MySQL高可用实践

王雪迎 著

清華大學出版社
北京

内 容 简 介

MySQL 数据库是现代软件系统基础应用的核心组成部分，如何保持 MySQL 的高可用性在系统应用中就变得极为重要。本书从复制、中间件、集群、存储、优化五个方面介绍 MySQL 高可用技术。

本书分为 16 章，内容包括异步复制、半同步复制、GTID 与复制、复制拓扑与性能、延迟复制与部分复制、组复制、MySQL Router、MySQL Fabric、MMM、MHA、Keepalived + LVS、Heartbeat + HAProxy、InnoDB Cluster、Galera Cluster、DRBD、优化案例等。

本书适合 DBA、系统管理运维、应用程序开发、架构师等 MySQL 相关技术人员作为参考书，也适合作为高等院校和培训学校相关专业的技术培训教材。

图书在版编目（CIP）数据

MySQL 高可用实践 / 王雪迎著.—北京：清华大学出版社，2020.6
ISBN 978-7-302-55554-4

Ⅰ. ①M…　Ⅱ. ①王…　Ⅲ. ①SQL 语言—程序设计　Ⅳ. ①TP311.132.3

中国版本图书馆 CIP 数据核字（2020）第 088369 号

责任编辑： 夏毓彦
封面设计： 王　翔
责任校对： 闫秀华
责任印制： 沈　露
出版发行： 清华大学出版社
　　网　　址：http://www.tup.com.cn，http://www.wqbook.com
　　地　　址：北京清华大学学研大厦 A 座　　邮　　编：100084
　　社 总 机：010-62770175　　邮　　购：010-62786544
　　投稿与读者服务：010-62776969，c-service@tup.tsinghua.edu.cn
　　质 量 反 馈：010-62772015，zhiliang@tup.tsinghua.edu.cn
印 装 者： 三河市君旺印务有限公司
经　　销： 全国新华书店
开　　本： 190mm×260mm　　**印　　张：** 29.75　　**字　　数：** 809 千字
版　　次： 2020 年 7 月第 1 版　　**印　　次：** 2020 年 7 月第 1 次印刷
定　　价： 99.00 元

产品编号：086804-01

前　　言

毋庸置疑，开源数据库 MySQL 是当今最流行的关系数据库系统之一，无论是在国内还是国外，都具有极高的人气。据统计，在 2019 年的全球数据库市场中，MySQL 以接近 40%的使用率高居榜首。在此前提下，作为基础技术支撑的核心组成部分，保持 MySQL 数据库的高可用性变得极为重要。这是我写这本书的第一个原因，希望读者在构建自己的 MySQL 高可用环境时，本书能提供一定的参考和帮助。

网上关于 MySQL 的资料比比皆是，但大部分都是针对单一具体问题的解答，例如如何配置 MySQL 双主复制等，全面系统讲述 MySQL 高可用这方面的资料并不多见。而且，这些资料大多只给出简单的实现方法，缺少相应的理论背景知识和分析验证过程。这是我写这本书的第二个原因，希望能填补缺口，让读者不但知其然，也知其所以然。

从 2000 年至今的 20 年里，我一直从事各种数据库相关的应用开发和管理工作，尤其是近年来一直跟 MySQL 数据库打交道。从最简单的单实例，到各种复制，再到集群以及与中间件的配合，在其使用上积累了一些经验和心得。这是我写这本书的第三个原因，希望将它们加以归纳和总结，汇集成书与读者分享。

本书内容

全书共 16 章，从复制、中间件、集群、存储、优化五个方面介绍 MySQL 高可用技术。

第 1~6 章讲述 MySQL 的复制，是 MySQL 高可用实现的基础。除了包括原生的异步复制、插件实现的半同步复制，以及 MySQL 5.7 版本引入的组复制等基本复制技术外，还讨论了几种常见的复制拓扑，以及各种复制的性能对比。在实际应用中，GTID、部分复制、延迟复制都是针对不同场景下常用的与复制密切相关的技术，这部分内容会分别用单独的章节加以说明。

第 7~12 章给出如何使用中间件构建 MySQL 高可用环境。具体包括 MySQL 自己的中间件产品 Router 和 Fabric，以及被广泛采用的 MMM、MHA、Keepalived + LVS、Heartbeat + HAProxy 等第三方 MySQL 高可用解决方案。每种产品使用一章的篇幅进行讲解。

第 13~14 章介绍两种 MySQL 集群：InnoDB Cluster 和 Galera Cluster。其中 InnoDB Cluster 是 MySQL 官方推出的集群产品，而 Galera Cluster 是包含在 MariaDB 中的 MySQL 多主集群。

第 15 章讲述利用 DRBD 技术实现 MySQL 高可用。前三部分介绍的 MySQL 高可用解决方案都是基于 MySQL 自带的各种复制技术。DRBD 的实现方式则完全不同，它本身与 MySQL 复制毫无关系，而是一个用软件实现的镜像块设备内容的存储复制解决方案。

在全书最后的第 16 章，引入四个 MySQL 高可用环境下的优化案例，分别是：快速生成数字辅助表；将 MySQL 去重优化到极致；重叠时间段问题之算法优化；快速安全删除大表。严格来说，本章内容并不特指高可用环境下的优化。但这些案例都取自我工作中的真实需求，具有一定的代表性和实用性，既有利于读者借鉴解决实际问题，同时在某种程度上也能增加阅读的趣味。

为力求严谨，书中所涉及产品或技术的知识绝大部分来源于相应的官方文档。正如 Linus 的那句话：“Talk is cheap，show me the code”，对于每一个过程或结论，都配有详细的实验步骤，用实例辅以佐证，帮助读者加深理解。

源码下载

本书示例源码请扫描右边二维码下载。如果有问题和建议，请联系 booksaga@163.com，邮件主题为“MySQL 高可用性实践”。

读者对象

本书所定位的读者是 DBA、系统管理运维、应用程序开发、架构师等 MySQL 相关的技术人员。所有的描绘场景与实验环境都是基于 Linux 操作系统的虚拟机。本书的读者应具有一定的 SQL、MySQL、Linux 基础。

致谢

在本书编写过程中，得到了很多人的帮助与支持。首先要感谢我所在的公司——优贝在线提供的平台和环境，感谢同事们工作中的鼎力相助。没有那里的工作氛围和团队，也就不会有这本书。感谢清华大学出版社图格事业部的老师和编辑们，他们的辛勤工作使得本书得以尽早与读者见面。感谢 CSDN 提供的技术分享平台，给我有一个将博客文章整理成书的机会。最后，感谢家人对我一如既往的支持。

由于水平有限，错漏之处在所难免，希望读者批评指正。

编者

2020 年 5 月

目　　录

第 1 章 异步复制

1.1 MySQL 异步复制简介

这里所说的复制，就是将来自一个 MySQL 数据库服务器（主库）的数据复制到一个或多个 MySQL 数据库服务器（从库）。传统的 MySQL 复制提供了一种简单的 Primary-Secondary（主-从）复制方法，默认情况下，复制是单向异步的。MySQL 支持两种复制方式：基于行的复制和基于语句的复制。这两种方式都是通过在主库上记录二进制日志（binlog）和在从库重放中继日志（relay-log）的方式来实现异步的数据复制。二进制日志或中继日志中的记录被称为事件。所谓异步包含两层含义：一是主库二进制日志的写入操作与将其发送到从库的操作是异步进行的；二是从库获取与重放日志事件是异步进行的。这意味着，在同一个时间点从库上的数据更新可能落后于主库，并且无法保证主库和从库之间的延迟间隔。

复制主库而增加的系统开销主要体现在启用二进制日志带来的 I/O，但是增加的开销并不大，MySQL 官方文档中称开启二进制日志会产生 1%的性能损耗。为了保证对历史事务的备份以及从介质失败中可以恢复过来，这点系统开销是非常必要的。除此之外，每个从库也会增加主库的一些负载（即系统开销），例如网络和 I/O。当从库读取主库的二进制日志时，就会产生一定的 I/O 开销。如果从一个主库复制到多个从库，唤醒多个复制线程发送二进制日志内容的开销就会累加。不过，所有这些复制带来的额外系统开销相对于各种应用对 MySQL 服务器造成的高负载来说都是微不足道的。

1.1.1 复制的用途

复制的用途主要体现在以下五个方面：

1. 横向扩展

通过复制可以将读操作指向从库来获得更好的读扩展。所有写入和更新都在主库上进行，但读取可能发生在一个或多个从库上。在这种读写分离模型中，主库专用于更新，显然比同时进行读写操作会有更好的写性能。需要注意的是，写操作并不适合通过复制来扩展。在“一主多从”架构中，写操作会被执行多次，正如“木桶效应”，整个系统的写入性能取决于写入最慢的那部分操作。

2. 负载均衡

通过 MySQL 复制可以将读操作分布到多个服务器上，实现对读密集型应用的优化。对于小规模的应用，可以简单地对机器名进行硬编码或者使用 DNS 轮询（将一个机器名指向多个 IP 地址）。当然也可以使用复杂的方法，例如使用 LVS 网络负载均衡器等，就能够很好地将负载分配到不同的 MySQL 服务器上。

3. 提高数据安全性

提高数据安全性可以从两方面来理解：其一，因为数据被复制到从库，并且从库可以暂停复制过程，所以可以在从库上执行备份操作而不会影响对应的主库；其二，当主库出现问题时，还有从库的数据可以被访问。但是，对备份来说，复制仅仅是一项有意义的技术补充，它既不是备份，也不能够取代备份。例如，当用户误删了一个表，而且此操作已经在从库上被复制执行，这种情况下只能用备份来恢复。

4. 提升高可用性

复制可以帮助应用程序避免 MySQL 单点故障，一个包含复制且设计良好的故障切换系统能够显著缩短宕机的时间。

5. 滚动升级

比较普遍的做法是，使用一个高版本的 MySQL 作为从库，保证在升级全部数据库实例之前，数据的查询能够在从库上按预期执行。在测试没有问题后，将高版本的 MySQL 切换为主库，并将应用连接至该主库，然后重新搭建高版本的从库。

1.1.2 复制如何工作

如前所述，MySQL 复制依赖二进制日志（binlog），想要理解复制的工作事项，就先要了解 MySQL 的二进制日志。

1. 二进制日志

二进制日志包含描述数据库更改的事件，如建表操作或对表数据的更改等。开启二进制日志有两个重要目的：

- 用于复制。主库上的二进制日志提供了要发送到从库的数据更改记录。主库将其二进制日志中包含的事件发送到从库，从库执行这些事件以对其本地数据进行相同的更改。
- 用于恢复。当出现介质错误，如磁盘故障时，数据恢复操作需要使用二进制日志。还原备份后，重新执行备份之后记录的二进制日志中的事件，最大限度地减少数据丢失。

不难看出，MySQL 二进制日志所起的作用与 Oracle 的归档日志类似。二进制日志只记录更新数据的事件，不记录 SELECT 或 SHOW 等语句。通过设置 log-bin 系统变量来开启二进制日志，MySQL 8 中这个系统变量默认是开启的。

二进制日志有 STATEMENT、ROW、MIXED 三种格式，通过 binlog-format 系统变量来设置：

- STATEMENT 格式，基于 SQL 语句的复制（Statement-Based Replication，SBR）。每一条会修改数据的 SQL 语句都会被记录到 binlog 中。这种格式的优点是不需要记录每行的数据变化，这样二进制日志会比较少，减少了磁盘 I/O，提高了性能。缺点是在某些情况下会导致主库与从库中的数据不一致，例如 last_insert_id()、now()等非确定性函数，以及用户自定义函数（User-Defined Function，UDF）等易出现问题。
- ROW 格式，基于行的复制（Row-Based Replication，RBR）。该格式不记录 SQL 语句的上下文信息，仅记录哪条数据被修改了，修改成了什么样子，能清楚地记录每一行数据的修改细节。这种格式的优点是不会出现某些特定情况下的存储过程、函数或触发器的调用和触发无法被正确复制的问题。缺点是通常会产生大量的日志，尤其像大表上执行 alter table 操作时会让日志暴涨。
- MIXED 格式，混合复制（Mixed-Based Replication，MBR）。它是 STATEMENT 和 ROW 这两种格式的混合体，默认使用 STATEMENT 格式保存二进制日志，对于 STATEMENT 格式无法正确复制的操作，会自动切换到基于 ROW 格式的复制操作，MySQL 会根据执行的 SQL 语句选择日志保存方式。

MySQL 8 默认使用 ROW 格式。二进制日志的存放位置最好设置到与 MySQL 数据目录不同的磁盘分区，以降低磁盘 I/O 的竞争，提升性能，并且在数据磁盘发生故障时还可以利用备份和二进制日志来恢复数据。

2. 复制步骤

总体来说，MySQL 复制有五个步骤：

步骤 01 在主库上把数据更改事件记录到二进制日志中。

步骤 02 从库上的 I/O 线程向主库询问二进制日志中的事件。

步骤 03 主库上的二进制日志转储（Binlog dump）线程向 I/O 线程发送二进制事件。

步骤 04 从库上的 I/O 线程将二进制日志事件复制到自己的中继日志中。

步骤 05 从库上的 SQL 线程读取中继日志中的事件，并将其重放到从库上。

图 1-1 详细描述了复制的细节。

第一步，是在主库上记录二进制日志。每次准备提交事务完成数据更新前，主库将数据更新的事件记录到二进制日志中。MySQL 会按事务提交的顺序而非每条语句的执行顺序来记录二进制日志。在记录二进制日志后，主库会告诉存储引擎可以提交事务了。

第二步，从库将主库的二进制日志复制到其本地的中继日志中。首先，从库会启动一个工作线程，称为 I/O 线程。I/O 线程与主库建立一个普通的客户端连接，然后在主库上启动一个特殊的二进制日志转储（Binlog dump）线程，它会读取主库上二进制日志中的事件，但不会对事件进行轮询。如果该线程追赶上了主库，它将进入睡眠状态，直到主库发送信号通知该线程有新的事件时才会被唤醒，从库 I/O 线程会将接收到的事件记录到中继日志中。

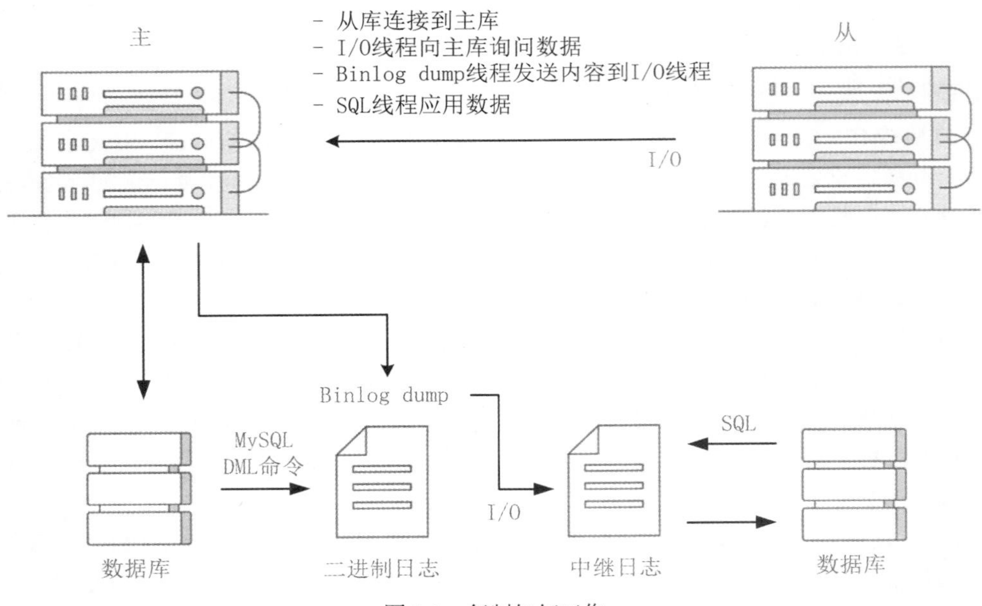

图 1-1　复制如何工作

从库的 SQL 线程执行最后一步，该线程从中继日志中读取事件并在从库上执行，从而实现从库数据的更新。当 SQL 线程追赶 I/O 线程时，中继日志通常已经在系统缓存中，所以读取中继日志的开销很低。SQL 线程执行的事件也可以通过 log_slave_updates 系统变量来决定是否写入其自己的二进制日志中，这可以用于级联复制的场景。

这种复制架构实现了获取事件和重放事件的解耦，允许这两个过程异步进行。也就是说 I/O 线程的执行能够独立于 SQL 线程的执行。但是，这种架构也限制了复制的过程，其中最重要的一点是在主库上并发更新的查询，到从库上通常只能串行化执行了，因为系统默认只有一个 SQL 线程来重放中继日志中的事件。在 MySQL 5.6 版本以后已经可以通过配置 slave_parallel_workers 等系统变量进行并行复制，在第 4 章讨论复制性能问题时会介绍并行复制的相关细节。

现在我们已经了解了 MySQL 复制是以二进制日志为基础，但是像 InnoDB 这样的事务引擎有自己的事务日志，如 ib_logfile，这些事务日志通常被称为重做日志（redo log）。作为背景知识，简单介绍一下 InnoDB 重做日志的作用。

对 InnoDB 的任何修改操作都会首先在称为缓冲池（InnoDB Buffer Pool）的内存页面上进行，然后这样的页面将被标记为脏页，并被放到专门的刷新列表上，后续将由主线程（Master Thread）或专门“刷脏页”的线程阶段性地将这些页面写入磁盘。这样做的好处是避免每次的写库操作都要操作磁盘，从而导致大量的随机 I/O 操作，阶段性地“刷脏页”可以将多次对页面的修改合并成一次 I/O 操作，同时异步写入也降低了访问时延。然而，如果在脏页还未刷入到磁盘时服务器就非正常关闭了，那么这些修改操作将会丢失，如果写入操作正在进行，甚至会由于损坏数据文件导致数据库不可用。为了避免上述问题的发生，InnoDB 将所有对页面的修改写入一个专门的文件，并在数据库启动时从此文件进行实例恢复，这个文件就是重做日志文件。每当有更新操作时，在数据页变更之前将操作写入重做日志，这样当发生掉电之类的情况时，系统可以在重启后继续工作。这就是所谓的预写日志（Write-Ahead Logging，WAL）。这种技术推迟了缓冲区页面的刷新，从而提升了数据库的吞吐量。同时由于重做日志的写操作是顺序 I/O，相对于写数据文件的随机 I/O 要快得多。大多数数据库系统都采用类似的技术实现。

聪明如你可能已经有了这样的疑问，在复制中二进制日志和重做日志如何协同工作？假设 InnoDB 写完重做日志后，服务异常关闭了。主库能够根据重做日志恢复数据，但由于二进制日志没写入，会导致从库同步时少了这个事务么？或者反之，二进制日志写成功，而重做日志没有写完，是否会导致从库执行事务而主库不执行？这些情况会不会造成主从数据不一致的问题呢？解决这些问题是 MySQL 的核心需求，让我们从 MySQL 基本架构说起。图 1-2 是 MySQL 的逻辑架构图。

在图 1-2 中，最上层用于处理客户端连接、授权认证、安全，等等。第二层架构是 MySQL 服务器层。大多数 MySQL 的核心功能都在这一层，包括查询解析、分析、优化、缓存以及全部内置函数，所有跨存储引擎的功能，如存储过程、触发器、视图等都在这一层实现。不出所料，二进制日志也在这一层实现。第三层包含了存储引擎，负责 MySQL 中数据的存储和提取。服务器通过 API 与存储引擎进行通信，存储引擎只是简单地响应上层服务器的请求。显然 InnoDB 的重做日志在这一层实现。

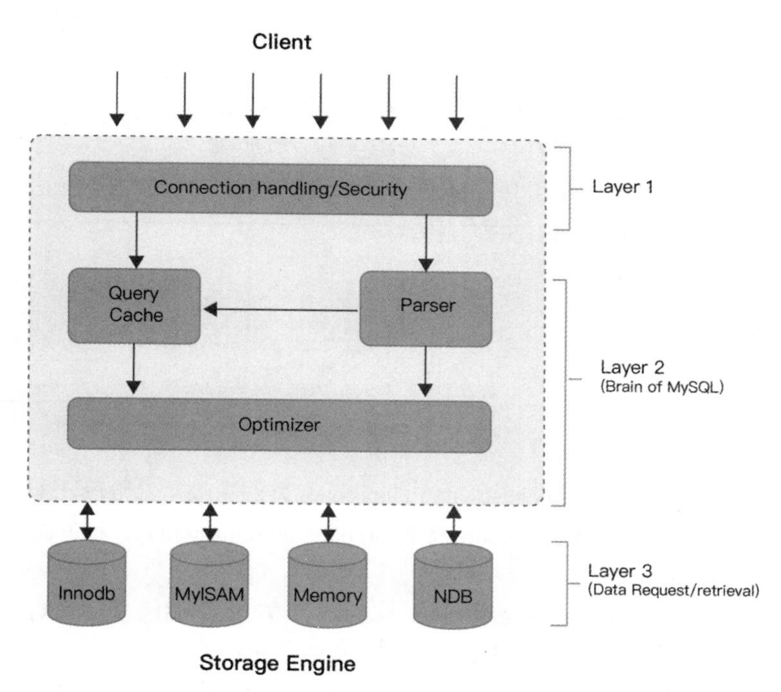

图 1-2　MySQL 服务器逻辑架构图

由于 MySQL 的事务日志包含二进制日志和重做日志，当发生崩溃恢复时，MySQL 主库通过重做日志进行恢复，而在主从复制的环境下，从库依据主节点的二进制日志进行数据同步。这样的架构对两种日志有两个基本要求：第一，保证二进制日志中存在的事务一定在重做日志中存在，也就是二进制日志里不会比重做日志里的事务多（可以少，因为重做日志里面记录的事务可能有部分没有提交，这些事务最终可能会被回滚）；第二，两种日志中事务的顺序一致，这也是很重要的一点。假设两者记录的事务顺序不一致，那么会出现类似于主库事务执行的顺序是 ta、tb、tc 和 td，但是二进制日志中记录的是 ta、tc、tb 和 td，也就是被复制到从库后主从数据不一致了。为了达到上述两个基本要求，MySQL 使用内部 XA 来实现。XA 是 eXtended Architecture 的缩写，是 X/Open 分布式事务定义的事务中间件与数据库之间的接口规范，其核心是两阶段提交（Two Phase Commit，2PC）。

1.1.3 两阶段提交

在两阶段提交协议中一般分为事务管理器（协调者）和若干事务执行器（参与者）两种角色。在 MySQL 内部实现的两阶段提交中，二进制日志充当了协调者角色，由它来通知 InnoDB 执行准备、提交或回滚。从实现角度分析，事务提交由准备阶段和提交阶段构成。提交流程和代码框架分别如图 1-3 和图 1-4 所示。

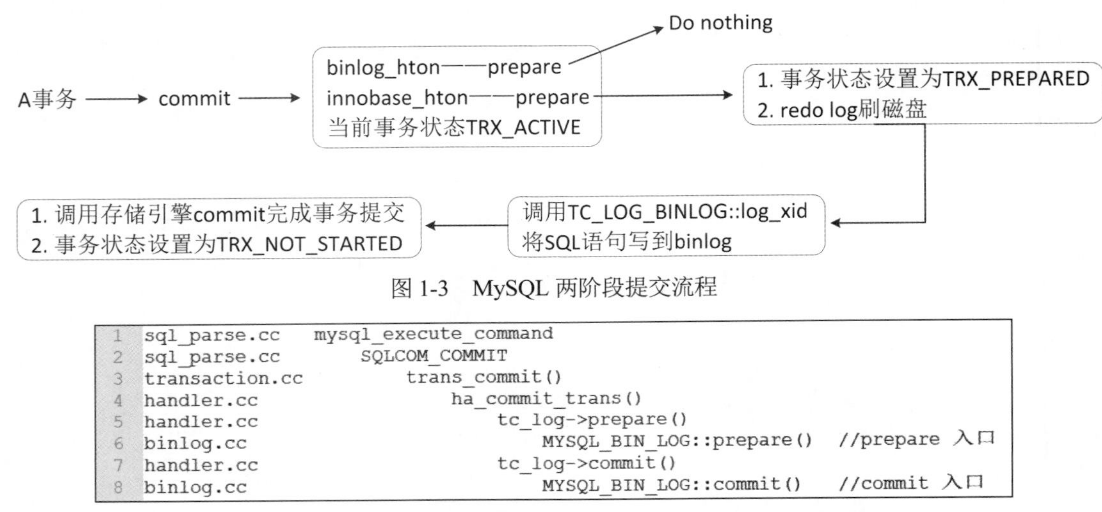

图 1-3 MySQL 两阶段提交流程

```
1 sql_parse.cc    mysql_execute_command
2 sql_parse.cc        SQLCOM_COMMIT
3 transaction.cc          trans_commit()
4 handler.cc                  ha_commit_trans()
5 handler.cc                      tc_log->prepare()
6 binlog.cc                           MYSQL_BIN_LOG::prepare()   //prepare 入口
7 handler.cc                      tc_log->commit()
8 binlog.cc                           MYSQL_BIN_LOG::commit()    //commit 入口
```

图 1-4 commit 命令的 MySQL 代码框架

（1）先调用 binglog_hton 和 innobase_hton 的 prepare 方法完成第一阶段，binlog_hton 的 prepare 方法实际上什么也没做，InnoDB 的 prepare 持有 prepare_commit_mutex，将重做日志刷磁盘，并将事务状态设为 TRX_PREPARED。

（2）如果事务涉及的所有存储引擎的 prepare 都执行成功，则调用 TC_LOG_BINLOG::log_xid 将事务（STATEMENT 格式或 ROW 格式）写到二进制日志中，此时，事务已经肯定要提交了。否则，调用 ha_rollback_trans 回滚事务，而事务实际上也不会写到二进制日志中。

（3）最后，调用引擎的 commit 完成事务的提交。实际上 binlog_hton->commit 什么也不会做（上一步已经将二进制日志写入磁盘），innobase_hton->commit 则会清除回滚信息，向重做日志中写入 COMMIT 标记，释放 prepare_commit_mutex，并将事务设为 TRX_NOT_STARTED 状态。

如果数据库系统发生崩溃，当重启数据库时会进行崩溃恢复操作。具体到代码层面，InnoDB 在恢复的时候，不同状态的事务，会进行不同的处理：

- 对于 TRX_COMMITTED_IN_MEMORY 的事务，清除回滚段后，将事务设为 TRX_NOT_STARTED。

- 对于 TRX_NOT_STARTED 的事务，表示事务已经提交，跳过。
- 对于 TRX_PREPARED 的事务，要根据二进制日志来决定事务是否提交，暂时跳过。
- 对于 TRX_ACTIVE 的事务，回滚。

简单来讲，当发生崩溃恢复时，数据库根据重做日志进行数据恢复，逐个查看每条重做条目的事务状态。根据图 1-3 的流程，如果已进行到 TRX_NOT_STARTED 阶段，也就是存储引擎 commit 阶段，那么说明重做日志和二进制日志是一致的，正常根据重做条目进行恢复即可。如果事务状态为 TRX_ACTIVE，没写到二进制日志中，就直接回滚。如果事务状态为 TRX_PREPARED，要分两种情况，先检查二进制日志是否已写入成功，如果没写入成功，那么就算是 TRX_PREPARED 状态也要回滚。如果写入成功了，那么就进行最后一步，调用存储引擎 commit，更改事务状态为 TRX_NOT_STARTED，也就是真正提交状态，可以用作数据恢复。

由此可见，MySQL 是以二进制日志的写入与否作为事务提交成功与否的标志，通过这种方式让 InnoDB 重做日志和 MySQL 服务器的二进制日志中的事务状态保持一致。两阶段提交很好地保持了数据一致性和事务顺序性。

了解了所有这些技术细节后，当初的疑问自然也就有了答案。假设在准备阶段（阶段 1）结束之后程序异常，此时没有写入二进制日志，则从库不会同步这个事务。主库上，崩溃恢复时重做日志中这个事务没有 trx_commit，因此会被回滚。逻辑上主从库都不会执行这个事务。假设在提交阶段（阶段 2）结束后程序异常，此时二进制日志已经写入，则从库会同步这个事务。主库上，根据重做日志能够正常恢复此事务。也就是说，若二进制日志写入完成，那么主从库都会正常提交事务，反之则主从库都回滚事务，都不会出现主从不一致的问题。

MySQL 通过 innodb_support_xa 系统变量控制 InnoDB 是否支持 XA 事务的 2PC，默认是 TRUE。如果关闭，则 InnoDB 在 prepare 阶段就什么也不做，这可能会导致二进制日志的顺序与 InnoDB 提交的顺序不一致，继而导致在恢复时或者从库上产生不同的数据。在 MySQL 8 中，innodb_support_xa 系统变量已被移除，也就是始终启用 InnoDB 对 XA 事务中两阶段提交的支持，就不再交由用户来选择了。

上述的 MySQL 两阶段提交流程并不是天衣无缝的，主从数据是否一致还与重做日志和二进制日志的写盘方式有关。innodb_flush_log_at_trx_commit 和 sync_binlog 系统变量分别控制两者的落盘策略。

- innodb_flush_log_at_trx_commit：有 0、1、2 共三个可选值。0 表示每秒进行一次刷新，但是每次事务提交不进行任何操作。每秒调用 fsync 使数据落到磁盘，不过这里需要注意如果底层存储有高速缓存（Cache），比如 RAID Cache，那么这时也不会真正落盘。由于一般 RAID 卡都带有备用电源，因此一般都认为此时数据是安全的。1 代表每次事务提交都会进行刷新，这是最安全的模式；2 表示每秒刷新，每次事务提交时不刷新，而是调用 write 将重做日志缓冲区中的内容刷到操作系统页面缓存中。从数据安全性和性能角度来比较这三种策略的优劣为：策略 1，因为每次事务提交都是重做日志落盘，所以最安全，但是由于 fsync 的次数增多会导致性能下降比较严重；策略 0，表示每秒刷新，每次事务提交都不进行任何操作，所以 MySQL 或操作系统崩溃时最多丢失一秒的事务；策略 2，相对于策略 0 来说多了每次事务提交时的一个 write 操作，此时数据虽然没有落盘，但是只要操作系统没有崩溃，即使 MySQL 崩溃，那么事务也不会丢失。

- sync_binlog：MySQL 在提交事务时调用 MYSQL_LOG::write 执行写二进制日志的操作，并根据 sync_binlog 决定是否进行刷新。默认值是 0，即不刷新，从而把控制权交给操作系统。如果设置为 1，则每次提交事务都会进行一次磁盘刷新。

这两个参数不同值的组合会带来不同的效果。两者都设置为 1，数据最安全，能保证主从一致，这也是 MySQL 8 的默认设置。innodb_flush_log_at_trx_commit 设置为非 1 时，假设在二进制日志写入完成后系统崩溃，则可能出现这样的情况：从库能够执行事务，但主库中 trx_prepare 的日志没有被写入到重做日志中，导致主库不执行事务，就出现了主从不一致的情况。同理，若 sync_binlog 设置为非 1 时，则可能导致二进制日志丢失（如操作系统异常崩溃），从而与 InnoDB 层面的数据不一致，体现在复制上，从库可能丢失事务。在数据一致性要求很高的场景下，建议就使用全部设置为 1 的默认设置。

1.2 搭建实验环境

理解了原理后，在着手配置 MySQL 异步复制前，先要安装 MySQL 数据库软件。本节介绍我们的实验环境以及在此环境中安装 MySQL 的具体步骤。

1.2.1 实验环境说明

1. 主机 IP

- 172.16.1.125（主）
- 172.16.1.126（从）
- 172.16.1.127（从）

2. 软件环境

- OS：CentOS Linux release 7.2.1511 (Core)
- MySQL：MySQL Community Server 8.0.16
- glibc：glibc-2.17-105.el7.x86_64

3. 硬件环境

三台虚拟机，每台的基本配置为：

- 双核双 CPU，Intel(R) Xeon(R) CPU E5-2420 0 @ 1.90GHz
- 8GB 物理内存，8GB Swap
- 100GB 物理硬盘

1.2.2 安装 MySQL

从 https://dev.mysql.com/downloads/mysql/下载二进制安装文件 mysql-8.0.16-linux-glibc2.12-

x86_64.tar.xz，相关选项如图 1-5 所示。

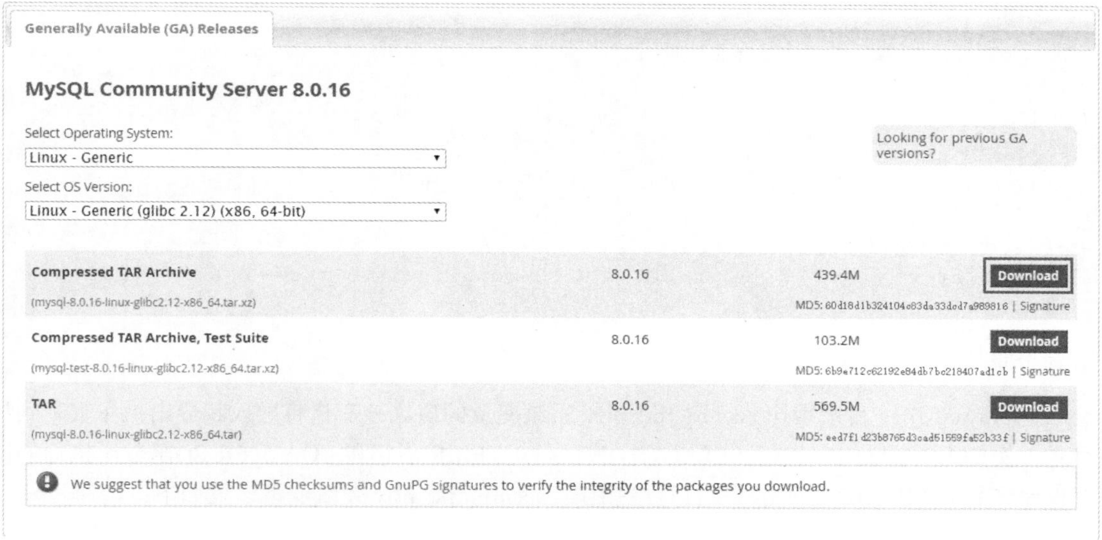

Generally Available (GA) Releases

MySQL Community Server 8.0.16

Select Operating System:
Linux - Generic

Select OS Version:
Linux - Generic (glibc 2.12) (x86, 64-bit)

Looking for previous GA versions?

Compressed TAR Archive (mysql-8.0.16-linux-glibc2.12-x86_64.tar.xz)	8.0.16	439.4M MD5: 60d18d1b324104c83da33dcd7a989816 | Signature	Download
Compressed TAR Archive, Test Suite (mysql-test-8.0.16-linux-glibc2.12-x86_64.tar.xz)	8.0.16	103.2M MD5: 6b9e712c62192e84db7bc218407ad1cb | Signature	Download
TAR (mysql-8.0.16-linux-glibc2.12-x86_64.tar)	8.0.16	569.5M MD5: eed7f1d23b8765d3cad51559fa52b33f | Signature	Download

We suggest that you use the MD5 checksums and GnuPG signatures to verify the integrity of the packages you download.

图 1-5　下载 mysql-8.0.16 安装包

然后以 root 用户按顺序执行下面的命令，在三台主机上安装 MySQL。

```
# 进入安装目录
cd /usr/local

# 从 tar 包中提取文件
tar xvf /home/mysql/mysql-8.0.16-linux-glibc2.12-x86_64.tar.xz

# 建立软链接
ln -s mysql-8.0.16-linux-glibc2.12-x86_64 mysql

# 进入 mysql 目录
cd mysql

# 建立 secure_file_priv 系统变量指向的目录
mkdir mysql-files

# 修改属主为 mysql
chown mysql:mysql mysql-files

# 修改目录权限
chmod 750 mysql-files

# mysql 系统初始化
bin/mysqld --initialize --user=mysql

# 建立 SSL/RSA 相关文件，如果不启用 SSL 连接，此步可省略
bin/mysql_ssl_rsa_setup
```

```
# 启动mysql服务器
bin/mysqld_safe --user=mysql &

# 连接mysql服务器
bin/mysql -u root -p

-- 修改root密码
alter user user() identified by "123456";

-- 创建一个新的mysql管理员账号
create user 'wxy'@'%' identified with mysql_native_password by '123456';
grant all on *.* to 'wxy'@'%' with grant option;
```

命令说明：

（1）mysql-files 目录用作 secure_file_priv 系统变量的值。该变量将导入和导出操作限制到特定目录。例如由 LOAD DATA、SELECT... INTO OUTFILE 语句和 LOAD_FILE()函数所执行的操作，仅允许具有 FILE 权限的用户执行这些操作。secure_file_priv 系统变量设置如下：

- 空字符串：变量不起作用，是不安全的设置。
- 目录名：MySQL 限制导入和导出操作仅用于该目录中的文件。目录必须已经存在，MySQL 不会创建它。
- NULL：MySQL 禁用导入和导出操作。

（2）mysqld --initialize 命令创建默认数据库并退出。在过程中会创建一个超级用户'root'@'localhost'，并为该用户生成一个随机密码。命令执行输出如下所示：

```
    [root@hdp2/usr/local/mysql]#bin/mysqld --initialize --user=mysql
    2019-05-05T06:31:58.956385Z 0 [System] [MY-013169] [Server]
/usr/local/mysql-8.0.16-linux-glibc2.12-x86_64/bin/mysqld (mysqld 8.0.16)
initializing of server in progress as process 10256
    2019-05-05T06:32:01.287093Z 5 [Note] [MY-010454] [Server] A temporary password
is generated for root@localhost: w1SN3pgRPL*D
    2019-05-05T06:32:02.901171Z 0 [System] [MY-013170] [Server]
/usr/local/mysql-8.0.16-linux-glibc2.12-x86_64/bin/mysqld (mysqld 8.0.16)
initializing of server has completed
    [root@hdp2/usr/local/mysql]#
```

MySQL 文档中说产生的临时密码会写到.err 日志文件里，但在本次安装中，这一步并没有生成.err 文件。安全的做法还是记下临时密码，因为在 initialize 情况下，临时密码默认标记为已过期，用户必须在第一次进入 MySQL 后首先修改 root 用户的密码。

（3）mysql_ssl_rsa_setup 程序将创建 SSL 证书和密钥文件，以及创建使用 SSL 进行安全连接所需的 RSA 密钥对文件。如果现有的 SSL 文件已过期，mysql_ssl_rsa_setup 也可用于创建新的 SSL 文件。mysql_ssl_rsa_setup 使用 openssl 命令，因此主机上必须安装有 OpenSSL。mysql_ssl_rsa_setup 检查数据目录中的以下 SSL 文件：

```
ca.pem
server-cert.pem
server-key.pem
```

只要存在这些文件之一，则mysql_ssl_rsa_setup不会创建任何SSL文件。否则，它会调用openssl来创建它们以及一些其他文件。启用SSL安全客户端连接时需要这些文件：

```
ca.pem                      自签名 CA 证书
ca-key.pem                  CA 私钥
server-cert.pem             服务器证书
server-key.pem              服务器私钥
client-cert.pem             客户端证书
client-key.pem              客户端私钥
```

之后mysql_ssl_rsa_setup检查数据目录中的以下RSA文件：

```
private_key.pem             私钥/公钥对的私有成员
public_key.pem              私钥/公钥对的公共成员
```

只要存在这些文件之一，则mysql_ssl_rsa_setup不会创建任何RSA文件。否则，它会调用openssl来创建它们。对于sha256_password或caching_sha2_password插件验证的账户不加密连接时，通过这些文件使用RSA进行安全密码交换。

mysql-8.0.16默认支持SSL加密连接：

```
mysql> show variables like 'have_ssl';
+---------------+-------+
| Variable_name | Value |
+---------------+-------+
| have_ssl      | YES   |
+---------------+-------+
1 row in set (0.00 sec)
```

（4）第一次进入MySQL后，执行任何命令都会报出以下错误：

```
ERROR 1820 (HY000): You must reset your password using ALTER USER statement
before executing this statement.
```

提示很明显，需要修改初始化时为用户'root'@'localhost'生成的临时密码。MySQL 8默认使用的认证插件是caching_sha2_password：

```
mysql> show variables like 'default_authentication_plugin';
+-------------------------------+-----------------------+
| Variable_name                 | Value                 |
+-------------------------------+-----------------------+
| default_authentication_plugin | caching_sha2_password |
+-------------------------------+-----------------------+
1 row in set (0.01 sec)
```

当用旧版本的MySQL客户端连接MySQL 8服务器时，可能报出以下错误：

```
ERROR 2059 (HY000): Authentication plugin 'caching_sha2_password' cannot be
loaded: /home/mysql/mysql-5.6.14/lib/plugin/caching_sha2_password.so: cannot
open shared object file: No such file or directory
```

可以使用两种方法解决此问题：第一种方法是在配置文件中设置default_authentication_plugin = 'mysql_native_password'，然后重启MySQL服务器使之生效。default_authentication_plugin是只读

系统变量，不能动态修改；第二种方法是在创建用户时，使用兼容新旧版本的认证方式，例如：

```
create user 'wxy'@'%' identified with mysql_native_password by '123456';
```

通过以下查询可以得到用户所对应的认证插件：

```
mysql> select host,user,plugin from mysql.user;
+-----------+------------------+-----------------------+
| host      | user             | plugin                |
+-----------+------------------+-----------------------+
| %         | wxy              | mysql_native_password |
| localhost | mysql.infoschema | caching_sha2_password |
| localhost | mysql.session    | caching_sha2_password |
| localhost | mysql.sys        | caching_sha2_password |
| localhost | root             | caching_sha2_password |
+-----------+------------------+-----------------------+
5 rows in set (0.00 sec)
```

1.3 配置异步复制

在本实验中分别针对空库、脱机、联机三种场景，配置“一主两从”的MySQL标准异步复制。只进行实例级别的全复制，不考虑对个别库表或使用过滤复制的情况。

1.3.1 空库

初始安装后，MySQL中还没有任何应用数据。此时事先配置好复制再投入使用，这是最理想也是最简单的一种情况，具体配置步骤如下：

步骤01 修改server_id系统变量。

```
-- 主库
set global server_id=1125;
-- 从库 1
set global server_id=1126;
-- 从库 2
set global server_id=1127;
```

要求在复制中所有MySQL实例的server_id都不相同，这里将三个实例的server_id分别配置为1125、1126、1127。server_id系统变量可以动态修改，这样做的好处是不需要重启实例，配置即可在新连接中生效。但是为了避免MySQL重启后配置信息丢失，还需要同时在/etc/my.cnf配置文件中设置server_id参数。

步骤02 查看主库二进制日志信息。

```
-- 主库
show master status;
```

因为是一个静态的空库，二进制信息此时不会发生变化，可以作为复制的起始点。本例中重启过 MySQL 实例，输出的信息为：

```
+----------+----------+-------------+-----------------+-------------------+
| File     | Position | Binlog_Do_DB  | Binlog_Ignore_DB| Executed_Gtid_Set |
+----------+----------+-------------+-----------------+-------------------+
| binlog.000003|    155 |              |                 |                   |
+----------+----------+-------------+-----------------+-------------------+
```

步骤03 在主库上建立复制专属用户。

```
-- 主库
create user 'repl'@'%' identified with mysql_native_password by '123456';
grant replication client,replication slave on *.* to 'repl'@'%';
```

前面介绍过 MySQL 8 的用户认证机制，这里仍然使用 mysql_native_password 方式。

步骤04 在从库创建主库信息。

```
change master to
     master_host='172.16.1.125',
     master_port=3306,
     master_user='repl',
     master_password='123456',
     master_log_file='binlog.000003',
     master_log_pos=155;
```

连接主库时，需要使用 change master to 提供到主库的连接选项，包括主机地址、端口、用户名、密码、二进制文件名、复制起始事件位置等信息。执行 change master to 语句后，在 mysql.slave_master_info 表中就会生成一条记录，此行为由 master_info_repository 系统变量控制。MySQL 8 中，该变量的默认值为 TABLE，即把与复制相关的主库信息记录到 mysql.slave_master_info 表中。随着复制的进行，表中的数据会随之更新。change master to 只是为 I/O 线程连接主库时提供连接参数，这条语句本身不会连接主库。以后启动 I/O 线程时，I/O 线程都会自动读取这条记录连接主库，不需要再执行 change master to 语句。

类似地，MySQL 8 默认会将中继日志的重放信息存到 mysql.slave_relay_log_info 表中。该行为由系统变量 relay_log_info_repository 控制。中继日志信息在首次启动复制时生成，并随着复制过程即时改变，这样 SQL 线程再次启动时就能从中获取到从中继日志的何处继续读取和执行。

步骤05 在从库启动复制并查看复制信息。

```
start slave;
show slave status\G
select * from mysql.user where user='repl'\G
```

刚才我们并没有在从库上创建 repl 用户，由于 create user 语句是在起始位置后执行的，因此可以正常复制到从库，查询 mysql.user 表即可确认。

start slave 语句会启动 I/O 线程和 SQL 线程，并创建一个到主库的普通客户端连接，此连接会伴随复制一直存在。该命令执行后，在主库的 processlist 中会看到类似如下的线程，它们是从库上的 I/O 线程所建立的，Binlog Dump 表示由 I/O 线程在主库上启动了 Binlog Dump 线程，每个连接到主库的从库对应一个线程，如 Id 32 和 56 是两个从库的连接线程：

```
    mysql> show processlist;
    ...
    | 32 | repl | hdp4:3723  | NULL | Binlog Dump | 328879 | Master has sent all binlog to slave; waiting for more updates | NULL |
    | 56 | repl | hdp3:57308 | NULL | Binlog Dump | 319204 | Master has sent all binlog to slave; waiting for more updates | NULL |
    ...
```

从库上的 processlist 中会看到类似如下的线程，Id 325 和 326 分别对应 I/O 线程与 SQL 线程：

```
    mysql> show processlist;
    ...
    | 325 | system user |  | NULL | Connect | 320408 | Waiting for master to send event | NULL |
    | 326 | system user |  | NULL | Query   | 320408 | Slave has read all relay log; ...| NULL |
    ...
```

从 show slave status 输出中可以查看复制的状态信息：

```
    mysql> show slave status\G
    *************************** 1. row ***************************
                   Slave_IO_State: Waiting for master to send event
                      Master_Host: 172.16.1.125
                      Master_User: repl
                      Master_Port: 3306
                    Connect_Retry: 60
                  Master_Log_File: binlog.000011
              Read_Master_Log_Pos: 155
                   Relay_Log_File: hdp3-relay-bin.000002
                    Relay_Log_Pos: 319
            Relay_Master_Log_File: binlog.000011
                 Slave_IO_Running: Yes
                Slave_SQL_Running: Yes
                  Replicate_Do_DB: 
              Replicate_Ignore_DB: 
               Replicate_Do_Table: 
           Replicate_Ignore_Table: 
          Replicate_Wild_Do_Table: 
      Replicate_Wild_Ignore_Table: 
                       Last_Errno: 0
                       Last_Error: 
                     Skip_Counter: 0
              Exec_Master_Log_Pos: 155
                  Relay_Log_Space: 526
                  Until_Condition: None
                   Until_Log_File: 
                    Until_Log_Pos: 0
               Master_SSL_Allowed: No
```

```
             Master_SSL_CA_File:
             Master_SSL_CA_Path:
                Master_SSL_Cert:
              Master_SSL_Cipher:
                 Master_SSL_Key:
          Seconds_Behind_Master: 0
  Master_SSL_Verify_Server_Cert: No
                  Last_IO_Errno: 0
                  Last_IO_Error:
                 Last_SQL_Errno: 0
                 Last_SQL_Error:
    Replicate_Ignore_Server_Ids:
               Master_Server_Id: 1125
                    Master_UUID: 8eed0f5b-6f9b-11e9-94a9-005056a57a4e
               Master_Info_File: mysql.slave_master_info
                      SQL_Delay: 0
            SQL_Remaining_Delay: NULL
        Slave_SQL_Running_State: Slave has read all relay log; waiting for more
updates
             Master_Retry_Count: 86400
                    Master_Bind:
        Last_IO_Error_Timestamp:
       Last_SQL_Error_Timestamp:
                 Master_SSL_Crl:
             Master_SSL_Crlpath:
             Retrieved_Gtid_Set:
              Executed_Gtid_Set:
                  Auto_Position: 0
           Replicate_Rewrite_DB:
                   Channel_Name:
             Master_TLS_Version:
         Master_public_key_path:
          Get_master_public_key: 0
              Network_Namespace:
  1 row in set (0.00 sec)
```

在从库上执行 show slave status 可以查看从库的状态，输出的信息非常多，其中除了那些描述 I/O 线程、SQL 线程状态的行，还有几个 log_file 和 pos 相关的行。理解这几行的意义至关重要，所以这里完整地描述它们：

- Master_Log_File：I/O 线程正在读取的 master binlog。
- Read_Master_Log_Pos：I/O 线程已经读取到 master binlog 的哪个位置。
- Relay_Log_File：SQL 线程正在读取和执行的 relay log。
- Relay_Log_Pos：SQL 线程已经读取和执行到 relay log 的哪个位置。
- Relay_Master_Log_File：SQL 线程最近执行的操作对应的是哪个 master binlog。
- Exec_Master_Log_Pos：SQL 线程最近执行的操作对应的是 master binlog 的哪个位置。

(Relay_Master_Log_File, Exec_Master_log_Pos)构成一个坐标，这个坐标表示从库上已经将主

库上的哪些数据重放到自己的实例中，它可以用于下一次执行 change master to 语句时指定的二进制日志坐标。与这个坐标相对应的是从库上 SQL 线程的中继日志坐标(Relay_Log_File, Relay_Log_Pos)。这两个坐标位置不同，但它们对应的数据是一致的。

还有一个延迟参数 Seconds_Behind_Master 需要说明一下，它的含义是 SQL 线程比 I/O 线程慢多少。如果主从之间的网络状况优良，那么从库的 I/O 线程读速度二进制日志和主库写二进制日志的速度基本一致，所以这个参数也用来描述“SQL 线程比主库慢多少”，也就是说从库比主库少多少数据，只不过衡量的单位是秒。需要注意的是，该参数的描述并不标准，只是在网速很好的时候做个大概的估计，很多情况下它的值都是 0，即使 SQL 线程比 I/O 线程慢了很多也是如此。

1.3.2 脱机

如果数据库已经存在应用数据，允许在一个可接受的脱机时间窗口进行复制，那么这种场景下常用的做法是先直接将主库的数据目录整体复制到从库，再启动从库复制操作，具体步骤如下：

步骤 01 在主库上建立复制专属用户：

```
-- 主库
create user 'repl'@'%' identified with mysql_native_password by '123456';
grant replication client,replication slave on *.* to 'repl'@'%';
```

步骤 02 停掉复制涉及的实例（“一主两从”三个实例都停止）：

```
mysqladmin -uroot -p123456 shutdown
```

步骤 03 将主库的数据目录整体复制到从库：

```
scp -r /usr/local/mysql/data/ 172.16.1.126:/usr/local/mysql
```

步骤 04 保证所有参与复制实例的 server-uuid 和 server_id 都不同。

这是非常重要的一步，相同的 server-uuid 或 server_id 会造成复制错误。从 MySQL 5.6 开始，用 128 位的 server_uuid 代替了原本的 32 位 server_id 的大部分功能。原因很简单，server_id 依赖于 my.cnf 的手工配置，有可能发生冲突。而自动产生 128 位 uuid 的算法可以保证所有的 MySQL uuid 都不会发生冲突。首次启动时 MySQL 会调用 generate_server_uuid() 自动生成一个 server_uuid，并且保存到数据目录下的 auto.cnf 文件中。这个文件目前存在的唯一目的就是保存 server_uuid。在 MySQL 再次启动时会读取 auto.cnf 文件，继续使用上次生成的 server_uuid。使用 show global variables like 'server_uuid'命令可以查看 MySQL 实例当前使用的 server_uuid，它是一个全局只读变量。全局唯一 server_uuid 的一个好处是，可以解决由 server_id 配置冲突导致的 MySQL 主从复制异常终止。从库向主库申请二进制日志时，会首先发送自己的 server_uuid，主库用从库发送的 server_uuid 代替 server_id 作为 kill_zombie_dump_threads 的参数，以终止冲突或者僵死的 BINLOG_DUMP 线程。

删除 data_dir/auto.cnf 文件，实例启动时会自动生成 server_uuid 的值。本例中删除了从库的 /usr/local/mysql/data/auto.cnf 文件，并编辑/etc/my.cnf 文件以修改 server_id，保证三个实例的配置互不相同。

步骤 01 重启实例（“一主两从”三个实例都启动）：

```
mysqld_safe --user=mysql &
```

步骤 02 查看主库二进制日志信息：

```
-- 主库
show master status;
```

步骤 03 使用上一步的输出在从库创建主库信息：

```
change master to
       master_host='172.16.1.125',
       master_port=3306,
       master_user='repl',
       master_password='123456',
       master_log_file='binlog.000004',
       master_log_pos=155;
```

步骤 04 在从库启动复制并查看复制信息：

```
start slave;
show slave status\G
```

此时在从库的 err 日志文件中可以看到复制正常启动的信息：

```
[mysql@hdp3~]$tail /usr/local/mysql/data/hdp3.err
...
    2019-05-10T02:56:17.333635Z 144 [System] [MY-010597] [Repl] 'CHANGE MASTER TO
FOR CHANNEL '' executed'. Previous state master_host='', master_port= 3306,
master_log_file='', master_log_pos= 4, master_bind=''. New state
master_host='172.16.1.125', master_port= 3306, master_log_file='binlog.000004',
master_log_pos= 155, master_bind=''.
    2019-05-10T02:56:25.235523Z 171 [Warning] [MY-010897] [Repl] Storing MySQL
user name or password information in the master info repository is not secure and
is therefore not recommended. Please consider using the USER and PASSWORD connection
options for START SLAVE; see the 'START SLAVE Syntax' in the MySQL Manual for more
information.
    2019-05-10T02:56:25.237851Z 171 [System] [MY-010562] [Repl] Slave I/O thread
for channel '': connected to master 'repl@172.16.1.125:3306',replication started
in log 'binlog.000004' at position 155
```

1.3.3 联机

脱机建立复制的需求太过理想化，大多数情况下，复制被要求在不影响线上业务的情况下联机创建，而且还要求对线上库的影响越小越好。例如，在复制过程中主库加锁会影响对主库的访问，因此通常不允许对主库加锁。这种场景下有两种备选的复制方案：使用 mysqldump 程序或使用如 XtraBackup 的第三方工具。这两种方案都有各自的适用场合。使用 mysqldump 联机建立复制的过程如下：

步骤 01 在主库上建立复制专属用户：

```
-- 主库
create user 'repl'@'%' identified with mysql_native_password by '123456';
grant replication client,replication slave on *.* to 'repl'@'%';
```

步骤 02 在从库上创建主库信息：

```
change master to
     master_host='172.16.1.125',
     master_port=3306,
     master_user='repl',
     master_password='123456';
```

注意在上面这条命令中并没有指定主库二进制日志的文件名和位置。

步骤 03 在从库用 mysqldump 建立复制：

```
mysqldump --single-transaction --all-databases --master-data=1 --host=
172.16.1.125 --default-character-set=utf8mb4 --user=wxy --password=123456
--apply-slave-statements | mysql -uroot -p123456 -h127.0.0.1
```

说明：

- --single-transaction 参数可以对 InnoDB 表执行非锁定导出。此选项将事务隔离模式设置为 REPEATABLE READ，并在转储数据之前向服务器发送 START TRANSACTION SQL 语句。它仅适用于 InnoDB 等事务表，因为它会在发出 START TRANSACTION 时转储数据库的一致状态，而不会阻塞任何应用程序。因此这里假定：①所有的应用数据表都使用 InnoDB 引擎；②所有非 InnoDB 表数据在备份过程中不会发生变化。
- --master-data 参数会导致转储输出包含类似 CHANGE MASTER TO MASTER_LOG_FILE = 'binlog.000004', MASTER_LOG_POS=1480; 的 SQL 语句，该语句指示主库的二进制日志坐标（文件名和偏移量）。如果选项值为 2，则 CHANGE MASTER TO 语句将写为 SQL 注释，因此仅提供信息，不会执行。如果参数值为 1，则该语句不会写为注释，并在重新加载转储文件时执行。如果未指定选项值，则默认值为 1。
- --apply-slave-statements 参数会在 CHANGE MASTER TO 语句之前添加 STOP SLAVE 语句，并在输出结尾处添加 START SLAVE 语句，用来自动开启复制。
- --default-character-set 参数指定默认字符集，MySQL 8 默认的字符集为 utf8mb4。
- 通过管道操作符，导出和导入一步进行，不需要中间落盘生成文件。

步骤 04 确认复制状态：

```
-- 从库
show slave status\G
```

mysqldump 方式的优点是可以进行部分复制，如在配置文件中定义 replicate-do-table=db1.*，则用这种方法可以只复制 db1 库而忽略其他复制事件。缺点是由于 mysqldump 会生成主库转储数据的 SQL 语句，实际是一种逻辑备份方式，所以速度较慢，不适用于大库。

联机建立复制的另一种可选方案是使用 XtraBackup。XtraBackup 是 Percona 公司的开源项目，用以实现 InnoDB 官方的热备份工具 InnoDB Hot Backup 类似的功能，它支持在线热备份，备份时不影响数据读写。到目前为止，最新的版本为 Percona XtraBackup 8.0.6，可以从 https://www.percona.com/downloads/下载安装包。XtraBackup 有很多功能和优点，例如支持全备份、增量备份；支持备份压缩；备份不影响数据读写、事务等。但是它也存在不足之处，例如不支持部分备份；不支持直接备份到磁带设备；不支持 Cloud Back；MyISAM 的备份也会阻塞等。不过瑕不掩瑜，这些小缺陷不影响 XtraBackup 成为一款流行的 MySQL 备份工具。另外，XtraBackup 只支持 Linux 平台。下面演示用 XtraBackup 联机搭建主从复制的过程，主库已经建立了用于执行复制的用户 repl。具体步骤如下：

步骤 01 在主库和从库安装 XtraBackup：

```
# 安装依赖包
yum -y install libev

# 安装 XtraBackup
rpm -ivh percona-xtrabackup-80-8.0.6-1.el7.x86_64.rpm
```

步骤 02 配置主库到从库的 SSH 免密连接：

```
# 主库执行
ssh-keygen
... 一路按回车键 ...
ssh-copy-id 172.16.1.126
```

步骤 03 停止从库，并清空从库的数据目录：

```
# 从库执行
mysqladmin -u root -p123456 shutdown

# 清空数据目录
rm -rf /usr/local/mysql/data/*
```

步骤 04 备份并传输：

```
# 主库执行
xtrabackup -uroot -p123456 --socket=/tmp/mysql.sock --no-lock --backup
--compress --stream=xbstream --parallel=4 --target-dir=./ | ssh mysql@172.16.1.126
"xbstream -x -C /usr/local/mysql/data/ --decompress"
```

这条命令用于连接主库，并行压缩流式备份，同时将备份通过管道操作符传输到从库，并直接解压缩到从库的数据目录。所有操作由一条命令来完成，不需要中间落盘生成文件。具体步骤如下：

步骤 01 在从库恢复备份：

```
# 应用日志
xtrabackup --prepare --target-dir=/usr/local/mysql/data/
```

步骤 02 查看二进制日志坐标：

```
cat /usr/local/mysql/data/xtrabackup_binlog_info
```

步骤 03 启动从库：

```
mysqld_safe --user=mysql &
```

步骤 04 登录从库启动复制：

```
-- 创建主库信息，其中的master_log_file和master_log_pos值来自第6步
change master to
      master_host='172.16.1.125',
      master_port=3306,
      master_user='repl',
      master_password='123456',
      master_log_file='binlog.000011',
      master_log_pos=155;

-- 启动复制
start slave;

-- 确认复制状态
show slave status\G
```

XtraBackup 是物理备份，性能比 mysqldump 高得多，而且对主库的影响极小，非常适用于从头联机创建高负载、大数据量、全实例从库的场景。

1.4 小 结

绝大部分 MySQL 高可用方案都离不开其原生的复制功能，因此作为全书的开篇，本章介绍了 MySQL 最基本的复制形式，异步复制。我们描述了复制的主要用途，复制的工作原理和流程。MySQL 复制以二进制日志 binlog 为基础，它是在 MySQL 服务器层实现的，而在存储引擎层有自己的事务日志。通过 XA 的两阶段提交，两种日志协调配合，既保证了实例级的崩溃恢复，又保证了主从库的数据一致性。

除理论外，我们还讲解了 MySQL 安装，以及配置异步复制的详细步骤。从最简单的空库，到理想情况下的脱机，再到现实中的联机，各种场景下配置 MySQL 复制的具体操作，都在本章中逐一说明。本章是学习 MySQL 高可用的起点，下一章我们将介绍 MySQL 的另一种复制形式——半同步复制。

第2章 半同步复制

2.1 MySQL 半同步复制简介

直到目前的最新版本，MySQL 默认依然使用异步复制策略。所谓的异步复制，指的是主库写二进制日志、从库的 I/O 线程读主库的二进制日志和写本地中继日志、从库的 SQL 线程重放中继日志，这三步操作都是异步进行的。如此选择的主要理由是出于性能的考虑，与同步复制相比，异步复制显然更快，同时能承载更高的吞吐量。不过，异步复制的缺点同样显而易见，它不能保证主从数据实时一致，也无法控制从库的延迟时间，因此它不适于要求主从数据强一致性的场景。例如，为了分解读写压力，同一个程序写主库读从库，且要求读到的数据与读主库的相同，异步复制无法满足这种强数据一致性的需求。异步复制的另一个问题是可能会有数据丢失，例如主库宕机时，已经提交的事务可能还没有传到从库上，如果此时强行进行主从切换，那么就可能导致新主库上的数据不完整。

于是在 MySQL 5.5 中就自然而然地引入了半同步复制，用作异步复制的可选替代方案，它具有以下特性：

- 从库在连接主库时表明它是否支持半同步复制。
- 如果在主库启用了半同步复制，并且至少有一个支持半同步复制的从库，则主库上执行事务提交的线程将等待，直到至少一个半同步从库确认已收到事务的所有事件（此时从库会向主库发送 ACK 信息，即确认信息。ACK 为 Acknowledgement 的缩写），或者直到发生超时。
- 只有在将事件写入其中继日志并刷新到磁盘后，从库才会确认收到事务的事件，即向主库发送 ACK。
- 如果在没有任何从库确认事务的情况下发生超时，则主库将退化为异步复制。当至少有一个半同步从库赶上时，主库恢复半同步复制。退化与恢复过程都是自动的。
- 必须在主库和从库上都启用半同步复制，否则使用异步复制。

当主库因等待来自从库的确认而阻塞时，它不会返回执行事务的会话。阻塞结束时，主库返回到会话，然后该会话可以继续执行其他语句。此时，事务已在主库提交，并且至少一个从库已确认其事件的接收。在继续之前，主库必须收到的确认从库的数量，可使用rpl_semi_sync_master_wait_for_slave_count系统变量进行配置，默认值为1。

不只是事务提交，事务回滚时主库也会发生同样的阻塞。MySQL同时支持多个数据库引擎，当一个事务中既包含事务表又包含非事务表时，回滚即使对事务表没有影响，二进制日志中也会记录非事务表的事件，因为对非事务表的修改无法回滚并且必须发送到从库。

在未以START TRANSACTION或SET autocommit = 0开启事务时，每条语句都自动隐式提交。使用半同步复制时，主库上的这类语句就像显式事务提交一样。

为了加深对半同步复制中“半”的理解，简单将其与异步和全同步复制进行比较：

- 异步复制：主库提交事务时，将事件写入它的二进制日志，而从库在准备就绪时请求它们。主库无需等待从库的ACK回复，直接提交事务并返回客户端。异步复制不确保所有事件都能到达从库，因而无法保证数据的强一致性。
- 全同步复制：当主库提交事务时，所有从库也将在主库返回执行事务的会话之前提交事务。这样做的缺点是完成事务可能会有很大的延迟。
- 半同步复制：介于异步复制和全同步复制之间。主库仅等待至少一个从库接收并记录事件。它不会等待所有从库都确认收到，并且从库只需要确认接收，而不是事件已在从库处理完成和提交。

与异步复制相比，半同步复制提供了改进的数据完整性，因为当提交成功返回时，已知数据至少存在于两个位置。半同步复制确实会对性能产生一些影响，因为需要等待从库，提交速度会变慢，延迟至少是将提交发送到从库，并等待从库确认收到的一个TCP/IP往返时间（RTT，Round-Trip Time）。这意味着半同步复制最好用于低延时的网络中。

2.2 性能提升

MySQL一直致力于提升半同步复制的性能，从以下三个方面便可见：

（1）支持发送二进制日志事件和接收ACK的异步化。

（2）控制主库接收确认从库的反馈数量。

（3）二进制日志互斥锁的改进。

2.2.1 ACK异步化

旧版本的半同步复制受限于Binlog Dump线程，原因是该线程承担了两个不同且又十分频繁的任务：传送二进制日志事件给从库；接收从库的ACK反馈信息。这两个任务是串行的，Binlog Dump线程必须等待从库返回之后才会传送下一个事件。Binlog Dump线程已然成为整个半同步复制性能的瓶颈。在高并发业务场景下，这样的机制会影响数据库整体的吞吐量。单个Binlog Dump

线程发送和接收的工作流程如图 2-1 所示。

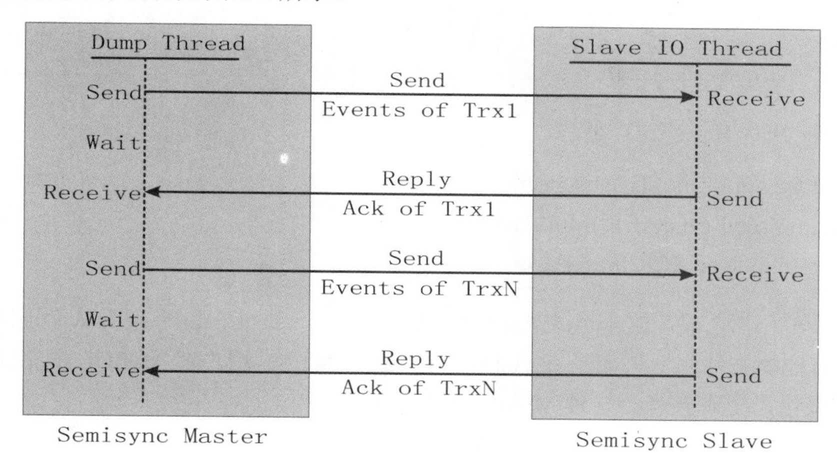

图 2-1　单个 Binlog Dump 线程发送和接收的工作流程

为了解决上述问题，在 MySQL 5.7.4 版本的半同步复制框架中，独立出一个 Ack Receiver 线程，专门用于接收从库返回的 ACK 请求，这将之前 Binlog Dump 线程的发送和接收工作分为了两个线程来处理。这样主库上有两个线程独立工作，可以同时发送二进制日志事件到从库，和接收从库的 ACK 信息。因此半同步复制得到了极大的性能提升。Binlog Dump 线程与 Ack Receiver 线程的工作流程如图 2-2 所示。

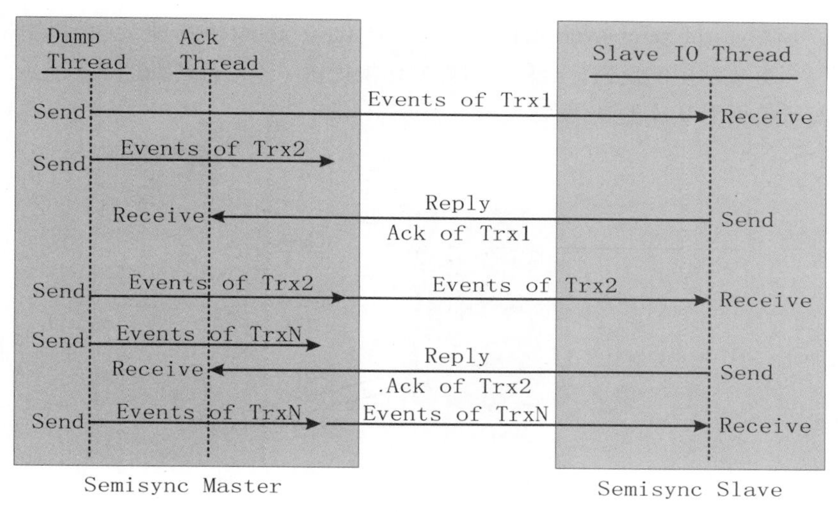

图 2-2　Binlog Dump 线程与 Ack Receiver 线程的工作流程

Ack Receiver 线程在主库启用半同步复制时创建，并在主库禁用半同步复制时销毁，它是自动创建和销毁的，因此不受用户控制。它的状态信息可以从 performance_schema 中查询到：

```
select name, type, processlist_state
 from performance_schema.threads
 where name like '%ack_receiver%';
+----------------------------+----------+-------------------------------+
|name                        |type      |processlist_state              |
```

```
+----------------------------+---------+--------------------------------+
|thread/semisync/Ack_receiver|BACKGROUND|Waiting for semi-sync ACK from 
slave|
+----------------------------+---------+--------------------------------+
```

Ack receiver 线程有以下三个状态:

- Waiting for semi-sync slave connection
- Waiting for semi-sync ACK from slave
- Reading semi-sync ACK from slave

在 MySQL 5.7.17 之前，这个 Ack Receiver 线程采用了 select 机制来监听从库返回的结果，然而 select 机制监控的文件句柄只能是 0~1024，当超过 1024 时，用户在 MySQL 的错误日志中会收到类似如下的报错，更有甚者会导致 MySQL 发生宕机。

```
semi-sync master failed on net_flush() before waiting for slave reply.
```

从 MySQL 5.7.17 版本开始，官方修复了这个 bug，开始使用 poll 机制来替换原来的 select 机制，从而可以避免上面的问题。其实 poll 调用本质上和 select 没有区别，只是 I/O 句柄数理论上没有了上限，因为它是基于链表来存储的。

2.2.2 控制从库反馈的数量

MySQL 5.7 新增了 rpl_semi_sync_master_wait_for_slave_count 系统变量，可以用来控制主库接收多少个从库写事务成功后的反馈，给高可用架构切换提供了灵活性。如图 2-3 所示，当该变量值为 2 时，主库需等待两个从库的 ACK。

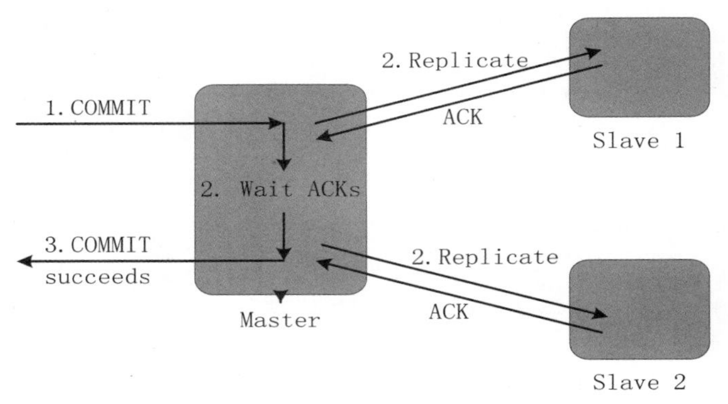

图 2-3 “一主两从”的半同步复制

使用这个功能，可以在不同机房部署主服务器和两个从服务器，并配置半同步复制以将事务复制到至少两个从库，以便在多个服务器一次性崩溃的情况下减少数据丢失的可能，从库越多，数据越安全。

2.2.3 二进制日志互斥锁的改进

旧版本半同步复制在主库提交二进制日志的写会话和 Binlog Dump 线程读取二进制日志的操作时，都会对二进制日志添加 binlog lock 互斥锁，用于保护二进制日志的读写安全。使用此互斥锁，二进制日志读写操作是安全的，但会导致二进制日志文件的读写串行化。不仅 Binlog Dump 线程和用户会话不能同时读写二进制日志，就连多个 Binlog Dump 线程本身也无法同时读写。每当一个会话正在读取或写入二进制日志文件时，所有其他会话都必须等待。如此顺序读写是一个瓶颈，尤其是当读写操作很慢时。串行化读写二进制日志如图 2-4 所示。

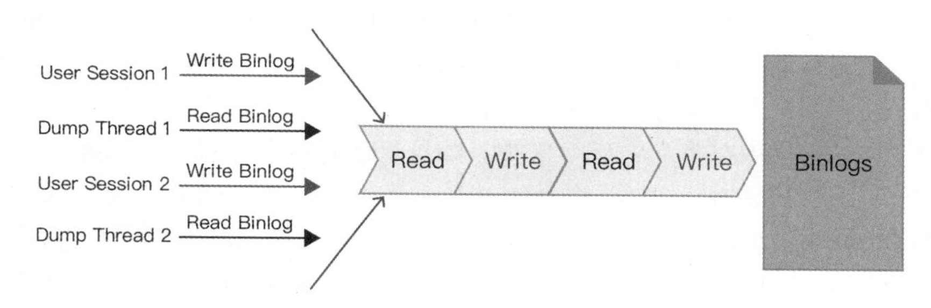

图 2-4 串行化读写二进制日志

MySQL 5.7.2 对 binlog lock 进行了以下两方面的优化:

- 从 Binlog Dump 线程中移除 binlog lock。
- 加入了安全边际以保证二进制日志的读安全。

二进制日志文件是一个仅用于追加二进制事件的日志文件，可以安全地从中读取没有锁定的二进制事件，因此可以从 Binlog Dump 线程中删除 binlog 锁。不使用 binlog 锁，而是为活动 binlog 维护安全读取边界（最大位置）。Binlog Dump 线程永远不会读取超过安全读取的边界。当到达边界时，它将等待边界更新。用户会话负责在追加了二进制事件后更新安全读取边界。改进后的二进制日志读写如图 2-5 所示。

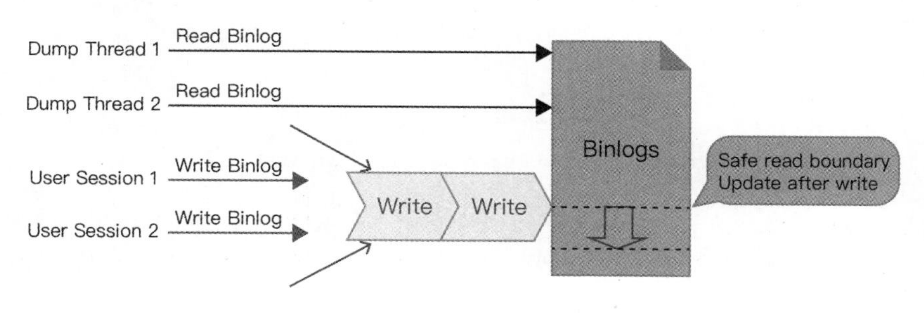

图 2-5 改进后的二进制日志读写

从图 2-5 中一目了然：

- 读取二进制日志事件时，Binlog Dump 线程不会相互阻塞。
- 正在写二进制日志事件的用户会话不会阻止 Binlog Dump 线程。

- 读取二进制日志事件的 Binlog Dump 不会阻塞用户会话。

因此，Binlog Dump 线程和用户会话都可以获得更好的吞吐量，尤其是在有很多从库时，这种改进非常显著。

2.3 数据一致性

半同步复制最主要的目标是保证主从数据强一致性。要想搞清楚这其中的原理和具体实现，需要研读 MySQL 源代码。MySQL semi-sync 半同步以插件方式引入，源代码在 plugin/semisync 目录下。阿里有一篇很好的文章（http://mysql.taobao.org/monthly/2017/04/01/），从官方 MySQL 5.7 源代码的层面分析了半同步复制的数据一致性问题。

在半同步方式中，主库在等待从库 ACK 时，如果超时则会退化为异步复制，这就可能导致数据丢失。主库等待提交确认的超时时间，由 rpl_semi_sync_master_timeout 参数控制，默认值为 10000 毫秒。在下面的分析中，假设 rpl_semi_sync_master_timeout 足够大，不会退化为异步方式。这里通过三个参数 rpl_semi_sync_master_wait_point、sync_binlog、sync_relay_log 的配置来对半同步复制进行数据一致性的分析。

2.3.1 rpl_semi_sync_master_wait_point 配置

（1）源代码剖析

```
plugin/semisync/semisync_master_plugin.cc

68 int repl_semi_report_binlog_sync(Binlog_storage_param *param,
69                                  const char *log_file,
70                                  my_off_t log_pos)
71 {
72   if (rpl_semi_sync_master_wait_point == WAIT_AFTER_SYNC)
73     return repl_semisync.commitTrx(log_file, log_pos);
74   return 0;
75 }

97 int repl_semi_report_commit(Trans_param *param)
   ...
102   if (rpl_semi_sync_master_wait_point == WAIT_AFTER_COMMIT &&
106     return repl_semisync.commitTrx(binlog_name, param->log_pos);
```

（2）设置为 WAIT_AFTER_COMMIT

rpl_semi_sync_master_wait_point 为 WAIT_AFTER_COMMIT 时，commitTrx 的调用在引擎层提交之后（由 ordered_commit 函数中的 process_after_commit_stage_queue 调用），如图 2-6 所示。即在等待从库 ACK 时，虽然没有返回当前客户端，但事务已经提交，其他客户端会读取到已提交的事务。如果从库还没有读到该事务在二进制日志中的事件，同时主库发生了崩溃，然后切换到从库。那么之前读到的事务就不见了，出现了幻读，如图 2-7 所示。

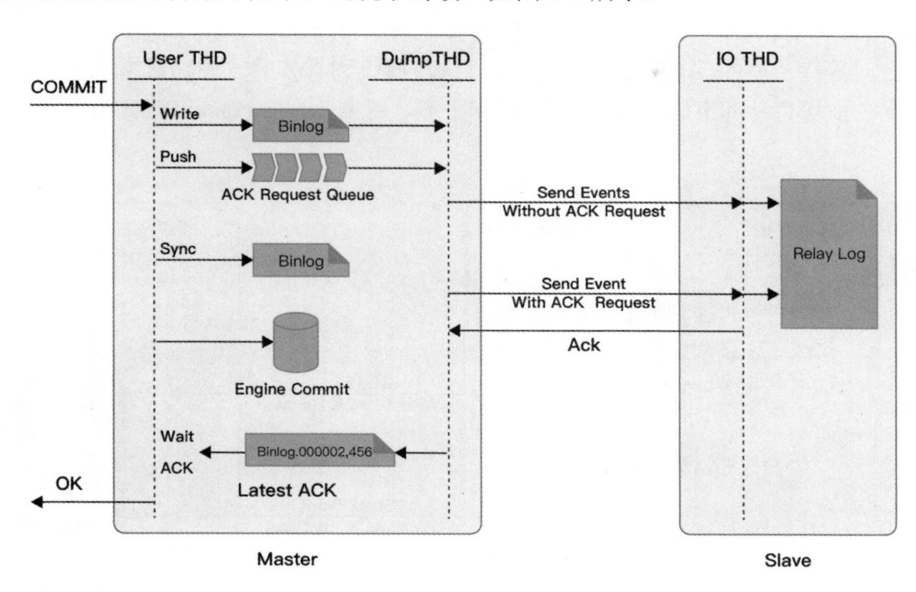

图 2-6　WAIT_AFTER_COMMIT 处理流程

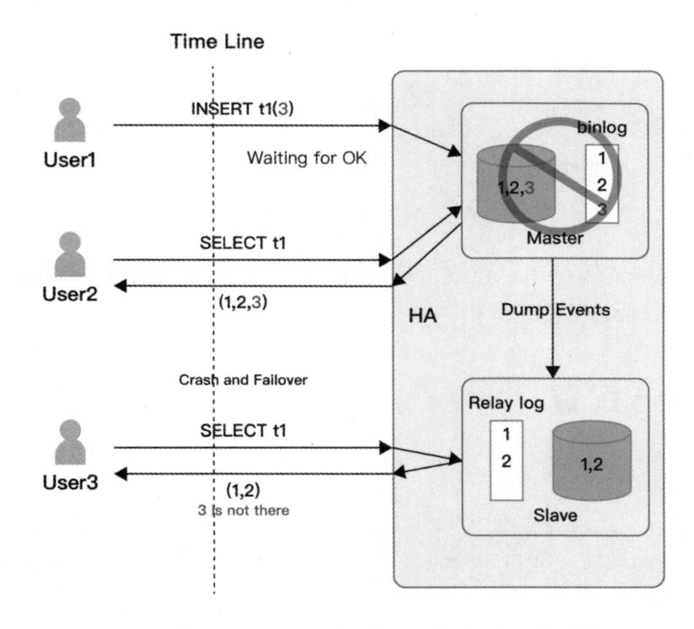

图 2-7　WAIT_AFTER_COMMIT 导致幻读

除了幻读，这种场景还有一个问题是，如果客户端会重新尝试把该事务提交到新的主库上，当宕机的主库重新启动后，以从库的身份重新加入到该主从结构中，那么此时就会发现，该事务在从库中被提交了两次，一次是之前作为主库的时候，一次是被新主库同步过来的，结果依然是主从数据不一致。

（3）设置为 WAIT_AFTER_SYNC

MySQL 针对上述问题，在 5.7.2 引入了 Loss-less Semi-Synchronous。在调用 binlog sync 之后，引擎层提交之前等待从库的 ACK。这样只有在确认从库收到事务的二进制日志事件后，事务才会提交。在提交之前等待从库 ACK，同时可以堆积事务，趋向 group commit，有利于提升性能，如图 2-8 所示。

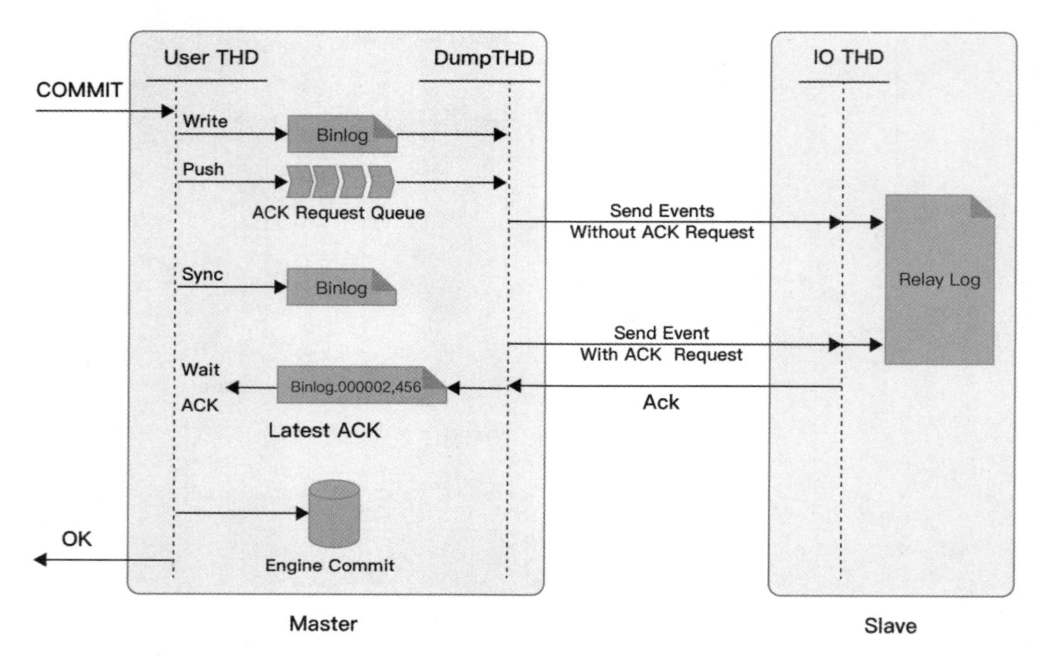

图 2-8 WAIT_AFTER_SYNC 处理流程

其实在图 2-8 的流程中依然存在着导致主从数据不一致，使主从同步失败的情形。进一步的说明可见下面章节对 sync_binlog 配置的分析。

2.3.2 sync_binlog 配置

（1）源代码剖析

```
sql/binlog.cc ordered_commit
     //当 sync_period(sync_binlog)为 1 时，在 sync 之后 update binlog end pos
9002  update_binlog_end_pos_after_sync= (get_sync_period() == 1);
      ...
9021    if (!update_binlog_end_pos_after_sync)
          //更新 binlog end position,dump 线程会发送更新后的 events
9022      update_binlog_end_pos();
      ...
        //
9057    std::pair<bool, bool> result= sync_binlog_file(false);
      ...
9061  if (update_binlog_end_pos_after_sync)
```

```
9062   {
       ...
9068       update_binlog_end_pos(tmp_thd->get_trans_pos());
9069   }

sql/binlog.cc sync_binlog_file
8618 std::pair<bool, bool>
8619 MYSQL_BIN_LOG::sync_binlog_file(bool force)
8620 {
8621   bool synced= false;
8622   unsigned int sync_period= get_sync_period();//sync_binlog 值
       //sync_period 为 0 不执行 sync 操作，非 0 值表示执行 sync 的调用次数
8623   if (force || (sync_period && ++sync_counter >= sync_period))
8624   {
```

（2）设置分析

当 sync_binlog 设置为 0 时，binlog sync 磁盘由操作系统负责。当不为 0 时，其数值为定期刷磁盘的 binlog commit group 次数。当 sync_binlog 值大于 1 时，sync binlog 操作可能并没有使 binlog 落盘。如果没有落盘，事务在提交前，主库掉电，然后恢复，那么这个时候该事务被回滚。但是，从库上可能已经收到了该事务的二进制日志事件并且执行了，这个时候就会出现从库事务比主库多的情况，主从同步失败。因此，如果要保持主从一致，则需要设置 sync_binlog 为 1。

WAIT_AFTER_SYNC 和 WAIT_AFTER_COMMIT 两图中 Send Events 的位置（图 2-6 和图 2-8），也可能导致主从数据不一致，出现同步失败的情况。实际在 rpl_semi_sync_master_wait_point 分析的图中是 sync binlog 大于 1 的情况。根据上面源代码，流程如图 2-9 所示。主库依次执行 flush binlog、update binlog position、sync binlog。如果主库在 update binlog position 后，且 sync binlog 前掉电，主库再次启动后原事务就会被回滚，但可能出现从库已经获取到事件，这也会导致从库数据比主库多的情况，结果主从同步失败。

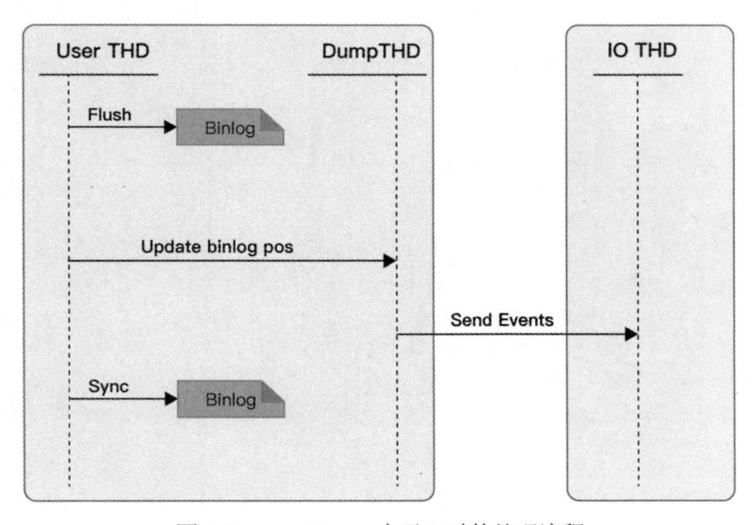

图 2-9 sync_binlog 大于 1 时的处理流程

由于上面的原因，sync_binlog 设置为 1 时，MySQL 会在 sync 后更新二进制日志坐标。流程

如图 2-10 所示。这时，对于每一个事务都需要 sync binlog，同时 sync binlog 和网络发送二进制日志事件会是一个串行的过程，性能会下降。

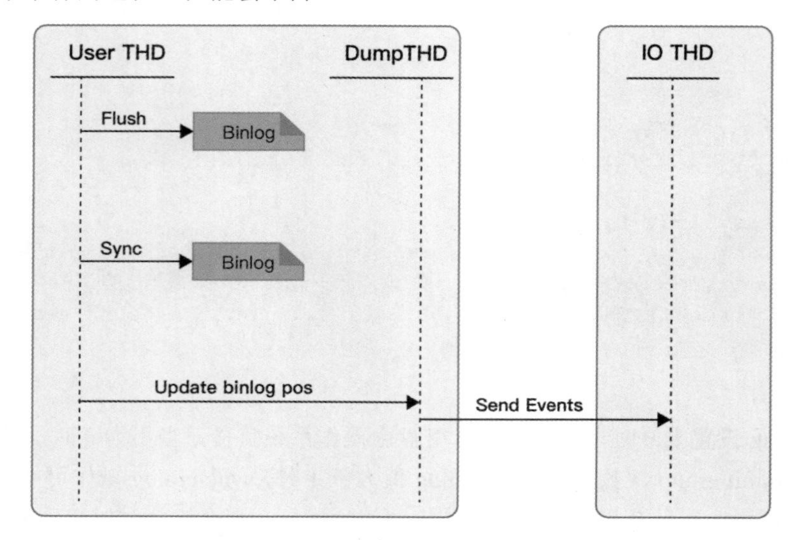

图 2-10　sync_binlog 等于 1 时的处理流程

2.3.3　sync_relay_log 配置

（1）源代码剖析

```
    sql/rpl_slave.cc handle_slave_io

    5764        if (queue_event(mi, event_buf, event_len))
                ...
    5771        if (RUN_HOOK(binlog_relay_io, after_queue_event,
    5772                     (thd, mi, event_buf, event_len, synced)))

    after_queue_event
    ->plugin/semisync/semisync_slave_plugin.cc repl_semi_slave_queue_event
    ->plugin/semisync/semisync_slave.cc ReplSemiSyncSlave::slaveReply

    queue_event
    ->sql/binlog.cc MYSQL_BIN_LOG::append_buffer(const char* buf, uint len,
Master_info *mi)
    ->sql/binlog.cc after_append_to_relay_log(mi);
    ->sql/binlog.cc flush_and_sync(0)
    ->sql/binlog.cc sync_binlog_file(force)
```

（2）设置分析

在从库的 I/O 线程中 get_sync_period 获得的是 sync_relay_log 的值。与 sync_binlog 对 sync 控制一样，当 sync_relay_log 不是 1 时，semisync 返回给主库的 position 可能没有刷新到磁盘。开启 GTID 时，在保证前面两个设置正确的情况下，sync_relay_log 不是 1 的时候，仅发生主库或从库的一次崩溃并不会造成数据丢失或者主从同步失败的情况。如果发生从库没有 sync relay log，主库事务提交，客户端观察到事务提交，然后从库崩溃，这样从库端就会丢失掉已经回复主库 ACK 的事务，如图 2-11 中所示的 GTID:xx:1-40。

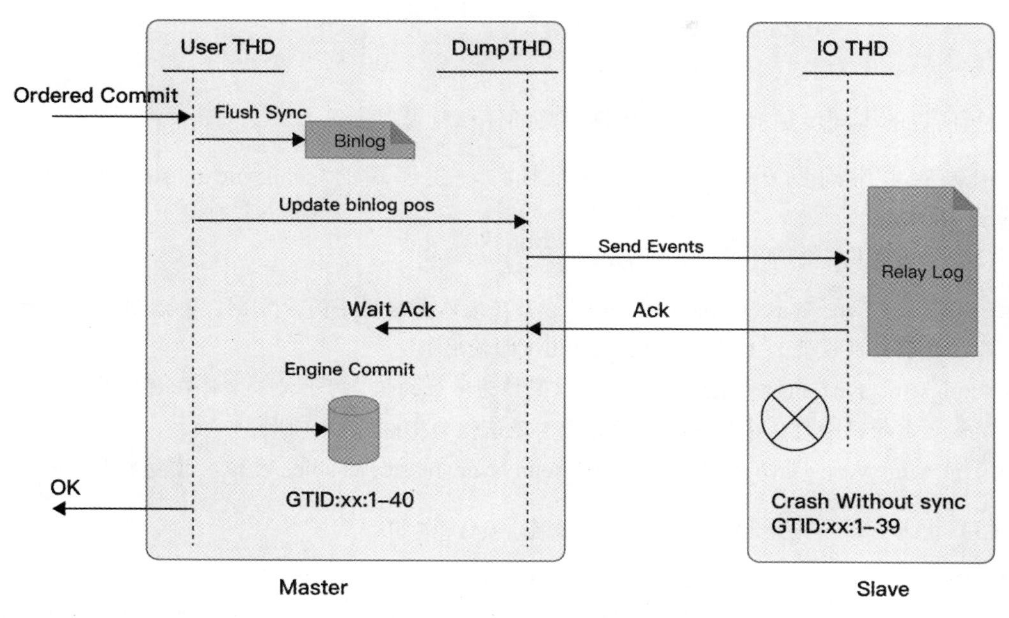

图 2-11　sync_relay_log 不为 1 时从库崩溃

但当从库再次启动时，会从主库同步丢失事务的二进制日志事件。如果没有来得及这样做主库就崩溃了，此时用户访问从库就会发现数据丢失，如图 2-12 所示。

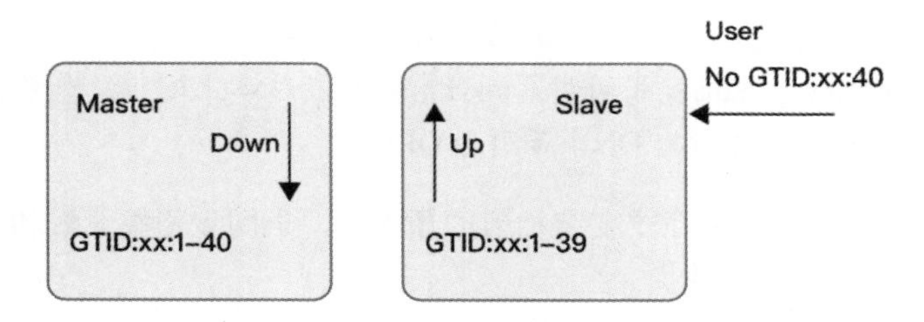

图 2-12　主库在从库同步前崩溃导致数据丢失

通过上面这个例子可知，MySQL 半同步复制如果要保证任意时刻发生一台机器宕机都不丢失数据，则需要同时设置 sync_relay_log 为 1。对 relay log 的 sync 操作是在 queue_event 中，每个事件都要 sync，所以 sync_relay_log 设置为 1 时，事务响应时间会受到影响，对于涉及数据比较多的事务延迟会增加很多。

这么多分析后我们不难发现，当前原生的 MySQL 主从复制要同时满足数据一致性、高可用和高性能，依然是力有不逮。

2.4 管理接口

这里所说的半同步复制管理接口包括相关插件和变量。

（1）实现半同步复制功能涉及两个插件，主库端的 semisync_master.so 和从库端的 semisync_slave.so。

（2）系统变量控制插件行为，例如：

- rpl_semi_sync_master_enabled：控制是否在主库上启用半同步复制。要启用或禁用插件，将此变量分别设置为 1 或 0，默认值为 0（即禁用）。
- rpl_semi_sync_master_timeout：一个以毫秒为单位的值，用于控制主库在超时并退化到异步复制之前等待来自从库确认提交的时间，默认值为 10000（10 秒）。
- rpl_semi_sync_slave_enabled：与 rpl_semi_sync_master_enabled 类似，控制启用从库的插件。

（3）状态变量用来反映半同步复制的状态信息，例如：

- Rpl_semi_sync_master_clients：半同步从库的数量。
- Rpl_semi_sync_master_status：半同步复制当前是否在主库上运行。如果已启用插件且未发生复制退化，则该值为 ON。如果未启用插件，或者由于提交确认超时，主服务器已回退到异步复制，则为 OFF。
- Rpl_semi_sync_master_no_tx：从库未成功确认的事务数。
- Rpl_semi_sync_master_yes_tx：从库成功确认的事务数。
- Rpl_semi_sync_slave_status：半同步复制当前是否在从库上运行。如果插件已启用且从库的 I/O 线程正在运行，则此值为 ON，否则为 OFF。

仅当使用 INSTALL PLUGIN 命令安装相应的插件后，半同步相关系统和状态变量才可用。

2.5 配置半同步复制

上一章中我们配置了“一主两从”的 MySQL 异步复制，下面在相同环境下配置半同步复制。

2.5.1 安装插件

半同步复制是使用插件实现的，因此必须将插件安装到 MySQL 服务器中才能使用它们。安装插件后，可通过与之关联的系统变量来控制它。要使用半同步复制，必须满足以下要求：

- 安装插件需要 MySQL 服务器支持动态加载。要验证这一点，检查 have_dynamic_loading 系统变量的值是否为 YES。MySQL 8 默认为 YES。

- 已经启动了异步复制。
- 半同步不支持多源复制（“一从多主”的情况）。

安装设置半同步复制，需要 REPLICATION_SLAVE_ADMIN 或 SUPER 权限。MySQL 发行版包括主、从端的半同步复制插件文件 semisync_master.so 和 semisync_slave.so，默认位于 MySQL 安装目录下的 lib/plugin 目录下，本例中为/usr/local/mysql/lib/plugin。也可以通过设置 plugin_dir 系统变量的值指定插件目录位置。

执行下面的 SQL 语句加载插件：

```
-- 在主库
install plugin rpl_semi_sync_master soname 'semisync_master.so';
-- 在每个从库
install plugin rpl_semi_sync_slave soname 'semisync_slave.so';
```

如果安装插件时 Linux 上出现类似于下面所示的错误，则必须安装 libimf：

```
mysql> install plugin rpl_semi_sync_master soname 'semisync_master.so';
ERROR 1126 (HY000): Can't open shared library
'/usr/local/mysql/lib/plugin/semisync_master.so'
(errno: 22 libimf.so: cannot open shared object file: No such file or directory)
```

可以通过网址 https://dev.mysql.com/downloads/os-linux.html 获取 libimf。检查 INFORMATION_SCHEMA.PLUGINS 表或使用 SHOW PLUGINS 语句验证插件安装，例如：

```
mysql> select plugin_name, plugin_status
    -> from information_schema.plugins
    -> where plugin_name like '%semi%';
+----------------------+---------------+
| plugin_name          | plugin_status |
+----------------------+---------------+
| rpl_semi_sync_master | ACTIVE        |
+----------------------+---------------+
1 row in set (0.00 sec)
```

检查相关系统变量的初始值：

```
-- 主库
mysql> show variables like '%semi%';
+-------------------------------------------+------------+
| Variable_name                             | Value      |
+-------------------------------------------+------------+
| rpl_semi_sync_master_enabled              | OFF        |
| rpl_semi_sync_master_timeout              | 10000      |
| rpl_semi_sync_master_trace_level          | 32         |
| rpl_semi_sync_master_wait_for_slave_count | 1          |
| rpl_semi_sync_master_wait_no_slave        | ON         |
| rpl_semi_sync_master_wait_point           | AFTER_SYNC |
+-------------------------------------------+------------+
6 rows in set (0.00 sec)

--从库
```

```
mysql> show variables like '%semi%';
+---------------------------------+-------+
| Variable_name                   | Value |
+---------------------------------+-------+
| rpl_semi_sync_slave_enabled     | OFF   |
| rpl_semi_sync_slave_trace_level | 32    |
+---------------------------------+-------+
2 rows in set (0.00 sec)
```

主库的相关系统变量：

- rpl_semi_sync_master_enabled：主库是否启用了半同步复制，默认为OFF。
- rpl_semi_sync_master_timeout：等待从库的ACK回复的超时时间，默认为10秒。
- rpl_semi_sync_master_trace_level：半同步复制时主库的调试级别。
- rpl_semi_sync_master_wait_for_slave_count：主库在超时时间内需要收到多少个ACK回复才认为此次提交成功，否则就降级为异步复制。该变量在MySQL 5.7.3版本才提供，在此之前的版本都默认为收到1个ACK则确认成功，且不可更改。MySQL 5.7.3之后该变量的默认值也是1。
- rpl_semi_sync_master_wait_no_slave：默认值为ON，当状态变量Rpl_semi_sync_master_clients中的值小于rpl_semi_sync_master_wait_for_slave_count时，Rpl_semi_sync_master_status依旧为ON，只有当事务提交后等待rpl_semi_sync_master_timeout超时后，Rpl_semi_sync_master_status才会变为OFF，即降级为异步复制；为OFF时，当状态变量Rpl_semi_sync_master_clients中的值小于rpl_semi_sync_master_wait_for_slave_count时，Rpl_semi_sync_master_status立即显示为OFF，即立即降级为异步复制。
- rpl_semi_sync_master_wait_point：控制主库上commit、接收ACK、返回消息给客户端的时间点。值为AFTER_SYNC或AFTER_COMMIT。该选项是MySQL 5.7.2版本才引入的，默认值为AFTER_SYNC。此版本之前，等价于使用了AFTER_COMMIT模式。

从库的相关系统变量：

- rpl_semi_sync_slave_enabled：从库是否开启半同步复制。
- rpl_semi_sync_slave_trace_level：从库的调试级别。

2.5.2 启用半同步复制

安装半同步复制插件后，默认情况下会禁用它。必须在主库和从库都启用插件才能使用半同步复制。如果仅启用一侧，则复制将是异步的。

1. 启用半同步复制

```
-- 主库
set global rpl_semi_sync_master_enabled = 1;

-- 从库
set global rpl_semi_sync_slave_enabled = 1;
```

以上的启动方式是在命令行执行的，也可写在配置文件中：

```
# 主库
plugin-load="rpl_semi_sync_master=semisync_master.so"
rpl_semi_sync_master_enabled=1

# 从库
plugin-load="rpl_semi_sync_slave=semisync_slave.so"
rpl_semi_sync_slave_enabled=1
```

在有的高可用架构下，master 和 slave 需同时启动，以便在切换后能继续使用半同步复制：

```
plugin-load="rpl_semi_sync_master=semisync_master.so;rpl_semi_sync_slave=
            semisync_slave.so"
rpl-semi-sync-master-enabled=1
rpl-semi-sync-slave-enabled=1
```

2. 重启从库上的 I/O 线程

```
stop slave io_thread;
start slave io_thread;
```

如果没有重启，则默认还是异步复制。重启后，从库会在主库上注册为启用半同步复制的从库角色。

3. 查看半同步是否在运行

```
-- 主库
mysql> show status like 'Rpl_semi_sync_master_status';
+------------------------------+-------+
| Variable_name                | Value |
+------------------------------+-------+
| Rpl_semi_sync_master_status  | ON    |
+------------------------------+-------+
1 row in set (0.01 sec)

-- 从库
mysql> show status like 'Rpl_semi_sync_slave_status';
+----------------------------+-------+
| Variable_name              | Value |
+----------------------------+-------+
| Rpl_semi_sync_slave_status | ON    |
+----------------------------+-------+
1 row in set (0.00 sec)
```

这两个变量常用来监控主从库是否运行在半同步复制模式下。至此，MySQL 半同步复制搭建完毕。

2.5.3　监控半同步复制

半同步复制功能的插件公开了几个状态变量，可以检查这些变量以确定其操作状态，它们仅当安装了半同步复制插件以后才可用。

```
-- 主库
mysql> show status like 'rpl_semi_sync%';
+--------------------------------------------+-------+
| Variable_name                              | Value |
+--------------------------------------------+-------+
| Rpl_semi_sync_master_clients               | 2     |
| Rpl_semi_sync_master_net_avg_wait_time     | 0     |
| Rpl_semi_sync_master_net_wait_time         | 0     |
| Rpl_semi_sync_master_net_waits             | 0     |
| Rpl_semi_sync_master_no_times              | 0     |
| Rpl_semi_sync_master_no_tx                 | 0     |
| Rpl_semi_sync_master_status                | ON    |
| Rpl_semi_sync_master_timefunc_failures     | 0     |
| Rpl_semi_sync_master_tx_avg_wait_time      | 0     |
| Rpl_semi_sync_master_tx_wait_time          | 0     |
| Rpl_semi_sync_master_tx_waits              | 0     |
| Rpl_semi_sync_master_wait_pos_backtraverse | 0     |
| Rpl_semi_sync_master_wait_sessions         | 0     |
| Rpl_semi_sync_master_yes_tx                | 0     |
+--------------------------------------------+-------+
14 rows in set (0.00 sec)

-- 从库
mysql> show status like 'rpl_semi_sync%';
+----------------------------+-------+
| Variable_name              | Value |
+----------------------------+-------+
| Rpl_semi_sync_slave_status | ON    |
+----------------------------+-------+
1 row in set (0.01 sec)
```

从库上只有一个半同步复制相关的状态变量 Rpl_semi_sync_slave_status，为 ON 时表示从库使用半同步复制，OFF 表示从库使用异步复制。

主库有 14 个半同步复制相关的状态变量：

- Rpl_semi_sync_master_clients：当前连接了多少个半同步从库。
- Rpl_semi_sync_master_net_avg_wait_time：主库等待从库回复的平均时间，以微秒为单位。此变量始终为 0，不推荐使用，并且将在以后的版本中删除。
- Rpl_semi_sync_master_net_wait_time：主库等待从库回复的总时间，以微秒为单位。此变量始终为 0，不推荐使用，并且将在以后的版本中删除。
- Rpl_semi_sync_master_net_waits：主库等待从库回复的总次数。
- Rpl_semi_sync_master_no_times：主库关闭半同步复制的次数。
- Rpl_semi_sync_master_no_tx：从库未成功确认的事务数。
- Rpl_semi_sync_master_status：为 ON 时表示主库使用半同步复制，为 OFF 时表示主库使用异步复制。

- Rpl_semi_sync_master_timefunc_failures：调用 gettimeofday 等时间函数时主库失败的次数。
- Rpl_semi_sync_master_tx_avg_wait_time：主库等待一个事务的平均时间，以微秒为单位。
- Rpl_semi_sync_master_tx_wait_time：主库等待事务的总时间，以微秒为单位。
- Rpl_semi_sync_master_tx_waits：主库等待事务的总次数。
- Rpl_semi_sync_master_wait_pos_backtraverse：主库等待事件的二进制日志次数低于之前等待事件的总次数。当事务等待回复的顺序与其二进制日志事件的写入顺序不同时，就会发生这种情况。
- Rpl_semi_sync_master_wait_sessions：当前等待从库回复的会话数。
- Rpl_semi_sync_master_yes_tx：从库成功确认的事务数。

2.6　测　试

2.6.1　正常提交事务

```
-- 主库
mysql> create database test;
Query OK, 1 row affected (0.00 sec)

mysql> use test;
Database changed

mysql> create table test.t1 (a int) engine=innodb;
Query OK, 0 rows affected (0.02 sec)

mysql> insert into t1 values(1);
Query OK, 1 row affected (0.02 sec)

mysql> show status like 'rpl_semi_sync%';
+--------------------------------------------+-------+
| Variable_name                              | Value |
+--------------------------------------------+-------+
| Rpl_semi_sync_master_clients               | 2     |
...
| Rpl_semi_sync_master_tx_avg_wait_time      | 758   |
| Rpl_semi_sync_master_tx_wait_time          | 2274  |
| Rpl_semi_sync_master_tx_waits              | 3     |
...
| Rpl_semi_sync_master_yes_tx                | 3     |
+--------------------------------------------+-------+
14 rows in set (0.00 sec)

-- 从库
mysql> select * from test.t1;
+------+
```

```
| a    |
+------+
|    1 |
+------+
1 row in set (0.00 sec)
```

MySQL 8 默认是每条语句自动提交。主库等待 3 个事务的确认，分别对应 create database、create table、insert 语句，平均每个事务等待确认的时长为 758 微秒。从库正常确认了 3 个事务。

2.6.2 回滚事务

```
-- 主库
mysql> show variables like 'autocommit';
+---------------+-------+
| Variable_name | Value |
+---------------+-------+
| autocommit    | OFF   |
+---------------+-------+
1 row in set (0.01 sec)

mysql> set session autocommit=0;                 -- 关闭自动提交，开启事务
Query OK, 0 rows affected (0.00 sec)

mysql> insert into t1 values(2);                 -- 向事务表插入记录
Query OK, 1 row affected (0.00 sec)

mysql> create table t2 (a int) engine=myisam;       -- 创建非事务表 t2
Query OK, 0 rows affected (0.01 sec)

mysql> insert into t1 values(3);                 -- 向事务表插入记录
Query OK, 1 row affected (0.00 sec)

mysql> insert into t2 values(3);                 -- 向非事务表插入记录
Query OK, 1 row affected (0.01 sec)

mysql> rollback;                           -- 回滚事务
Query OK, 0 rows affected, 1 warning (0.00 sec)

mysql> show warnings;
+---------+-----+--------------------------------------------------------------------
----+
| Level   | Code| Message                                                            |
+---------+-----+--------------------------------------------------------------------
----+
| Warning | 1196|Some non-transactional changed tables couldn't be rolled back|
+---------+-----+--------------------------------------------------------------------
----+
1 row in set (0.00 sec)
```

```
mysql> select * from t1;
+-------+
| a     |
+-------+
|     1 |
|     2 |
+-------+
2 rows in set (0.00 sec)

mysql> select * from t2;
+-------+
| a     |
+-------+
|     3 |
+-------+
1 row in set (0.00 sec)

mysql> show status like 'rpl_semi_sync%';
+--------------------------------------------+-------+
| Variable_name                              | Value |
+--------------------------------------------+-------+
| Rpl_semi_sync_master_clients               | 2     |
...
| Rpl_semi_sync_master_tx_avg_wait_time      | 691   |
| Rpl_semi_sync_master_tx_wait_time          | 4149  |
| Rpl_semi_sync_master_tx_waits              | 6     |
...
| Rpl_semi_sync_master_yes_tx                | 6     |
+--------------------------------------------+-------+
14 rows in set (0.00 sec)

-- 从库
mysql> show variables like 'autocommit';
+---------------+-------+
| Variable_name | Value |
+---------------+-------+
| autocommit    | ON    |
+---------------+-------+
1 row in set (0.01 sec)

mysql> select * from test.t1;
+-------+
| a     |
+-------+
|     1 |
|     2 |
+-------+
2 rows in set (0.00 sec)

mysql> select * from test.t2;
```

```
+-------+
| a     |
+-------+
|    3 |
+-------+
1 row in set (0.01 sec)
mysql>
```

从上面的测试可以得出以下结论：

- set session autocommit=0 语句不被复制。
- DDL 语句会触发一个 commit，自动提交 DDL 语句本身及其之前所有未提交的事务。
- 非事务表不能回滚。
- Rpl_semi_sync_master_tx_waits 和 Rpl_semi_sync_master_yes_tx 都增加了 3，分别对应的 3 个事务为：insert into t1 values(2)、create table t2 (a int) engine=myisam、insert into t2 values(3)。

2.6.3 rpl_semi_sync_master_wait_no_slave 与从库数量

（1）rpl_semi_sync_master_wait_no_slave 为 ON 时从库数小于 rpl_semi_sync_master_wait_for_slave_count

```
-- 关闭两个从库的复制
mysql> stop slave;
Query OK, 0 rows affected (0.01 sec)

mysql> show status like 'rpl_semi_sync%';    -- 查看当前是否启用半同步复制
+----------------------------+-------+
| Variable_name              | Value |
+----------------------------+-------+
| Rpl_semi_sync_slave_status | OFF   |
+----------------------------+-------+
1 row in set (0.00 sec)

-- 主库
mysql> show status like 'Rpl_semi_sync_master_clients'; -- 查看当前半同步从库数
+------------------------------+-------+
| Variable_name                | Value |
+------------------------------+-------+
| Rpl_semi_sync_master_clients | 0     |
+------------------------------+-------+
1 row in set (0.00 sec)

mysql> show status like 'Rpl_semi_sync_master_status';-- 查看是否启用半同步复制
+-----------------------------+-------+
| Variable_name               | Value |
+-----------------------------+-------+
| Rpl_semi_sync_master_status | ON    |
+-----------------------------+-------+
```

```
1 row in set (0.00 sec)

mysql> insert into t1 values(3);
Query OK, 1 row affected (0.00 sec)

mysql> commit;
Query OK, 0 rows affected (10.00 sec)

mysql> select * from t1;
+-------+
| a     |
+-------+
|    1 |
|    2 |
|    3 |
+-------+
3 rows in set (0.00 sec)

mysql> show status like 'Rpl_semi_sync_master_status';
+-----------------------------+-------+
| Variable_name               | Value |
+-----------------------------+-------+
| Rpl_semi_sync_master_status | OFF   |
+-----------------------------+-------+
1 row in set (0.01 sec)
```

可以看到，主库提交后等待了10秒（rpl_semi_sync_master_timeout的默认值）才完成，而且此时主库已经降级为异步复制。

```
-- 启动一个从库的复制
mysql> start slave;
Query OK, 0 rows affected (0.00 sec)

mysql> select * from test.t1;
+-------+
| a     |
+-------+
|    1 |
|    2 |
|    3 |
+-------+
3 rows in set (0.00 sec)

mysql> show status like 'rpl_semi_sync%';
+----------------------------+-------+
| Variable_name              | Value |
+----------------------------+-------+
| Rpl_semi_sync_slave_status | ON    |
+----------------------------+-------+
1 row in set (0.01 sec)
```

```
-- 主库
mysql> show variables like 'rpl_semi_sync_master_wait_for_slave_count';
+-------------------------------------------+-------+
| Variable_name                             | Value |
+-------------------------------------------+-------+
| rpl_semi_sync_master_wait_for_slave_count | 1     |
+-------------------------------------------+-------+
1 row in set (0.00 sec)

mysql> show status like 'Rpl_semi_sync_master_clients';
+------------------------------+-------+
| Variable_name                | Value |
+------------------------------+-------+
| Rpl_semi_sync_master_clients | 1     |
+------------------------------+-------+
1 row in set (0.00 sec)

mysql> show status like 'Rpl_semi_sync_master_status';
+-----------------------------+-------+
| Variable_name               | Value |
+-----------------------------+-------+
| Rpl_semi_sync_master_status | ON    |
+-----------------------------+-------+
1 row in set (0.00 sec)

mysql> insert into t1 values(4);
Query OK, 1 row affected (0.00 sec)

mysql> commit;
Query OK, 0 rows affected (0.01 sec)

-- 从库
mysql> select * from test.t1;
+------+
| a    |
+------+
|    1 |
|    2 |
|    3 |
|    4 |
+------+
4 rows in set (0.00 sec)
```

当 Rpl_semi_sync_master_clients 大于等于 rpl_semi_sync_master_wait_for_slave_count 时，主库立即恢复为半同步复制，并发送那些未被复制的 binlog。

（2）rpl_semi_sync_master_wait_no_slave 为 OFF 时，从库数小于 rpl_semi_sync_master_wait_for_slave_count。

```
-- 关闭 rpl_semi_sync_master_wait_no_slave
mysql> set global rpl_semi_sync_master_wait_no_slave=off;
```

```
Query OK, 0 rows affected (0.00 sec)

-- 关闭所有从库的复制
mysql> stop slave;
Query OK, 0 rows affected (0.01 sec)

-- 主库
mysql> show status like 'Rpl_semi_sync_master_clients';--查看当前半同步的从库数
+------------------------------+-------+
| Variable_name                | Value |
+------------------------------+-------+
| Rpl_semi_sync_master_clients | 0     |
+------------------------------+-------+
1 row in set (0.00 sec)

mysql> show status like 'Rpl_semi_sync_master_status';
+-----------------------------+-------+
| Variable_name               | Value |
+-----------------------------+-------+
| Rpl_semi_sync_master_status | OFF   |
+-----------------------------+-------+
1 row in set (0.00 sec)

mysql> insert into t1 values(5);
Query OK, 1 row affected (0.00 sec)

mysql> commit;
Query OK, 0 rows affected (0.00 sec)

mysql> select * from t1;
+------+
| a    |
+------+
|    1 |
|    2 |
|    3 |
|    4 |
|    5 |
+------+
5 rows in set (0.00 sec)
```

可以看到，当 rpl_semi_sync_master_wait_no_slave 设置为 OFF 时，一旦 Rpl_semi_sync_master_clients 小于 rpl_semi_sync_master_wait_for_slave_count，Rpl_semi_sync_master_status 立即变为 OFF，即立即降为异步复制。

2.7 小 结

为了解决主从数据一致性的问题，MySQL 早在 5.5 版本就引入了半同步复制。经过支持发送二进制日志事件和接收 ACK 的异步化，控制主库接收确认从库的反馈数量，二进制日志互斥锁改进等一系列的优化，使得半同步复制在保证数据一致性的同时，性能有了大幅提升。

在本章中，我们还展示了 rpl_semi_sync_master_wait_point、sync_binlog、sync_relay_log 这三个配置参数在 MySQL 5.7 中实现的源代码，详细分析了不同参数值对数据一致性产生的影响。在需要强一致性的场景中，sync_binlog 和 sync_relay_log 都应设置为 1，而 rpl_semi_sync_master_wait_point 应设置为 WAIT_AFTER_SYNC。不过，这样做会牺牲一定的性能，因而需要权衡考虑。

最后，我们介绍了配置半同步复制的步骤，并且测试了主库提交事务、回滚事务对从库的影响，以及 rpl_semi_sync_master_wait_no_slave 参数所起的作用。通过设置相关的配置参数可以控制半同步复制的行为，同样通过查询相关的状态参数可以查看半同步复制的当前状态，它们被统称为半同步复制的管理接口。下一章将讲述一个与 MySQL 复制相关的重要主题——GTID。

第3章

GTID与复制

MySQL 复制中使用的事务类型有以下两种：

- GTID 事务：在二进制日志中每个 GTID 事务始终都以 Gtid_log_event 开头。可以使用 GTID 或使用文件名和位置来定位 GTID 事务。
- 匿名事务：MySQL 8 的二进制日志中的每个匿名事务都以 Anonymous_gtid_log_event 开头，不分配 GTID。匿名事务只能使用文件名和位置来定位。

在 GTID 出现之前，在“一主多从”的复制拓扑结构中，如果主库宕机，则需要从多个从库选择一个作为新主库。这个过程比较复杂，没有一种直截了当的方法找到其他从库对应的新主库二进制日志位置。通常的做法是先要寻找每个从库复制原主库的最后语句，然后找到新主库中包含该语句的二进制日志文件，其中该语句后的第一个事件位置即为连接新主库的二进制位置。主要难点在于不存在一个唯一标识指出“复制原主库的最后语句”，于是后来的 MySQL 中就出现了 GTID 的概念。

3.1 GTID 简介

3.1.1 什么是 GTID

全局事务标识符 GTID 的英文全称为 Global Transaction Identifier，是在整个复制环境中对一个事务的唯一标识。它是 MySQL 5.6 加入的一个强大特性，目的在于能够实现主从自动定位和切换，而不像以前需要指定文件和位置。使用 GTID 复制时，在主库上提交事务时创建事务对应的 GTID，从库在应用中继日志时用 GTID 识别和跟踪每个事务。在启动新从库或因故障转移到新主库时，可以使用 GTID 来标识复制的位置，极大地简化了这些任务。由于 GTID 的复制完全基于事务，因此只要在主库上提交的所有事务也在从库上提交，两者之间的一致性就能得到保证。GTID 支持基于语句或基于行的复制格式，但为了获得最佳效果，MySQL 建议使用基于行的格式。GTID 始终保留在主库和从库上，这意味着可以通过检查它的二进制日志来确定应用源于哪一个从库的何种事务。而且，一旦在指定库上提交了具有给定 GTID 的事务，则该库将忽略具有相同 GTID 的任何后续事务。因此，在主库上提交的事务只会在从库上应用一次，这也有助于保证一致性。

3.1.2 GTID 的格式与存储

1. 单个 GTID

GTID 与主库上提交的每个事务相关联。此标识符不仅对发起事务的库是唯一的，而且在给定复制拓扑结构中的所有库中都是唯一的。GTID 是由冒号分隔的一对坐标来表示的，例如：

```
8eed0f5b-6f9b-11e9-94a9-005056a57a4e:23
```

前一部分是主库的 server_uuid，后面一部分是主库上按提交事务的顺序确定的序列号，提交的事务序号从 1 开始。上面的 GTID 表示：具有 8eed0f5b-6f9b-11e9-94a9-005056a57a4e 的服务器上提交的第 23 个事务具有此 GTID。MySQL 5.6 后使用自动生成的 128 位 server_uuid 以避免冲突。数据目录下的 auto.cnf 文件用来保存 server_uuid。MySQL 启动的时候会读取 auto.cnf 文件，如果没有读取到则会生成一个 server_id，并保存到 auto.cnf 文件中。

在主库上提交客户端事务时，如果事务已写入二进制日志，则会为其分配新的 GTID，保证为客户事务生成单调递增且没有间隙的 GTID。如果未将客户端事务写入二进制日志（例如，因为事务已被过滤掉，或者事务是只读的），则不会在源服务器上为其分配 GTID。从库上复制的事务保留与主库上事务相同的 GTID。即使从库上未开启二进制日志，GTID 也会被保存。MySQL 系统表 mysql.gtid_executed 用于保存 MySQL 服务器上应用的所有事务的 GTID，但存储在当前活动二进制日志文件中的事务除外。

GTID 的自动跳过功能意味着一旦在指定服务器上提交了具有给定 GTID 的事务，则该服务器将忽略使用相同 GTID 执行的任何后续事务（这种情况是可能发生的，如手工设置了 gtid_next 时）。这有助于保证主从一致性，因为在主库上提交的事务在从库上应用不超过一次。如果具有给定 GTID 的事务已开始在服务器上执行但尚未提交或回滚，则任何在该服务器上启动具有相同 GTID 的并发事务都将被阻止。服务器既不执行并发事务也不将控制权返回给客户端。一旦先前的事务提交或回滚，就可以继续执行在同一个 GTID 上被阻塞的并发会话。如果是回滚，则一个并发会话继续执行事务，并且在同一个 GTID 上阻塞的任何其他并发会话仍然被阻止。如果是提交，则所有并发会话都将被阻止，并自动跳过事务的所有语句。mysqlbinlog 输出中的 GTID_NEXT 包含事务的 GTID，用于标识复制中的单个事务。

下面做三个简单实验来验证 GTID 的自动跳过功能。

实验 1：验证自动跳过

（1）准备初始数据：

```
use test;
create table t1(a int);
create table t2(a int);
```

```
insert into t1 values(1),(2);
insert into t2 values(1),(2);
commit;
```

（2）查看当前 GTID：

```
mysql> show master status\G
*************************** 1. row ***************************
             File: binlog.000027
         Position: 34614
     Binlog_Do_DB:
 Binlog_Ignore_DB:
Executed_Gtid_Set: 8eed0f5b-6f9b-11e9-94a9-005056a57a4e:1-356
1 row in set (0.00 sec)
```

（3）将 GDIT 设置为已经执行过的值，再执行事务：

```
mysql> set gtid_next = '8eed0f5b-6f9b-11e9-94a9-005056a57a4e:356';
Query OK, 0 rows affected (0.00 sec)
mysql> truncate table test.t1;
Query OK, 0 rows affected (0.00 sec)
mysql> select * from test.t1;
+------+
| a    |
+------+
|    1 |
|    2 |
+------+
2 rows in set (0.00 sec)
mysql> set gtid_next = automatic;
Query OK, 0 rows affected (0.00 sec)
```

可以看到，服务器已经执行了 GTID 为 356 的事务，后续相同 GTID 的事务都被自动跳过，虽然 truncate 语句没有报错，但并未执行，数据无变化。

实验 2：验证两个相同 GTID 事务，事务 1 提交，事务 2 被跳过

（1）准备两个 SQL 脚本 s1.sql 和 s2.sql，gtid_next 是一个没用过的新值。

s1.sql 内容如下：

```
set gtid_next='8eed0f5b-6f9b-11e9-94a9-005056a57a4e:357';
begin;
delete from test.t1 where a=1;
select sleep(10);
commit;
set gtid_next=automatic;
```

s2.sql 内容如下：

```
set gtid_next='8eed0f5b-6f9b-11e9-94a9-005056a57a4e:357';
begin;
delete from test.t2 where a=1;
commit;
```

```
set gtid_next=automatic;
```

（2）在会话 1 执行 s1.sql，并且在其 sleep 期间，在会话 2 执行 s2.sql：

```
-- 会话 1
mysql -uroot -p123456 test < s1.sql
-- 会话 2
mysql -uroot -p123456 test < s2.sql
```

（3）查询数据：

```
mysql> select * from t1;
+------+
| a    |
+------+
|    2 |
+------+
1 row in set (0.00 sec)

mysql> select * from t2;
+------+
| a    |
+------+
|    1 |
|    2 |
+------+
2 rows in set (0.00 sec)
```

可以看到，事务 1 提交前，事务 2 被阻塞。事务 1 提交后，具有相同 GTID 的事务 2 被跳过。

实验 3：验证两个相同 GTID 事务，事务 1 回滚，事务 2 提交

（1）准备两个 SQL 脚本 s1.sql 和 s2.sql，gtid_next 是一个没用过的新值。

s1.sql 内容如下：

```
set gtid_next='8eed0f5b-6f9b-11e9-94a9-005056a57a4e:360';
begin;
delete from test.t1 where a=2;
select sleep(10);
rollback;
set gtid_next=automatic;
```

s2.sql 内容如下：

```
set gtid_next='8eed0f5b-6f9b-11e9-94a9-005056a57a4e:360';
begin;
delete from test.t2 where a=1;
commit;
set gtid_next=automatic;
```

（2）在会话 1 执行 s1.sql，并且在其 sleep 期间，在会话 2 执行 s2.sql：

```
-- 会话 1
mysql -uroot -p123456 test < s1.sql
```

```
-- 会话2
mysql -uroot -p123456 test < s2.sql
```

（3）查询数据：

```
mysql> select * from t1;
+------+
| a    |
+------+
|    2 |
+------+
1 row in set (0.00 sec)

mysql> select * from t2;
+------+
| a    |
+------+
|    2 |
+------+
1 row in set (0.00 sec)
```

可以看到，事务1回滚前，事务2被阻塞。事务1回滚后，具有相同GTID的事务2被提交。

2. GTID集

GTID集是包括一个或多个单个GTID或GTID范围的集合。源自同一个服务器的一系列GTID可以折叠为单个表达式，例如：

```
8eed0f5b-6f9b-11e9-94a9-005056a57a4e:1-321
```

上面的示例表示源自server_uuid为8eed0f5b-6f9b-11e9-94a9-005056a57a4e服务器的第1到第321个事务。源自同一个服务器的多个单GTID或GTID范围可以同时包含在由冒号分隔的单个表达式中，例如：

```
8eed0f5b-6f9b-11e9-94a9-005056a57a4e:1-3:11:47-49
```

GTID集可以包括单个GTID和GTID范围的任意组合，甚至它可以包括源自不同服务器的GTID。例如一个存储在从库gtid_executed系统变量中的GTID集可能如下：

```
565a6b0a-6f05-11e9-b95c-005056a5497f:1-20,
8eed0f5b-6f9b-11e9-94a9-005056a57a4e:1-321
```

表示该从库已从两个主库应用了事务，也有可能是在从库执行的写操作。当从库变量返回GTID集时，UUID按字母顺序排列，并且数值间隔按升序合并。

MySQL服务器中很多地方都用到GTID集，例如：gtid_executed和gtid_purged系统变量存储的值是GTID集；START SLAVE的UNTIL SQL_BEFORE_GTIDS和UNTIL SQL_AFTER_GTIDS子句的值是GTID集；内置函数GTID_SUBSET()和GTID_SUBTRACT()需要GTID集作为输入等。

3. mysql.gtid_executed表

mysql.gtid_executed表结构如下：

```
mysql> desc mysql.gtid_executed;
```

```
+----------------+-----------+------+-----+---------+-------+
| Field          | Type      | Null | Key | Default | Extra |
+----------------+-----------+------+-----+---------+-------+
| source_uuid    | char(36)  | NO   | PRI | NULL    |       |
| interval_start | bigint(20) | NO  | PRI | NULL    |       |
| interval_end   | bigint(20) | NO  |     | NULL    |       |
+----------------+-----------+------+-----+---------+-------+
3 rows in set (0.00 sec)
```

mysql.gtid_executed 表记录的是服务器上已经执行事务的 GTID。三个字段分别表示发起事务的服务器 UUID、UUID 集的起始和结束事务 ID。对于单个 GTID，后两个字段的值相同。

mysql.gtid_executed 表供 MySQL 服务器内部使用。当从库禁用二进制日志时用该表记录 GTID，或者当二进制日志丢失时，可从该表查询 GTID 状态。RESET MASTER 命令将重置 mysql.gtid_executed 表，清空表数据。和所有系统表一样，用户不要修改该表。

仅当 gtid_mode 设置为 ON 或 ON_PERMISSIVE 时，GTID 才存储在 mysql.gtid_executed 表中。存储的 GTID 值取决于是否启用二进制日志：

- 对于从库，如果禁用了二进制日志记录（skip-log-bin）或 log_slave_updates，则服务器将在该表中存储每个事务的 GTID。
- 如果启用了二进制日志记录，当刷新二进制日志或重启服务器时，服务器都会将当前二进制日志中所有事务的 GTID 写入 mysql.gtid_executed 表。这种情况适用于主库或启用了二进制日志记录的从库。

启用二进制日志记录时，mysql.gtid_executed 表并不保存所有已执行事务 GTID 的完整记录，该信息由 gtid_executed 全局系统变量的值提供，每次提交事务后更新。如果服务器意外停止，则当前二进制日志文件中的 GTID 集不会保存在 mysql.gtid_executed 表中。在 MySQL 实例恢复期间，这些 GTID 将从二进制日志文件添加到表中。即使服务器处于只读模式，MySQL 服务器也可以写入 mysql.gtid_executed 表，这样二进制日志文件仍然可以在只读模式下轮转。如果无法访问 mysql.gtid_executed 表时进行二进制日志文件轮转，则继续使用二进制日志文件存储 GTID，同时在服务器上记录警告信息：

```
2019-06-03T09:37:07.777423Z 287633 [Warning] [MY-010015] [Repl] Gtid table is
not ready to be used. Table 'mysql.gtid_executed' cannot be opened.
```

前面已经提到，mysql.gtid_executed 表的记录可能并不是完整的已执行 GTID，而且有不可访问的可能性，例如误删除此表，因此建议始终通过查询@@global.gtid_executed 来确认 MySQL 服务器的 GTID 状态，而不是查询 mysql.gtid_executed 表。mysql.gtid_executed 表可能随着事务量的增多而快速膨胀，存储了源自同一个服务器的大量不同的单个 GTID，这些 GTID 构成一个范围，例如：

```
+--------------------------------------+----------------+--------------+
| source_uuid                          | interval_start | interval_end |
+--------------------------------------+----------------+--------------+
| 8eed0f5b-6f9b-11e9-94a9-005056a57a4e |              1 |          329 |
| 8eed0f5b-6f9b-11e9-94a9-005056a57a4e |            330 |          330 |
| 8eed0f5b-6f9b-11e9-94a9-005056a57a4e |            331 |          331 |
```

```
| 8eed0f5b-6f9b-11e9-94a9-005056a57a4e |          332 |          332 |
| 8eed0f5b-6f9b-11e9-94a9-005056a57a4e |          333 |          333 |
| 8eed0f5b-6f9b-11e9-94a9-005056a57a4e |          334 |          334 |
...
```

为了节省空间，MySQL 服务器定期压缩 mysql.gtid_executed 表，方法是将每个这样的行集替换为跨越整个事务标识符间隔的单行，如下所示：

```
+--------------------------------------+----------------+--------------+
| source_uuid                          | interval_start | interval_end |
|--------------------------------------+----------------+--------------|
| 8eed0f5b-6f9b-11e9-94a9-005056a57a4e |       1        |     334      |
...
```

通过设置 gtid_executed_compression_period 系统变量，可以控制压缩表之前允许的事务数，从而控制压缩率。此变量的默认值为 1000，指的是在每 1000 次事务之后执行表的压缩。把 gtid_executed_compression_period 设置为 0，将不执行压缩。注意，启用二进制日志时不使用 gtid_executed_compression_period 的值，并在每个二进制日志轮转时压缩 mysql.gtid_executed 表。mysql.gtid_executed 表的压缩由名为 thread/sql/compress_gtid_table 的专用前台线程执行。此线程未在 SHOW PROCESSLIST 的输出中列出，但可以从 performance_schema.threads 中查询到：

```
mysql> select * from performance_schema.threads where name like '%gtid%'\G
*************************** 1. row ***************************
          THREAD_ID: 44
               NAME: thread/sql/compress_gtid_table
               TYPE: FOREGROUND
     PROCESSLIST_ID: 6
   PROCESSLIST_USER: NULL
   PROCESSLIST_HOST: NULL
     PROCESSLIST_DB: NULL
PROCESSLIST_COMMAND: Daemon
   PROCESSLIST_TIME: 438302
  PROCESSLIST_STATE: Suspending
   PROCESSLIST_INFO: NULL
   PARENT_THREAD_ID: 1
               ROLE: NULL
       INSTRUMENTED: YES
            HISTORY: YES
    CONNECTION_TYPE: NULL
       THREAD_OS_ID: 73199
     RESOURCE_GROUP: SYS_default
1 row in set (0.00 sec)
```

通常该线程都处于暂停状态，只有当满足条件时被唤醒，如达到 gtid_executed_compression_period 或发生了二进制日志轮转（如 flush logs 等）时。

下面做个简单实验展示一下 reset master 的作用和影响。

（1）查看从库当前已经执行的 GTID 和二进制日志：

```
show master status\G
```

```
show variables like 'gtid%';
select * from mysql.gtid_executed;
show slave status\G
```

（2）查询结果如下：

```
mysql> show master status\G
*************************** 1. row ***************************
             File: binlog.000004
         Position: 195
     Binlog_Do_DB:
 Binlog_Ignore_DB:
Executed_Gtid_Set: 8eed0f5b-6f9b-11e9-94a9-005056a57a4e:1-6
1 row in set (0.00 sec)

mysql> show variables like 'gtid%';
+----------------------------------+----------------------------------------
----+
| Variable_name                    | Value                                  |
+----------------------------------+----------------------------------------
----+
| gtid_executed                    | 8eed0f5b-6f9b-11e9-94a9-005056a57a4e:1-6 |
| gtid_executed_compression_period | 1000                                   |
| gtid_mode                        | ON                                     |
| gtid_next                        | AUTOMATIC                              |
| gtid_owned                       |                                        |
| gtid_purged                      |                                        |
+----------------------------------+----------------------------------------
----+
6 rows in set (0.01 sec)

mysql> select * from mysql.gtid_executed;
+--------------------------------------+----------------+--------------+
| source_uuid                          | interval_start | interval_end |
+--------------------------------------+----------------+--------------+
| 8eed0f5b-6f9b-11e9-94a9-005056a57a4e |    1           |      6       |
+--------------------------------------+----------------+--------------+
1 row in set (0.00 sec)

mysql> show slave status\G
...
Retrieved_Gtid_Set: 8eed0f5b-6f9b-11e9-94a9-005056a57a4e:1-6
 Executed_Gtid_Set: 8eed0f5b-6f9b-11e9-94a9-005056a57a4e:1-6
...
1 row in set (0.00 sec)
```

所有查询显示已经执行的 GTID 均为 8eed0f5b-6f9b-11e9-94a9-005056a57a4e:1-6。

查看当前的 binlog 结果如下：

```
[mysql@hdp4/usr/local/mysql/data]$more binlog.index
./binlog.000001
```

```
./binlog.000002
./binlog.000003
./binlog.000004
[mysql@hdp4/usr/local/mysql/data]$ls -lt binlog.*
-rw-r----- 1 mysql mysql   64 Jun  5 14:43 binlog.index
-rw-r----- 1 mysql mysql  195 Jun  5 14:43 binlog.000004
-rw-r----- 1 mysql mysql  239 Jun  5 14:43 binlog.000003
-rw-r----- 1 mysql mysql  239 Jun  5 14:43 binlog.000002
-rw-r----- 1 mysql mysql 1569 Jun  5 14:43 binlog.000001
[mysql@hdp4/usr/local/mysql/data]$
```

当前从库有4个binlog文件。

（1）在从库执行reset master。

（2）再次执行（1）的查询。可以看到所有查询的gtid_executed都置空，binlog文件只有binlog.000001一个。说明reset master命令会清空gtid_executed变量和mysql.gtid_executed表，并会只保留一个初始的binlog文件。

（3）在主库上执行一些更新。

```
use test;
create table t1(a int);
insert into t1 select 1;
```

（4）再次执行（1）的查询。可以看到mysql.gtid_executed表中没有记录，其他查询都已显示出新执行GTID的值，复制正常。说明mysql.gtid_executed不记录当前binlog中的GTID。

（5）从库执行flush logs。在从库上执行flush logs后，mysql.gtid_executed表中存储了从reset master到flush logs之间binlog中的GTID。

从以上步骤看到，从库上执行reset master只是清空从库的gtid_executed，随着复制的继续，其gtid_executed的值也将随之变化，对复制和主从数据一致性没有影响。下面继续实验，看一下在主库上执行reset master会产生哪些影响。

（6）在主库上执行以下语句：

```
use test;
delimiter //
create procedure p1(a int)
begin
   declare i int default 1;
   while i<=a do
      insert into t1 values (i);
      set i=i+1;
   end while;
end;
//
delimiter ;

call p1(10000);
```

（7）在上一步执行期间，开启一个新会话在主库上执行reset master。

（8）查看从库的复制状态。从 show slave status 的输出中可以看到复制的 I/O 线程已停止，并报以下错误：

```
    Last_IO_Errno: 13114
    Last_IO_Error: Got fatal error 1236 from master when reading data from binary
log: 'I/O error reading log event; the first event '' at 4, the last event read
from './binlog.000001' at 201303, the last byte read from './binlog.000001' at
201303.'
```

由于主库正在执行事务中间进行了 reset master，从库无法读取主库的二进制日志而报错。更有甚之，这些二进制日志的丢失是永久性的，结果很可能需要从头重建复制。由此实验得出的结论是，作为一条基本原则，不要随意在主库上执行 reset master，这样做极有可能导致复制停止或造成主从数据不一致等严重后果，而且不易恢复。

3.2 GTID 生命周期

3.2.1 典型事务的 GTID 生命周期

典型事务的 GTID 生命周期包括以下步骤：

步骤 01 客户端事务在主库上执行并提交，此事务被分配一个 GTID，该 GTID 由主服务器的 UUID 和此服务器上尚未使用的最小非零事务序列号组成。GTID 作为 Gtid_log_event 紧接在事务本身之前，与事务本身一起被写入主库的二进制日志，这是一个原子操作（即不可分割的操作）。如果未将客户端事务写入二进制日志，例如因为事务已被过滤掉，或者事务是只读的，则不会为其分配 GTID。轮转二进制日志或关闭 MySQL 实例时，都会将写入之前二进制日志文件的所有事务的 GTID 写入 mysql.gtid_executed 表。

步骤 02 如果为事务分配了 GTID，则将 GTID 添加到主库 gtid_executed 系统变量 @@global.gtid_executed 的 GTID 集合中，这一步将在事务提交后进行，并且与事务处理本身不是一个原子操作。gtid_executed 系统变量包含所有已提交事务的 GTID 集，是应用事务的完整记录，并在复制中用作表示服务器状态的标记。mysql.gtid_executed 表不包含当前二进制日志文件中的最新 GTID 记录。

步骤 03 在将二进制日志数据传输到从库并存储在从库的中继日志中之后，从库读取 GTID 并将其设置为 gtid_next 系统变量的值。这告诉从库必须使用此 GTID 记录下一个事务。

步骤 04 在处理事务本身之前，从库首先读取和检查复制事务的 GTID，不仅保证没有先前的事务具有此 GTID，还保证没有其他会话已经读取此 GTID、尚未提交相关事务。因此，如果多个客户端同时提交同一个 GTID 事务，则服务器只允许其中一个执行。从库的 gtid_owned 系统变量 @@global.gtid_owned 显示当前正在使用的 GTID 以及拥有它的线程 ID。如果已经使用了该 GTID，通过自动跳过功能忽略该事务，并且不会引发错误。

步骤 05 如果 GTID 尚未使用，则从库应用复制的事务。gtid_next 设置为主库已分配的 GTID，从库不会为此事务生成新的 GTID，而是使用存储在 gtid_next 中的 GTID。

步骤 06 如果在从库上启用了二进制日志记录，则与主库操作类似。GTID 会在提交时作为 Gtid_log_event 原子写入其二进制日志。当轮转二进制日志或关闭 MySQL 实例时，都会将之前写入二进制日志文件的所有事务的 GTID 写入 mysql.gtid_executed 表。

步骤 07 如果从库禁用二进制日志记录，则通过将 GTID 直接写入 mysql.gtid_executed 表保留 GTID。MySQL 会在事务中附加一条语句，将 GTID 插入该表中。从 MySQL 8.0 开始，此操作对于 DDL 语句和 DML 语句都是原子操作。在这种情况下，mysql.gtid_executed 表是从库上应用事务的完整记录。

步骤 08 从库提交复制事务后，GTID 将被添加到从库 gtid_executed 系统变量 @@global.gtid_executed 的 GTID 集合中，这步将在事务应用后进行，并且与事务处理本身不是一个原子操作。

主库上过滤掉的客户端事务未分配 GTID，因此它们不会添加到 gtid_executed 系统变量中的事务集中，也不会添加到 mysql.gtid_executed 表中。但是，在从库上过滤掉的复制事务的 GTID 是持久化的。如果在从库上启用了二进制日志，则过滤掉的事务将作为 Gtid_log_event 写入其二进制日志，后跟仅包含 BEGIN 和 COMMIT 语句的空事务。如果禁用二进制日志，则已过滤掉的事务的 GTID 将写入 mysql.gtid_executed 表。为过滤掉的事务保留 GTID 可确保将 mysti.gtid_executed 表和 gtid_executed 系统变量中的 GTID 用 GTID 集表示。它还确保如果从库重新连接到主库，不会再次检索过滤掉的事务。

在主库或单线程复制的从库上，GTID 从 1 开始单向递增且没有间隙。但在多线程复制的从库（slave_parallel_workers> 0）上，可以并行应用事务，因此复制的事务可能无序提交，除非设置了 slave_preserve_commit_order = 1。在发生这种情况时，gtid_executed 系统变量中的 GTID 集合将包含多个 GTID 范围，它们之间可能存在间隙。多线程复制从库上的间隙仅发生在最近应用的事务中，并在复制过程中填充。当使用 STOP SLAVE 语句停止复制线程时，将应用正在进行的事务以填补空白。如果发生异常关闭，例如服务器故障或使用 KILL 语句停止复制线程，则可能依然存在间隙。下一章将详细讨论多线程复制。

下面实验中将演示 GTID 存在间隙的情况。

（1）从库开启多线程复制：

```
    set global slave_parallel_workers=8;
    stop slave;
    start slave;
    show processlist;
```

在最后的输出中可以看到 8 个复制线程：

```
    ...
    |1124|system user|    |NULL|Connect|2|Waiting for an event from
Coordinator|NULL|
    |1125|system user|   |NULL|Connect|2|Waiting for an event from
Coordinator|NULL|
    |1126|system user|   |NULL|Connect|2|Waiting for an event from
Coordinator|NULL|
    |1127|system user|   |NULL|Connect|2|Waiting for an event from
Coordinator|NULL|
```

```
|1128|system user|   |NULL|Connect|2|Waiting for an event from
Coordinator|NULL|
|1129|system user|   |NULL|Connect|2|Waiting for an event from
Coordinator|NULL|
|1130|system user|   |NULL|Connect|2|Waiting for an event from
Coordinator|NULL|
|1131|system user|   |NULL|Connect|2|Waiting for an event from
Coordinator|NULL|
+----+-----------+----+----+-------+-+--------------------------------+-----+
12 rows in set (0.00 sec)
```

（2）在主库上执行一个可以并行复制的长操作。

因为并行复制默认是按数据库分配线程的，所以会建立多个库表：

```
create database db1;
create database db2;
create database db3;
create database db4;
create database db5;
create database db6;
create database db7;
create database db8;
create table db1.t1(a int);
create table db2.t1(a int);
create table db3.t1(a int);
create table db4.t1(a int);
create table db5.t1(a int);
create table db6.t1(a int);
create table db7.t1(a int);
create table db8.t1(a int);

use test;
delimiter //
create procedure p1(a int)
begin
  declare i int default 1;
  while i<=a do
    insert into db1.t1 values (i);
    insert into db2.t1 values (i);
    insert into db3.t1 values (i);
    insert into db4.t1 values (i);
    insert into db5.t1 values (i);
    insert into db6.t1 values (i);
    insert into db7.t1 values (i);
    insert into db8.t1 values (i);
    set i=i+1;
  end while;
end;
//
delimiter ;
call p1(5000);
```

（3）在上一步正在执行过程中杀掉从库的mysqld进程，模拟异常宕机：

```
ps -ef | grep mysqld | grep -v grep | awk {'print $2'} | xargs kill -9
```

（4）启动从库，不自动启动复制：

```
mysqld_safe --defaults-file=/etc/my.cnf --skip-slave-start
--slave_parallel_workers=8 &
```

（5）查看从库的GTID间隙：

```
mysql> show variables like 'gtid_executed'\G
*************************** 1. row ***************************
Variable_name: gtid_executed
        Value:
8eed0f5b-6f9b-11e9-94a9-005056a57a4e:1-42171:42173-42179:42181-42187:42189-421
95:42197-42203:42205-42211:42213-42219:42221-42227:42229-42235:42237-42243:422
45-42251:42253-42259:42261-42267:42269-42275:42277-42283:42285-42291:42293-422
99:42301-42307:42309-42315:42317-42323:42325-42331:42333-42339:42341-42347:423
49-42355:42357-42363:42365-42371:42373-42379:42381-42387:42389-42395:42397-424
03:42405-42411:42413-42419:42421-42427:42429-42435:42437-42443:42445-42451:424
53-42459:42461-42467:42469-42475:42477-42483:42485-42491:42493-42499:42501-425
07:42509-42515:42517-42523:42525-42531:42533-42539:42541-42547:42549-42555:425
57-42563:42565-42571:42573-42579:42581-42587:42589-42595:42597-42603:42605-426
11:42613:42615-42619:42621:42623-42627:42629:42631-42635:42637:42639-42643
1 row in set (0.01 sec)
```

GTID范围的输出是排序的，可以看到42172、42180、42188、42196……这些GTID没有出现在gtid_executed变量中，这些就是GTID间隙。查询各个库的记录数（已经执行的事务）也是各不相同的。

（6）启动从库的复制，检查复制情况：

```
start slave;
```

当所有事务都执行完后，再次查看gtid_executed系统变量，已经合并为一个GTID范围，所有间隙都已经被填充：

```
mysql> show variables like 'gtid_executed'\G
*************************** 1. row ***************************
Variable_name: gtid_executed
        Value: 8eed0f5b-6f9b-11e9-94a9-005056a57a4e:1-50021
1 row in set (0.00 sec)
```

从show slave status的输出和各个库表的记录数，也能确认复制正常。通过这个简单的实验可以看到，启用并行复制的从库，在复制期间从库实例异常终止会产生GTID间隙，但在实例重启后复制会自动填充GTID间隙，最终达到主从数据一致。

3.2.2 GTID分配

典型情况是服务器为已提交的事务生成新的GTID。写入二进制日志的每个数据库更改（DDL

或 DML）都会分配一个 GTID。这包括自动提交的更改以及使用 BEGIN 或 START TRANSACTION 和 COMMIT 语句提交的更改。当数据库以及非表数据库对象，例如过程、函数、触发器、事件、视图、用户、角色，在创建、更改或删除时都会分配 GTID。授权语句和非事务表的更新也会分配 GTID。

当二进制日志中的生成语句自动删除表时，会为该语句分配 GTID。例如，当具有打开临时表的用户会话断开连接时，将会自动删除临时表，或者使用 MEMORY 存储引擎的表在服务器启动后会自动删除。

未写入二进制日志的事务不会被分配 GTID。这包括回滚的事务，或在禁用二进制日志时执行的事务，或指定 sql_log_bin=0 时执行的事务，或空事务（begin;commit;）等。

XA 事务为事务的 XA PREPARE 阶段和事务的 XA COMMIT 或 XA ROLLBACK 阶段分配了单独的 GTID。XA 事务的准备阶段是持久化的，以便用户可以在发生故障时将其提交或回滚。因此，事务的两个部分是分开复制的，因此两个阶段必须有自己单独的 GTID。

在以下特殊情况下，单个语句可以生成多个事务，因此会分配多个 GTID：

- 调用存储过程时，为过程提交的每个更新事务生成一个 GTID。
- 多表 DROP TABLE 语句中包含任何不支持原子 DDL 存储引擎的表（如 myisam）或临时表时，会生成多个 GTID。

注意，触发器内的语句和触发它的语句是在一个事务中，因此不会单独分配 GTID。MySQL 不支持类似 Oracle 自治事务的功能。

3.2.3 gtid_next 系统变量

gtid_next 是会话级系统变量。默认情况下，对于在用户会话中提交的新事务，服务器会自动生成并分配新的 GTID。在从库上应用事务时，将保留来自原始服务器的 GTID。可以通过设置 gtid_next 系统变量的会话值来更改此行为：

- 当 gtid_next 设置为默认值 AUTOMATIC，并且事务已提交并写入二进制日志时，服务器会自动生成并分配新的 GTID。如果由于其他原因而回滚事务或未将事务写入二进制日志，则服务器不会生成和分配 GTID。
- 如果将 gtid_next 设置为有效的单个 GTID，服务器会将该 GTID 分配给下一个事务。只要事务提交，就会将此 GTID 分配并添加到 gtid_executed。

在将 gtid_next 设置为特定 GTID 并且已提交或回滚事务之后，必须在任何其他语句之前发出显式 SET @@SESSION.gtid_next 语句。如果不想分配更多 GTID，可以将此选项值的值设置回 AUTOMATIC。

```
    mysql> show variables like 'gtid%';
    +------------------------------+------------------------------------------
----+
    |Variable_name                |Value                                    |
    +------------------------------+------------------------------------------
----+
```

```
|gtid_executed                |8eed0f5b-6f9b-11e9-94a9-005056a57a4e:1-50057|
|gtid_executed_compression_period|1000                                   |
|gtid_mode                   |ON                                |
|gtid_next                   |AUTOMATIC                               |
|gtid_owned                   |                                 |
|gtid_purged                  |                                 |
+-------------------------------+---------------------------------------------
----+
6 rows in set (0.00 sec)

mysql> set gtid_next='8eed0f5b-6f9b-11e9-94a9-005056a57a4e:50058';
Query OK, 0 rows affected (0.00 sec)

mysql> show variables like 'gtid%';
+-------------------------------+---------------------------------------------
----+
|Variable_name                |Value                                |
+-------------------------------+---------------------------------------------
----+
|gtid_executed                |8eed0f5b-6f9b-11e9-94a9-005056a57a4e:1-50057|
|gtid_executed_compression_period|1000                                   |
|gtid_mode                   |ON                                |
|gtid_next                   |8eed0f5b-6f9b-11e9-94a9-005056a57a4e:50058 |
|gtid_owned                   |8eed0f5b-6f9b-11e9-94a9-005056a57a4e:50058 |
|gtid_purged                  |                                 |
+-------------------------------+---------------------------------------------
----+
6 rows in set (0.00 sec)

mysql> begin;commit;
Query OK, 0 rows affected (0.00 sec)
Query OK, 0 rows affected (0.00 sec)

mysql> show variables like 'gtid%';
+-------------------------------+---------------------------------------------
----+
|Variable_name                |Value                                |
+-------------------------------+---------------------------------------------
----+
|gtid_executed                |8eed0f5b-6f9b-11e9-94a9-005056a57a4e:1-50058|
|gtid_executed_compression_period|1000                                   |
|gtid_mode                   |ON                                |
|gtid_next                   |8eed0f5b-6f9b-11e9-94a9-005056a57a4e:50058 |
|gtid_owned                   |                                 |
|gtid_purged                  |                                 |
+-------------------------------+---------------------------------------------
----+
6 rows in set (0.01 sec)

mysql> create table t1(a int);
```

```
    ERROR 1837 (HY000): When @@SESSION.GTID_NEXT is set to a GTID, you must
explicitly set it to a different value after a COMMIT or ROLLBACK. Please check
GTID_NEXT variable manual page for detailed explanation. Current
@@SESSION.GTID_NEXT is '8eed0f5b-6f9b-11e9-94a9-005056a57a4e:50058'.

    mysql> set gtid_next=automatic;
    Query OK, 0 rows affected (0.00 sec)

    mysql> create table t1(a int);
    Query OK, 0 rows affected (0.01 sec)
```

正如前面所讲，从库的 SQL 线程应用复制事务时使用此技术，将@@SESSION.gtid_next 设置为在源服务器上分配给事务的 GTID。这意味着保留来自原始服务器的 GTID，而不是由从库生成和分配的新 GTID。即使从库禁用 log_bin 或 log_slave_updates，或者事务是空操作或在从库上过滤掉时，GTID 也会添加到从库上的 gtid_executed 中。

客户端可通过在执行事务之前将@@SESSION.gtid_next 设置为特定 GTID 来模拟复制的事务。mysqlbinlog 使用此技术生成二进制日志的转储，客户端可以重放该转储以保留 GTID。通过客户端提交的模拟复制事务完全等同于通过复制应用程序线程提交的复制事务，事后是无法区分它们的。

```
[mysql@hdp2/usr/local/mysql/data]$mysqlbinlog  --base64-output=decode-rows
 binlog.000001 | tail -15
/*!80001 SET @@session.original_commit_timestamp=1559800983100268*//*!*/;
/*!80014 SET @@session.original_server_version=80016*//*!*/;
/*!80014 SET @@session.immediate_server_version=80016*//*!*/;
SET @@SESSION.GTID_NEXT= '8eed0f5b-6f9b-11e9-94a9-005056a57a4e:50059'/*!*/;
# at 13622355
#190606 14:03:03 server id 1125  end_log_pos 13622465 CRC32 0xbf6bf581
Query  thread_id=184    exec_time=0    error_code=0    Xid = 601312
SET TIMESTAMP=1559800983/*!*/;
/*!80013 SET @@session.sql_require_primary_key=0*//*!*/;
create table t1(a int)
/*!*/;
SET @@SESSION.GTID_NEXT= 'AUTOMATIC' /* added by mysqlbinlog */ /*!*/;
DELIMITER ;
# End of log file
/*!50003 SET COMPLETION_TYPE=@OLD_COMPLETION_TYPE*/;
/*!50530 SET @@SESSION.PSEUDO_SLAVE_MODE=0*/;
[mysql@hdp2/usr/local/mysql/data]$
```

3.2.4 gtid_purged 系统变量

gtid_purged 是全局系统变量。@@GLOBAL.gtid_purged 中的 GTID 集包含已在服务器上提交，但在服务器上的任何二进制日志文件中不存在的所有事务的 GTID。gtid_purged 是 gtid_executed 的子集。以下类别的 GTID 位于 gtid_purged 中：

- 第一种情况：在从库上禁用二进制日志记录时提交的复制事务的 GTID。
- 第二种情况：已清除的二进制日志文件中事务的 GTID。

- 第三种情况：通过语句 SET @@GLOBAL.gtid_purged 明确添加到集合中的 GTID。

第一种情况：

```
[mysql@hdp4~]$mysqladmin -uroot -p123456 shutdown
mysqladmin: [Warning] Using a password on the command line interface can be insecure.
[mysql@hdp4~]$mysqld_safe --defaults-file=/etc/my.cnf --skip-log-bin &
[1] 97160
[mysql@hdp4~]$2019-06-06T06:25:56.483366Z mysqld_safe Logging to '/usr/local/mysql/data/hdp4.err'.
2019-06-06T06:25:56.544557Z mysqld_safe Starting mysqld daemon with databases from /usr/local/mysql/data

[mysql@hdp4~]$mysql -uroot -p123456 -e "show variables like 'gtid_purged'"
mysql:[Warning] Using a password on the command line interface can be insecure.
+---------------+-------+
| Variable_name | Value |
+---------------+-------+
| gtid_purged   |       |
+---------------+-------+

... 主库执行更新 ...

[mysql@hdp4~]$mysql -uroot -p123456 -e "show variables like 'gtid_purged'"
mysql:[Warning] Using a password on the command line interface can be insecure.
+---------------+------------------------------------------------+
| Variable_name | Value                                          |
+---------------+------------------------------------------------+
| gtid_purged   | 8eed0f5b-6f9b-11e9-94a9-005056a57a4e:50060 |
+---------------+------------------------------------------------+
[mysql@hdp4~]$
```

第二种情况：

```
mysql> show binary logs;
+---------------+-----------+-----------+
| Log_name      | File_size | Encrypted |
+---------------+-----------+-----------+
| binlog.000001 |  13683049 | No        |
+---------------+-----------+-----------+
1 row in set (0.00 sec)

mysql> show variables like 'gtid_purged';
+---------------+-------+
| Variable_name | Value |
+---------------+-------+
| gtid_purged   |       |
+---------------+-------+
1 row in set (0.00 sec)
```

```
mysql> flush logs;
Query OK, 0 rows affected (0.01 sec)

mysql> purge master logs to 'binlog.000002';
Query OK, 0 rows affected (0.01 sec)

mysql> show variables like 'gtid_purged';
+---------------+------------------------------------------------+
| Variable_name | Value                                          |
+---------------+------------------------------------------------+
| gtid_purged   | 8eed0f5b-6f9b-11e9-94a9-005056a57a4e:1-50060  |
+---------------+------------------------------------------------+
1 row in set (0.01 sec)
```

第三种情况：

```
mysql> show variables like 'gtid%';
+----------------------------------+------------------------------------------------
----+
|Variable_name                     |Value                                           |
+----------------------------------+------------------------------------------------
----+
|gtid_executed                     |...b-6f9b-11e9-94a9-005056a57a4e:50060-50061|
|gtid_executed_compression_period|1000                                           |
|gtid_mode                         |ON                                              |
|gtid_next                         |AUTOMATIC                                       |
|gtid_owned                        |                                                |
|gtid_purged                       |8eed0f5b-6f9b-11e9-94a9-005056a57a4e:50060   |
+----------------------------------+------------------------------------------------
----+
6 rows in set (0.01 sec)

mysql> set gtid_purged='8eed0f5b-6f9b-11e9-94a9-005056a57a4e:50060-50061';
ERROR 1229 (HY000): Variable 'gtid_purged' is a GLOBAL variable and should be
set with SET GLOBAL

mysql> set global
gtid_purged='8eed0f5b-6f9b-11e9-94a9-005056a57a4e:50060-50061';
ERROR 3546 (HY000): @@GLOBAL.GTID_PURGED cannot be changed: the added gtid set
must not overlap with @@GLOBAL.GTID_EXECUTED

mysql> set global gtid_purged='8eed0f5b-6f9b-11e9-94a9-005056a57a4e:50061';
ERROR 3546 (HY000): @@GLOBAL.GTID_PURGED cannot be changed: the new value must
be a superset of the old value

mysql> set global
gtid_purged='8eed0f5b-6f9b-11e9-94a9-005056a57a4e:1-50059';
ERROR 3546 (HY000): @@GLOBAL.GTID_PURGED cannot be changed: the new value must
be a superset of the old value

mysql> set global
```

```
gtid_purged='8eed0f5b-6f9b-11e9-94a9-005056a57a4e:1-50059:50060';
    Query OK, 0 rows affected (0.00 sec)

    mysql> show variables like 'gtid%';
    +--------------------------------+-----------------------------------------
-----+
    |Variable_name                   |Value                                     |
    +--------------------------------+-----------------------------------------
-----+
    |gtid_executed                   |8eed0f5b-6f9b-11e9-94a9-005056a57a4e:1-50061|
    |gtid_executed_compression_period|1000                                        |
    |gtid_mode                       |ON                                          |
    |gtid_next                       |AUTOMATIC                                   |
    |gtid_owned                      |                                            |
    |gtid_purged                     |8eed0f5b-6f9b-11e9-94a9-005056a57a4e:1-50060|
    +--------------------------------+-----------------------------------------
-----+
    6 rows in set (0.01 sec)
```

可以更改gtid_purged的值，以便在服务器上记录已应用某个GTID集中的事务，尽管它们不存在于服务器上的任何二进制日志中。将GTID添加到gtid_purged时，它们也会添加到gtid_executed中。下面来看一个相对极端的例子。

（1）从库清除二进制日志和gtid_executed信息：

```
reset master;
stop slave;
reset slave all;
show variables like 'gtid%';
```

RESET MASTER会导致gtid_purged和gtid_executed的全局值重置为空字符串。最后的输出为：

```
mysql> show variables like 'gtid%';
+----------------------------------+-----------+
| Variable_name                    | Value     |
+----------------------------------+-----------+
| gtid_executed                    |           |
| gtid_executed_compression_period | 1000      |
| gtid_mode                        | ON        |
| gtid_next                        | AUTOMATIC |
| gtid_owned                       |           |
| gtid_purged                      |           |
+----------------------------------+-----------+
6 rows in set (0.00 sec)
```

（2）重置复制：

```
change master to
     master_host = '172.16.1.125',
    master_port = 3306,
    master_user = 'repl',
```

```
        master_password = '123456',
        master_auto_position = 1;
    start slave;
    show slave status\G
```

最后的输出中会显示以下错误：

```
    mysql> show slave status\G
    ...
    Last_Errno: 1007
    Last_Error: Error 'Can't create database 'test'; database exists' on query.
Default database: 'test'. Query: 'create database test'
    ...
    Retrieved_Gtid_Set: 8eed0f5b-6f9b-11e9-94a9-005056a57a4e:1-10005
     Executed_Gtid_Set:
    ...
    1 row in set (0.00 sec)
```

可以看到，从主库读到的 GTID 已经到了 10005，但没有已经执行的 GTID。实际上这些事务都已经在从库中应用了，只是由于 reset master 而没有留下执行的痕迹，因此要从 1 开始执行，重复执行事务会造成这样的错误。

（3）将所有已读的 GTID 都标记为已执行，然后重启复制：

```
    set global gtid_purged='8eed0f5b-6f9b-11e9-94a9-005056a57a4e:1-10005';
    stop slave;
    start slave;
    show slave status\G
    show variables like 'gtid%';
```

从 show slave status 的输出中可以看到复制已恢复正常，最后的输出为：

```
    mysql> show variables like 'gtid%';
    +--------------------------------+------------------------------------------
-----+
    |Variable_name                  | Value                                    |
    +--------------------------------+------------------------------------------
-----+
    |gtid_executed                  | 8eed0f5b-6f9b-11e9-94a9-005056a57a4e:1-10005|
    |gtid_executed_compression_period| 1000                                     |
    |gtid_mode                      | ON                                       |
    |gtid_next                      | AUTOMATIC                                |
    |gtid_owned                     |                                          |
    |gtid_purged                    | 8eed0f5b-6f9b-11e9-94a9-005056a57a4e:1-10005|
    +--------------------------------+------------------------------------------
-----+
    6 rows in set (0.00 sec)
```

服务器启动时，将初始化 gtid_executed 和 gtid_purged 系统变量中的 GTID 集。每个二进制日志文件都以事件 Previous_gtids_log_event 开头，该事件包含所有先前二进制日志文件中的 GTID 集，由前一个文件的 Previous_gtids_log_event 中的 GTID 和前一个文件中每个 Gtid_log_event 的 GTID

组成。最旧和最新的二进制日志文件中 Previous_gtids_log_event 的内容用于计算服务器启动时的 gtid_executed 和 gtid_purged 的 GTID 集：

- gtid_executed 是最新二进制日志文件中 Previous_gtids_log_event 中的 GTID、该二进制日志文件中事务的 GTID 和存储在 mysql.gtid_executed 表中的 GTID 这三者的并集。此 GTID 集包含服务器上已使用或显式添加到 gtid_purged 的所有 GTID（无论它们当前是否位于服务器上的二进制日志文件中）。这个 GTID 集不包括当前正在服务器上处理事务的 GTID（@@GLOBAL.gtid_owned）。
- gtid_purged 的计算方法是首先添加最新二进制日志文件 Previous_gtids_log_event 中的 GTID，再添加该二进制日志文件中事务的 GTID。此步提供当前或曾经记录在服务器上的二进制日志中的 GTID 集（gtids_in_binlog）。然后从 gtids_in_binlog 中减去最旧的二进制日志文件中的 Previous_gtids_log_event 中的 GTID。此步骤提供当前记录在服务器上的二进制日志中的 GTID 集（gtids_in_binlog_not_purged）。最后，从 gtid_executed 中减去 gtids_in_binlog_not_purged。结果是服务器上已经执行，但当前未记录在服务器上的二进制日志文件中的 GTID 集，此结果用于初始化 gtid_purged。

下面用一个例子来说明 gtid_executed 和 gtid_purged 的计算过程。

```
mysql> show variables like 'gtid%';
+----------------------------------+--------------------------------------------+
|Variable_name                     |Value                                       |
+----------------------------------+--------------------------------------------+
|gtid_executed                     |8eed0f5b-6f9b-11e9-94a9-005056a57a4e:1-11006|
|gtid_executed_compression_period|1000                                          |
|gtid_mode                         |ON                                          |
|gtid_next                         |AUTOMATIC                                   |
|gtid_owned                        |                                            |
|gtid_purged                       |8eed0f5b-6f9b-11e9-94a9-005056a57a4e:1-10005|
+----------------------------------+--------------------------------------------+
6 rows in set (0.01 sec)
```

服务器重启后，gtid_executed 的值为 1-11006，gtid_purged 值为 1-10005，下面倒推这些数是怎么得来的。

当前有三个二进制日志文件，最旧的是 binlog.000001，最新的是 binlog.000003：

```
[mysql@hdp4/usr/local/mysql/data]$ls -lt binlog.*
-rw-r----- 1 mysql mysql    195 Jun  6 16:02 binlog.000003
-rw-r----- 1 mysql mysql     48 Jun  6 16:02 binlog.index
-rw-r----- 1 mysql mysql  275358 Jun  6 16:01 binlog.000002
-rw-r----- 1 mysql mysql    199 Jun  6 16:00 binlog.000001
[mysql@hdp4/usr/local/mysql/data]$
```

binlog.000001 的 Previous-GTIDs 为空，文件本身也没有 GTID：

```
[mysql@hdp4/usr/local/mysql/data]$mysqlbinlog --base64-output=decode-rows
binlog.000001
/*!50530 SET @@SESSION.PSEUDO_SLAVE_MODE=1*/;
/*!50003 SET @OLD_COMPLETION_TYPE=@@COMPLETION_TYPE,COMPLETION_TYPE=0*/;
```

```
    DELIMITER /*!*/;
    # at 4
    #190606 15:52:18 server id 1127  end_log_pos 124 CRC32 0x57a0d989 tart: binlog
v 4, server v 8.0.16 created 190606 15:52:18 at startup
    ROLLBACK/*!*/;
    # at 124
    #190606 15:52:18 server id 1127  end_log_pos 155 CRC32 0x69663b35 evious-GTIDs
    # [empty]
    # at 155
    #190606 16:00:15 server id 1127  end_log_pos 199 CRC32 0x08490198  Rotate to
binlog.000002  pos: 4
    SET @@SESSION.GTID_NEXT= 'AUTOMATIC' /* added by mysqlbinlog */ /*!*/;
    DELIMITER ;
    # End of log file
    /*!50003 SET COMPLETION_TYPE=@OLD_COMPLETION_TYPE*/;
    /*!50530 SET @@SESSION.PSEUDO_SLAVE_MODE=0*/;
    [mysql@hdp4/usr/local/mysql/data]$
```

binlog.000002 的 Previous-GTIDs 由 binlog.000001 的 Previous-GTIDs 和 binlog.000001 本身的 GTID 组成，由于两者都为空，因此 binlog.000002 的 Previous-GTIDs 也为空：

```
    [mysql@hdp4/usr/local/mysql/data]$mysqlbinlog  --base64-output=decode-rows
     binlog.000002 | head -10
    /*!50530 SET @@SESSION.PSEUDO_SLAVE_MODE=1*/;
    /*!50003 SET @OLD_COMPLETION_TYPE=@@COMPLETION_TYPE,COMPLETION_TYPE=0*/;
    DELIMITER /*!*/;
    # at 4
    #190606 16:00:15 server id 1127  end_log_pos 124 CRC32 0x66692a6f tart: binlog
v 4, server v 8.0.16 created 190606 16:00:15
    # at 124
    #190606 16:00:15 server id 1127 end_log_pos 155 CRC32 0x2c439049 Previous-GTIDs
    # [empty]
    # at 155
    #190606 16:00:35 server id 1125 end_log_pos 239 CRC32 0x4bace6c6
GTID last_committed=0 sequence_number=1 rbr_only=no original_committed_timesta
mp=1559808035672038 immediate_commit_timestamp=1559808035637464 transaction_le
ngth=180
    [mysql@hdp4/usr/local/mysql/data]$
```

binlog.000002 本身的 GTID 为 10006-11006：

```
    [mysql@hdp4/usr/local/mysql/data]$mysqlbinlog  --base64-output=decode-rows
     binlog.000002 > binlog.000002.txt
    [mysql@hdp4/usr/local/mysql/data]$grep @@SESSION.GTID_NEXT
binlog.000002.txt | head -1
    SET @@SESSION.GTID_NEXT= '8eed0f5b-6f9b-11e9-94a9-005056a57a4e:10006'/*!*/;
    [mysql@hdp4/usr/local/mysql/data]$grep @@SESSION.GTID_NEXT
binlog.000002.txt | tail -2
    SET @@SESSION.GTID_NEXT= '8eed0f5b-6f9b-11e9-94a9-005056a57a4e:11006'/*!*/;
    SET @@SESSION.GTID_NEXT= 'AUTOMATIC' /* added by mysqlbinlog */ /*!*/;
    [mysql@hdp4/usr/local/mysql/data]$
```

binlog.000003 的 Previous-GTIDs 由 binlog.000002 的 Previous-GTIDs 和 binlog.000002 本身的 GTID 组成，所以 binlog.000003 的 Previous-GTIDs 为 10006-11006：

```
[mysql@hdp4/usr/local/mysql/data]$mysqlbinlog --base64-output=decode-rows
 binlog.000003
/*!50530 SET @@SESSION.PSEUDO_SLAVE_MODE=1*/;
/*!50003 SET @OLD_COMPLETION_TYPE=@@COMPLETION_TYPE,COMPLETION_TYPE=0*/;
DELIMITER /*!*/;
# at 4
#190606 16:02:02 server id 1127 end_log_pos 124 CRC32 0x9443c747 Start: binlog
v 4, server v 8.0.16 created 190606 16:02:02 at startup
# Warning: this binlog is either in use or was not closed properly.
ROLLBACK/*!*/;
# at 124
#190606 16:02:02 server id 1127 end_log_pos 195 CRC32 0xbcd9d46f Previous-GTIDs
# 8eed0f5b-6f9b-11e9-94a9-005056a57a4e:10006-11006
SET @@SESSION.GTID_NEXT= 'AUTOMATIC' /* added by mysqlbinlog */ /*!*/;
DELIMITER ;
# End of log file
/*!50003 SET COMPLETION_TYPE=@OLD_COMPLETION_TYPE*/;
/*!50530 SET @@SESSION.PSEUDO_SLAVE_MODE=0*/;
[mysql@hdp4/usr/local/mysql/data]$
```

binlog.000003 本身没有 GTID。mysql.gtid_executed 的记录为：

```
mysql> select * from mysql.gtid_executed;
+--------------------------------------+----------------+--------------+
| source_uuid                          | interval_start | interval_end |
+--------------------------------------+----------------+--------------+
| 8eed0f5b-6f9b-11e9-94a9-005056a57a4e |              1 |        10005 |
| 8eed0f5b-6f9b-11e9-94a9-005056a57a4e |          10006 |        11006 |
+--------------------------------------+----------------+--------------+
2 rows in set (0.00 sec)
```

按照 gtid_executed 的计算方法，gtid_executed 为 10006-11006 和 1-11006 的并集，于是得出 1-11006。

gtid_purged 的计算过程如下：

```
gtids_in_binlog_not_purged = gtids_in_binlog - binlog.000001 的 Previous-GTIDs
                           = gtids_in_binlog
gtids_in_binlog = binlog.000003 的 Previous-GTIDs + binlog.000003 本身的 GTID
                = binlog.000003 的 Previous-GTIDs
                = 10006-11006
gtid_purged = gtid_executed - gtids_in_binlog_not_purged
            = 1-11006 - 10006-11006
            = 1-10005
```

3.3 GTID 自动定位

GTID 是用来代替传统复制的方法，GTID 复制与普通复制模式的最大不同在于，启动和恢复复制时能够自动定位，而不需要指定二进制日志文件名和位置。配置非 GTID 复制时，需要在 CHANGE MASTER TO 语句中包含 MASTER_LOG_FILE 或 MASTER_LOG_POS 选项，用于指示从主库复制的开始点。但对于 GTID，从库不需要此非本地数据，其与主库同步的所有信息都直接从复制数据流中获取，因此不需要指定这些选项。要使用基于 GTID 的复制启动从库，推荐启用 MASTER_AUTO_POSITION 选项。

默认情况下禁用 MASTER_AUTO_POSITION 选项。如果在从库上启用了多源复制，则需要为每个适用的复制通道设置该选项。设置 MASTER_AUTO_POSITION=0 会使从库恢复为基于文件的复制，这时必须指定 MASTER_LOG_FILE 或 MASTER_LOG_POS 选项。当从库启用 GTID（GTID_MODE = ON、ON_PERMISSIVE 或 OFF_PERMISSIVE）并使用 MASTER_AUTO_POSITION 选项时，将激活自动定位以连接到主库。主库必须设置 GTID_MODE = ON 才能使连接成功。

在初始握手中，从库向主库发送一个 GTID 集，其中包含已经收到、已提交或两者都已完成的事务。此 GTID 集等于@@GLOBAL.gtid_executed 系统变量与 select received_transaction_set from performance_schema.replication_connection_status 查询结果的并集。主库会比较其二进制日志中记录的所有事务和从库发来的 GTID 集合，并将不包括在从库发送的 GTID 集中的事务全部发送给从库。自动跳过功能可确保同一事务不会应用两次。如果从库缺失的 GTID 已经被主库清除（purge），则复制中断，主库将错误 ER_MASTER_HAS_PURGED_REQUIRED_GTIDS 发送给从库。主库错误日志的 ER_FOUND_MISSING_GTIDS 警告消息中将列出丢失事务的 GTID。从库无法自动解决此问题，尝试在不启用 MASTER_AUTO_POSITION 选项的情况下重新连接主库只会导致已清除事务在从库上的丢失。可以考虑修改主库上的 binlog_expire_logs_seconds 系统参数值（默认为 2592000 秒，即 30 天），以确保不再发生还需要使用二进制日志时它却已经被提前清除的情况。下面模拟一下这个场景：

```
-- 从库停止复制
stop slave;

-- 主库进行更新
truncate table t1;

-- 主库修改 binlog 文件名，模拟事务丢失
mysql -uroot -p123456 -e "show master status;"
mv binlog.000001 binlog.000001.bak

-- 从库启动复制
start slave;
show slave status\G
```

会看到 1236 错误：

```
    Last_IO_Errno: 13114
    Last_IO_Error: Got fatal error 1236 from master when reading data from binary
log: 'Cannot replicate because the master purged required binary logs. Replicate
the missing transactions from elsewhere, or provision a new slave from backup.
Consider increasing the master's binary log expiration period. To find the missing
transactions, see the master's error log or the manual for GTID_SUBTRACT.'
```

主库的错误日志中会显示如下信息：

```
    2019-06-11T00:18:14.500248Z 207 [ERROR] [MY-010958] [Server] Could not open
     log file.
    2019-06-11T00:18:14.500299Z 207 [Warning] [MY-011809] [Server] Cannot
replicate to server with server_uuid='565a6b0a-6f05-11e9-b95c-005056a5497f'
because the present server has purged required binary logs. The connecting server
needs to replicate the missing transactions from elsewhere, or be replaced by a
new server created from a more recent backup. To prevent this error in the future,
consider increasing the binary log expiration period on the present server. The
missing transactions are '8eed0f5b-6f9b-11e9-94a9-005056a57a4e:1-11007'.
```

主库二进制日志修改成正确的文件名，重启从库复制后恢复正常：

```
    -- 主库
    mv binlog.000001.bak binlog.000001

    -- 从库
    stop slave;
    start slave;
    show slave status\G
```

如果在事务交换期间发现从库已经在 GTID 中接收或提交了事务，但主库本身没有它们的记录，则复制停止，主库将错误 ER_SLAVE_HAS_MORE_GTIDS_THAN_MASTER 发送给从库。当没有配置 sync_binlog=1 的主库遇到电源故障或操作系统崩溃，此时尚未同步，而导致二进制日志文件的已提交事务未被从库接收，就会发生这种情况。如果主库重新提交事务，可能导致主库和从库对不同的事务使用相同的 GTID，这时只能根据需要对各个事务以手动方式解决冲突，例如手工设置 gtid_next。如果问题仅仅是主库缺少事务，则可以进行主从切换，允许它跟上复制拓扑结构中的其他服务器，然后在需要时再次将其设置为主库。可见 sync_binlog=1 对于主从数据一致性至关重要，这也是 MySQL 8 的默认配置值。下面模拟一下这个场景：

```
    -- 主库 reset master
    reset master;

    -- 从库重启复制
    stop slave;
    start slave;
    show slave status\G
```

会看到以下错误：

```
    Last_IO_Errno: 13114
    Last_IO_Error: Got fatal error 1236 from master when reading data from binary
log: 'Slave has more GTIDs than the master has, using the master's SERVER_UUID.
```

```
This may indicate that the end of the binary log was truncated or that the last
binary log file was lost, e.g., after a power or disk failure when sync_binlog !=
1. The master may or may not have rolled back transactions that were already replica'
```

重新配置从库以恢复复制：

```
reset master;
stop slave;
reset slave all;
change master to
     master_host = '172.16.1.125',
     master_port = 3306,
     master_user = 'repl',
     master_password = '123456',
     master_auto_position = 1;
start slave;
```

3.4 配置 GTID 复制

前面解释了许多 GTID 的原理以及在 MySQL 复制中所起的作用，并且进行了很多实验加以辅助说明。本节演示如何从头开始一步步配置 GTID 复制，实验环境同 1.2 节。这里只讨论在联机情况下进行配置，因为相对于空库或脱机等理想情况，联机配置复制的需求更为典型和常见。

3.4.1 联机配置 GTID 复制

依照第 1 章讨论异步复制时得出的结论，本实验使用 XtraBackup 工具进行联机数据备份。主、从服务器已经进行了以下配置：

- 在主库上建立复制专属用户。
- 在主、从库上安装 XtraBackup。
- 配置主库到从库的 SSH 免密码连接。
- 停止作为从库的 MySQL 实例，并清空其数据目录。

这些作为配置 MySQL 复制的前置步骤，具体操作参考 1.3.3 小节。下面说明联机配置 GTID 复制的步骤：

步骤 01 检查主库中是否有不支持 GTID 的操作：

```
set global enforce_gtid_consistency=warn;
```

让服务器在正常工作负载下运行一段时间并监控错误日志，最好包含一天负载最高的时间段，有条件建议观察几天。如果此步骤导致错误日志中出现任何警告，需要调整应用程序，使其仅使用与 GTID 兼容的功能，并且不能生成与 GTID 相关的任何警告。这是一个重要步骤，在进行下一步之前，必须确保错误日志中未生成警告。

步骤 02 在主库联机设置 GTID 相关参数：

```
set global enforce_gtid_consistency=true;
set global gtid_mode=off_permissive;
set global gtid_mode=on_permissive;
set global gtid_mode=on;
```

enforce-gtid-consistency 启用后，MySQL 服务器通过仅允许执行使 GTID 安全的语句来强制 GTID 一致性。在启用基于 GTID 的复制之前，必须将此选项设置为 true。enforce_gtid_consistency 的可配置值为：

- false：允许事务违反 GTID 一致性。
- true：不允许事务违反 GTID 一致性。
- warn：允许事务违反 GTID 一致性，但在这种情况下会生成警告。

当 enforce_gtid_consistency 设置为 true 时，只能使用 GTID 安全的语句，例如如下操作不能与此选项一起使用：

- CREATE TABLE ... SELECT 语句。
- 事务内的 CREATE TEMPORARY TABLE 或 DROP TEMPORARY TABLE 语句。
- 更新事务和非事务表的事务或语句。

enforce_gtid_consistency 仅在语句进行二进制日志记录时生效。如果在服务器上禁用了二进制日志记录，或者由于过滤器删除了语句而未将语句写入二进制日志，则不会对未记录的语句检查或强制执行 GTID 一致性。

在包含 gtid_mode 系统变量的所有 MySQL 版本中，它都可以设置成 on 或 off。MySQL 5.7.6 版本之后，gtid_mode 提供了两个新的选项分别为 on_permissive 和 off_permissive。当 gtid_mode = on 时，无法复制匿名事务，而当 gtid_mode = off 时，只能复制匿名事务。当 gtid_mode = off_permissive 时，新事务是匿名的，同时允许复制的事务是 GTID 或匿名事务。当 gtid_mode = on_permissive 时，新事务使用 GTID，同时允许复制事务为 GTID 或匿名事务。这意味着可以拥有一个复制拓扑结构，其中包含使用匿名和 GTID 事务的服务器。例如，具有 gtid_mode = on 的主库可以有使用 gtid_mode = on_permissive 从库。gtid_mode 在主、从库上的兼容性以及能否使用自动定位如表 3-1 所示，每个条目的含义如下：

- Y：主库和从库的 gtid_mode 兼容。
- N：主库和从库的 gtid_mode 不兼容。
- *：自动定位可与此组合一起使用。

表 3-1　gtid_mode 兼容性

Slave Master	OFF	OFF_PERMISSIVE	ON_PERMISSIVE	ON
OFF	Y	Y	N	N
OFF_PERMISSIVE	Y	Y	Y	Y*
ON_PERMISSIVE	Y	Y	Y	Y*
ON	N	N	Y	Y*

联机设置 gtid_mode 时，只能基于 OFF、OFF_PERMISSIVE、ON_PERMISSIVE、ON 的顺序一次改变一步。例如，如果 gtid_mode 当前设置为 OFF_PERMISSIVE，则可以更改为 OFF 或 ON_PERMISSIVE，但不能直接更改为 ON，否则会报以下错误：

```
ERROR 1788 (HY000): The value of @@GLOBAL.GTID_MODE can only be changed one
 step at a time: OFF <-> OFF_PERMISSIVE <-> ON_PERMISSIVE <-> ON. Also note
that this value must be stepped up or down simultaneously on all servers. See the
Manual for instructions.
```

这样实现是为了确保服务器能够正确处理从匿名事务更改为 GTID 事务的过程，此过程中可能同时包含这两种模式的事务。在 gtid_mode = on 和 gtid_mode = off 之间切换时，GTID 状态（也就是 gtid_executed 的值）是持久化的，因此不管 gtid_mode 的类型如何更改，都可确保始终保留服务器应用的 GTID 集。无论当前选择的 gtid_mode 如何，与 GTID 相关的字段都会显示正确的信息。显示 GTID 集的字段，例如 performance_schema.replication_connection_status 表中的 gtid_executed、gtid_purged、received_transaction_set 以及 show slave status 的 GTID 相关结果，在没有 GTID 时返回空字符串。显示单个 GTID 的字段，如 performance_schema.replication_applier_status_by_worker 表中的 current_transaction，在未使用 GTID 时显示 ANONYMOUS。从库使用 gtid_mode = on 复制提供了自动定位的功能。

当前选择的 gtid_mode 也会影响 gtid_next 变量。表 3-2 显示了服务器对 gtid_mode 和 gtid_next 不同值的行为。每个条目的含义如下：

- ANONYMOUS：生成匿名事务。
- Error：生成错误并且无法执行 SET GTID_NEXT。
- UUID:NUMBER：使用指定的 UUID:NUMBER 生成 GTID。
- New GTID：使用自动生成的数字生成 GTID。

表 3-2 gtid_mode 和 gtid_next 不同值的行为

gtid_mode gtid_next	AUTOMATIC binary log on	AUTOMATIC binary log off	ANONYMOUS	UUID:NUMBER
OFF	ANONYMOUS	ANONYMOUS	ANONYMOUS	Error
OFF_PERMISSIVE	ANONYMOUS	ANONYMOUS	ANONYMOUS	UUID:NUMBER
ON_PERMISSIVE	New GTID	ANONYMOUS	ANONYMOUS	UUID:NUMBER
ON	New GTID	ANONYMOUS	Error	UUID:NUMBER

当二进制日志关闭且 gtid_next 设置为 AUTOMATIC 时，不会生成 GTID，这与先前版本的行为一致。注意，为了保证主从数据一致性和实例恢复的性能，在 MySQL 8 中作为一项基本原则，除非有特殊需求，与复制相关的其他系统变量最好保持默认值，包括但不限于下面所列出的系统变量：

- autocommit = ON
- log_bin = ON
- log_slave_updates = ON

- innodb_flush_log_at_trx_commit = 1
- sync_binlog = 1
- master_info_repository = TABLE
- relay_log_info_repository = TABLE
- relay_log_recovery = OFF
- binlog_gtid_simple_recovery = ON
- innodb_replication_delay = 0

（1）使用 xtrabackup 备份并传输：

```
    xtrabackup -uroot -p123456 --socket=/tmp/mysql.sock --no-lock --backup
    --compress --stream=xbstream --parallel=4 --target-dir=./ | ssh
mysql@172.16.1.126 "xbstream -x -C /usr/local/mysql/data/ --decompress"
```

使用 xtrabackup 的一个好处是，不必考虑从 gtid_mode=off 到 gtid_mode=on 的过程中是否存在正在进行的匿名事务。xtrabackup 执行的是物理备份，主库上无论是匿名事务还是 GTID 事务，最终数据文件和二进制日志文件都会被复制到从库，并且在恢复备份、启动实例和复制后，可以使用 GTID 的自动定位功能找到初始复制点。

（2）在从库中恢复备份：

```
xtrabackup --prepare --target-dir=/usr/local/mysql/data/
```

（3）在从库的配置文件中添加以下选项：

```
server_id=1126                      # 服务器 ID
read_only=on                        # 从库只读
gtid_mode=on                        # 启用 GTID
enforce-gtid-consistency=true       # 强制 GTID 一致
```

（4）启动从库：

```
mysqld_safe --defaults-file=/etc/my.cnf &
```

（5）在从库启动复制：

```
change master to
     master_host = '172.16.1.125',
    master_port = 3306,
    master_user = 'repl',
    master_password = '123456',
    master_auto_position = 1;
start slave;
show slave status\G
```

可以在 slave status 中看到复制正在进行，Retrieved_Gtid_Set 和 Executed_Gtid_Set 不断增加，Seconds_Behind_Master 逐渐缩小至 0。

（6）将 GTID 参数添加到主库的配置文件中：

```
gtid_mode=on
enforce-gtid-consistency=true
```

至此完成联机配置GTID复制。

3.4.2 联机更改复制模式

如果已经在未启用GITD的情况下配置了主从复制，可以联机将复制模式修改为GTID以及自动定位。由于整个过程不需要停止MySQL实例，因此这种方式适合在生产环境中使用。开始前确保MySQL服务器满足以下前提条件：

- 复制拓扑结构中的所有服务器都必须使用MySQL 5.7.6或更高版本。除非拓扑结构中的所有服务器都使用此版本，否则无法在任何单个服务器上联机启用GTID事务。
- 所有服务器的gtid_mode默认设置为OFF。

以下过程可以随时暂停，之后再恢复，这使得该过程具有容错能力。如果过程中出现任何不相关的错误，可以先暂停过程解决问题，然后再从停止的地方继续。但至关重要的一点是，在继续下一步之前必须完成之前的步骤。联机改为GTID复制的步骤如下：

步骤01 在每台服务器上执行：set global enforce_gtid_consistency=warn; 保证所有操作都与GTID兼容，并且确保错误日志中没有GTID的相关警告。

步骤02 在每台服务器上执行：set global enforce_gtid_consistency=true;。

步骤03 在每台服务器上执行：set global gtid_mode=off_permissive; 哪台服务器首先执行此语句无关紧要，重要的是在开始下一步之前，在所有服务器完成此步骤。

步骤04 在每台服务器上执行：set global gtid_mode=on_permissive;。

步骤05 在每台服务器上，等待状态变量ongoing_anonymous_transaction_count为0。检查方法：show status like 'ongoing_anonymous_transaction_count';。

步骤06 如果二进制日志还用于复制以外的其他目的，如基于时间点的恢复等，需要在执行flush logs后备份二进制日志文件。包含匿名事务的二进制日志在下一步执行之后无法使用。完成此步后，确保拓扑中的任何位置都不存在匿名事务。

步骤07 在每台服务器上执行：set global gtid_mode=on;。

步骤08 在每个从库上执行以下操作：

```
stop slave;
change master to master_auto_position = 1;
start slave;
```

步骤09 在每台服务器上，将gtid-mode = on和enforce_gtid_consistency=true添加到my.cnf配置文件。

现在可以保证所有事务都具有GTID（步骤5或更早生成的事务已经过处理），已经改为GTID复制模式。

联机将GTID事务复制改为匿名事务复制模式的过程基本是上述步骤的逆过程，唯一不同的是等待记录事务复制的方式。

步骤01 在每个从库上执行以下操作，xxx可以从show slave status输出的Master_Log_File和

Read_Master_Log_Pos 获得：

```
stop slave;
show slave status\G
change master to
      master_auto_position = 0,
      master_log_file = 'xxx',
      master_log_pos = xxx;
start slave;
```

步骤 02 在每台服务器上执行：set global gtid_mode=on_permissive;。

步骤 03 在每台服务器上执行：set global gtid_mode=off_permissive;。

步骤 04 在每台服务器上，等待变量@@GLOBAL.GTID_OWNED 等于空字符串。检查方法：select @@global.gtid_owned; 此时复制正常进行，但 slave status 中的 Retrieved_Gtid_Set 和 Executed_Gtid_Set 值不再变化。

步骤 05 如果二进制日志还用于复制以外的其他目的（如基于时间点的恢复等），执行 FLUSH LOGS 后再备份二进制日志文件。包含 GTID 事务的二进制日志在下一步执行之后无法使用。完成此步骤后，确保拓扑结构中的任何位置都不存在 GTID 事务。

步骤 06 在每台服务器上执行：set global gtid_mode=off;。

步骤 07 在每台服务器上执行：set global enforce_gtid_consistency=false;。

步骤 08 在每台服务器上，在 my.cnf 配置文件中设置 gtid-mode = off 和 enforce_gtid_consistency=false。

3.4.3 GTID 相关系统变量

前面已经在各个地方分散介绍过很多 GTID 相关的系统变量，如 gtid_executed、gtid_next、gtid_purged、gtid_mode 等，下面对 MySQL 8 中的重要 GTID 系统变量加以简单整理。

- binlog_gtid_simple_recovery：布尔类型全局变量，控制 MySQL 启动时从哪些 binlog 文件中寻找 GTID，默认值为 true。当 binlog_gtid_simple_recovery=true 时，初始化 gtid_executed 和 gtid_purged 值时只读取最旧和最新的 binlog 文件，否则需要遍历所有 binlog 文件。
- enforce_gtid_consistency：枚举类型全局变量，指示是否强制 GTID 数据一致性，有效值为 OFF、ON 和 WARN，默认值为 OFF。
- gtid_executed：全局和会话级别都可以用，用来保存已经执行过的 GTID 集合。
- gtid_executed_compression_period：整型全局变量，指示压缩 mysql.gtid_executed 表之前允许的事务数，默认值为 1000。使用二进制日志时该值不起作用，而是在每个二进制日志轮转时压缩 mysql.gtid_executed 表。
- gtid_mode：枚举类型全局变量，控制是否启用 GTID 功能，有效值为 OFF、OFF_PERMISSIVE、ON_PERMISSIVE 和 ON，默认值为 OFF。联机设置时，只能按顺序一步步修改。
- gtid_next：会话级枚举变量，用于指定如何获取下一个 GTID。有效值为 AUTOMATIC、ANONYMOUS 和 UUID:NUMBER，默认值为 AUTOMATIC。
- gtid_owned：内部使用的只读变量，指示当前正在使用的 GTID 以及拥有它的线程 ID。

- gtid_purged：全局变量，设置已经执行但在 binlog 中被清除的 GTID 集合，是 gtid_executed 的子集。

3.5 GTID 运维

每个 GTID 唯一标识构成事务的一组二进制日志事件，在二进制日志中跟踪 GTID 事务与其事件集之间的映射。应用连接到数据库时，MySQL 服务器自动跳过之前已处理的 GTID 事务，此行为对于自动复制定位和正确的故障转移至关重要。启用 GTID 也给运维带来了一些改变。

3.5.1 跳过一个事务

传统基于 GTID 二进制位置的复制中，从库由于某些错误导致复制中断时，一个可能的解决方案是设置 sql_slave_skip_counter 全局系统变量，跳过导致错误的事件，然后重启复制。但是，在启用 GTID 后，执行的单位由事件变为事务，因此该方法不再有效（slave_skip_errors 仍然可用），并会报出以下错误：

```
mysql> set global sql_slave_skip_counter=1;
ERROR 1858 (HY000): sql_slave_skip_counter can not be set when the server is running with @@GLOBAL.GTID_MODE = ON. Instead, for each transaction that you want to skip, generate an empty transaction with the same GTID as the transaction
```

从错误信息可以看到，GTID 跳过事务的方法是注入一个空事务，具体步骤如下：

步骤 01 定位出错事务的 GTID。我们需要获得从库执行的最后一个事务，方法有：show slave status \G 中的 Executed_Gtid_Set，show global variables like '%gtid%'; 中的 gtid_executed，show master status;中的 Executed_Gtid_Set。

步骤 02 将会话级系统变量 gtid_next 设置为上一步的 GTID，如：set gtid_next = '8eed0f5b-6f9b-11e9-94a9-005056a57a4e:980058'。注意 gtid_next 的值只能是单个 GTID。

步骤 03 注入空事务：begin;commit;。

步骤 04 重启复制：set gtid_next='automatic'; start slave; 重启复制前需要将 gtid_next 设置为默认值 'automatic'。下面是跳过多个事务的一个例子：

```
stop slave;
set gtid_next='8eed0f5b-6f9b-11e9-94a9-005056a57a4e:980055';
begin;commit;
set gtid_next='8eed0f5b-6f9b-11e9-94a9-005056a57a4e:980056';
begin;commit;
set gtid_next='8eed0f5b-6f9b-11e9-94a9-005056a57a4e:980057';
begin;commit;
set gtid_next='automatic';
start slave;
```

3.5.2　mysqldump 导出

使用 mysqldump 受 set-gtid-purged 选项影响，set-gtid-purged 选项设置为 AUTO（默认值）或 ON 时的输出如下所示：

```
[mysql@hdp3~]$mysqldump --single-transaction --all-databases --master-data=2
 --host=172.16.1.125 --user=repl --password=123456
-- MySQL dump 10.13  Distrib 8.0.16, for linux-glibc2.12 (x86_64)
...
SET @MYSQLDUMP_TEMP_LOG_BIN = @@SESSION.SQL_LOG_BIN;
SET @@SESSION.SQL_LOG_BIN= 0;

--
-- GTID state at the beginning of the backup
--

SET @@GLOBAL.GTID_PURGED=/*!80000 '+'*/
'8eed0f5b-6f9b-11e9-94a9-005056a57a4e:1-980059';

--
-- Position to start replication or point-in-time recovery from
--

-- CHANGE MASTER TO MASTER_LOG_FILE='binlog.000022',
MASTER_LOG_POS=209837996;
SET @@SESSION.SQL_LOG_BIN = @MYSQLDUMP_TEMP_LOG_BIN;
...
-- Dump completed on 2019-06-13 10:36:35
[mysql@hdp3~]$
```

开始部分的 SET @@SESSION.SQL_LOG_BIN= 0 用于防止导入数据时基于本地服务器生成新的 GTID。接着 GTID_PURGED 被设置为备份时刻已经执行过的 GTID 事务，该操作将会初始化 mysql.gtid_executed 表、gtid_purge 变量及 gtid_executed 变量。当 mysqldump 命令加入 --set-gtid-purged=off 选项时，则输出中不会加入 SQL_LOG_BIN= 0 和 GTID_PURGED 的设置。如果要将数据导入从库进行初始化，则不能设置--set-gtid-purged=off。下面是这个选项的含义：

```
--set-gtid-purged[=name]
                Add 'SET @@GLOBAL.GTID_PURGED' to the output. Possible
     values for this option are ON, OFF and AUTO. If ON is
     used and GTIDs are not enabled on the server, an error is
     generated. If OFF is used, this option does nothing. If
     AUTO is used and GTIDs are enabled on the server, 'SET
     @@GLOBAL.GTID_PURGED' is added to the output. If GTIDs
     are disabled, AUTO does nothing. If no value is supplied
     then the default (AUTO) value will be considered.
```

细心的读者到这里可能心生疑问：为初始化从库数据，命令行使用了--all-databases 选项。mysql.gtid_executed 表会不会被重建，进而通过 GTID_PURGED 设置的 mysql.gtid_executed 表会重新改变，重启数据库后读取 mysql.gtid_executed 表可能获得错误 GTID 集合导致复制错误？答案也在 mysqldump 的输出中。

首先，如果从库实例的 mysql 库存在，则不会删除重建：

```
    ...
    -- CHANGE MASTER TO MASTER_LOG_FILE='binlog.000022',
MASTER_LOG_POS=209837996;

    --
    -- Current Database: 'mysql'
    --

    CREATE DATABASE /*!32312 IF NOT EXISTS*/ 'mysql' /*!40100 DEFAULT CHARACTER
SET utf8mb4 COLLATE utf8mb4_0900_ai_ci */ /*!80016 DEFAULT ENCRYPTION='N' */;

    USE 'mysql';
    ...
```

其次，如果 mysql.gtid_executed 表存在，则不会删除重建。最后，如果该表不存在，则创建它，但不会向其加载数据。由此得出结论，除非手工删除了 mysql.gtid_executed 表，否则不会因它造成复制问题，至少 MySQL 8 是这样。具体如下所示：

```
    ...
    --
    -- Table structure for table `gtid_executed`
    --

    /*!40101 SET @saved_cs_client     = @@character_set_client */;
     SET character_set_client = utf8mb4 ;
    CREATE TABLE IF NOT EXISTS 'gtid_executed' (
      'source_uuid' char(36) NOT NULL COMMENT 'uuid of the source where the
transaction was originally executed.',
      'interval_start' bigint(20) NOT NULL COMMENT 'First number of interval.',
      'interval_end' bigint(20) NOT NULL COMMENT 'Last number of interval.',
      PRIMARY KEY ('source_uuid','interval_start')
    ) /*!50100 TABLESPACE 'mysql' */ ENGINE=InnoDB DEFAULT CHARSET=utf8mb4
COLLATE=utf8mb4_0900_ai_ci;
    /*!40101 SET character_set_client = @saved_cs_client */;

    --
    -- Table structure for table 'help_category'
    --
    ...
```

3.5.3 主从切换

这里分三种情况进行讨论：从库只读、从库读写并且有全部写操作的二进制日志、从库读写但写操作的二进制日志不全。

1. 从库只读

这种情况从库（新主库）没有执行过本地的事务，只需执行正常切换：

```
-- 从库（新主库）
stop slave;
reset slave all;

-- 主库（新从库）
change master to
     master_host = '172.16.1.126',
     master_port = 3306,
     master_user = 'repl',
     master_password = '123456',
     master_auto_position = 1;
start slave;
```

新主库会生成自己的 GTID 事务，此时会出现两个 server_uuid 对应的 GTID：

```
mysql> select @@global.gtid_executed;
+------------------------------------------------------------------------------------
-----------------+
| @@global.gtid_executed                                                            |
+------------------------------------------------------------------------------------
-----------------+
| 53442434-8bfa-11e9-bc15-005056a50f77:1-2,
8eed0f5b-6f9b-11e9-94a9-005056a57a4e:1-980059 |
+------------------------------------------------------------------------------------
-----------------+
1 row in set (0.00 sec)
```

2. 从库读写并且有全部写操作的二进制日志

```
-- 从库（新主库）
drop table test.t1;
create table test.t1(a int);
insert into test.t1 select 100;

stop slave;
reset slave all;

-- 主库（新从库）
change master to
     master_host = '172.16.1.125',
     master_port = 3306,
     master_user = 'repl',
     master_password = '123456',
     master_auto_position = 1;
start slave;
```

此时从 show slave status 的输出中可以看到：

```
Retrieved_Gtid_Set: 8eed0f5b-6f9b-11e9-94a9-005056a57a4e:980060-980062
```

```
    Executed_Gtid_Set: 53442434-8bfa-11e9-bc15-005056a50f77:1-2,
8eed0f5b-6f9b-11e9-94a9-005056a57a4e:1-980062
```

刚才从库执行的三个本地事务，在新从库上正常复制。因为本地事务与复制事务 GTID 的 server_uuid 部分不同，只要 binlog 保留完整，从库上的写操作在主从切换后可以自动复制到新的从库上，与匿名复制相比明显方便许多。

3. 从库读写但写操作的二进制日志不全

```
-- 从库（新主库）
drop table test.t1;
create table test.t1(a int);
insert into test.t1 select 100;

stop slave;
reset slave all;
flush logs;

# 模拟binlog文件丢失
mv binlog.000022 binlog.000022.bak

-- 主库（新从库）
change master to
    master_host = '172.16.1.125',
     master_port = 3306,
     master_user = 'repl',
     master_password = '123456',
     master_auto_position = 1;
start slave;
```

此时在 show slave status 的输出中报错如下：

```
Last_IO_Errno: 13114
Last_IO_Error: Got fatal error 1236 from master when reading data from binary
log: 'Could not open log file'
```

真实环境中要是遇到这种情况建议还是重建从库。二进制日志文件默认的保留时间是 30 天（binlog_expire_logs_seconds = 2592000）。一般来说从库的写操作通常是为保留一些报表结果或临时数据，这些操作的最早时间很大可能已超过三十天，在这之后进行主从切换就会出现问题。这也是建议从库 readonly（只读）的原因之一。如果确实要执行比如加索引等不影响数据的操作，则可以在执行前将 sql_log_bin 变量设置为 0，这样不会增加本地 GTID：

```
set sql_log_bin=0;
create index idx1 on test.t1(a);
```

但还是要强调，从库最好始终 readonly。

3.6　GTID 限制

1. 涉及非事务存储引擎的更新

使用 GTID 时，一条语句或一个事务中，不能对非事务性存储引擎如 MyISAM 表和事务存储引擎（如 InnoDB 表）同时更新，因为这种混合引擎同时更新可能导致将多个 GTID 分配给同一个事务。下面两组命令都会报同样的错误。

```
use test;
create table t_myisam(a int) engine=myisam;
create table t_innodb(a int) engine=innodb;
update t_myisam, t_innodb set t_myisam.a=1, t_innodb.a=1;

begin;
insert into t_myisam select 1;
insert into t_innodb select 1;
update t_myisam set a=2;
update t_innodb set a=2;
commit;
```

错误信息：

```
ERROR 1785 (HY000): Statement violates GTID consistency: Updates to
non-transactional tables can only be done in either autocommitted statements
or single-statement transactions, and never in the same statement as updates to
transactional tables.
```

在 MySQL 8 中，这个限制并没有多大影响，因为包括系统表在内都是 InnoDB 表，默认已经没有 myisam 表了，除非用户建表时显示定义。

2. CREATE TABLE ... SELECT 语句

CREATE TABLE ...使用基于 GTID 的复制时不允许使用 SELECT 语句。当 binlog_format 设置为 STATEMENT 时，CREATE TABLE ... SELECT 语句作为一个具有单一 GTID 的事务记录在二进制日志中。但如果使用 ROW 格式，则该语句将记录为具有两个 GTID 的两个事务。如果主服务器使用 STATEMENT 格式而从服务器使用 ROW 格式，则从服务器将无法正确处理事务，因此 GTID 不允许使用 CREATE TABLE ... SELECT 语句来防止出现这种情况。

```
mysql> show variables like 'binlog_format';
+---------------+-------+
| Variable_name | Value |
+---------------+-------+
| binlog_format | ROW   |
+---------------+-------+
1 row in set (0.00 sec)

mysql> create table t2 as select * from t1;
```

```
ERROR 1786 (HY000): Statement violates GTID consistency: CREATE TABLE ... SELECT.

mysql> set binlog_format=statement;
Query OK, 0 rows affected (0.00 sec)

mysql> create table t2 as select * from t1;
ERROR 1786 (HY000): Statement violates GTID consistency: CREATE TABLE...SELECT.
```

3. 临时表

当 binlog_format 设置为 STATEMENT，在服务器上启用 GTID 时，不能在事务、过程、函数或触发器内使用 CREATE TEMPORARY TABLE 和 DROP TEMPORARY TABLE 语句。如果设置了 autocommit = 1，则可以在使用 GTID 时在这些上下文之外使用它们。从 MySQL 8.0.13 开始，当 binlog_format 设置为 ROW 或 MIXED 且启用 GTID 时，允许在事务、过程、函数或触发器内使用 CREATE TEMPORARY TABLE 和 DROP TEMPORARY TABLE 语句。这些语句不会写入二进制日志，因此不会复制到从库。

```
mysql> show variables like 'binlog_format';
+---------------+-------+
| Variable_name | Value |
+---------------+-------+
| binlog_format | ROW   |
+---------------+-------+
1 row in set (0.01 sec)

mysql> begin;
Query OK, 0 rows affected (0.00 sec)

mysql> create temporary table tmp1 select * from t1;
Query OK, 1 row affected (0.00 sec)
Records: 1  Duplicates: 0  Warnings: 0

mysql> commit;
Query OK, 0 rows affected (0.00 sec)

mysql> set binlog_format=statement;
ERROR 3745 (HY000): Changing @@session.binlog_format is disallowed when the session has open temporary table(s). You could wait until these temporary table(s) are dropped and try again.

mysql> drop temporary table tmp1;
Query OK, 0 rows affected (0.00 sec)

mysql> set binlog_format=statement;
Query OK, 0 rows affected (0.00 sec)

mysql> create temporary table tmp1 select * from t1;
Query OK, 1 row affected (0.00 sec)
Records: 1  Duplicates: 0  Warnings: 0
```

```
mysql> drop temporary table tmp1;
Query OK, 0 rows affected (0.00 sec)

mysql> begin;
Query OK, 0 rows affected (0.00 sec)

mysql> create temporary table tmp1 select * from t1;
ERROR 3748 (HY000): Statement violates GTID consistency: CREATE TEMPORARY TABLE
and DROP TEMPORARY TABLE are not allowed inside a transaction or inside a procedure
in a transactional context when @@session.binlog_format=STATEMENT.
```

要防止执行会导致基于 GTID 的复制失败的语句，必须在启用 GTID 时使用--enforce-gtid-consistency 选项启动所有服务器。这会导致前面讨论的语句失败并显示错误。

4. 忽略服务器

使用 GTID 时，不推荐使用 CHANGE MASTER TO 语句的 IGNORE_SERVER_IDS 选项，因为已经应用的事务会自动被忽略。在启动基于 GTID 的复制之前，需要检查并使用 CHANGE MASTER TO IGNORE_SERVER_IDS = ()清除之前在相关服务器上设置的所有忽略的服务器 ID 列表。可以为各个通道发出的 SHOW SLAVE STATUS 语句显示被忽略的服务器 ID 列表。如果没有，则 Replicate_Ignore_Server_Ids 字段为空。

5. GTID 模式和 mysqldump

可以将使用 mysqldump 创建的转储导入到启用了 GTID 模式的 MySQL 服务器中，前提是目标服务器的二进制日志中没有重叠的 GTID。

```
[mysql@hdp3/usr/local/mysql/data]$mysqldump --single-transaction --all
-databases --master-data=2 --host=172.16.1.125 --user=wxy --password=123456
| mysql -uroot -p123456
mysql: [Warning] Using a password on the command line interface can be insecure.
mysqldump: [Warning] Using a password on the command line interface can be
insecure.
Warning: A partial dump from a server that has GTIDs will by default include
the GTIDs of all transactions, even those that changed suppressed parts of the
database. If you don't want to restore GTIDs, pass --set-gtid-purged=OFF. To make
a complete dump, pass --all-databases --triggers --routines --events.
ERROR 3546 (HY000) at line 26: @@GLOBAL.GTID_PURGED cannot be changed: the added
gtid set must not overlap with @@GLOBAL.GTID_EXECUTED
mysqldump: Got errno 0 on write

[mysql@hdp3/usr/local/mysql/data]$mysql -uroot -p123456 -e "reset master;"
mysql: [Warning] Using a password on the command line interface can be insecure.
[mysql@hdp3/usr/local/mysql/data]$mysqldump --single-transaction
--all-databases --master-data=2 --host=172.16.1.125 --user=wxy --password=123456
| mysql -uroot -p123456
mysql: [Warning] Using a password on the command line interface can be insecure.
mysqldump: [Warning] Using a password on the command line interface can be
insecure.
```

```
    Warning: A partial dump from a server that has GTIDs will by default include
the GTIDs of all transactions, even those that changed suppressed parts of the
database. If you don't want to restore GTIDs, pass --set-gtid-purged=OFF. To make
a complete dump, pass --all-databases --triggers --routines --events.
    [mysql@hdp3/usr/local/mysql/data]$
```

3.7 GTID 集合运算函数

3.7.1 GTID 内置函数

MySQL 8 包含 GTID_SUBSET、GTID_SUBTRACT、WAIT_FOR_EXECUTED_GTID_SET 和 WAIT_UNTIL_SQL_THREAD_AFTER_GTIDS 这 4 个内置函数，用于 GTID 集合的基本运算。

- GTID_SUBSET(set1,set2)：给定两个 GTID 集 set1 和 set2，set1 是 set2 的子集返回 true，否则返回 false。
- GTID_SUBTRACT(set1,set2)：给定两个 GTID 集 set1 和 set2，返回 set1 与 set2 的差集。
- WAIT_FOR_EXECUTED_GTID_SET(gtid_set[, timeout])：等到服务器应用了包含在 gtid_set 中的所有事务。如果指定可选的 timeout 值（秒数），超时会使函数停止等待而退出。
- WAIT_UNTIL_SQL_THREAD_AFTER_GTIDS(gtid_set[, timeout][,channel])：与 WAIT_FOR_EXECUTED_GTID_SET 类似，但针对单个启动的复制通道。

3.7.2 用户自定义函数

用户可以在自定义函数中调用这些内置函数，实现一些常用的 GTID 集合运算，下面是 MySQL 8 文档中的几个例子。

- 如果两个 GTID 集相同，函数返回非零值。

```
    create function gtid_is_equal(gtid_set_1 longtext, gtid_set_2 longtext)
    returns int deterministic
      return gtid_subset(gtid_set_1, gtid_set_2) and gtid_subset(gtid_set_2,
gtid_set_1);
```

- 如果两个 GTID 集不相交，函数返回非零值。

```
    create function gtid_is_disjoint(gtid_set_1 longtext, gtid_set_2 longtext)
    returns int deterministic
      return gtid_subset(gtid_set_1, gtid_subtract(gtid_set_1, gtid_set_2));
```

- 如果两个 GTID 集不相交，则函数返回非零，sum 是两个集的并集。

```
    create function gtid_is_disjoint_union(gtid_set_1 longtext, gtid_set_2
                                            longtext, sum longtext)
    returns int deterministic
```

```
    return gtid_is_equal(gtid_subtract(sum, gtid_set_1), gtid_set_2) and
        gtid_is_equal(gtid_subtract(sum, gtid_set_2), gtid_set_1);
```

- 函数返回格式化的 GTID 集。没有空格且没有重复，UUID 按字母顺序排列，间隔按数字顺序排列。

```
create function gtid_normalize(g longtext)
returns longtext deterministic
return gtid_subtract(g, '');
```

- 函数返回两个 GTID 集的并集。

```
create function gtid_union(gtid_set_1 longtext, gtid_set_2 longtext)
returns longtext deterministic
  return gtid_normalize(concat(gtid_set_1, ',', gtid_set_2));
```

- 函数返回两个 GTID 集的交集。

```
create function gtid_intersection(gtid_set_1 longtext, gtid_set_2 longtext)
returns longtext deterministic
  return gtid_subtract(gtid_set_1, gtid_subtract(gtid_set_1, gtid_set_2));
```

- 函数返回两个 GTID 集的对称差集。

```
create function gtid_symmetric_difference(gtid_set_1 longtext, gtid_set_2
                                           longtext)
returns longtext deterministic
  return gtid_subtract(concat(gtid_set_1, ',', gtid_set_2),
gtid_intersection(gtid_set_1, gtid_set_2));
```

- 函数返回除去指定 UUID 的 GTID 集。

```
create function gtid_subtract_uuid(gtid_set longtext, uuid text)
returns longtext deterministic
  return gtid_subtract(gtid_set, concat(uuid, ':1-', (1 << 63) - 2));
```

- 函数返回指定 UUID 的 GTID 集。

```
create function gtid_intersection_with_uuid(gtid_set longtext, uuid text)
returns longtext deterministic
  return gtid_subtract(gtid_set, gtid_subtract_uuid(gtid_set, uuid));
```

3.7.3 使用示例

1. 验证从库的复制是否最新

内置函数 GTID_SUBSET 和 GTID_SUBTRACT 可用于检查从库是否应用了主库的每个事务。使用 GTID_SUBSET 执行此检查，在从库上执行以下命令：

```
master_gtid_executed=`mysql -uwxy -p123456 -h172.16.1.125 -N -e "select
                     replace(@@global.gtid_executed,char(10),'')"`
slave_gtid_executed=`mysql -uwxy -p123456 -N -e "select replace
```

```
(@@global.gtid_executed,char(10),'')"`
    sql="select gtid_subset('$master_gtid_executed', '$slave_gtid_executed')"
    mysql -uwxy -p123456 -e "$sql"
```

如果返回 0，则 master_gtid_executed 中的某些 GTID 不存在于 slave_gtid_executed 中，因此从库不是最新的。要使用 GTID_SUBTRACT 执行检查，在从库上执行以下命令：

```
    master_gtid_executed=`mysql -uwxy -p123456 -h172.16.1.125 -N -e "select
                        replace(@@global.gtid_executed,char(10),'')"`
    slave_gtid_executed=`mysql -uwxy -p123456 -N -e "select replace
(@@global.gtid_executed,char(10),'')"`
    sql="select gtid_subtract('$master_gtid_executed', '$slave_gtid_executed')"
    mysql -uwxy -p123456 -e "$sql"
```

返回 master_gtid_executed 中但未在 slave_gtid_executed 中的 GTID。如果返回值不为空，则从库不是最新的。

2. 验证 mysqldump 导出和导入

自定义函数 GTID_IS_EQUAL、GTID_IS_DISJOINT 和 GTID_IS_DISJOINT_UNION 可用于验证涉及多个数据库和服务器的备份、还原操作。此示例中，server1 包含数据库 db1，server2 包含数据库 db2。目标是将数据库 db2 复制到 server1，server1 上的结果应该是两个数据库的并集。过程是使用 mysqldump 备份 server2，然后在 server1 上恢复此备份。

如果 mysqldump 的选项--set-gtid-purged 设置为 ON 或默认值为 AUTO 时，则程序的输出包含 SET @@GLOBAL.gtid_purged 语句，该语句将 server2 中的 gtid_executed 集添加到 server1 上的 gtid_purged 集。gtid_purged 集包含已在服务器上提交但在服务器上的任何二进制日志文件中不存在的所有事务的 GTID。将数据库 db2 复制到 server1 时，必须将 server2 上提交的事务的 GTID（不在 server1 上的二进制日志文件中）添加到 server1 的 gtid_purged 集中，以使这些 GTID 集不在 server1 上重建。

- 使用 GTID_IS_EQUAL 验证备份操作是否为 SET @@GLOBAL.gtid_purged 语句计算了正确的 GTID 集。在 server2 上，从 mysqldump 输出中提取该语句，并将 GTID 集存储到本地变量中，例如$gtid_purged_set。然后执行以下语句：

```
server2> SELECT GTID_IS_EQUAL($gtid_purged_set, @@GLOBAL.gtid_executed);
```

如果结果为 1，则两个 GTID 集相等，并且已正确计算该集。

- 使用 GTID_IS_DISJOINT 验证 mysqldump 输出中设置的 GTID 与 server1 上的 gtid_executed 集不重叠。如果存在任何重叠，则在将数据库 db2 复制到 server1 时会出现错误。将 mysqldump 输出中的 gtid_purged 集提取并存储到如上所述的局部变量中，然后执行以下语句：

```
server1> SELECT GTID_IS_DISJOINT($gtid_purged_set, @@GLOBAL.gtid_executed);
```

如果结果为 1，则两个 GTID 集之间没有重叠，因此不存在重复的 GTID。

- 使用 GTID_IS_DISJOINT_UNION 验证还原操作是否导致 server1 上的 GTID 状态正确。在恢复备份之前，在 server1 上，通过执行以下语句获取现有的 gtid_executed 集：

```
server1> SELECT @@GLOBAL.gtid_executed;
```

将结果存储在本地变量$original_gtid_executed 中。还将 gtid_purged 集存储在局部变量中。当 server2 的备份已恢复到 server1 上时，执行以下语句以验证 GTID 状态：

```
server1> SELECT GTID_IS_DISJOINT_UNION($original_gtid_executed,
                        $gtid_purged_set,
                        @@GLOBAL.gtid_executed);
```

如果结果为 1，则已验证来自 server1 的原始 gtid_executed 集（$original_gtid_executed）和从 server2 添加的 gtid_purged 集（$gtid_purged_set）没有重叠，并且 server1 上已更新的 gtid_executed 集现在包含来自 server1 的前一个 gtid_executed 集加上来自 server2 的 gtid_purged 集，这是所需的结果。确保在 server1 上进行任何进一步的事务之前执行此检查，否则 gtid_executed 集中的新事务将会失败。

3. 手工选择作为新主库的从库

自定义函数 GTID_UNION 可用于从一组复制从库中识别最新的从库，以便在主库意外停止后执行手动切换。如果某些从库遇到复制延迟，则此函数可用于计算最新的从库，而无需等待所有从库应用完其现有的中继日志，从而最大限度地缩短主从切换时间。该函数可以返回每个从库上的 gtid_executed 集合与从库接收的事务集合的并集，后者记录在 performance_schema.replication_connection_status 表中。可以比较这些结果，以查找哪个从库的事务记录是最新的，即使并非所有事务都已提交。

在每个复制从库上，通过发出以下语句来计算完整的事务记录：

```
SELECT GTID_UNION(RECEIVED_TRANSACTION_SET, @@GLOBAL.gtid_executed)
 FROM performance_schema.replication_connection_status
 WHERE channel_name = 'name';
```

然后比较每个从库的结果，选择具有最新事务记录的从库用作新主库。

4. 检查从库上的异常事务

自定义函数 GTID_SUBTRACT_UUID 可用于检查从库是否只接收到源自其指定主库的事务。对于单个主库的复制，执行以下语句，server_uuid_of_master 是主库的 server_uuid：

```
SELECT GTID_SUBTRACT_UUID(@@GLOBAL.gtid_executed, server_uuid_of_master);
```

如果结果不为空，则返回的事务为不是源自指定主库的异常事务。对于多主库复制拓扑结构中的从库，重复该功能，例如：

```
SELECT GTID_SUBTRACT_UUID(GTID_SUBTRACT_UUID(@@GLOBAL.gtid_executed,
                             server_uuid_of_master_1),
                             server_uuid_of_master_2);
```

如果结果不为空，则返回的事务是来自非指定主库的异常事务。

5. 验证复制拓扑结构中的服务器是否执行过本地事务

自定义函数 GTID_INTERSECTION_WITH_UUID 可用于验证服务器是否执行过本地事务。可以在服务器上发出以下语句来检查：

```
SELECT GTID_INTERSECTION_WITH_UUID(@@GLOBAL.gtid_executed, my_server_uuid);
```

6. 在多主复制设置中验证附加从库

假设 master1 和 master2 为双主复制，互为主从，同时 master2 还有自己的从库 slave3，如图 3-1 所示。

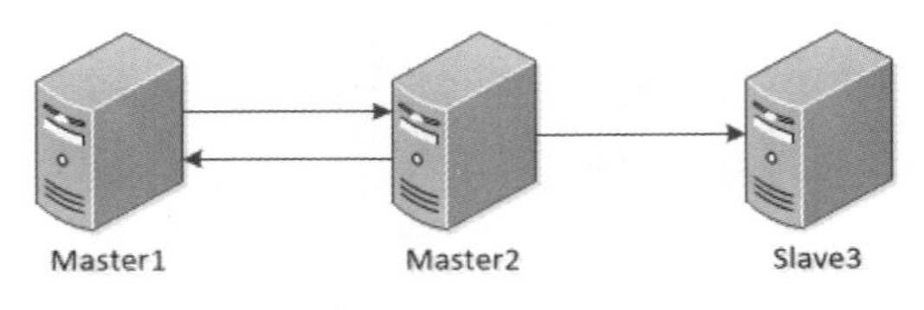

图 3-1　多主复制

如果 master2 配置了 log_slave_updates = ON，那么 slave3 也将接收并应用 master1 的事务，如果 master2 使用 log_slave_updates = OFF，则不会这样做。在这种情况下，自定义函数 GTID_INTERSECTION_WITH_UUID 可用于标识 master2 发起的事务，丢弃 master2 从 master1 复制的事务。然后可以使用内置函数 GTID_SUBSET 将结果与 slave3 上的 gtid_executed 集进行比较。如果 slave3 与 master2 保持同步，则 slave3 上的 gtid_executed 设置包含交集中的所有事务（源自 master2 的事务）。

要执行此检查，可将 master2 的 gtid_executed、master2 的 server_uuid 和 slave3 的 gtid_executed 集存储到客户端变量中，例如：

```
master2_gtid_executed := 'mysql -h master2 -N -e "SELECT
                         @@GLOBAL.gtid_executed;" '
master2_server_uuid := 'mysql -h master2 -N -e "SELECT @@GLOBAL.server_uuid;"'
slave_gtid_executed := 'mysql -h slave3 -N -e "SELECT @@GLOBAL.gtid_executed;"'
```

然后使用 GTID_INTERSECTION_WITH_UUID 和 GTID_SUBSET 将这些变量作为输入，例如在 slave3 上执行：

```
SELECT GTID_SUBSET(GTID_INTERSECTION_WITH_UUID(
$master2_gtid_executed,$master2_server_uuid),$slave_gtid_executed);
```

来自 master2 的服务器标识符（$master2_server_uuid）与 GTID_INTERSECTION_WITH_UUID 一起使用，以识别并返回源自 master2 的 gtid_executed 集合中的那些 GTID，省略源自 master1 的 GTID。然后使用 GTID_SUBSET 将得到的 GTID 集与从库上所有已执行 GTID 的集合进行比较。如果此语句返回非零（true），则来自 master2 的所有已识别的 GTID（第一个集输入）也位于从库的 gtid_executed 集（第二个集输入）中，这意味着从库已复制源自 master2 的所有事务。

3.8 小　结

引入 GTID 是 MySQL 的一个重大改进，它涉及到实现的方方面面。对于复制来说，最重要的是能够自动确定初始复制位置，不需要手动在从库上配置主库的二进制位置，极大地简化了维护工作。本章用比较长的篇幅，全面介绍了 GTID 的存储格式、生命周期、自动定位、配置、运维、限制和集合运算函数，对于每个要点都用实验或例子加以说明。下一章将讨论不同的复制拓扑结构和影响复制性能的因素。

第 4 章

复制拓扑结构与性能

4.1　复制拓扑结构

可以在任意主从库之间建立复杂的复制拓扑结构，如普通的一主一（多）从、双（多）主复制、级联复制，MySQL 5.7.2 版本后新增的多源复制，在特殊场景下使用的 Blackhole 引擎与日志服务器等等。复制中的 MySQL 服务器须要遵循以下基本原则：

- 拓扑结构中的每个服务器必须有一个唯一的 server_id 和 server_uuid。
- 一个主库可以有多个从库，或者说一个从库可以有多个兄弟从库。
- 如果打开了 log_slave_updates 选项，一个从库就可以把它对应主库上的数据变化传播给自己。

4.1.1　一主一（多）从

1. 一级主从

一主一从的基本配置是最简单的拓扑结构，而一主多从的拓扑结构和基本配置也同样简单，因为从库之间根本没有交互，它们仅仅是连接到同一个主库而已。图 4-1 显示了这种结构。

尽管这是非常简单的拓扑结构，但它非常灵活，能满足多种需求。为保证数据一致性和主从切换，从库原则上应为只读，下面是从库的一些用途：

- 为不同的角色使用不同的从库，例如为满足读写分离需求，在从库添加不同于主库的适合读操作的索引。不要忘记在从库执行任何写操作前设置 set sql_log_bin=0。
- 将一台从库只作为备用主库，除了复制没有其他数据传输。
- 将一台从库放到远程数据中心，用作灾备容错。
- 延迟复制一个从库，用以恢复用户错误。
- 使用其中一个从库，作为备份或测试使用。

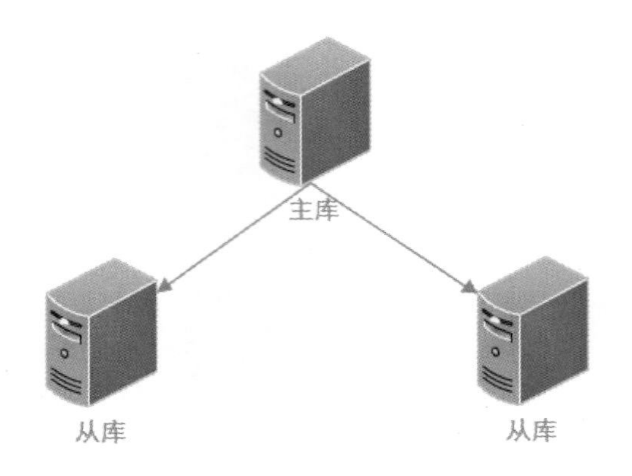

图 4-1　一主二从的拓扑结构

这种拓扑结构流行的原因是它避免了很多其他拓扑结构的复杂性。例如，可以方便地比较不同从库的重放事件在主库二进制日志中的位置，当然如果启用 GTID 就更简单了，支持自动定位。换句话说，如果在同一个逻辑点停止所有从库的复制，它们正在读取的是主库上同一个日志文件的相同物理位置。这是个很好的特性，可以减轻管理员的许多工作，例如把从库提升为主库。

这种特性只存在于兄弟从库之间。在没有直接的主从或者兄弟关系的服务器上去比较日志文件的位置要复杂很多。例如树形复制或分布式主库，就很难计算出复制事件的逻辑顺序。

2. 级联主从

如果正在将主库复制到大量从库中，不管是把数据分发到不同的地方，还是提供更高的读性能，使用级联复制都能更好地管理，如图 4-2 所示。

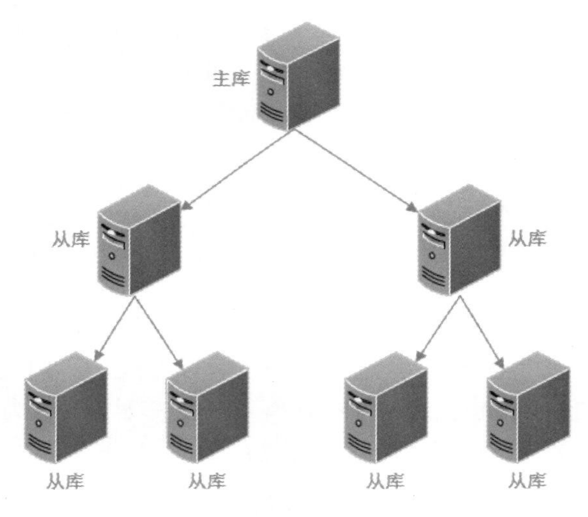

图 4-2　级联主从拓扑结构

这种设计的好处是减轻了主库的负担，将读负载分发到多个从库。缺点是中间层出现的任何错误都会影响到多台服务器。如果每个从库和主库直接相连就不会存在这样的问题。同时中间层次越多，处理故障就会越复杂和困难。

4.1.2　双（多）主复制

1. 主动-主动模式下的双主复制

双主复制包含两台 MySQL 服务器，每一个都被配置为对方的主库和从库，换句话说，它们是一对主库。图 4-3 显示了该拓扑结构。

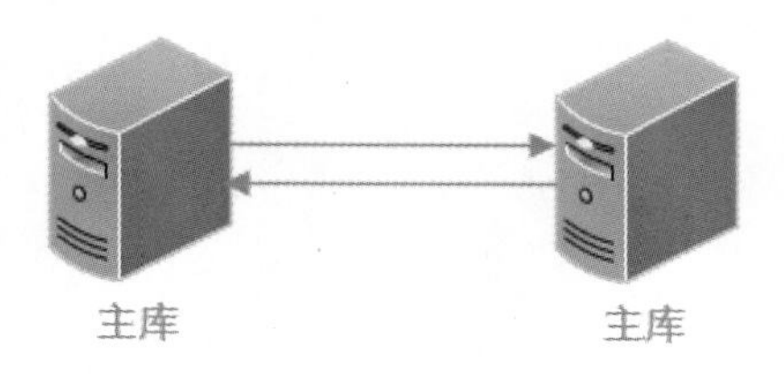

图 4-3　对称双主拓扑结构

主动-主动模式指的是两台服务器对于应用均可读写，通常用于特殊目的。例如一个可能的应用场景是两个处于不同地理位置的办公室，并且都需要一份可写的数据拷贝。这种配置最大的问题是如何解决冲突，两个可写的互主服务器导致的问题非常多。这通常发生在两台服务器同时修改一行记录，或同时在两台服务器上向一个包含 auto_increment 列的表里插入数据。这些问题会经常发生，而且不易解决，因此并不推荐这种模式。下面演示两个常见的问题。

（1）在两台机器更新顺序不同导致数据不一致或 1032 错误：

```
-- 主库 1
mysql> select * from t1;
+----+
| a |
+----+
| 1 |
+----+
1 row in set (0.00 sec)

-- 主库 2
mysql> select * from t1;
+----+
| a |
+----+
| 1 |
+----+
1 row in set (0.00 sec)

-- 主库 2 延迟复制，模拟两个主库不同的执行顺序
stop slave;
change master to master_delay = 10;
start slave;

-- 主库 1
set binlog_format='statement';
```

```
update t1 set a=a+1;

-- 主库 2 在复制之前（10 秒之内）执行
set binlog_format='statement';
update t1 set a=a*2;

-- 10 秒之后查询
-- 主库 1
mysql> select * from t1;
+------+
| a    |
+------+
|    4 |
+------+
1 row in set (0.00 sec)

-- 主库 2
mysql> select * from t1;
+------+
| a    |
+------+
|    3 |
+------+
1 row in set (0.00 sec)
```

复制没有报告任何错误，但两个库的数据已经不一致。主库 1 上先执行的 a=a+1，紧接着由于复制执行 a=a*2，最终结果是 4。主库 2 上由于延迟复制，先执行 a=a*2，10 秒后执行复制的 a=a+1，最终结果是 3。此实验是在 binlog_format='statement'下进行的，如果设置 binlog_format='row'，则两个库（首先是主库 1，10 秒后是主库 2）都会报 1032 错误，show slave status 中显示的错误如下：

```
    Last_SQL_Errno: 1032
    Last_SQL_Error: Could not execute Update_rows event on table test.t1; Can't
find record in 't1', Error_code: 1032; handler error HA_ERR_END_OF_FILE; the event's
master log binlog.000001, end_log_pos 2534
```

报 1032 的原因是应用复制时记录已经发生改变，找不到更新时间点的数据行。

（2）由 auto_increment 引起的 1062 错误：

```
-- 主库 1
use test;
create table t1(a int auto_increment primary key);

delimiter //
create procedure p1(a int)
begin
   declare i int default 1;
   while i<=a do
      insert into t1(a) select null;
       set i=i+1;
   end while;
```

```
end;
//
delimiter ;
call p1(1000);

-- 主库 2，在主库 1 执行过程期间同时在主库 2 执行
call p1(1000);
```

show slave status 中显示如下错误：

```
Last_SQL_Errno: 1062
Last_SQL_Error: Could not execute Write_rows event on table test.t1; Duplicate
entry '366' for key 'PRIMARY', Error_code: 1062; handler error
HA_ERR_FOUND_DUPP_KEY; the event's master log binlog.000001, end_log_pos 101521
```

因为本机插入的数据与复制插入的数据产生冲突而报 1062 错误。通过在两个服务器设置不同的 auto_increment_offset 和 auto_increment_increment，可以让 MySQL 自动为 insert 语句选择不互相冲突的值，稍微增加了点安全性。

```
-- 主库 1
set auto_increment_offset=1;
set auto_increment_increment=2;
call p1(1000);

-- 主库 2，在主库 1 执行过程期间同时在主库 2 执行
set auto_increment_offset=2;
set auto_increment_increment=2;
call p1(1000);
```

主库 1 上插入单数，主库 2 插入双数，复制与本机数据不冲突。过程执行完后，两个库都插入了 2000 条数据，但默认配置 innodb_autoinc_lock_mode=2 会造成序列值不连续。

```
-- 主库 1
mysql> select count(*),min(a),max(a) from t1;
+----------+--------+--------+
| count(*) | min(a) | max(a) |
+----------+--------+--------+
|     2000 |      1 |   2414 |
+----------+--------+--------+
1 row in set (0.00 sec)

-- 主库 2
mysql> select count(*),min(a),max(a) from t1;
+----------+--------+--------+
| count(*) | min(a) | max(a) |
+----------+--------+--------+
|     2000 |      1 |   2414 |
+----------+--------+--------+
1 row in set (0.00 sec)
```

可以看到复制正常，两台服务器数据一致。但这只极端理想的场景：从空表开始插入数据，

并且配置复制时没有联机访问。如果在配置双主复制时已经有数据，情况将复杂得多。同时允许向两台主库写入很危险，极易造成复制错误或数据不一致。数据不同步还仅仅是开始。当正常的复制发生错误停止了，但应用仍然在同时向两台服务器写入数据，这时候会发生什么呢？不能简单地把数据从一台服务器复制到另外一台，因为这两台机器上需要复制的数据都可能发生了变化。解决这个问题将非常困难。总的来说，允许向两个服务器上同时写入所带来的麻烦远远大于其带来的好处。只要作为从库可写，就存在主从数据不一致的风险。

2. 主动-被动模式下的双主复制

这是前面描述的双主拓扑结构的变体，主要区别在于其中的一台服务器是只读的被动服务器。这种拓扑结构能够避免之前讨论的问题，也是构建容错性和高可用性系统的强大方式。两个服务器从硬件到操作系统再到 MySQL 配置都应该完全相同。为便于故障转移，只读最好由客户端应用保证，通过设置以下系统变量强制只读仅作为可选项。

```
set global read_only=1;
set global super_read_only=1;
```

这种方式使得反复切换主动和被动服务器非常方便，因为服务器的配置是对称的，故障转移和故障恢复相对容易。它也允许用户在不关闭服务器的情况下执行维护、优化、升级操作系统（或者应用程序、硬件等）或其他任务。

例如，执行 alter table 操作可能会锁住整个表，阻塞对表的读写，这可能会花费很长时间并导致服务中断。

```
use test;
create table t1(a int auto_increment primary key);
insert into t1 select -1;
commit;

-- session 1
set autocommit=0;
insert into t1 select null;

-- session 2
alter table t1 add column (b int);

-- session 3
update t1 set a=-2 where a=-1;

-- session 4
show processlist;
...
| 170 | ... | 17 | Waiting for table metadata lock | alter table t1 add column
(b int) |
| 171 | ... |  9 | Waiting for table metadata lock | update t1 set a=-2 where
a=-1    |
| 172 | ... |  0 | init                            | show processlist          |
...
7 rows in set (0.00 sec)
```

可以看到，如果在执行 alter table 时，表上有未提交的事务，alter table 本身和其后的所有 DML 都会等待 table metadata lock，而不论这些后续的 DML 操作的是哪些行，因为 metadata lock 是一个表级锁。当 session 1 的事务提交或回滚，session 2 才能得以执行。高并发场景下，在线 DDL 极有可能造成灾难性的后果。一种暴力的解决方案是，先 kill 掉所有 sleep 的 MySQL 线程，紧接着执行 alter table，这样不会因为 metadata lock 而卡住后面的 DML。在这个例子中，如果 session 2 可以先得到执行，即使操作需要很长时间，也不会让后面的 DML 等待。脚本文件的内容可能为：

```
#!/bin/bash
source ~/.bashrc

rm -rf /tmp/kill.sql
mysql -u root -p123456 -P3306 -h127.0.0.1 -e "select * into outfile
'/tmp/kill.sql' from (select concat('kill ',id,';') from
information_schema.processlist where command='sleep' union all select 'set
sql_log_bin=0;' union all select 'alter table test.t1 add column (b int);') t;"

mysql -u root -p123456 -P3306 -h127.0.0.1 < /tmp/kill.sql
```

注意，将所有 sleep 的线程都杀掉这个操作会导致没有提交的事务回滚，是有风险的，需要根据业务场景选择操作。如果希望进行更精细化的控制，下面的查询是有用的：

```
-- 查询事务对应的线程
select t1.trx_id,
       t1.trx_state,
       t1.trx_query,
       t2.id,
       t2.state,
       t2.command
  from information_schema.innodb_trx t1,
       information_schema.processlist t2
 where t1.trx_mysql_thread_id = t2.id\G

-- 查询正在执行的事务：
select * from information_schema.innodb_trx;

-- 查看正在锁的事务
select * from information_schema.innodb_locks;

-- 查看等待锁的事务
select * from information_schema.innodb_lock_waits;
```

主库上大表的 DDL 操作可能引起从库的复制延时变长。在不影响数据一致性的前提下，如 drop、truncate 等，一种可能的解决方案是执行 DDL 前先设置 set sql_log_bin=0，让大的 DDL 操作不写入 binlog，从而不会复制到从库，之后再在从库手动执行一遍。

然而，在主动-被动模式的双主配置下，在线 DDL 变得更具可操作性。可以先停止主动服务器上的复制线程，这样就不会复制被动服务器上执行的任何更新。然后在被动服务器上执行 alter table 操作，之后交换角色，最后在先前的主动服务器上启动复制线程。这个服务器将会读取中继日志并执行相同的 alter 语句。这可能花费很长时间，但不要紧，因为该服务器没有为任何活跃查询提供服务。假设 A、B 库配置了双主复制，A 为主动库提供服务，B 为被动库进行复制。如果需要在一个大表上增加字段，可能的操作步骤如下：

步骤 01 A 库执行 stop slave，此时 A 不会复制 B 的更新。

步骤 02 B 库执行 alter table，B 此时仍然复制来自 A 的更新。

步骤 03 交互角色，B 变为主动提供读写服务，A 变为被动，这意味着应用连接需要切换到 B。

步骤 04 A 库执行 start slave，此时 A 将重放 B 上的 alter table 语句和其他更新。

整个过程不停库，只需修改应用连接的数据库即可。如果使用 VIP 技术，则应用不需要做任何修改，原本可能导致服务中断的 DDL 操作将对应用完全透明。下面的过程用于验证第 2 步操作中，B 上的 alter table 不会阻塞它对 A 的复制。

```
-- 1. A 停止复制
stop slave;

-- 2. B 上执行一个长时间的 alter table 操作
alter table t1 add column (b int);

-- 3. 在上一步执行过程中，A 上操作同一个表
call p1(1000000);

-- 4. B 确认复制状态和线程状态
show slave status\G
show processlist;
select max(a) from t1;

-- 5. 前面的步骤都执行完后，A 启动复制
start slave;
```

show slave status 的 Read_Master_Log_Pos 和 Exec_Master_Log_Pos 不停改变，show processlist 中的 State 没有任何锁，t1 表的数据一直处于更新状态，说明 B 对 A 的复制不会被其上的 alter table 阻塞。注意，如果在第 2 步执行前从库被修改的表上有未提交的事务（从主库复制过来），依然会阻塞第 2 步执行，但情况要比在主库上缓解很多。其一是因为从库默认为单线程复制，不是并发执行方式，事务应该很快被提交。其次是从库可以设置成 autocommit=on，这也会缩短 alter table 语句被阻塞的时间。调换以上步骤 2 和 3 的执行顺序，可以验证 B 对 A 的复制同样也不会阻塞其上的 alter table 语句执行。

上面的步骤并非无懈可击，成立的前提是 alter table 与复制的语句兼容，否则会导致复制错误。但通常来说都是先修改数据库表结构，再升级应用程序，这样看来此前提成立是自然的，问题并不大。下面的过程只是演示一种出错的情况。

```
-- 1. A 停止复制
stop slave;

-- 2. A 上执行一个长时间的操作
call p1(1000000);

-- 3. 在上一步执行过程中，B 上 alter table 同一个表
```

```
alter table t1 add column b int,drop column a;

-- 4. B 确认复制状态和线程状态
show slave status\G
```

由于 t1.a 列被删除，添加了一列 b，而列 a 与列 b 的数据类型不兼容，导致 B 库上的复制报错：

```
Last_Errno: 1677
Last_Error: Column 0 of table 'test.t1' cannot be converted from type 'bigint' to type 'int(11)'
```

执行下面的修复后复制继续：

```
alter table t1 change b a bigint auto_increment primary key;
stop slave;
start slave;
```

让我们看看主动服务器上更新时会发生什么事情。更新被记录到二进制日志中，通过复制传递给被动服务器的中继日志中。被动服务器重放中继日志里的查询，如果启用了 log_slave_updates 选项，它还会将复制事件记录到自己的二进制日志中。由于复制事件的服务器 ID 与主动服务器相同，因此主动服务器将忽略这些事件，通过这种方式避免复制死循环。设置主动-被动的双主拓扑结构在某种意义上类似于创建一个热备份，但可以使用这个“备份”来提高性能，例如，用它来执行读操作、备份、轮换维护以及升级等。

3. 拥有从库的双主拓扑结构

另外一种相关的配置是为每个主库增加一个从库，如图 4-4 所示。

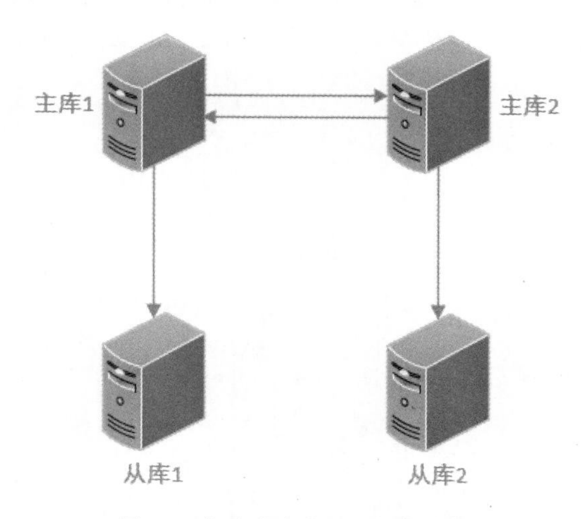

图 4-4　拥有从库的双主拓扑结构

这种配置的优点是增加了冗余，对于不同地理位置的复制拓扑结构，能够消除站点单点失效的问题。也可以像平常一样，将读查询分配到从库上。如果为了故障转移在本地使用双主拓扑结构，这种配置同样有用。当主库失效时，有两种不同的处理方式，一是用从库代替主库，二是把从库指向另一个不同的主库。以图 4-4 为例，假设主库 1 失效，采用第一种方式，需要将从库 1 提升为新的主库 1，修改主库 2 的复制配置，指向新的主库 1，并将新主库指向主库 2，保持双主配置。如果采用第二种方式，只需要将从库 1 指向主库 2，但这样拓扑结构已从双主变为一主两从。

4. 环形复制

如图 4-5 所示，双主拓扑结构实际上是环形结构的一种特例。环形结构可以有三个或更多的主库。每个服务器都是在它之前的服务器的从库，是在它之后的服务器的主库。这种拓扑结构也称为环形复制（Circular Replication）。

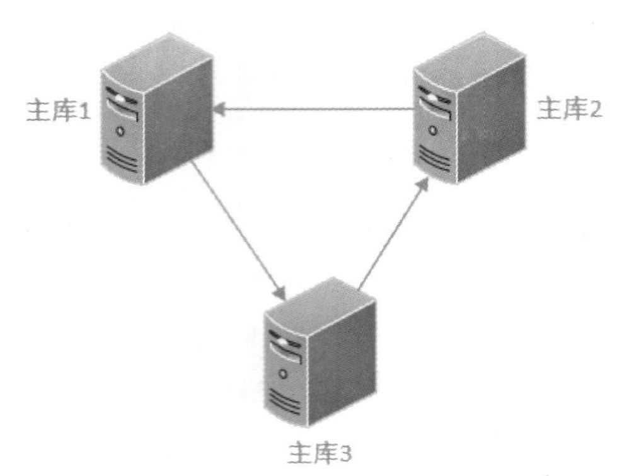

图 4-5　环形复制拓扑结构

环形拓扑结构没有双主拓扑结构的一些优点，例如对称配置和简单的故障转移，并且完全依赖于环上的每一个可用节点，这大大增加了整个系统失效的几率。如果从环中移除一个节点，这个节点发起的事件就会陷入无限循环：它将永远绕着服务器循环。因为唯一可以根据服务器 ID 将其过滤的服务器是创建这个事件的服务器。下面的步骤可以模拟这种场景，M1、M2、M3 构成的三个主库的环形复制拓扑结构，M1 复制 M3、M3 复制 M2、M2 复制 M1。

```
    -- 1. M1 停止 sql_thread 线程
    stop slave sql_thread;

    -- 2. M2 停止 sql_thread 线程
    stop slave sql_thread;

    -- 3. M3 进行更新
    insert into test.t1 values (1);
    commit;

    -- 4. M3 停库
    mysqladmin -uroot -p123456 shutdown

    -- 5. M1 启动 sql_thread 线程，此时 M3 的更新复制到 M1
    start slave sql_thread;

    -- 6. M1 复制 M2，此时原环形复制拓扑结构中移除了 M3，其中 master_log_file 和
master_log_pos 从 M2 的 show master status 的输出得到

    stop slave;
    change master to
         master_host = '172.16.1.126',
```

```
        master_port = 3306,
        master_user = 'repl',
        master_password = '123456',
        master_auto_position = 0,
        master_log_file='binlog.000002',
        master_log_pos=664210;
    start slave;

    -- 7. M2 启动 sql_thread 线程，此时 M2 复制了来自 M3 的更新，并继续传递给 M1，复制陷入死
循环。在 M1、M2 上查询 test.t1，可以看到记录不停地增长
    start slave sql_thread;
```

如果三个主库都启用 GTID 复制，则以上过程不会陷入死循环，因为复制不再通过 server_id 过滤本地事件，而是通过 server_uuid 复制事务。总的来说，环形拓扑结构非常脆弱，应该尽量避免。可以通过为每个节点增加从库的方式来减少环形复制拓扑结构的风险，如图 4-6 所示。但这仅仅防范了服务器失效的风险，断电或者其他一些影响到网络连接的问题都可能破坏整个环。

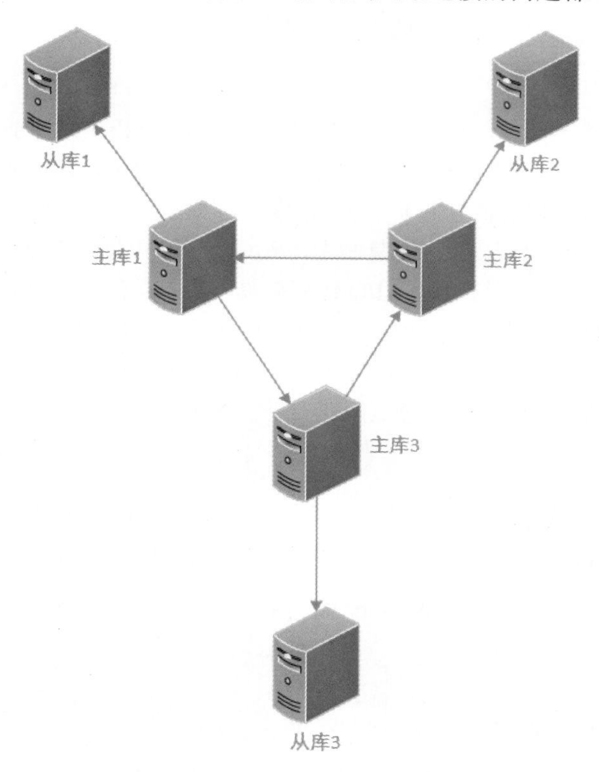

图 4-6 拥有从库的环形复制拓扑结构

4.1.3 多源复制

MySQL 5.7.6 版本开始支持多源复制（Multi-Source Replication）。多源复制不同于多主复制，多主复制指的是在复制拓扑结构中存在多个主库，而多源复制是指一个从库可以同时从多个主库进行复制。图 4-7 所示为“两主一从”的多源复制拓扑结构。

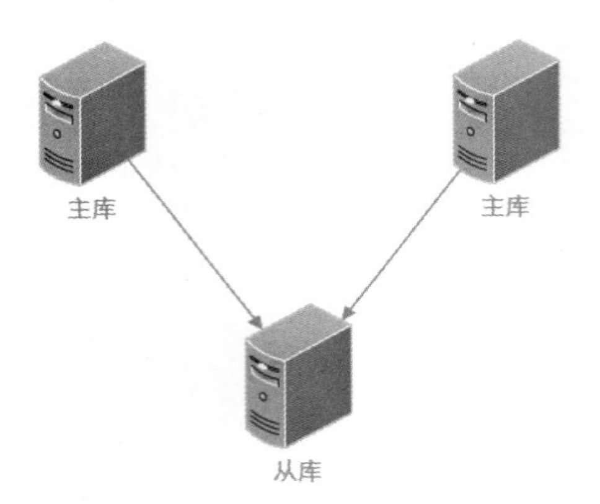

图 4-7 多源复制拓扑结构

多源复制可用于将来自多台服务器的数据合并或备份到单台服务器，比如合并表分片。应用事务时，多源复制不会检测或解决任何冲突，如果需要，这些任务将留给应用程序去实现。在多源复制拓扑结构中，从库为每个接收事务的主库创建复制通道。

1. 复制通道

复制通道是一个字符串，表示从主库到从库的复制路径。为提供与先前版本的兼容性，MySQL 服务器在启动时自动创建一个默认通道，其名称为空字符串（""）。这个通道始终存在，不能被用户创建或销毁。如果没有创建其他具有非空名称的通道，则复制语句仅作用于默认通道，以便旧版从库的所有复制语句按预期运行。在多源复制中，从库打开多个命名通道，每个命名通道都有自己的中继日志和复制线程。一旦复制通道的 I/O 线程接收到事务，它们就会被添加到通道对应的中继日志文件中并传递给 SQL 线程，这使得每个通道能独立运行。复制通道还与主机名和端口关联，可以将多个通道分配给主机名和端口的相同组合。在 MySQL 8.0 中，添加到一个从库的最大通道数为 256。每个复制通道独立配置，必须具有唯一非空名称。

2. 配置

多源复制拓扑结构至少需要配置两个主库和一个从库。可以将多源复制中的主库配置为使用基于全局事务标识符 GTID 的复制或基于二进制日志位置的复制。配置多源复制的步骤如下：

步骤 01 将从库的 master_info_repository 和 relay_log_info_repository 选项设置为 TABLE：

```
stop slave;
set global master_info_repository = 'table';
set global relay_log_info_repository = 'table';
```

这是 MySQL 8.0 的默认值。多源复制拓扑结构中的从库需要使用表存储主库二进制日志和本身中继日志的信息，多源复制与基于文件的存储库不兼容。现在不推荐将这两个参数设置为“file”。

步骤 02 将主库添加到从库：

```
change master to
     master_host = '172.16.1.125',
     master_port = 3306,
     master_user = 'repl',
```

```
        master_password = '123456',
        master_auto_position = 1
    for channel 'master-125';

change master to
        master_host = '172.16.1.126',
        master_port = 3306,
        master_user = 'repl',
        master_password = '123456',
        master_auto_position = 1
    for channel 'master-126';
```

这里使用GTID复制，设置“两主一从”的多源复制。CHANGE MASTER TO语句通过使用FOR CHANNEL子句将新主库添加到复制通道。多源复制与自动定位兼容。

步骤03 启动从库复制：

```
-- 启动所有线程所有通道的复制
start slave;

-- 启动所有通道的io_thread线程
start slave io_thread;

-- 启动所有通道的sql_thread线程
start slave sql_thread;

-- 启用单个通道
start slave for channel 'master_125';
start slave io_thread for channel 'master_125';
start slave sql_thread for channel 'master_125';
```

停止复制命令与启动复制类似，只是把start换成stop。同样重置也可以选择重置所有通道和单一通道：

```
reset slave;
reset slave for channel 'master_125';
```

3. 监控

监控可以使用performance_schema.replication*表，这些表的第一列都是channel_name。注意SHOW VARIABLES语句不适用于多个复制通道。这些变量的信息已迁移到复制性能表。在具有多个通道的拓扑结构中，使用SHOW VARIABLES语句仅显示默认通道的状态。

```
-- 查询特定通道的连接状态
mysql> select * from replication_connection_status where
channel_name='master-125'\G
*************************** 1. row ***************************
                     CHANNEL_NAME: master-125
                       GROUP_NAME:
                      SOURCE_UUID: 8eed0f5b-6f9b-11e9-94a9-005056a57a4e
                        THREAD_ID: 10421
                    SERVICE_STATE: ON
```

```
                    COUNT_RECEIVED_HEARTBEATS: 41
                     LAST_HEARTBEAT_TIMESTAMP: 2019-06-24 16:21:31.583443
                    RECEIVED_TRANSACTION_SET:
                         LAST_ERROR_NUMBER: 0
                         LAST_ERROR_MESSAGE:
                       LAST_ERROR_TIMESTAMP: 0000-00-00 00:00:00.000000
                           LAST_QUEUED_TRANSACTION:
     LAST_QUEUED_TRANSACTION_ORIGINAL_COMMIT_TIMESTAMP: 0000-00-00
00:00:00.000000
    LAST_QUEUED_TRANSACTION_IMMEDIATE_COMMIT_TIMESTAMP: 0000-00-00
00:00:00.000000
       LAST_QUEUED_TRANSACTION_START_QUEUE_TIMESTAMP: 0000-00-00 00:00:00.000000
        LAST_QUEUED_TRANSACTION_END_QUEUE_TIMESTAMP: 0000-00-00 00:00:00.000000
                       QUEUEING_TRANSACTION:
      QUEUEING_TRANSACTION_ORIGINAL_COMMIT_TIMESTAMP: 0000-00-00
00:00:00.000000
     QUEUEING_TRANSACTION_IMMEDIATE_COMMIT_TIMESTAMP: 0000-00-00
00:00:00.000000
       QUEUEING_TRANSACTION_START_QUEUE_TIMESTAMP: 0000-00-00 00:00:00.000000
    1 row in set (0.00 sec)
```

使用 SHOW SLAVE STATUS FOR CHANNEL 监控特定通道的状态，如果不加 FOR CHANNEL 子句，则返回所有复制通道的状态，每个通道一行。

4. 测试

```
-- 主库 1
mysql> insert into test.t1 values(125);
Query OK, 1 row affected (0.01 sec)

-- 主库 2
mysql> insert into test.t1 values(126);
Query OK, 1 row affected (0.01 sec)

-- 从库
mysql> select * from test.t1;
+------+
| a    |
+------+
|    1 |
|  125 |
|  126 |
+------+
3 rows in set (0.00 sec)

-- 主库 1
mysql> truncate table test.t1;
Query OK, 0 rows affected (0.01 sec)

-- 从库
mysql> select * from test.t1;
```

```
Empty set (0.00 sec)
```

两个主库新增的数据都复制到从库，但只在一个主库清空表，从库表所有数据全部被清空。因此使用多源复制要避免多个主库具有同名的数据库表。

```
-- 主库 1
mysql> create user 'u1'@'%' identified by '123456';
Query OK, 0 rows affected (0.01 sec)

-- 主库 2
mysql> create user 'u1'@'%' identified by '123456';
Query OK, 0 rows affected (0.01 sec)

-- 从库
mysql> show slave status\G
```

通道 master-125 复制状态正常，但 master-126 报错：

```
    Last_SQL_Errno: 1396
    Last_SQL_Error: Error 'Operation CREATE USER failed for 'u1'@'%'' on query.
Default database: 'test'. Query: 'CREATE USER 'u1'@'%' IDENTIFIED WITH
'caching_sha2_password' AS
'$A$005$*B_B^@}R;15egC4\nYdRPGtaEXbF.jB36e2UpAZEoXEPck87oeMl4j8rO6iu5''
```

创建用户的时候报告 1396 错误，原因是 mysql 库中已经有了这个用户。恢复复制的过程如下：

（1）停止从库通道 master-126 的复制：

```
stop slave for channel 'master-126';
```

（2）在从库上确认出错的事务 ID：

```
show slave status for channel 'master-126'\G
...
Retrieved_Gtid_Set: 53442434-8bfa-11e9-bc15-005056a50f77:1008-1009
          Executed_Gtid_Set: 53442434-8bfa-11e9-bc15-005056a50f77:1-1008,
6a739bf0-961d-11e9-8dd8-005056a5497f:1-1885,
8eed0f5b-6f9b-11e9-94a9-005056a57a4e:1-24240
...
```

可以看到，从库从 53442434-8bfa-11e9-bc15-005056a50f77 接收到事务 1009，但只执行到 1008，所以确定报错的事务为：53442434-8bfa-11e9-bc15-005056a50f77:1009。

（3）在从库上注入一个空事务跳过错误：

```
set gtid_next='53442434-8bfa-11e9-bc15-005056a50f77:1009';
begin;commit;
set gtid_next=automatic;
start slave for channel 'master-126';
```

对于 mysql 库，建议使用 REPLICATE_IGNORE_DB 将其屏蔽掉：

```
stop slave;
change replication filter replicate_ignore_db = (mysql);
start slave;
```

在主库上对 mysql 库进行操作时，需要加 use mysql，否则不会进行过滤。

4.1.4 Blackhole 引擎与日志服务器

1. Blackhole 存储引擎与复制

MySQL 8 中 show engines 命令返回存储引擎如下：

```
mysql> show engines;
+--------------------+---------+------------------------------------+------
+------+------+
| Engine             | Support | Comment                            | T... | XA   | S... |
+--------------------+---------+------------------------------------+------
+------+------+
| FEDERATED          | NO      | Federated MySQL storage engine     | NULL | NULL |
NULL |
| MEMORY             | YES     | Hash based, stored in memory...  | NO   | NO   | NO   |
| InnoDB             | DEFAULT | Supports transactions, ...       | YES  | YES  | YES  |
| PERFORMANCE_SCHEMA | YES     | Performance Schema               | NO   | NO   | NO   |
| MyISAM             | YES     | MyISAM storage engine            | NO   | NO   | NO   |
| MRG_MYISAM         | YES     | Collection of identical MyISAM ...| NO   | NO   |
NO   |
| BLACKHOLE          | YES     | /dev/null storage engine ...     | NO   | NO   | NO   |
| CSV                | YES     | CSV storage engine               | NO   | NO   | NO   |
| ARCHIVE            | YES     | Archive storage engine           | NO   | NO   | NO   |
+--------------------+---------+--------------------+------+------+------+
9 rows in set (0.00 sec)
```

像 MyISAM 和 InnoDB 一样，blackhole 是另一种 MySQL 引擎。该引擎的功能可谓名副其实，任何写入到此引擎的数据均会被丢弃掉，不做实际存储，和 Linux 中的 /dev/null 文件所起的作用类似。创建一个 blackhole 的表时，MySQL 服务器在数据库目录创建一个.frm 表定义文件，没有其他文件关联到这个表。虽然 blackhole 表不存储任何数据，但它却能够接收并重放二进制日志，如果启用了 log_slave_updates，它也能把复制向下传播，如同普通的级联复制拓扑结构一样。

当从库足够多时，会对主库造成很大的负载。每个从库会在主库上创建一个线程执行 binlog dump 命令，该命令读取二进制文件中的数据并将其发送给从库。每个从库都会重复这样的工作，它们不会共享 binlog dump 的资源。如果从库很多，并且有大的事件时，例如 binlog_format 为 statement 时执行一次很大的 load data infile 操作，主库的负载会显著上升，甚至可能由于从库同时请求同样的事件因耗尽内存而崩溃。另一方面，如果从库请求的数据不在文件系统的缓存中，可能会导致大量的磁盘检索，这同样会影响主库的性能并增加锁的竞争。

因此，如果需要多个从库，一个好办法是从主库移除负载并利用 blackhole 进行分发，即所谓的分发主库。分发主库实际上也是一个从库，它唯一的目的就是提取和提供主库的二进制日志。多个从库连接到分发主库，这使原来的主库摆脱了负担，如图 4-8 所示。

很难说当主库数据达到多少时需要一个分发主库。按照通用准则，如果主库接近满负载，不应该为其建立 10 个以上的从库。如果只有少量写操作，或者只复制其中一部分表，则主库可以提供更多的复制。如果需要，可以使用多个分发主库向大量从库进行复制，或者使用级联的分发主库。

对于跨数据中心的复制，设置 slave_compressed_protocol 能节约一些主库带宽。该变量是全局系统变量，默认值为 off，可以动态设置。

还可以通过分发主库实现其他目的，如对二进制日志事件执行过滤和重放规则。这比在每个从库上重复进行日志记录、重放和过滤要高效得多。使用 blackhole 存储引擎可以支持更多从库。虽然会在分发主库执行查询，但代价极小，因为 blackhole 表中没有任何数据。

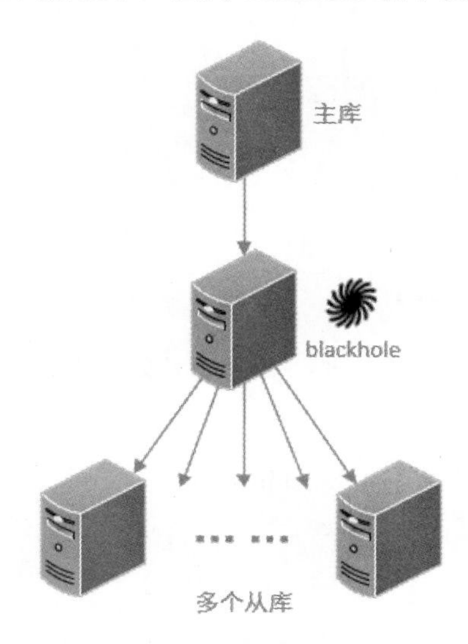

图 4-8 blackhole 引擎

一个常见的问题是如何确保分发服务器上的每个表都是 blackhole 存储引擎。如果在主库创建了一个表并指定了不同的存储引擎呢？确实，不管何时在从库上使用不同的存储引擎总会导致同样的问题。通常的解决方案是设置服务器的默认存储引擎：

```
default_storage_engine=blackhole
```

这只会影响那些没有指定存储引擎的 create table 的语句。如果有一个无法控制的应用，这种拓扑结构可能会非常脆弱。可以设置 disabled_storage_engines 禁用多个存储引擎。该系统变量为只读，只能通过配置文件修改，并重启 MySQL 服务器使之生效。下面演示如何在 MySQL 8 上联机搭建一个 blackhole 的分发主库。

- 服务器角色分配：

172.16.1.125：主库。假设为生产主库，可以在以下整个过程中存在负载。

172.16.1.126：blackhole 分发主库。一个初始化的 MySQL 服务器。

172.16.1.127：从库。

- MySQL 服务器配置：

```
172.16.1.125:
[mysqld]
server_id=1125
gtid_mode=ON
enforce-gtid-consistency=true
```

```
172.16.1.126:
[mysqld]
server_id=1126
gtid_mode=ON
enforce-gtid-consistency=true
default_storage_engine=blackhole
default_tmp_storage_engine=blackhole
disabled_storage_engines='innodb'
secure_file_priv='/tmp'

172.16.1.127:
server_id=1127
gtid_mode=ON
enforce-gtid-consistency=true
```

其他配置使用默认值。启用 GTID 复制，三台 MySQL 服务器均已创建复制账号。

（1）初始化 blackhole 分发主库

在 126 执行内容如下的脚本文件 init_blackhole.sh。

```
source ~/.bashrc

# 全量导入主库，无数据
mysqldump --single-transaction --all-databases --host=172.16.1.125 -d
--user=wxy --password=123456 | mysql -uroot -p123456

# 修改所有表的存储引擎为 blackhole
rm -rf /tmp/black.sql
mysql -uroot -p123456 -e "
select concat('alter table ', table_schema, '.', table_name, '
engine=''blackhole''', ';')
 from information_schema.tables
 where table_schema not in
('information_schema','mysql','performance_schema','sys')
  and table_type='BASE TABLE' into outfile '/tmp/black.sql';"

# 在执行的 SQL 文件第一行加入 sql_log_bin=0，否则下级从库也会执行
sed -i '1i\set sql_log_bin=0;' /tmp/black.sql
mysql -uroot -p123456 < /tmp/black.sql
```

（2）初始化从库

因为是联机配置复制，使用 xtrabackup 初始化从库。

```
# 将主库备份到从库，在 125 执行
xtrabackup -uroot -p123456 --socket=/tmp/mysql.sock --no-lock --backup
--compress --stream=xbstream --parallel=4 --target-dir=./ | ssh mysql@172.16.1.127
"xbstream -x -C /usr/local/mysql/data/ --decompress"

# 从库执行应用日志，在 127 执行
xtrabackup --prepare --target-dir=/usr/local/mysql/data/
```

```
# 启动从库，在 127 执行
mysqld_safe --defaults-file=/etc/my.cnf &
```

（3）启动复制

```
-- 在 126 执行
change master to
       master_host = '172.16.1.125',
       master_port = 3306,
       master_user = 'repl',
       master_password = '123456',
       master_auto_position = 1;
start slave;
show slave status\G

-- 在 127 执行
change master to
       master_host = '172.16.1.126',
       master_port = 3306,
       master_user = 'repl',
       master_password = '123456',
       master_auto_position = 1;
start slave;
show slave status\G
```

至此完成了分发主库复制拓扑结构的搭建。

2. 日志服务器

使用 MySQL 复制的另一种用途是创建没有数据的日志服务器。它唯一的目的就是更加容易重放或过滤二进制日志事件。假设有一组二进制日志或中继日志，可能从备份或者一台崩溃的服务器上获取，希望能够重放这些日志中的事件。最容易想到的是通过 mysqlbinlog 命令行工具从其中提取出事件，但更加方便和高效的方法是配置一个没有任何应用数据的 MySQL 实例，并使其认为这些二进制日志是它拥有的。因为无须执行二进制日志，所以日志服务器也就不需要任何数据，它的目的仅仅是将复制事件提供给别的服务器。

我们来看看该策略是如何工作的。假设日志被命名为 binlog.000001、binlog.000002 等等，将这些日志放到日志服务器的日志文件夹中，假设为/usr/local/mysql/data。然后在启动服务器前编辑 my.cnf 文件：

```
log_bin = /usr/local/mysql/data/binlog
log_bin_index = /usr/local/mysql/data/binlog.index
```

服务器不会自动发现日志文件，因此还需要更新日志的索引文件，在 Linux 中可用下面的命令来完成：

```
/bin/ls -1 /usr/local/mysql/data/binlog.[0-9]* >
/usr/local/mysql/data/binlog.index
```

确保运行 MySQL 的账户能够读写日志索引文件。现在可以启动日志服务器并通过 show master

logs 命令来确保其找到日志文件。当主库失效但二进制日志尚存，可以设置一个日志服务器，把从库指向它，然后让所有从库赶上主库的失效点。

相比于使用 mysqlbinlog 来实现恢复，日志服务器有优势主要体现在：

- 速度快，因为无须将语句从日志导出来并传给 MySQL。
- 可以观察到复制过程。
- 容易处理错误，如跳过执行失败的语句。
- 便于过滤复制事件。

4.2 复制性能

我们可以将复制的时间分为两部分：一是事件从主库到从库的传输时间；二是事件在从库上的执行时间。事件在主库上记录二进制日志后到传递到从库的时间理论上非常快，因为它只取决于网络速度。MySQL 二进制日志的 dump 线程不是通过轮询方式请求事件，而是由主库来通知从库新的事件的，因为前者低效且缓慢。从主库读取一个二进制日志事件是一个阻塞型网络调用，当主库记录事件后，马上就开始发送。因此可以说，只要 I/O 线程被唤醒并且能够通过网络传输数据，事件就会很快到达从库。但是，如果网络很慢并且二进制日志事件很大，记录二进制日志和在从库上执行的延迟可能会非常明显。如果查询需要执行很长时间而网络很快，通常可以认为重放时间占据了更多的复制时间开销。

本节主要从日志持久化、组提交与多线程复制，以及新增的 WRITESET 特性三个方面，讨论对复制性能产生的影响。我们先简要介绍每种特性的基础知识，然后针对不同情况进行测试，最后由测试结果得出结论。所有测试均基于 GTID 的标准主从异步复制。

4.2.1 测试规划

这里使用的思路是：记录主库加压前后的 GTID，得到从库需要执行的事务数。然后在从库上执行复制，记录执行时间，得到从库的每秒执行事务数（Transaction Per Second，TPS），以作为衡量复制性能的指标。测试目的在于对比不同情况下复制的性能，而不是针对测量绝对值进行优化。主库加压使用 tpcc-mysql 基准测试工具。

1. 测试环境

测试环境如下，已经配置好 GTID 异步复制。

主库：172.16.1.125
从库：172.16.1.126
MySQL 版本：8.0.16

测试通用参数：

主库：

server_id=1125
gtid_mode=ON
enforce-gtid-consistency=true
innodb_buffer_pool_size=4G

从库：

server_id=1126
gtid_mode=ON
enforce-gtid-consistency=true
innodb_buffer_pool_size=4G

2. tpcc-mysql 测试前准备

TPC-C 是专门针对联机事务处理过程（On-Line Transaction Processing，OLTP）的规范，tpcc-mysql 则是 Percona 公司基于 TPC-C 衍生出来的产品，专用于 MySQL 基准测试，下载地址为 https://github.com/Percona-Lab/tpcc-mysql。这里使用 tpcc-mysql 只是为了给主库加压。使用 tpcc-mysql 开始测试前需要完成以下准备工作，所有步骤均在主库上执行。

（1）安装

```
cd tpcc-mysql-master/src
make
```

（2）建立测试库

```
mysql -uroot -p123456 -e "create database tpcc_test;"
```

（3）建表和索引

```
cd tpcc-mysql-master
mysql -uroot -p123456 -Dtpcc_test < create_table.sql
mysql -uroot -p123456 -Dtpcc_test < add_fkey_idx.sql
```

（4）生成数据，-w 参数指定建立的仓库数

```
tpcc_load -h127.0.0.1 -d tpcc_test -u root -p "123456" -w 10
```

（5）备份测试库

```
mysqldump --databases tpcc_test -uroot -p123456 --set-gtid-purged=off >
tpcc_test.sql
```

为在同等环境下进行比较，每次测试前都要重新生成测试库中的表、索引和数据，因此这里做一个测试库的逻辑备份。一定要加--set-gtid-purged=off，因为将备份导入主库时，需要在从库通过复制同时生成。

下面是每次测试在从库执行的自动化脚本：

```
# 初始化 tpcc 数据
mysql -uwxy -p123456 -h172.16.1.125 < tpcc_test.sql

# 读取主库的二进制坐标
```

```
    read master_file master_pos < <(mysql -uwxy -p123456 -h172.16.1.125 -e "show
master status;" --skip-column-names | awk '{print $1,$2}')

    # 从库初始化 tcpp 数据结束后停止复制
    mysql -uwxy -p123456 -e "select
master_pos_wait('$master_file',$master_pos);stop slave;"

    # 取得从库起始的 GTID
    read start_gtid < <(mysql -uwxy -p123456 -e "show variables like
'gtid_executed';" --skip-column-names | awk '{print $2}' | sed "s/\\\n//g")

    # 主库执行压测，10 个仓库，32 个并发线程，预热 1 分钟，压测 5 分钟
    tpcc_start -h172.16.1.125 -d tpcc_test -u wxy -p "123456" -w 10 -c 32 -r 60
-l 300 > tpcc_test.log 2>&1

    # 读取主库的二进制坐标
    read master_file master_pos < <(mysql -uwxy -p123456 -h172.16.1.125 -e "show
master status;" --skip-column-names | awk '{print $1,$2}')

    # 从库复制开始时间
    start_time=`date '+%s'`

    # 从库执行复制
    mysql -uwxy -p123456 -e "start slave;select
master_pos_wait('$master_file',$master_pos);"

    # 从库复制结束时间
    end_time=`date '+%s'`

    # 复制执行时长
    elapsed=$(($end_time - $start_time))

    # 取得从库结束的 GTID
    read end_gtid < <(mysql -uwxy -p123456 -e "show variables like 'gtid_executed';"
--skip-column-names | awk '{print $2}' | sed "s/\\\n//g")

    # 取得从库执行的事务数
    read start end < <(mysql -uwxy -p123456 -e "select
gtid_subtract('$end_gtid','$start_gtid');" --skip-column-names | awk -F: '{print
$2}' | awk -F- '{print $1,$2}')
    trx=$(($end - $start + 1))

    # 计算从库和主库的 TPS
    Slave_TPS=`expr $trx / $elapsed`
    Master_TPS=`expr $trx / 360`

    # 打印输出
    echo "TRX: $trx" "Elapsed: $elapsed" "Slave TPS: $Slave_TPS" "Master TPS:
$Master_TPS"
```

4.2.2　sync_binlog 与 innodb_flush_log_at_trx_commit

sync_binlog 控制 MySQL 服务器将二进制日志同步到磁盘的频率，可取值 0、1 和 N，MySQL 8 的默认值为 1。innodb_flush_log_at_trx_commit 控制提交时是否将 InnoDB 重做日志同步到磁盘，可取值 0、1 和 2，MySQL 8 的默认值为 1。关于这两个参数已经在“第 1 章 异步复制”中详细讨论过，这里就不再赘述。对于复制而言，sync_binlog 为 0 可能造成从库丢失事务，innodb_flush_log_at_trx_commit 为 0 可能造成从库比主库事务多。而从性能角度来看，设置为双 1 的性能最差，设置为双 0 的性能最好。权衡数据安全与性能，一般建议主库都设置为双 1，根据应用场景，从库也可以设置成其他组合来提升性能。

表 4-1 所示为从库上 sync_binlog 和 innodb_flush_log_at_trx_commit 四种组合设置的测试结果。

表 4-1　日志持久化性能对比

sync_binlog	innodb_flush_log_at_trx_commit	事务数	复制执行时间（秒）	从库 TPS	主库 TPS
0	0	183675	330	556	510
0	1	184177	498	369	511
1	0	183579	603	304	509
1	1	183020	683	267	508

测试中主库执行了一共 360 秒（预热+压测），TPS 为 510。从表中可以明显看到这两个参数的不同组合对复制性能的影响。当从库仅为单线程复制时，只有双 0 的设置在执行时间和 TPS 上优于主库，其他组合都会造成复制延迟。

4.2.3　组提交与多线程复制

MySQL 5.6 支持多线程从库（Multi-Threaded Slave，MTS）复制，但太过局限。它只实现了基于 schema 的多线程从库复制，使不同数据库下的 DML 操作可以在从库并行重放，这样设计的复制效率并不高。如果用户实例仅有一个库，那么就无法实现并行重放，甚至性能会比原来的单线程更差，而单库多表是比多库多表更为常见的一种情况。MySQL 5.7 的多线程从库复制是基于组提交来实现的，不再有基于 schema 的多线程从库复制的限制。

1. 组提交

从 MySQL 5.6 开始同时支持 InnoDB redo log 和 binlog 组提交，并且默认为启用组提交，因而大大提高了 MySQL 的事务处理性能。和很多 RDBMS 一样，MySQL 为了保证事务处理的一致性和持久性，使用了预写日志（Write Ahead Log，WAL）机制，也就是在对数据文件进行修改之前，必须将修改先记录到日志中。redo log 就是一种 WAL 的应用，每次事务提交时，不用同步刷新磁盘数据文件，只需要同步刷新 redo log 就够了。相比写数据文件时的随机 I/O，写 redo log 时的顺序 I/O 能够提高事务提交的速度。redo log 的刷盘操作是最终影响 MySQL TPS 的瓶颈。为了缓解这一问题所造成的影响，MySQL 使用了 redo log 组提交，将多个 redo log 刷盘操作合并成一个操作。

为了保证 redo log 和 binlog 的数据一致性，MySQL 使用了两阶段提交（prepare 阶段和 commit 阶段），由 binlog 作为事务的协调者。而引入两阶段提交使得 binlog 又成为了性能的瓶颈，于是 MySQL 5.6 增加了 binlog 的组提交，目的同样是将 binlog 的多个刷盘操作合并成一个操作。结合 redo log 本身已经实现的组提交，将提交过程分成 Flush stage、Sync stage 和 Commit stage 三个阶段完成组提交，最大化每次刷盘的收益，弱化磁盘瓶颈。每个阶段都有各自的队列，使每个会话的事务进行排队，提高并发性能。

（1）Flush 阶段

- 首先获取队列中的事务组，将 redo log 中 prepare 阶段的数据刷盘。
- 将 binlog 数据写入文件系统缓存，并不能保证数据库崩溃时 binlog 不丢失。
- Flush 阶段队列的作用是提供了 redo log 的组提交。
- 如果在这一步完成后数据库崩溃，由于协调者 binlog 中不保证有该组事务的记录，因此 MySQL 可能会在重启后回滚该组事务。

（2）Sync 阶段

- 将 binlog 缓存 sync 到磁盘，sync_binlog=1 时该队列中所有事务的 binlog 将永久写入磁盘。
- 为了增加一组事务中的事务数量，提高刷盘收益，MySQL 使用两个参数控制获取队列事务组的时机:
  - binlog_group_commit_sync_delay = N：在等待 N 微秒后，开始事务刷盘。
  - binlog_group_commit_sync_no_delay_count = N：如果队列中的事务数达到 N 个，就忽视 binlog_group_commit_sync_delay 的设置，直接开始刷盘。
- Sync 阶段队列的作用是支持 binlog 的组提交。
- 如果在这一步完成后数据库崩溃，由于协调者 binlog 中已经有了事务记录，因此 MySQL 会在重启后通过 Flush 阶段中 redo log 刷盘的数据继续进行事务的提交。

（3）Commit 阶段

- 首先获取队列中的事务组。
- 依次将 redo log 中已经 prepare 的事务在存储引擎层提交，清除回滚信息，向 redo log 中写入 COMMIT 标记。
- Commit 阶段不用刷盘，如上所述，Flush 阶段中的 redo log 刷盘已经足够保证数据库崩溃时的数据安全了。
- Commit 阶段队列的作用是承接 Sync 阶段的事务，完成最后的引擎提交，使得 Sync 可以尽早地处理下一组事务，最大化组提交的效率。

Commit 阶段会受到参数 binlog_order_commits 的影响，当该参数为 OFF 时，不保证 binlog 和事务提交的顺序一致，因为此时允许多个线程发出事务提交指令。正是基于同样的原因，可以防止逐个事务提交成为吞吐量的瓶颈，性能会有少许提升。多数情况下，存储引擎的提交指令与 binlog 不同序无关紧要，因为多个单独事务中执行的操作，无论提交顺序如何变化都应该产生一致的结果。但也不是绝对的，例如会影响 XtraBackup 工具的备份。XtraBackup 会从 innodb page 中获取最后提交事务的 binlog 位置信息，binlog_order_commits=0 时事务提交顺序和 binlog 顺序可能不一致，这样此位置前可能存在部分 prepare 状态的事务，这些事务在备份恢复后会因回滚而丢失。binlog_order_commits 的默认值为 ON，此时存储引擎的事务提交指令将在单个线程上串行化，使得事务始终以与写入二进制日志相同的顺序提交。

“[图解 MySQL]MySQL 组提交”（https://yq.aliyun.com/articles/617776?utm_content=m_1000008043）是一篇 MySQL 组提交的图解说明，以更形象的方式解释了组提交的原理。

2. 多线程复制

MySQL 5.6 开始出现基于 schema 的多线程从库复制，言而言之就是主库上不同数据库上的 DML 可以在从库上并行重放。因为大多数生产环境依然习惯于单库多表的架构，这种情况下 MTS 依然还是单线程的效果。MySQL 5.7 实现了基于组提交的多线程从库复制，它的思想简单易懂：主库上同一个组提交的事务可以在从库并行重放，这些事务之间没有任何冲突，由存储引擎的 ACID 来保证。为了与 5.6 版本兼容，5.7 版本引入了新的变量 slave_parallel_type，可以配置为下面两个值之一：

- DATABASE：默认值，基于 schema 的多线程从库复制方式。
- LOGICAL_CLOCK：基于组提交的多线程从库复制方式。

那么从库如何知道事务是否在一组中呢？MySQL 5.7 的设计方式是将组提交信息存放在二进制日志的 GTID_EVENT 中：

```
    [mysql@hdp2/usr/local/mysql/data]$mysqlbinlog binlog.000064 | grep
last_committed | awk '{print $11, $12}' | head -10
    last_committed=0 sequence_number=1
    last_committed=0 sequence_number=2
    last_committed=0 sequence_number=3
    last_committed=0 sequence_number=4
    last_committed=0 sequence_number=5
    last_committed=0 sequence_number=6
    last_committed=0 sequence_number=7
    last_committed=0 sequence_number=8
    last_committed=0 sequence_number=9
    last_committed=0 sequence_number=10
    [mysql@hdp2/usr/local/mysql/data]$
```

last_committed 表示事务提交时上次事务提交的编号。事务在 prepare 阶段获取相同的 last_committed 而且相互不影响，最终会作为一组进行提交。如果事务具有相同的 last_committed，表示这些事务都在一组内，可以进行并行重放。例如上述 last_committed 为 0 的 10 个事务在从库是可以进行并行重放的。这个机制是 Commit-Parent-Based Scheme 的实现方式。

sequence_number 是事务计数器。记录在 GTID_EVENT 中的 sequence_number 和 last_committed 使用的是相对当前二进制日志文件的值。即每个二进制日志文件中事务的 last_commited 起始值为 0，sequence_number 为 1。由于在二进制日志文件切换时，需要等待上一个文件的事务执行完，因此这里记录相对值并不会导致冲突事务并行执行。

由于在 MySQL 中写入的是基于锁的并发控制，因此所有在主库同时处于 prepare 阶段且未提交的事务就不会存在锁冲突，从库就可以并行执行。Commit-Parent-Based Scheme 使用的就是这个原理，简单描述如下：

- 主库上有一个全局计数器（Global Counter）。每一次存储引擎提交之前，计数器值就会增加。
- 主库上，事务进入 prepare 阶段之前，全局计数器的当前值会被存储在事务中，这个值称为此事务的 commit-parent。
- 主库上，commit-parent 会在事务的开头被存储在 binlog 中。
- 从库上，如果两个事务有同一个 commit-parent，它们就可以并行执行。

此 commit-parent 就是在 binlog 中看到的 last_committed。如果 commit-parent 相同，即 last_committed 相同，则被视为同一组，可以并行重放。

Commit-Parent-Based Scheme 的问题在于会降低复制的并行程度，如图 4-9 所示。

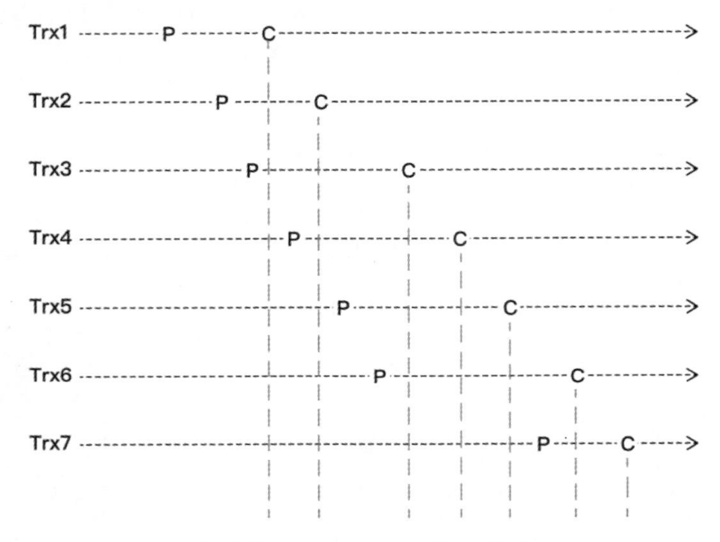

图 4-9 Commit-Parent-Based Scheme

从图 4-9 中可以看到，每一个水平线代表一个事务，时间从左到右。P 表示事务在进入 prepare 阶段之前读到的 commit-parent 值的那个时间点，可以简单视为加锁时间点。C 表示事务增加了全局计数器值的那个时间点，可以简单视为释放锁的时间点。P 对应的 commit-parent 是取自所有已经执行完的事务的最大的 C 所对应的 sequence_number。举例来说，Trx4 的 P 对应的 commit-parent 是 Trx1 的 C 所对应的 sequence_number。因为这个时候 Trx1 已经执行完，但是 Trx2 还未执行完。Trx5 的 P 所对应的 commit-parent 是 Trx2 的 C 所对应的 sequence_number。Trx6 的 P 所对应的 commit-parent 也是 Trx2 的 C 所对应的 sequence_number。

Trx5 和 Trx6 具有相同的 commit-parent，在进行重放时，Trx5 和 Trx6 可以并行执行。Trx4 和 Trx5 不能并行执行，Trx6 和 Trx7 也不能并行执行，因为它们的 commit-parent 不同。需要注意的是，在同一时段，Trx4 和 Trx5、Trx6 和 Trx7 分别持有它们各自的锁，事务互不冲突，所以在从库上并行执行不会有问题。针对这种情况，为了进一步增加并行度，MySQL 对并行复制的机制做了改进，提出了一种新的并行复制方式：Lock-Based Scheme，使同时持有各自锁的事务可以在从库并行执行。

Lock-Based Scheme 定义了一个称为 lock interval 的概念，表示一个事务持有锁的时间间隔。

假设有两个事务 Trx1 和 Trx2，Trx1 先于 Trx2，那么当且仅当 Trx1 和 Trx2 的 lock interval 有重叠，则可以并行执行。换言之，若 Trx1 结束自己的 lock interval 早于 Trx2 开始自己的 lock interval，则不能并行执行。如图 4-10 所示，L 表示 lock interval 的开始点，C 表示 lock interval 的结束。

对于 C（lock interval 的结束点），MySQL 会给每个事务分配一个逻辑时间戳（Logical Timestamp），命名为 transaction.sequence_number。此外，MySQL 会获取全局变量 global.max_committed_transaction，表示所有已经结束 lock interval 的事务的最大的 sequence_number。对于 L（lock interval 的开始点），MySQL 会把 global.max_committed_transaction 分配给一个变量，并取名为 transaction.last_committed。transaction.sequence_number 和 transaction.last_committed 这两个时间戳都会存放在 binlog 中，就是前面看到的 last_committed 和 sequence_number。

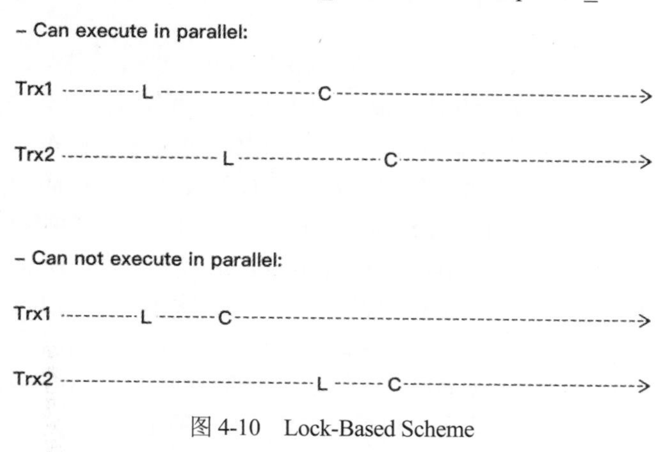

图 4-10　Lock-Based Scheme

根据以上分析得出，只要事务和当前执行事务的 lock interval 都存在重叠部分，就可以在从库并行执行。图 4-9 中，Trx3、Trx4、Trx5 和 Trx6 这四个事务可以并行执行，因为 Trx3 的 sequence_number 大于 Trx4、Trx5 和 Trx6 的 last_committed，即它们的 lock interval 存在重叠部分。当 Trx3、Trx4 和 Trx5 执行完成之后，Trx6 和 Trx7 可以并发执行，因为 Trx6 的 sequence_number 大于 Trx7 的 last_committed，即两者的 lock interval 存在重叠部分。Trx5 和 Trx7 不能并发执行，因为 Trx5 的 sequence_number 小于 Trx7 的 last_committed，即两者的 lock interval 不存在重叠部分。

可以通过以下命令粗略查看并发度：

```
    [mysql@hdp2/usr/local/mysql/data]$mysqlbinlog binlog.000064 | grep -o
'last_committed.*' | sed 's/=/ /g' | awk '{print $4-$2-1}' | sort -g | uniq -c
       1693 0
       4795 1
       8174 2
      11378 3
    ...
          1 49
          1 50
          1 56
          1 120
```

第一列为事务数量，第二列表示这些事务能与它们之前的多少个事务并行执行。例如有 1693 个事务不能与之前的事务并发，必须等到所有前面的事务完成之后才能开始，但并不表示不能和后面的事务并行执行。当前事务无法判断能否与后面的事务并行执行，只能与前面事务的sequence_number 比较，得出自己是否可以并发执行。

仅仅设置为 LOGICAL_CLOCK 还会存在问题，因为此时在从库上应用事务是无序的，和 relay log 中记录的事务顺序可能不一样。在这种情况下，从库的 GTID 会产生间隙，事务可能在某个时刻主从是不一致的，但是最终会一致，满足最终的一致性。相同记录的修改，会按照顺序执行，这由事务隔离级来保证。不同记录的修改，可以产生并行，并无数据一致性的风险。这大概也是 slave_preserve_commit_order 参数默认为 0 的原因之一。

如果要保证事务是按照 relay log 中记录的顺序来重放，需要设置参数 slave_preserve_commit_order=1，这要求从库启用 log_bin 和 log_slave_updates，并且 slave_parallel_type 设置为 LOGICAL_CLOCK。

启用 slave_preserve_commit_order 后，正在执行的 worker 线程将等待，直到所有先前的事务提交后再提交。当复制线程正在等待其他 worker 线程提交其事务时，它会将其状态报告为等待提交前一个事务。使用此模式，多线程复制的重放顺序与主库的提交顺序保持一致。

slave_parallel_workers 参数控制并行复制 worker 线程的数量。若将 slave_parallel_workers 设置为 0，则退化为单线程复制。如果 slave_parallel_workers=N（N>0），则单线程复制中的 SQL 线程将转为 1 个 coordinator 线程和 N 个 worker 线程，coordinator 线程负责选择 worker 线程执行事务的二进制日志。例如将 slave_parallel_workers 设置为 1，则 SQL 线程转化为 1 个 coordinator 线程和 1 个 worker 线程，也是单线程复制。然而，与 slave_parallel_workers=0 相比，多了一次 coordinator 线程的转发，因此 slave_parallel_workers=1 的性能反而比设置为 0 还要差。MySQL 8 中 slave_parallel_workers 参数可以动态设置，但需要重启复制才能生效。

LOGICAL_CLOCK 多线程复制为了准确性和实现的需要，其中 lock interval 实际获得的区间比理论值要窄，会导致原本一些可以并行的事务在从库上没有并行执行。当使用级联复制时，LOGICAL_CLOCK 可能会使离主库越远的从库并行度越小。

3. 多线程复制测试

从库增加以下配置参数：

```
sync_binlog = 1
innodb_flush_log_at_trx_commit = 1
slave_preserve_commit_order = 1
slave_parallel_type = LOGICAL_CLOCK
```

表 4-2 所示为从库上 slave_parallel_workers 分别设置为 2、4、8 和 16 的测试结果。

表 4-2 基于 LOGICAL_CLOCK 的多线程复制性能对比

slave_parallel_workers	事务数	复制执行时间（秒）	从库 TPS	主库 TPS
2	183717	460	399	510
4	183248	396	462	509
8	182580	334	546	507
16	183290	342	535	509

测试中主库执行了一共 360 秒（预热+压测），TPS 为 509。从表中可以看到，在实验负载场景下，多线程复制的性能明显高于单线程复制的性能。slave_parallel_workers=8 时性能最好，当

worker 数量增加到 16 时，性能反而比设置为 8 时差。太多线程会增加线程间同步的开销，因此 slave_parallel_workers 值并非越大越好，需要根据实际负载进行测试来确定其最佳值，通常建议 4~8 个 worker 线程。

4.2.4 基于 WriteSet 的多线程复制

基于组提交 LOGICAL_CLOCK 多线程复制机制在每组提交的事务足够多，即业务量足够大时表现较好。不过，在很多实际业务中，虽然事务间的 lock interval 没有重叠部分，但是这些事务操作的往往是不同的数据行，就不会有锁冲突，因而是可以并行执行的，但 LOGICAL_CLOCK 的实现无法使这部分事务进行并行重放。为了解决这个问题，MySQL 在 5.7.22 版本中推出了基于 WriteSet 的并行复制。简单来说，WriteSet 并行复制的思想是：不同事务的记录不重叠，则都可在从库上并行重放。可以看到并行的力度从事务级细化为记录级。

1. WriteSet 对象

MySQL 中用 WriteSet 对象来记录每个数据行，从源代码来看 WriteSet 就是每条记录求得的哈希值（必须启用 ROW 格式的二进制日志），具体算法如下：

```
WriteSet=hash(index_name | db_name | db_name_length | table_name |
table_name_length | value | value_length)
```

上述公式中的 index_name 只记录唯一索引，主键也是唯一索引。如果有多个唯一索引，则每条记录会产生对应多个 WriteSet 值。另外，value 会分别计算原始值和带有字符集排序规则（Collation）值的两种 WriteSet，所以一条记录可能有多个 WriteSet 对象。

新产生的 WriteSet 对象会插入到 WriteSet 哈希表，哈希表的大小由参数 binlog_transaction_dependency_history_size 设置，默认为 25000。内存中保留的哈希行数达到此值后，将清除历史记录。

2. 实现原理

基于 WriteSet 的复制优化了主库组提交的实现，体现主库端 last_committed 的定义也变了。原来一组事务是指拥有同一个 parent_commit 的事务，在二进制日志中记录为同一个 last_committed。基于 WriteSet 的方式中，last_committed 的含义是保证冲突事务（更新相同记录的事务）不能拥有同样的 last_committed 值，事务执行的并行度进一步提高。

当事务每次提交时，会计算修改的每个行记录的 WriteSet 值，然后查找哈希表中是否已经存在同样的 WriteSet。若无，WriteSet 插入到哈希表，写入二进制日志的 last_committed 值不变。上一个事务的 last_committed 与当前事务的 last_committed 相等，意味着它们可以作为一组进行提交。若有，更新哈希表对应的 WriteSet 值为 sequence_number，并且写入到二进制日志的 last_committed 值也更新为 sequence_number。上一个事务的 last_committed 与当前事务的 last_committed 必然不同，表示事务发生冲突了，必须等待之前的事务提交后才能执行。

从库端的逻辑跟以前一样没有变化，last_committed 相同的事务可以并行执行。

要使用 WriteSet 方式组提交，需要把 binlog_transaction_dependency_tracking 参数设置为 WRITESET。binlog_transaction_dependency_tracking 参数用于指定主库确定哪些事务可以作为一组

提交的方法，有三个可选值：

- COMMIT_ORDER：依赖事务提交的逻辑时间戳，这个值是默认值。如果事务更新的表上没有主键和唯一索引，也使用该值。这是 MySQL 5.7 所使用的方式。
- WRITESET：更新不同记录的事务（不冲突）都可以并行化。
- WRITESET_SESSION：与 WRITESET 的区别是 WRITESET_SESSION 需要保证同一个会话内事务的先后顺序。消除了从库中某一时刻可能看到主库从未出现过的数据库状态的问题。

从下面这个简单的实验可以直观地看到 COMMIT_ORDER 与 WRITESET 的区别。

```
drop table if exists t1;
create table t1 (a int primary key);
insert into t1 values (1), (2);

flush logs;
set global binlog_transaction_dependency_tracking  = WRITESET;
update t1 set a=10 where a=1;
update t1 set a=20 where a=2;

set global binlog_transaction_dependency_tracking  = COMMIT_ORDER;
update t1 set a=1 where a=10;
update t1 set a=2 where a=20;
```

查看二进制日志：

```
    [mysql@hdp2/usr/local/mysql/data]$mysqlbinlog binlog.000002 --base64-output
=decode-rows -v | grep -e 'last_committed' -A4 -e 'UPDATE' | grep -v "# original\|#
immediate\|/*!" | awk '{if ($1!="###") {print $11, $12} else {print $0}}'
    last_committed=0 sequence_number=1

    ### UPDATE 'test'.'t1'
    ### WHERE
    ###   @1=1
    ### SET
    ###   @1=10

    last_committed=0 sequence_number=2

    ### UPDATE 'test'.'t1'
    ### WHERE
    ###   @1=2
    ### SET
    ###   @1=20

    last_committed=2 sequence_number=3

    ### UPDATE 'test'.'t1'
    ### WHERE
    ###   @1=10
    ### SET
```

```
###   @1=1

last_committed=3 sequence_number=4

### UPDATE 'test'.'t1'
### WHERE
###   @1=20
### SET
###   @1=2
[mysql@hdp2/usr/local/mysql/data]$
```

第一个事务和第二个事务的 last_committed 都是 0。虽然这两个事务的 lock_interval 没有重叠部分，但是它们修改的是不同的数据行，不存在事务冲突，因此它们的 last_committed 相同，可以作为一组并行提交。

当设置 global binlog_transaction_dependency_tracking 为 COMMIT_ORDER 时，第三个事务和第四个事务的 last_committed 分别为 2 和 3。这两个事务的 lock_interval 没有重叠部分，即使更新的行不冲突，它们的 last_committed 也不相同，不能作为同一组并行提交。

与 WriteSet 相关的另一个参数是 transaction_write_set_extraction。该参数用于定义计算 WriteSet 使用的哈希算法。如果用于多线程复制，必须将此变量设置为 XXHASH64，这也是默认值。如果设置为 OFF，则 binlog_transaction_dependency_tracking 只能设置为 COMMIT_ORDER。如果 binlog_transaction_dependency_tracking 的当前值为 WRITESET 或 WRITESET_SESSION，则无法更改 transaction_write_set_extraction 的值。

3. WriteSet 多线程复制测试

主库增加以下配置参数：

```
binlog_transaction_dependency_tracking  = WRITESET
transaction_write_set_extraction      = XXHASH64
```

从库增加以下配置参数：

```
sync_binlog = 1
innodb_flush_log_at_trx_commit = 1
slave_preserve_commit_order = 1
slave_parallel_type = LOGICAL_CLOCK
```

表 4-3 所示为从库 slave_parallel_workers 分别设置为 2、4、8、16 和 32 的测试结果。

表 4-3　基于 WriteSet 的多线程复制性能对比

slave_parallel_workers	事务数	复制执行时间（秒）	从库 TPS	主库 TPS
2	209237	515	406	581
4	207083	438	472	575
8	207292	364	569	575
16	205060	331	619	569
32	201488	340	592	559

测试中主库执行了一共 360 秒（预热+压测），TPS 平均为 572，同等场景下比 COMMIT_ORDER

高出 12%。当有 16 个复制线程时，从库 TPS 达到峰值 619，比有 8 个复制线程时的从库 TPS 高出 13%。

MySQL 的复制延迟是一直被诟病的问题之一，从以上三组测试得出了目前解决延迟最普遍的三种方法：

- 如果负载和数据一致性要求都不是太高，则从库可以采用单线程复制 + 安全参数设置为双 0。这种模式同样拥有不错的表现，一般压力均可应付。
- 如果主库的并发量很高，那么基于 order-commit 模式的多线程复制可以有很好的表现。
- 基于 WriteSet 的模式是目前并发度最高的多线程复制，基本可以满足大部分应用场景。如果并发量非常高，或是要求从库与主库的延迟降至最低，则可以采取这种方式。

4.3 小 结

可以把 MySQL 复制设计成不同的拓扑结构，每种拓扑结构都有其适用的场景。“一主多从”能满足大多数应用的需求，它的简单性有利于减轻维护工作的负担。不推荐使用多主可写的配置，因为这种拓扑结构极易产生数据冲突，并且不好修复。MySQL 复制总的原则是把从库设置为只读。多源复制是指“一从多主”的情况，通过通道实现，但使用场景并不多见。blackhole 是一个特殊的引擎，它不存储任何数据，可用作复制分发服务器或 binlog 日志服务器。

MySQL 不断更新的版本在不断地提升复制性能，减少复制延迟。本章我们提供了一个自动化测试的 shell 脚本，利用 tpcc-mysql 工具，在实验环境下测试不同 sync_binlog 与 innodb_flush_log_at_trx_commit 设置、基于 order-commit 的多线程复制以及基于 WriteSet 的多线程复制的 TPS，并以此为基础，最后给出了简单的适用性指南。下一章将介绍与此截然相反的概念——延迟复制与部分复制。

第5章

延迟复制与部分复制

5.1 延迟复制

上一章我们介绍了几种减少复制延迟的方法，本章将反其道而行之，介绍人为延长复制滞后时间的方法。

5.1.1 延迟复制简介

通常情况下 MySQL 复制都很快，但 MySQL 默认的复制依然存在延迟。另一方面，有时却需要特意增加复制的延迟。设想这样一种应用场景，用户在主库上误删了一个表，并且该操作很快被复制到从库。当用户发现这个错误时，从库早就完成了该事件重放。此时主库和从库都没有那个被误删的表了，如何恢复？如果有备份，可以幸运地从备份恢复，丢失的数据量取决于备份的新旧以及从备份时间点到表被删除时间点之间该表上数据的变化量。如果没有备份呢？在这种情况下，延迟复制作为一种恢复数据的备选方案，或许可以帮上忙。如果在发现问题时，从库还没有来得及重放相应的中继日志，那么就有机会在从库获得该表，继而进行恢复。当然还有一些其他数据恢复方案，例如已经存在类似 Oracle 闪回技术（Flashback）在 MySQL 上的实现，该实现方式为解析相应的二进制日志事件，生成反向的 SQL 语句。这些程序多为个人作品，并没有被加入 MySQL 发行版本中，因此在易用性、适用性、可靠性等方面还不能与原生的功能相提并论。

MySQL 支持延迟复制，以便从库故意执行比主库晚至少在指定时间间隔的事务。在 MySQL 8.0 中，延迟复制的方法取决于两个时间戳：immediate_commit_timestamp 和 original_commit_timestamp。如果复制拓扑结构中的所有服务器都运行 MySQL 8.0.1 或更高版本，则使用这些时间戳测量延迟。如果从库未使用这些时间戳，则执行 MySQL 5.7 的延迟复制。

复制延迟默认为 0 秒，使用 CHANGE MASTER TO MASTER_DELAY = N 语句将延迟设置为 N 秒，从主库接收的事务比主库上的提交至少晚 N 秒才在从库上执行。每个事务发生延迟（不是以前 MySQL 版本中的事件），实际延迟仅强制在 gtid_log_event 或 anonymous_gtid_log_event 事件上。二进制日志中的每个 GTID 事务始终都以 Gtid_log_event 开头，匿名事务没有分配 GTID，MySQL 确保日志中的每个匿名事务都以 Anonymous_gtid_log_event 开头。对于事务中的其他事件，不会对它们施加任何等待时间，而是立即执行。注意，START SLAVE 和 STOP SLAVE 立即生效并忽略任何延迟，RESET SLAVE 将延迟重置为 0。

例如，下面将实验环境中“一主两从”半同步复制中的一个从库设置为延迟 60 秒复制：

```
mysql> change master to master_delay = 60;
ERROR 3085 (HY000): This operation cannot be performed with a running slave sql thread; run STOP SLAVE SQL_THREAD FOR CHANNEL '' first.

mysql> stop slave sql_thread;
Query OK, 0 rows affected (0.00 sec)

mysql> change master to master_delay = 60;
Query OK, 0 rows affected (0.01 sec)

mysql> start slave sql_thread;
Query OK, 0 rows affected (0.00 sec)
```

联机设置延迟复制时，需要先停止 sql_thread 线程。现在主库执行一个事务，观察从库的变化：

```
-- 主库
mysql> create table test.t3(a int);
Query OK, 0 rows affected (0.01 sec)

-- 从库
mysql> desc test.t3;
ERROR 1146 (42S02): Table 'test.t3' doesn't exist

mysql> desc test.t3;
+-------+---------+------+-----+---------+-------+
| Field | Type    | Null | Key | Default | Extra |
+-------+---------+------+-----+---------+-------+
| a     | int(11) | YES  |     | NULL    |       |
+-------+---------+------+-----+---------+-------+
1 row in set (0.00 sec)
```

主库上建立了一个表 test.t3，DDL 语句自成一个事务。60 秒后，从库上才出现该表。

从库上 performance_schema 模式下的 replication_applier_configuration.desired_delay 列显示使用 master_delay 选项配置的延迟，replication_applier_status.remaining_delay 列显示剩余的延迟秒数。

```
-- 从库
mysql> select desired_delay from performance_schema.replication_applier_configuration;
+---------------+
| desired_delay |
+---------------+
|            60 |
+---------------+
```

```
1 row in set (0.00 sec)

mysql> select remaining_delay from performance_schema.replication_applier
_status;
+-----------------+
| remaining_delay |
+-----------------+
|        NULL |
+-----------------+
1 row in set (0.00 sec)

-- 主
mysql> drop table test.t3;
Query OK, 0 rows affected (0.02 sec)

-- 从
mysql> select remaining_delay from performance_schema.replication_applier
_status;
+-----------------+
| remaining_delay |
+-----------------+
|          54 |
+-----------------+
1 row in set (0.00 sec)

mysql> select remaining_delay from performance_schema.replication_applier
_status;
+-----------------+
| remaining_delay |
+-----------------+
|          23 |
+-----------------+
1 row in set (0.00 sec)

mysql> select remaining_delay from performance_schema.replication_applier
_status;
+-----------------+
| remaining_delay |
+-----------------+
|        NULL |
+-----------------+
1 row in set (0.00 sec)
```

延迟复制可用于多种目的：

- 防止用户在主库上误操作。延迟复制时，可以将延迟的从库回滚到错误之前的时间。
- 测试滞后时系统的行为方式。例如，在应用程序中，延迟可能是由从库设备上的重负载引起的。但是，生成此负载级别可能很困难。延迟复制可以模拟滞后而无需模拟负载。它还可用于调试与从库滞后相关的条件。4.1.2 小节的示例中已经用到过此方法。

- 检查数据库过去的快照，而不必重新加载备份。例如，通过配置延迟为一周的从库，如果需要看一下几天前数据库的样子，可以检查延迟的从库。

5.1.2 延迟复制时间戳

MySQL 8.0 提供了一种新方法，用于测量复制拓扑中的延迟，或称复制滞后。该方法取决于与写入二进制日志的每个事务的 GTID 相关联的以下时间戳：

- original_commit_timestamp：将事务写入（提交）到主库二进制日志的时间戳。
- immediate_commit_timestamp：将事务写入（提交）到从库二进制日志的时间戳。

mysqlbinlog 的输出以两种格式显示这些时间戳，从 epoch 开始的微秒和 TIMESTAMP 格式，后者基于用户定义的时区以获得更好的可读性。例如：

```
    #190516 15:12:18 server id 1125 end_log_pos 239 CRC32 0xc1ebcb7c
    Anonymous_GTID  last_committed=0  sequence_number=1  rbr_only=no
    original_committed_timestamp=1557990738835397  immediate_commit_timestamp=
1557990738838735  transaction_length=192
    # original_commit_timestamp=1557990738835397 (2019-05-16 15:12:18.835397 CST)
    # immediate_commit_timestamp=1557990738838735 (2019-05-16 15:12:18.838735
CST)
    /*!80001 SET @@session.original_commit_timestamp=1557990738835397*//*!*/;
    /*!80014 SET @@session.original_server_version=80016*//*!*/;
    /*!80014 SET @@session.immediate_server_version=80016*//*!*/;
    SET @@SESSION.GTID_NEXT= 'ANONYMOUS'/*!*/;
    # at 239
```

通常，original_commit_timestamp 在应用事务的所有副本上始终相同。在主从复制中，主库二进制日志中事务的 original_commit_timestamp 始终与其 immediate_commit_timestamp 相同。在从库的中继日志中，事务的 original_commit_timestamp 和 immediate_commit_timestamp 与主库的二进制日志中的相同，而在其自己的二进制日志中，事务的 immediate_commit_timestamp 对应于从库提交事务的时间。

在组复制设置中，当原始主服务器是组的成员时，将在事务准备好提交时生成 original_commit_timestamp。再具体地说，当事务在原始主服务器上执行完成并且它的写集已经准备好发送给该组的所有成员以进行认证时，就会生成 original_commit_timestamp。因此，相同的 original_commit_timestamp 被复制到所有服务器的应用事务，并且每个服务器使用 immediate_commit_timestamp 在其自己的二进制日志中存储本地提交时间。

组复制中独有的视图更改事件是一种特殊情况。包含该事件的事务由每个服务器生成，但共享相同的 GTID。因此，这种事务不是先在主服务器中执行，然后复制到该组的其他成员，而是该组的所有成员都执行并应用相同的事务。由于没有原始主服务器，因此这些事务的 original_commit_timestamp 设置为零。关于组复制的概念和原理将在下一章详细说明。

5.1.3 监控延迟复制

在 MySQL 8 之前的版本中，监控复制的延迟（滞后）最常用的方法之一是依赖于 show slave status 输出中的 seconds_behind_master 字段。但是，当使用比传统主从复制更复杂的复制拓扑结构，例如组复制时，此度量标准不再适用。MySQL 8 中添加的 immediate_commit_timestamp 和 original_commit_timestamp 可提供有关复制延迟的更精细的信息。要监控支持这些时间戳的复制延迟，推荐使用以下 performance_schema 模式中的表。

- replication_connection_status：与主服务器连接的当前状态，提供有关连接线程排队到中继日志中的最后和当前事务的信息。
- replication_applier_status_by_coordinator：协调器线程的当前状态，仅在使用多线程复制时显示该信息，提供有关协调器线程缓冲到工作队列的最后一个事务的信息，以及当前正在缓冲的事务。
- replication_applier_status_by_worker：应用从主服务器接收事务线程的当前状态，提供有关 SQL 线程或使用多线程复制时每个工作线程应用的事务信息。

使用这些表，可以监控相应线程处理的最后一个事务以及该线程当前正在处理的事务信息，包括：

- 事务的 GTID。
- 从库中继日志中检索事务的 original_commit_timestamp 和 immediate_commit_timestamp。
- 线程开始处理事务的时间。
- 对于上次处理的事务，线程处理它的完成时间。

除 Performance Schema 表之外，show slave status 的输出还有三个字段与延迟复制有关：

- SQL_Delay：非负整数，表示使用 CHANGE MASTER TO MASTER_DELAY = N 配置的复制延迟，以秒为单位。与 performance_schema.replication_applier_configuration.desired_delay 值相同。
- SQL_Remaining_Delay：当 Slave_SQL_Running_State 等待主库执行事件后的 MASTER_DELAY 秒时，该字段包含一个整数，表示延迟剩余的秒数。在其他时候，此字段为 NULL。与 performance_schema.replication_applier_status.remaining_delay 值相同。
- Slave_SQL_Running_State：一个字符串，指示 SQL 线程的状态，类似于 Slave_IO_State。该值与 SHOW PROCESSLIST 显示的 SQL 线程的 State 值相同。

当从库的 SQL 线程在执行事件之前等待延迟时，SHOW PROCESSLIST 将其状态值显示为：Waiting until MASTER_DELAY seconds after master executed event。

5.2 部分复制

到目前为止，我们讨论的都是 MySQL 实例级的复制，复制拓扑结构中的所有服务器都包含整个实例的全部数据集，主库的任何数据变化都会原封不动地在从库进行重放。本节将介绍另一种不同的复制——部分复制。

5.2.1 部分复制简介

如果主库未将修改数据的 SQL 语句或变化的数据行写入其二进制日志，则不会复制该事件。如果主库记录了二进制日志并将其中的事件发送到从库，从库也可以自己确定是执行它还是忽略它。这就是实现 MySQL 部分复制的两种方式。

在主库上，可以使用--binlog-do-db 和--binlog-ignore-db 选项来控制要在二进制日志中记录更改的数据库。但是不应该使用这些选项来控制复制哪些数据库和表，推荐的方法是在从库上使用过滤来控制从库上执行的事件。在从库端，是否执行接收事件的决定是根据从库上启用的--replicate-* 选项来做出的。在 MySQL 5.7 及以后版本中，可以使用 CHANGE REPLICATION FILTER 语句动态地设置由这些选项指定的过滤器，而不需要重启 MySQL 实例。无论是使用--replicate-* 选项在启动时创建还是通过 CHANGE REPLICATION FILTER 运行从库，管理此类过滤器的规则都是相同的。注意，复制过滤器不能用于组复制，因为在某些服务器上过滤事务会使组无法就一致状态达成协议。

默认时没有--replicate-* 选项，从库执行接收的所有事件，这是最简单的情况。否则，结果取决于给定的特定选项。首先检查数据库级选项（--replicate-do-db，--replicate-ignore-db），如果未使用任何数据库级选项，则继续检查可能正在使用的任何表级选项，未匹配的选项不会被执行。对于仅影响数据库的语句（即 CREATE DATABASE，DROP DATABASE 和 ALTER DATABASE），数据库级选项始终优先于任何--replicate-wild-do-table 选项。换句话说，对于此类语句，当且仅当没有适用的数据库级选项时，才会检查--replicate-wild-do-table 选项。

为了更容易地确定选项集合会产生什么影响，建议避免混合使用“do”和“ignore”选项、通配符或非通配符选项。如果指定了任何--replicate-rewrite-db 复制映射选项，则在测试--replicate-* 过滤规则之前应用它们。所有复制过滤选项都遵循相同的区分字母大小写规则，这些规则适用于 MySQL 服务器中其他位置的数据库和表的名称，包括 lower_case_table_names 系统变量的效果。

5.2.2 评估库级复制选项

在评估复制选项时，从库首先检查是否存在适用的--replicate-do-db 或--replicate-ignore-db 选项。使用--binlog-do-db 或--binlog-ignore-db 时，过程类似，只是在主库上检查选项。检查匹配的数据库取决于正在处理的事件的二进制日志格式。如果使用 ROW 格式，则要更改数据的数据库就是要检查的数据库。如果使用 STATEMENT 格式记录了语句，则默认数据库（使用 USE 语句指定）就是要检查的数据库。来看下面的实验。

（1）在从库设置过滤器为 replicate_do_db=(db2)：

```
-- 从库
mysql> stop slave sql_thread;
Query OK, 0 rows affected (0.01 sec)
```

```
mysql> change replication filter replicate_do_db=(db2);
Query OK, 0 rows affected (0.00 sec)

mysql> start slave sql_thread;
Query OK, 0 rows affected (0.01 sec)
```

（2）主库设置默认数据库为 db1，然后删除 db2.t1：

```
-- 主库
mysql> set binlog_format=statement;
Query OK, 0 rows affected (0.00 sec)

mysql> use db1;
Database changed

mysql> drop table db2.t1;
Query OK, 0 rows affected (0.01 sec)
```

（3）检查从库的复制执行情况：

```
-- 从库
mysql> desc db2.t1;
+-------+---------+------+-----+---------+-------+
| Field | Type    | Null | Key | Default | Extra |
+-------+---------+------+-----+---------+-------+
| a     | int(11) | YES  |     | NULL    |       |
+-------+---------+------+-----+---------+-------+
1 row in set (0.01 sec)
```

从库上并没有删除 db2.t1。原因是在 STATEMENT 格式，过滤器没有匹配默认的数据库 db1。

（4）改变主的默认数据库为 db2，然后创建表 db1.t1：

```
-- 主库
mysql> use db2;
Database changed

mysql> create table db1.t1(a int);
Query OK, 0 rows affected (0.03 sec)
```

（5）检查从库的复制执行情况：

```
-- 从库
mysql> desc db1.t1;
+-------+---------+------+-----+---------+-------+
| Field | Type    | Null | Key | Default | Extra |
+-------+---------+------+-----+---------+-------+
| a     | int(11) | YES  |     | NULL    |       |
+-------+---------+------+-----+---------+-------+
1 row in set (0.00 sec)
```

因为过滤器匹配了默认数据库，所以语句在从库上执行。

（6）将主库的二进制日志格式改为 ROW，再进行测试：

```
-- 主库
mysql> use db1;
Database changed

mysql> set binlog_format=row;
Query OK, 0 rows affected (0.00 sec)

mysql> create table db2.t1(a int);
Query OK, 0 rows affected (0.02 sec)
```

此时从库已经存在 db2.t1，并且 replicate_do_db=(db2)，按照文档的说法，此时会执行复制，预想的结果是因为从库上的表已经存在而报错，然而并没有报错。

```
-- 主库
mysql> drop table db2.t1;
Query OK, 0 rows affected (0.02 sec)

mysql> create table db2.t1(a varchar(5));
Query OK, 0 rows affected (0.01 sec)

mysql> insert into db2.t1 values('aaa');
Query OK, 1 row affected (0.01 sec)
```

当主库删除表 db2.t1，而从库却没删除。主库再建立新表 db2.t1，与从库已存在的 db2.t1 结构不兼容。向主库的 db2.t1 插入记录后，从库的复制就报错了：

```
    Last_SQL_Error: Column 0 of table 'db2.t1' cannot be converted from type
'varchar(20(bytes))' to type 'int(11)'
```

可以看到，当默认数据库与 replicate_do_db 不相同时，create table 和 drop table 语句不会被复制到从库，但 DML 语句会正常复制。注意，行格式只记录 DML 语句，即使 binlog_format = ROW，DDL 语句也始终记录为语句。因此，始终根据基于语句的复制规则筛选所有 DDL 语句。这意味着必须使用 USE 语句显式地选择默认数据库，以便应用 DDL 语句。数据库级过滤选项的检查流程如图 5-1 所示。

重要的是，此阶段通过的语句尚未实际执行，在检查了所有表级选项（如果有）之后，结果允许执行该语句，语句才会真正执行。二进制日志选项的检查步骤简单描述如下：

步骤01 是否有--binlog-do-db 或--binlog-ignore-db 选项?
是，继续第 2 步；否，记录语句并退出。

步骤02 是否有默认数据库（USE 选择了任何数据库）?
是，继续第 3 步；否，忽略语句并退出。

步骤03 有一个默认数据库。是否有--binlog-do-db 选项?
是，它们中的任何一个都匹配数据库吗?
是，记录该语句并退出；否，忽略语句并退出。
否，继续执行第 4 步。

步骤04 是否有任何--binlog-ignore-db 选项与数据库匹配?
是，忽略该语句并退出；否，记录语句并退出。

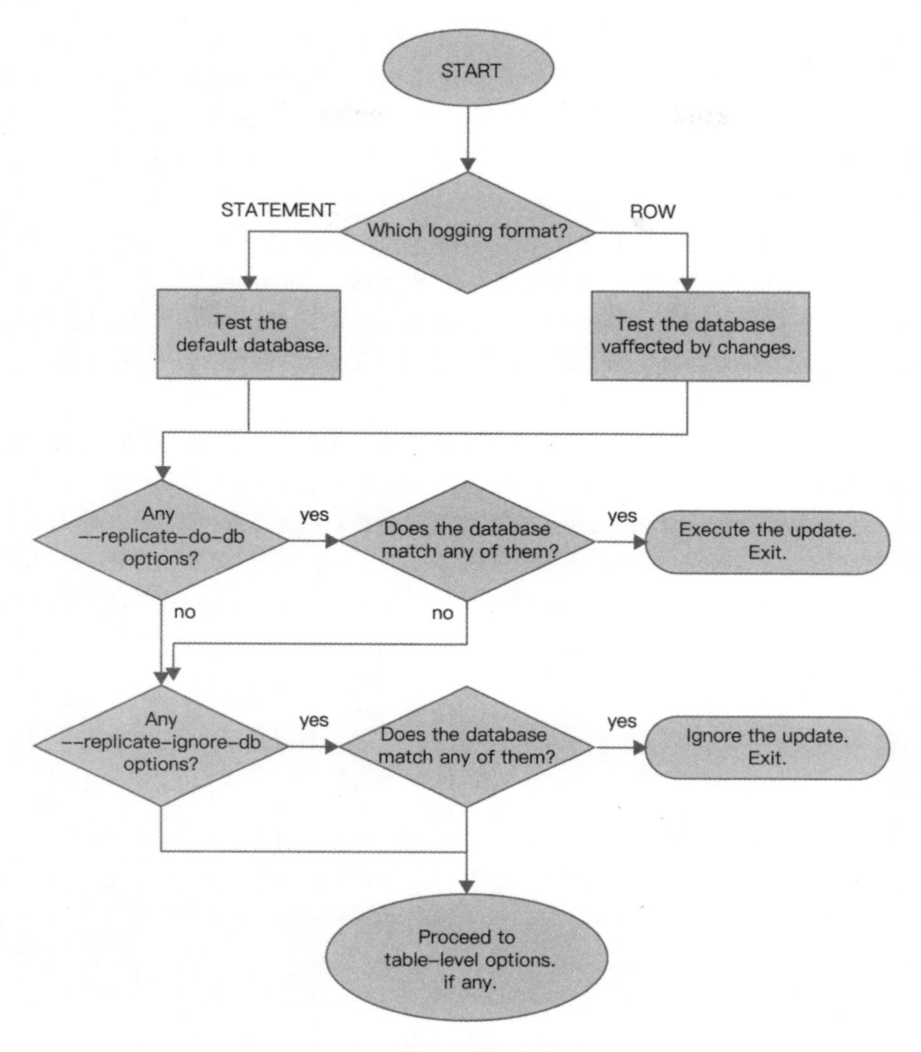

图 5-1 数据库级过滤选项的检查流程

在确定 CREATE DATABASE、ALTER DATABASE 和 DROP DATABASE 语句是记录还是忽略时，正在创建、更改或删除的数据库将替换默认数据库。--binlog-do-db 有时可能意味着"忽略其他数据库"。例如，使用基于语句的日志记录时，仅使用--binlog-do-db = sales 运行的服务器不会写入默认数据库与 sales 不同的二进制日志语句。使用具有相同选项的基于行的日志记录时，服务器仅记录那些更改 sales 库数据的更新。

5.2.3 评估表级复制选项

仅当满足以下两个条件之一时，从库才会检查并评估表选项：

- 没有数据库选项。
- 有数据库选项但与语句不匹配。

作为初始规则，如果主库启用了基于语句的复制并且语句出现在存储函数内，则从库执行语句并退出。对于基于语句的复制，复制事件表示语句，构成给定事件的所有更改都与单个 SQL 语句相关联。对于基于行的复制，每个事件表示单个表行中的更改，因此单个语句如 UPDATE mytable SET mycol = 1 可能会产生许多基于行的事件。从事件角度来看，检查表选项的过程对于基于行和基于语句的复制都是相同的。

到达表级选项检查时，如果没有表选项，从库简单地执行所有事件。如果有任何--replicate-do-table 或--replicate-wild-do-table 选项，则事件必须匹配其中一个才能执行，否则它会被忽略。如果存在任何--replicate-ignore-table 或--replicate-wild-ignore-table 选项，则执行所有事件，但匹配任何这些选项的事件除外。图 5-2 详细地描述了表级选项评估过程，它的起点是数据库级选项评估的结束点，如上一节中的图 5-1 所示。

如果单个 SQL 语句中同时含有--replicate-do-table 或--replicate-wild-do-table 选项包含的表，以及--replicate-ignore-table 或--replicate-wild-ignore-table 选项包含的另一个表，若语句是使用 binlog_format = ROW 记录的 DML 语句，那么更新的表和忽略的表都可以按预期复制，该更新的行更新，该忽略的行忽略。如果是基于语句的复制，无论是 DDL 还是 DML 语句，若匹配了--replicate-do-table 中的表，那么语句将被复制，包括--replicate-ignore-table 中的表也会更新。

下面简单验证一下表级的复制过滤规则。

（1）从库中设置表级复制过滤：

```
    mysql> stop slave sql_thread;
    Query OK, 0 rows affected (0.00 sec)

    mysql> change replication filter replicate_do_table = (db1.t1),
replicate_ignore_table = (db1.t2);
    Query OK, 0 rows affected (0.00 sec)

    mysql> start slave sql_thread;
    Query OK, 0 rows affected (0.01 sec)
```

（2）在主库上执行更新，并在从库检查复制情况：

```
    -- 主库
    mysql> create database db1;
    Query OK, 1 row affected (0.01 sec)

    mysql> create table db1.t1(a int);
    Query OK, 0 rows affected (0.02 sec)

    mysql> create table db1.t2(a int);
    Query OK, 0 rows affected (0.03 sec)

    mysql> insert into db1.t1 values (1);
    Query OK, 1 row affected (0.01 sec)

    mysql> insert into db1.t2 values (1);
    Query OK, 1 row affected (0.00 sec)
```

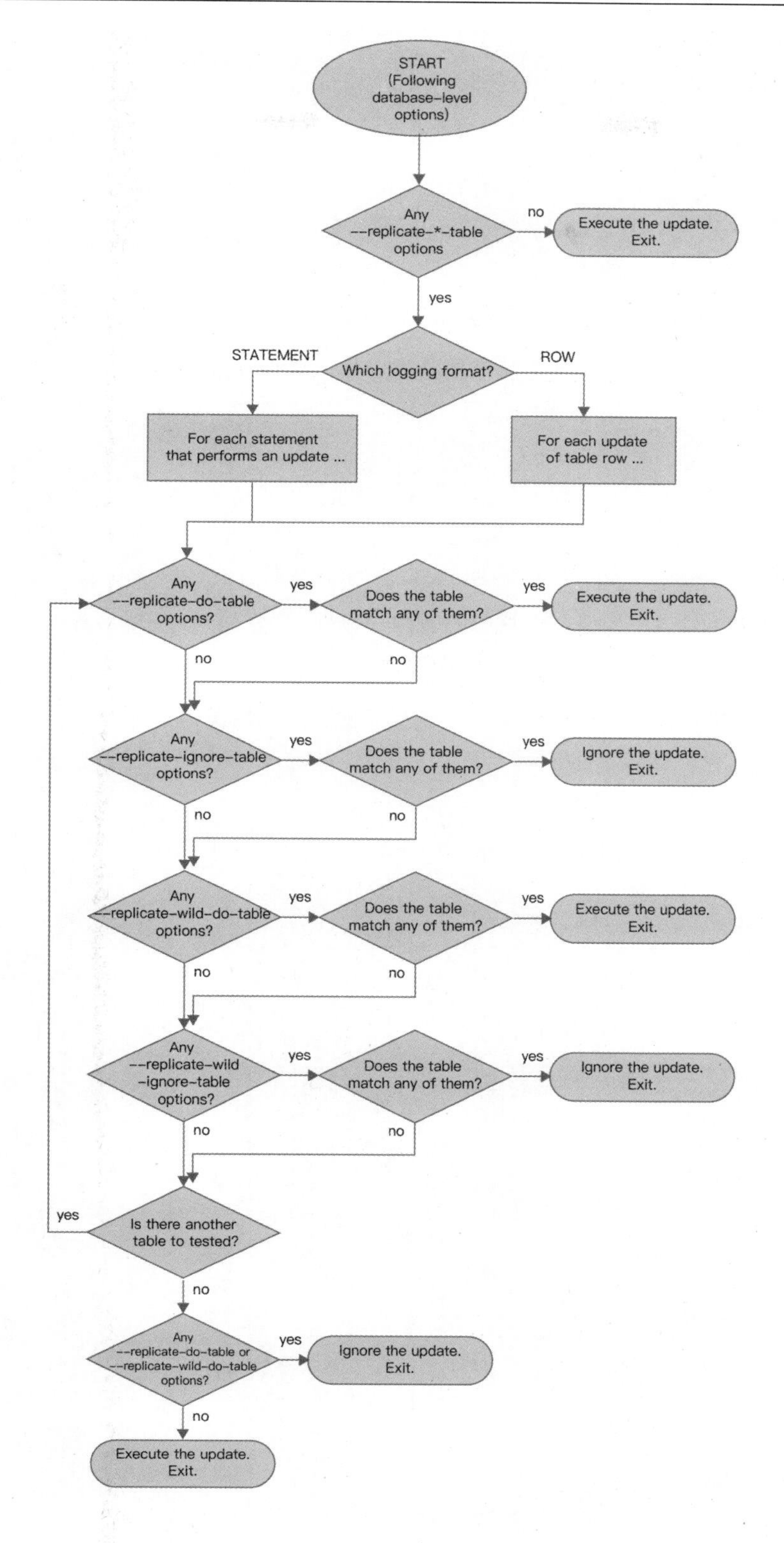

图 5-2　表级选项评估流程

```
-- 从库
mysql> select * from db1.t1;
+-------+
| a     |
+-------+
|     1 |
+-------+
1 row in set (0.00 sec)
```

可以看到，create database 以及和表匹配的 DDL、DML 语句正常复制。

（3）在从库执行同样地语句使主从库的数据一致，以便继续实验：

```
-- 从库
mysql> create table db1.t2(a int);
Query OK, 0 rows affected (0.02 sec)

mysql> insert into db1.t2 values (1);
Query OK, 1 row affected (0.01 sec)
```

在 MySQL 主从复制中，为保证主从库的数据一致性，通常将从库设置为只读（read_only=on），这里只是为了方便后续实验才在从库执行写操作。

（4）在主库上执行正常的单表更新，并在从库检查复制情况：

```
-- 主库
mysql> update db1.t1 set a=2;
Query OK, 1 row affected (0.00 sec)
Rows matched: 1  Changed: 1  Warnings: 0

mysql> update db1.t2 set a=2;
Query OK, 1 row affected (0.01 sec)
Rows matched: 1  Changed: 1  Warnings: 0

-- 从库
mysql> select * from db1.t1;
+-------+
| a     |
+-------+
|     2 |
+-------+
1 row in set (0.00 sec)

mysql> select * from db1.t2;
+-------+
| a     |
+-------+
|     1 |
+-------+
1 row in set (0.00 sec)
```

符合预期，db1.t1 正常复制，db1.t2 被忽略。

（5）在一句更新语句中同时包含 replicate_do_table 与 replicate_ignore_table 中的表：

```
-- 主库
mysql> update db1.t1 t1, db1.t2 t2 set t1.a=3, t2.a=3;
Query OK, 2 rows affected (0.00 sec)
Rows matched: 2  Changed: 2  Warnings: 0

-- 从库
mysql> select * from db1.t1;
+------+
| a    |
+------+
|    3 |
+------+
1 row in set (0.00 sec)
mysql> select * from db1.t2;
+------+
| a    |
+------+
|    1 |
+------+
1 row in set (0.00 sec)
```

在 binlog_format=row 时，两个表的复制符合预期，db1.t1 正常复制，db1.t2 被忽略。将二进制日志格式换成 statement 再试试。

```
-- 主库
mysql> set binlog_format=statement;
Query OK, 0 rows affected (0.00 sec)

mysql> update db1.t1 t1, db1.t2 t2 set t1.a=4, t2.a=4;
Query OK, 2 rows affected (0.00 sec)
Rows matched: 2  Changed: 2  Warnings: 0

-- 从库
mysql> select * from db1.t1;
+------+
| a    |
+------+
|    4 |
+------+
1 row in set (0.00 sec)

mysql> select * from db1.t2;
+------+
| a    |
+------+
|    4 |
+------+
1 row in set (0.00 sec)
```

这个语句还是复制成功了，包括 replicate_ignore_table 中的 db1.t2，也正常更新了。

（6）在主库同时删除 db1.t1 和 db1.t2 表：

```
-- 主库
mysql> drop table db1.t1,db1.t2;
Query OK, 0 rows affected (0.04 sec)

-- 从库
mysql> select * from db1.t1;
ERROR 1146 (42S02): Table 'db1.t1' doesn't exist

mysql> select * from db1.t2;
ERROR 1146 (42S02): Table 'db1.t2' doesn't exist
```

从库复制成功了，db1.t2 也被删除了。

5.2.4 复制规则应用

本小节提供一些有关复制过滤选项不同组合的说明和用法示例。表 5-1 给出了复制过滤规则类型的一些典型组合。

表 5-1 复制过滤类型

条件（选项类型）	结果
没有--replicate-*选项	从库执行从主库接收的所有事件
有--replicate-*-db 选项，但没有表选项	从库使用数据库选项接收或忽略事件。它执行这些选项允许的所有事件，因为没有表的限制
有--replicate-*-table 选项，但没有数据库选项	由于没有数据库过滤条件，因此在数据库检查阶段接收所有事件。从库仅根据表选项执行或忽略事件
数据库和表选项的组合	从库使用数据库选项接收或忽略事件。然后，它根据表选项评估这些选项允许的所有事件。这样有时会导致结果看似违反直觉，根据使用的是基于语句还是基于行的复制，结果可能会有所不同

下面是一个更复杂的示例，我们检查基于语句和基于行的设置的结果。假设主库上有两个表 db1.t1 和 db2.t2，并且从库在运行时只有以下选项：

```
replicate-ignore-db = db1
replicate-do-table  = db2.t2
```

执行下面的步骤初始化复制。

步骤 01 从库去掉以前的过滤规则：

```
mysql> stop slave sql_thread;
Query OK, 0 rows affected (0.00 sec)

mysql> change replication filter replicate_do_table = (),
replicate_ignore_table = ();
Query OK, 0 rows affected (0.00 sec)
```

```
mysql> start slave sql_thread;
Query OK, 0 rows affected (0.01 sec)
```

步骤 02 主库建立对象：

```
mysql> create database db1;
Query OK, 1 row affected (0.00 sec)

mysql> create database db2;
Query OK, 1 row affected (0.01 sec)

mysql> create table db1.t1(a int);
Query OK, 0 rows affected (0.02 sec)

mysql> create table db2.t2(a int);
Query OK, 0 rows affected (0.02 sec)
```

步骤 03 从库增加过新的滤规则：

```
mysql> stop slave sql_thread;
Query OK, 0 rows affected (0.01 sec)

mysql> change replication filter replicate_ignore_db = (db1),
replicate_do_table = (db2.t2);
Query OK, 0 rows affected (0.00 sec)

mysql> start slave sql_thread;
Query OK, 0 rows affected (0.00 sec)
```

现在我们在主库上执行以下语句：

```
mysql> set binlog_format=statement;
Query OK, 0 rows affected (0.00 sec)

mysql> use db1;
Database changed

mysql> insert into db2.t2 values (1);
Query OK, 1 row affected (0.00 sec)
```

查看从库：

```
mysql> select * from db2.t2;
Empty set (0.00 sec)
```

从库的 db2.t2 表没有数据。USE 语句使 db1 成为默认数据库，与--replicate-ignore-db 选项匹配，因此忽略 INSERT 语句，不检查表选项。

用 row 方式再执行一遍：

```
mysql> set binlog_format=row;
Query OK, 0 rows affected (0.00 sec)
```

```
mysql> use db1;
Database changed

mysql> insert into db2.t2 values (1);
Query OK, 1 row affected (0.00 sec)
```

查看从库：

```
mysql> select * from db2.t2;
+------+
| a    |
+------+
|    1 |
+------+
1 row in set (0.00 sec)
```

这回 db2.t2 复制了一条数据。使用基于行的复制时，默认数据库对从库读取数据库选项的方式没有影响。因此，USE 语句对如何处理--replicate-ignore-db 选项没有影响。此选项指定的数据库与 INSERT 语句更改数据的数据库不匹配，因此从库继续检查表选项。--replicate-do-table 指定的表与要更新的表匹配，并插入行。

5.2.5 部分复制示例

在某些情况下，可能只有一个主库服务器，并且希望将不同的数据库复制到不同的从库服务器。例如，可能希望将不同的销售数据分发到不同的部门，以帮助在数据分析期间分散负载。如图 5-3 所示，将主库的 db1 复制到从库 1，db2 复制到从库 2。

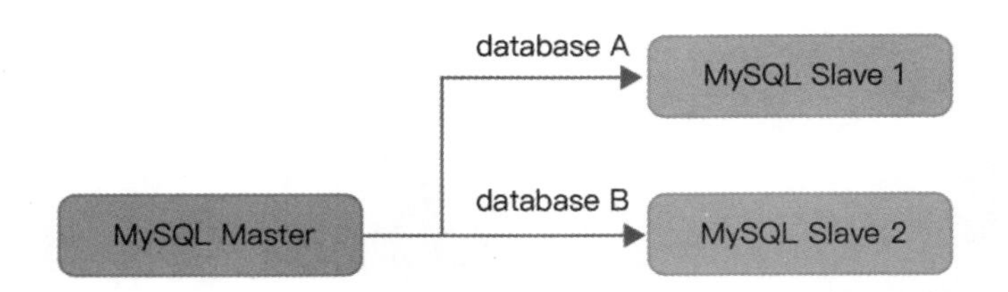

图 5-3 将主库上不同的数据库复制到不同的从库

实现时可以先配置正常的“一主两从”复制，然后通过在每个从库上使用--replicate-wild-do-table 配置选项来限制每个从库执行的事件。注意，在使用基于语句的复制时，不应将--replicate-do-db 用于此目的，基于语句的复制会导致此选项的影响会因当前所选的数据库而有所差异。这也适用于混合格式复制，因为此时可能使用基于语句的格式复制某些更新。

在从库 1 执行：

```
stop slave sql_thread;
change replication filter replicate_wild_do_table=('db1.%');
start slave sql_thread;
```

在从库 2 执行：

```
stop slave sql_thread;
```

```
change replication filter replicate_wild_do_table=('db2.%');
start slave sql_thread;
```

此配置中的每个从库从主库接收整个二进制日志，但仅执行二进制日志中--replicate-wild-do-table选项所包含的数据库和表的那些事件。

测试：

```
-- 主库
create database db1;
create database db2;
create table db1.t1(a int);
create table db2.t2(a int);
insert into db1.t1 select 1;
insert into db2.t2 select 2;

-- 从库1
mysql> select * from db1.t1;
+-------+
| a     |
+-------+
|     1 |
+-------+
1 row in set (0.00 sec)

mysql> select * from db2.t2;
ERROR 1049 (42000): Unknown database 'db2'

-- 从库2
mysql> select * from db1.t1;
ERROR 1049 (42000): Unknown database 'db1'

mysql> select * from db2.t2;
+-------+
| a     |
+-------+
|     2 |
+-------+
1 row in set (0.00 sec)
```

数据如预期复制，db1和db2的数据分别复制到从库1和从库2。下面看一下routine的复制情况。

```
-- 主库
delimiter //
create procedure db1.p1 ()
begin
select 1;
end;
//
delimiter ;

-- 从库1
```

```
mysql> call db1.p1();
+---+
| 1 |
+---+
| 1 |
+---+
1 row in set (0.00 sec)

Query OK, 0 rows affected (0.00 sec)

-- 从库 2
mysql> call db1.p1();
ERROR 1305 (42000): PROCEDURE db1.p1 does not exist

mysql> show slave status\G
*************************** 1. row ***************************
...
           Replicate_Do_DB:
         Replicate_Ignore_DB:
          Replicate_Do_Table:
       Replicate_Ignore_Table:
      Replicate_Wild_Do_Table: db2.%
  Replicate_Wild_Ignore_Table:
                 Last_Errno: 1049
                 Last_Error: Error 'Unknown database 'db1'' on query. Default
database: ''. Query: 'CREATE DEFINER=`wxy`@`%` PROCEDURE `db1`.`p1`()
begin
select 1;
end'
...
Last_SQL_Errno: 1049
              Last_SQL_Error: Error 'Unknown database 'db1'' on query. Default
database: ''. Query: 'CREATE DEFINER=`wxy`@`%` PROCEDURE `db1`.`p1`()
begin
select 1;
end'
...
1 row in set (0.00 sec)
```

在主库上的 db1 建立存储过程，从库 1 正常复制，但从库 2 却报错了，它还是执行了复制，只是因为缺少 db1 数据库而复制报错。可见，replicate_wild_do_table 只对表起作用，而对于 routine 无效，主库上所有库的 routine 都会在所有从库执行复制。

如果在复制开始之前有必须同步到从库的数据，则可以将所有数据同步到每个从库，然后在从库删除不想保留的数据库或表。或者使用 mysqldump 为每个数据库创建单独的转储文件，并在每个从库上加载相应的转储文件，例如：

```
# 从库 1
mysqldump --single-transaction --databases db1 --master-data=1
--host=172.16.1.125 --user=wxy --password=123456 --apply-slave-statements | mysql
```

```
-uroot -p123456

    # 从库 2
    mysqldump --single-transaction --databases db2 --master-data=1
--host=172.16.1.125 --user=wxy --password=123456 --apply-slave-statements | mysql
-uroot -p123456
```

这两种方式都可以联机进行，并且对主库的影响不大。实例复制与 mysqldump 的联机复制具体步骤参见“第 1 章 异步复制”。

5.3 主从切换

有时需要把从库指向一个新的主库。例如滚动升级服务器，或者主库出现问题时需要把一台从库转换成主库。可以使用 CHANGE MASTER TO 语句告诉从库连接新的主库。从库不检查主库上的数据库是否与从库上现有的数据库兼容，它只是从新主库二进制日志中的指定位置开始读取和执行事件。这种主从角色转换可以粗略地分为计划内和计划外两种。下面分别说明如何在两种应用场景下进行主从切换，假设主库定位使用默认的二进制位置方式。

5.3.1 计划内主从切换

计划内主从切换简单说就是事先有准备的维护性操作，通常需要执行以下步骤：

步骤 01 停止当前主库的所有写操作。如果可以，最好能关闭所有的客户端连接。

步骤 02 通过 flush tables with read lock 在主库上停止所有活跃的写入，这一步是可选的。也可以在主库上设置 read_only 选项。从这一刻开始，应该禁止向即将被替换的主库做任何写入。因为一旦它不是主库，写入就意味着数据不一致。注意，即使设置 read_only 也不会阻止当前已经存在的事务继续提交。为了更好地保证这一点，可以“kill”所有打开的事务，这将会真正地结束所有写入。例如可以编写以下内容的 kill_mysql_session.sh 脚本：

```
    #!/bin/bash
    source ~/.bashrc

    rm -rf /tmp/kill.sql
    mysql -u root -p123456 -P3306 -h127.0.0.1 -e "select * into outfile '/tmp/
kill.sql' from (select 'set global read_only=on;' union all select concat('kill
',id,';') from information_schema.processlist where command='sleep' ) t; "

    mysql -u root -p123456 -P3306 -h127.0.0.1 < /tmp/kill.sql
```

之后就可以执行 kill_mysql_session.sh 杀掉会话。较新版本的 MySQL，需要在配置文件中设置 secure_file_priv 参数，并重启 MySQL 后才能执行数据导出操作。

步骤 03 选择一个从库作为新的主库，并确保它已经完全跟上主库，即执行完所有中继日志。

步骤04 确保新主库和旧主库的数据一致，可选。例如在两个库上执行“mysqldump -uroot --skip-dump-date | md5sum”，检查校验和是否相同。

步骤05 在新主库上执行 stop slave。

步骤06 在新主库上执行 reset slave all，使其断开与旧主库的连接。

步骤07 执行 show master status 记录新主库的二进制日志位置。

步骤08 确保其他从库已经追上旧主库。

步骤09 关闭旧主库。

步骤10 如果需要，在新主库上执行 set global read_only=off，使其可写。

步骤11 在每台从库上执行 change master to 语句，使用前面步骤 07 中获得的二进制位置来指向新主库。

步骤12 将客户端连接到新主库。

5.3.2 计划外切换

当主库崩溃时，需要提升一个从库来替代它。如果只有一个从库，那别无选择，只能使用这台从库。如果有超过一个的从库，就需要做一些额外工作。对主从拓扑结构中的从库进行提升的过程可以简单描述如下：

- 让所有从库执行完其从崩溃前的旧主库获得的中继日志。
- 选择并设置新主库。
- 查找其他所有从库最后执行的事件，在新主库上对应的二进制位置。
- 其他所有从库重置复制，按上一步获得的二进制位置连接到新主库，启动新复制。

这个过程中隐藏着很多细节，因此用一个具体的例子进行详细说明。假设一个标准的 MySQL 8“一主两从”复制拓扑结构，主库标记为 M（172.16.1.125），两个从库分别标记为 S1（172.16.1.126）和 S2（172.16.1.127）。全部使用 MySQL 8 默认的复制相关配置：

```
log_bin=ON
binlog_format=ROW
log_slave_updates=ON
gtid_mode=OFF
# 两个从库
read_only=ON
```

我们在这个复制场景下，模拟主库服务器不可用时，如何提升一个从库成为新主库，并将其他从库指向新主库，目标是保证最少的事务丢失。实验步骤如下：

步骤01 在 M 上更新一些数据：

```
create database db1;
use db1;
create table t1(a int);
insert into t1 values (1),(2),(3);
update t1 set a=3 where a=1;
delete from t1 where a=3;
```

```
insert into t1 select 1; -- 用于查看同样的 SQL 语句，它们的事件是否一样
insert into t1 select 1;
insert into t1 select 1;
commit;
```

步骤 02 停止 S2 复制，模拟 S2 是落后的从库：

```
stop slave;
```

步骤 03 刷新 S1 的日志：

```
flush logs;
```

由于服务器重启、不同的配置、日志轮转或者 flush logs 命令等原因，会造成同一事件在不同的服务器上有不同的二进制日志位置，这一步模拟此种情况。

步骤 04 在 M 再更新一些数据：

```
insert into t1 values (10),(11),(12);
delete from t1 where a=1;
commit;
```

此时 S2 的复制已经停止，上面的数据更新对它不可知。S1 正常复制，以此人为模拟两个从库的快慢之分。

步骤 05 停止 M，模拟主库损坏：

```
mysqladmin -uwxy -p shutdown
```

步骤 06 启动 S2 的复制：

```
start slave;
```

步骤 07 查看当前两个从库的线程状态、复制状态和当前数据：

```
show processlist;
show slave status\G
select * from db1.t1;
```

S1 上的显示如下：

```
mysql> show processlist;
+------+------------+-------+------+--------+-------+----------+--------+
| Id   | User          | Host    | db   | Command| Time  | State       | Info      |
+------+----------+-------+------+--------+-------+--------+----------+
|    4 | event_scheduler | localhost | NULL | Daemon | 15228 | Waiting on ...|
NULL       |
| 7614 | wxy          | localhost | NULL | Query  |     0 | starting      | show
pro ...|
| 7619 | system user   |         | NULL | Connect| 9519 | Reconnecting..| NULL |
| 7620 | system user    |          | NULL | Query  |  59 | Slave has ... | NULL |
+------+-------------+--------+------+--------+-----+----------+--------+
4 rows in set (0.00 sec)
```

```
    mysql> show slave status\G
    *************************** 1. row ***************************
               Slave_IO_State: Reconnecting after a failed master event read
                  Master_Host: 172.16.1.125
                  Master_User: repl
                  Master_Port: 3306
                Connect_Retry: 60
              Master_Log_File: binlog.000015
          Read_Master_Log_Pos: 2918
               Relay_Log_File: hdp3-relay-bin.000003
                Relay_Log_Pos: 837
        Relay_Master_Log_File: binlog.000015
             Slave_IO_Running: Connecting
            Slave_SQL_Running: Yes
    ...
        Seconds_Behind_Master: NULL
    Master_SSL_Verify_Server_Cert: No
                Last_IO_Errno: 2003
                Last_IO_Error: error reconnecting to master
'repl@172.16.1.125:3306' - retry-time: 60  retries: 1
               Last_SQL_Errno: 0
               Last_SQL_Error:
    Replicate_Ignore_Server_Ids:
             Master_Server_Id: 1125
                  Master_UUID: 8eed0f5b-6f9b-11e9-94a9-005056a57a4e
             Master_Info_File: mysql.slave_master_info
                    SQL_Delay: 0
          SQL_Remaining_Delay: NULL
      Slave_SQL_Running_State: Slave has read all relay log; waiting for more
updates
    ...
    1 row in set (0.00 sec)

    mysql> select * from db1.t1;
    +------+
    | a    |
    +------+
    |    2 |
    |   10 |
    |   11 |
    |   12 |
    +------+
    4 rows in set (0.00 sec)
```

S2 上的显示如下：

```
    mysql> show processlist;
    +------+-----------------+-----------+------+---------+-------+-----------
---+--------------+| Id   | User            | Host      | db   | Command| Time  |
State        |
Info         |+------+-----------------+-----------+------+---------+-------+-----
```

```
----------+-------------+|   6 | event_scheduler | localhost | NULL | Daemon | 85864
| Waiting on ...| NULL       |
    | 48967| wxy          | localhost | mysql| Query  |    0 | starting     | show
pro ...|
    | 57199| system user    |          | NULL | Connect|   72 | Connecting ...|
NULL      |
    | 57200| system user   |         | NULL | Query  |   72 | Slave has ... | NULL |
    +------+-----------------+-----------+------+--------+-------+-----------
---+-------------+4 rows in set (0.00 sec)

    mysql> show slave status\G
    *************************** 1. row ***************************
               Slave_IO_State: Connecting to master
                  Master_Host: 172.16.1.125
                  Master_User: repl
                  Master_Port: 3306
                Connect_Retry: 60
              Master_Log_File: binlog.000015
          Read_Master_Log_Pos: 2356
               Relay_Log_File: hdp4-relay-bin.000002
                Relay_Log_Pos: 2520
        Relay_Master_Log_File: binlog.000015
             Slave_IO_Running: Connecting
            Slave_SQL_Running: Yes
    ...
        Seconds_Behind_Master: NULL
    Master_SSL_Verify_Server_Cert: No
                Last_IO_Errno: 2003
                Last_IO_Error: error connecting to master 'repl@172.16.1.125:3306'
- retry-time: 60  retries: 2
               Last_SQL_Errno: 0
               Last_SQL_Error:
     Replicate_Ignore_Server_Ids:
             Master_Server_Id: 1125
                  Master_UUID: 8eed0f5b-6f9b-11e9-94a9-005056a57a4e
             Master_Info_File: mysql.slave_master_info
                    SQL_Delay: 0
          SQL_Remaining_Delay: NULL
      Slave_SQL_Running_State: Slave has read all relay log; waiting for more
updates
    ...
    1 row in set (0.00 sec)

    mysql> select * from db1.t1;
    +------+
    | a    |
    +------+
    |    2 |
    |    1 |
    |    1 |
```

```
|    1 |
+------+
4 rows in set (0.00 sec)
```

S1 和 S2 的 SQL 线程状态均为“Slave has read all relay log; waiting for more updates”，这说明两个从库都已经完成了所有中继日志的重放。S1 的 Relay_Master_Log_File 和 Exec_Master_Log_Pos 分别是 binlog.000015 和 2918，S2 的 Relay_Master_Log_File 和 Exec_Master_Log_Pos 分别是 binlog.000015 和 2356，这说明 S1 更接近于原来的主库 M，应当将 S1 提升为新主库。从表 db1.t1 的数据也可以明显看到 S1 的数据更新。

步骤 08 提升 S1 为新主库：

```
stop slave;
reset slave all;
set global read_only=off;
```

同时去掉配置文件中的 read_only，以免重启库时忘记更改配置：

```
sed -i 's/^read_only/#&/' /etc/my.cnf
```

步骤 09 找到 S2 上最后的重放事件，在 S1 上对应的二进制位置。

- 首先在 S2 上查看最后的二进制位置：

```
mysql> show master status;
+---------------+----------+--------------+------------------+------------
-------+
| File          | Position | Binlog_Do_DB | Binlog_Ignore_DB | Executed_Gtid_Set |
+---------------+----------+--------------+------------------+------------
-------+
| binlog.000006 |     9620 |              |                  |                   |
+---------------+----------+--------------+------------------+------------
-------+
1 row in set (0.00 sec)
```

- 然后在当前 binlog 文件 binlog.000006 里查找最后的重放事件，在 S2 上执行：

```
mysqlbinlog --base64-output=decode-rows --verbose /usr/local/mysql/data
/binlog.000006
```

结果显示如下：

```
...
# at 9347
#190528 14:14:24 server id 1125 end_log_pos 9433 CRC32 0xd6fcb00c
Anonymous_GTID last_committed=36 sequence_number=37 rbr_only=yes original_comm
itted_timestamp=1559024064330837 immediate_commit_timestamp=1559024064349703 t
ransaction_length=273
/*!50718 SET TRANSACTION ISOLATION LEVEL READ COMMITTED*//*!*/;
# original_commit_timestamp=1559024064330837 (2019-05-28 14:14:24.330837 CST)
# immediate_commit_timestamp=1559024064349703 (2019-05-28 14:14:24.349703
CST)
```

```
/*!80001 SET @@session.original_commit_timestamp=1559024064330837*//*!*/;
/*!80014 SET @@session.original_server_version=80016*//*!*/;
/*!80014 SET @@session.immediate_server_version=80016*//*!*/;
SET @@SESSION.GTID_NEXT= 'ANONYMOUS'/*!*/;
# at 9433
#190528 14:14:24 server id 1125  end_log_pos 9502 CRC32
0xf10333d5    Query    thread_id=8    exec_time=0    error_code=0
SET TIMESTAMP=1559024064/*!*/;
BEGIN
/*!*/;
# at 9502
#190528 14:14:24 server id 1125  end_log_pos 9549 CRC32
0x0fb24522    Table_map: `db1`.`t1` mapped to number 104
# at 9549
#190528 14:14:24 server id 1125  end_log_pos 9589 CRC32
0x6af67d3e    Write_rows: table id 104 flags: STMT_END_F
### INSERT INTO `db1`.`t1`
### SET
###   @1=1
# at 9589
#190528 14:14:24 server id 1125  end_log_pos 9620 CRC32 0x763d087a    Xid =
211
COMMIT/*!*/;
SET @@SESSION.GTID_NEXT= 'AUTOMATIC' /* added by mysqlbinlog */ /*!*/;
DELIMITER ;
# End of log file
/*!50003 SET COMPLETION_TYPE=@OLD_COMPLETION_TYPE*/;
/*!50530 SET @@SESSION.PSEUDO_SLAVE_MODE=0*/;
[mysql@hdp4~]$
```

最后一个事件在原主库上的提交时间戳是 original_committed_timestamp=1559024064330837，正如前面讨论延迟复制时所述，这个时间戳是主库的原始提交时间，单位精确到微秒，在所有从库上都相同。因此可以通过它将不同从库上相同的事件联系起来。

- 最后查找 S1 上的 binlog 中“original_committed_timestamp=1559024064330837”对应的文件名和偏移量，在 S1 上执行:

```
while read LINE
do
    filename=/usr/local/mysql/data/${LINE:2}
    echo $filename
    mysqlbinlog --base64-output=decode-rows --verbose $filename | grep -A30
-n "original_committed_timestamp=1559024064330837"
done  < /usr/local/mysql/data/binlog.index
```

结果显示如下：

```
[mysql@hdp3~]$while read LINE
> do
> filename=/usr/local/mysql/data/${LINE:2}
> echo $filename
```

```
    > mysqlbinlog --base64-output=decode-rows --verbose $filename | grep -A30 -n
"original_committed_timestamp=1559024064330837"
    > done  < /usr/local/mysql/data/binlog.index
    /usr/local/mysql/data/binlog.000011
    ...
    /usr/local/mysql/data/binlog.000022
    /usr/local/mysql/data/binlog.000023
    194:#190528 14:14:24 server id 1125  end_log_pos 2193 CRC32 0x75e3db63
Anonymous_GTID
last_committed=8 sequence_number=9 rbr_only=yes original_committed_timestamp=1
559024064330837
immediate_commit_timestamp=1559024064361609 transaction_length=273
    195-/*!50718 SET TRANSACTION ISOLATION LEVEL READ COMMITTED*//*!*/;
    196-# original_commit_timestamp=1559024064330837 (2019-05-28 14:14:24.330837
CST)
    197-# immediate_commit_timestamp=1559024064361609 (2019-05-28
14:14:24.361609 CST)
    198-/*!80001 SET
@@session.original_commit_timestamp=1559024064330837*//*!*/;
    199-/*!80014 SET @@session.original_server_version=80016*//*!*/;
    200-/*!80014 SET @@session.immediate_server_version=80016*//*!*/;
    201-SET @@SESSION.GTID_NEXT= 'ANONYMOUS'/*!*/;
    202-# at 2193
    203-#190528 14:14:24 server id 1125  end_log_pos 2262 CRC32 0xb047eaf4  Query
thread_id=8 exec_time=0 error_code=0
    204-SET TIMESTAMP=1559024064/*!*/;
    205-BEGIN
    206-/*!*/;
    207-# at 2262
    208-#190528 14:14:24 server id 1125  end_log_pos 2309 CRC32 0xd9f43e3d
Table_map: 'db1'. 't1' mapped to number 91
    209-# at 2309
    210-#190528 14:14:24 server id 1125  end_log_pos 2349 CRC32 0x69e276a4
Write_rows: table id 91 flags: STMT_END_F
    211-### INSERT INTO 'db1'. 't1'
    212-### SET
    213-###   @1=1
    214-# at 2349
    215-#190528 14:14:24 server id 1125  end_log_pos 2380 CRC32 0xee93c6db  Xid =
44
    216-COMMIT/*!*/;
    217-# at 2380
    218-#190528 14:14:48 server id 1126  end_log_pos 2424 CRC32 0xe9f46f61  Rotate
to binlog.000024  pos: 4
    219-SET @@SESSION.GTID_NEXT= 'AUTOMATIC' /* added by mysqlbinlog */ /*!*/;
    220-DELIMITER ;
    221-# End of log file
    222-/*!50003 SET COMPLETION_TYPE=@OLD_COMPLETION_TYPE*/;
    223-/*!50530 SET @@SESSION.PSEUDO_SLAVE_MODE=0*/;
    /usr/local/mysql/data/binlog.000024
```

```
[mysql@hdp3~]$
```

可以看到，S2 上最后一个事务在 S1 上对应的位置为 binlog.000023、2380，至此已经确定了 S2 连接新主库（change master to）的位置。

步骤 10 将 S2 连接到新主库：

```
stop slave;
reset slave all;
change master to
    master_host='172.16.1.126',
    master_port=3306,
    master_user='repl',
    master_password='123456',
    master_log_file='binlog.000023',
    master_log_pos=2380;
```

步骤 11 在 S2 启动复制，查看复制状态，验证数据：

```
start slave;
show slave status\G
select * from db1.t1;
```

复制状态显示如下，可以看到 S2 从新主库的复制一切正常：

```
mysql> show slave status\G
*************************** 1. row ***************************
               Slave_IO_State: Waiting for master to send event
                  Master_Host: 172.16.1.126
                  Master_User: repl
                  Master_Port: 3306
                Connect_Retry: 60
              Master_Log_File: binlog.000024
          Read_Master_Log_Pos: 721
               Relay_Log_File: hdp4-relay-bin.000004
                Relay_Log_Pos: 929
        Relay_Master_Log_File: binlog.000024
             Slave_IO_Running: Yes
            Slave_SQL_Running: Yes
...
        Seconds_Behind_Master: 0
Master_SSL_Verify_Server_Cert: No
                Last_IO_Errno: 0
                Last_IO_Error:
               Last_SQL_Errno: 0
               Last_SQL_Error:
...
1 row in set (0.00 sec)
```

查询 db1.t1 表的数据显示如下，事务已经执行到 M 崩溃的时间点：

```
mysql> select * from db1.t1;
```

```
+------+
| a    |
+------+
|    2 |
|   10 |
|   11 |
|   12 |
+------+
4 rows in set (0.00 sec)
```

至此，新主库提升和其他从库的重新指向已经完成，复制拓扑结构中所有库的数据变化都已经追赶到最近，新的主从库复制正常进行，客户端应用可以连接到新主库。

5.4 小 结

延迟复制最大的用处是修正用户误操作引发的问题，只要在延迟时间段内被及时发现，就可以用从库的数据加以恢复，起到一个备份的作用。部分复制也是一个常用的技术手段。例如在我们的生产环境中，为简化应用程序开发，逻辑上需要在一条 SQL 查询语句中关联不同 MySQL 实例中的表，就是使用部分复制加以实现的。本章最后还用一个示例，演示了在“一主多从”复制拓扑结构的情况下，主库意外宕机后手工执行主从切换的详细步骤。下一章将介绍 MySQL 原生提供的多主复制架构——组复制。

第6章

组复制

6.1 组复制基本原理

MySQL Group Replication（MGR）是 MySQL 5.7.17 版本引入的一个服务器插件，可用于创建高可用、可扩展、容错的复制拓扑结构。组复制可以在单主模式下操作，其中只有一个服务器接受更新，这个单主是系统自动选举出来的。对于高级用户，也可以部署为多主模式，其中所有服务器都可以接受更新。内置的组成员服务可以在任何给定的时间点保持组的视图一致，并可供所有服务器使用。服务器加入或离开组时，视图也会相应更新。当服务器宕机，故障检测机制会检测到此情况并通知组——它的视图已更改。这些都是自动进行的。组复制对属于同一组的服务器自动进行协调。对于要提交的事务，组成员必须就全局事务序列中给定事务的顺序达成一致。提交或回滚事务由每个服务器单独完成，但所有服务器都必须做出相同的决定。

6.1.1 MySQL 复制技术

在深入了解 MySQL 组复制的细节之前，先介绍一些它产生的背景以及工作原理，以帮助理解组复制，以及传统异步复制、半同步复制和组复制之间的区别。

1. 主从复制

传统的 MySQL 复制提供了一种简单的主从复制方法。有一个主库，一个或多个从库。主库执行并提交事务，然后通过二进制日志将事务相关的事件异步发送到从库，以便重放。这是一个无共享系统，默认情况下所有服务器都拥有完整的数据副本。

半同步复制为异步复制协议添加了一个同步步骤。这意味着主库在提交时等待至少一个从库确认它已收到该事务，才会继续提交操作。

图 6-1 和图 6-2 分别表示 MySQL 异步复制协议以及它的半同步变体。箭头表示服务器之间或服务器与客户端之间发送消息的方向。

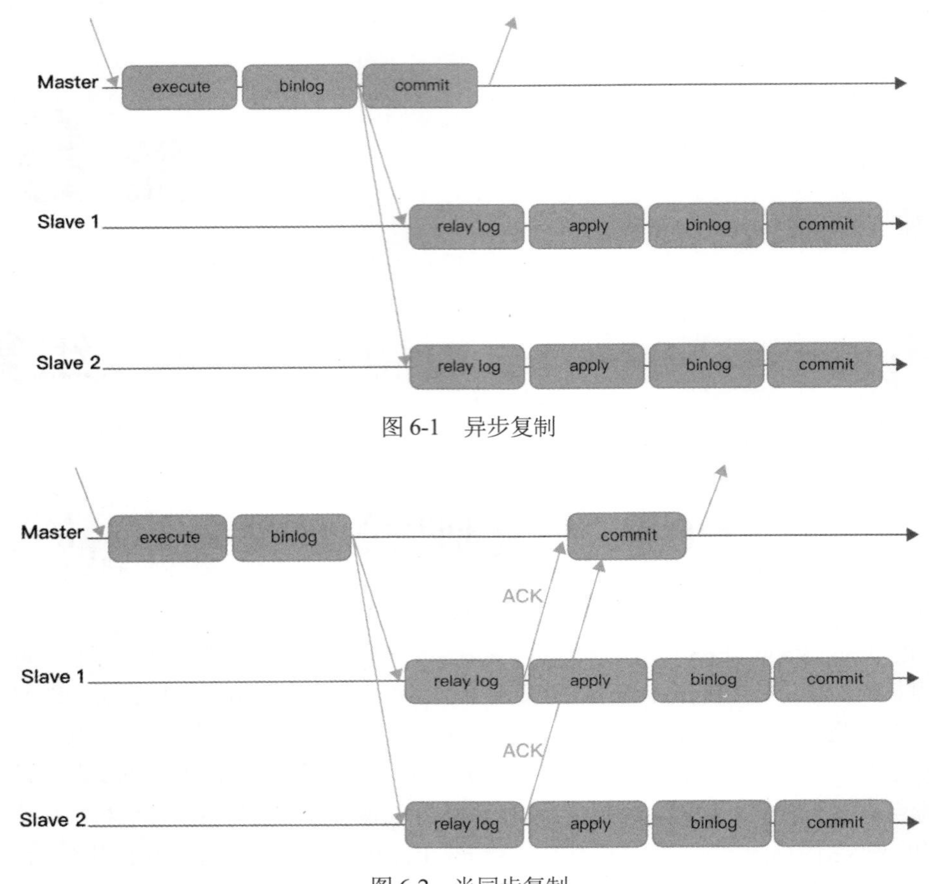

图 6-1　异步复制

图 6-2　半同步复制

2. 组复制

组复制由多个服务器构成，它们通过传递消息进行交互。组中每个服务器独立地执行事务，但是所有读写事务只有在得到组批准后才会提交。只读事务在组内不需要协调，因此立即提交。对于任何读写事务，当事务准备好在始发服务器处提交时，服务器以原子方式广播写入值（更改的行）和对应的写入集（更新行的唯一标识符），然后将该事务加入全局事务列表。最终所有服务器都以相同的顺序接收并应用相同的事务集，所以它们在组内保持一致。

不同服务器上并发执行的事务之间可能存在冲突。组复制在认证（certify）过程中检查并发事务的写集来检测这种冲突。如果在不同服务器上执行的两个并发事务更新同一行，则存在冲突。解决方案是先到事务提交，后到事务回滚，即按顺序第一个事务在所有服务器提交，而第二个事务在原始服务器上回滚并在组中的其他服务器中删除。这实际上体现的是多主分布式事务的首个提交获胜原则。组复制同样是一种无共享复制方案，其中每个服务器都拥有整个数据副本。图 6-3 描述了 MGR 协议。

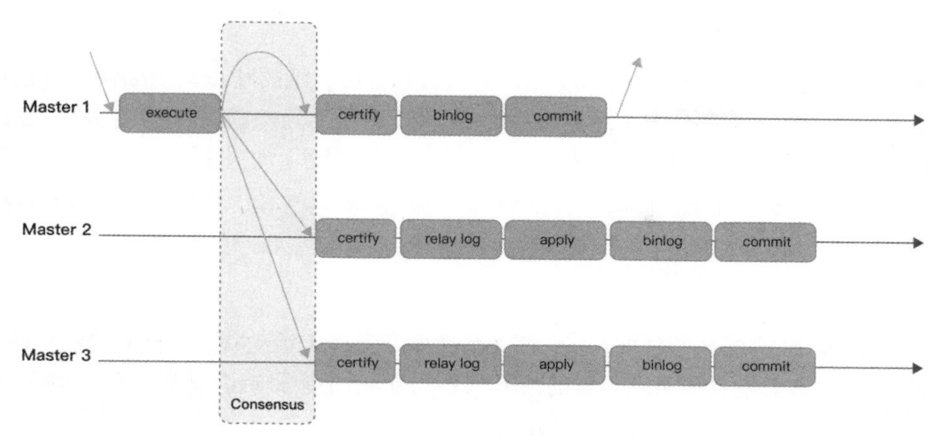

图 6-3　MGR 协议

6.1.2　组复制使用场景

组复制可用来创建具有冗余的容错系统。某些服务器发生故障时，根据失败服务器的数量，可能会降低性能或可伸缩性，但只要不是全部或大多数的服务器都发生故障，就能保证数据库服务持续可用。必须将连接到崩溃服务器的客户端重定向或转移到其他服务器。组复制本身并不会解决数据库连接重定向问题，更适合处理此问题的是如 MySQL Router 式的中间件。以下是组复制的典型使用场景：

- 弹性伸缩：服务器的数量能够动态增加或减少，并且尽可能减小副作用，例如云数据库服务。
- 高可用分片：分片是实现写扩展的流行方法。可使用组复制来实现高可用分片，其中每个分片映射到复制组。
- 主从复制的替代方案：在某些情况下，使用单个主服务器会使其成为热点，写入整个组会更具可扩展性。

6.1.3　组复制相关服务

1. 故障检测

组复制包括一种故障检测机制，能够查找并报告哪些服务器已经宕机。当服务器 A 在给定时间内没有收到服务器 B 的消息时，因发生超时并引发怀疑，如果组认为这种怀疑是真实情况，那么组会认定服务器 B 确实宕机了。这意味着组中的其余成员将采取协调策略来排除宕机的成员。

如果一个服务器与组的其余部分隔离，它会怀疑所有其他服务器都已失败，但由于无法与该组达成协议（因为无法获得多数票），因此其怀疑并没有结果。当服务器以这种方式与组隔离时，它无法执行任何本地事务。

2. 组成员服务

组成员服务定义了组中哪些服务器在线，在线服务器列表通常称为视图。组中的每台服务器都具有一致的视图，其中的服务器是在给定时刻参与该组的成员。当有服务器加入或离开组时，会触发视图的更改。

组成员离开组可分主动离开与被动离开。主动离开会启动组的动态重新配置，会触发所有其他成员必须在没有该服务器的情况下就新视图达成一致。被动离开，例如服务器意外停止或断网时，故障检测机制会建议重新配置组，这需要组中大多数成员的同意。如果无法达成协议，为阻止出现脑裂（Split-brain，即大脑分裂）问题，导致系统无法动态更改配置，这就意味着需要管理员介入解决此问题。

3. 容错

组复制是以 Paxos 分布式算法实现的，需要多数服务器处于活动状态才能获得足够的票数，从而做出决定。这会直接影响系统可以容忍的故障机之数量，但不会影响组复制自身及其整体的功能。容忍 f 台故障机所需的服务器数量 n 至少为：n = 2 * f + 1。

在实际应用中为了容忍一台故障机，组必须具有至少三台服务器，因为此时如果一台服务器发生了故障，仍然有两台服务器构成多数，允许系统自动做出决策并继续提供服务。但是，如果第二台服务器又失败了，那么该组就只剩下一台服务器，此时整个系统就会陷入阻塞，因为没有多数票可以做出决定。

6.1.4 组复制插件体系结构

组复制是一个 MySQL 插件，它以现有的 MySQL 复制架构为基础，利用二进制日志、基于行的日志记录和全局事务标识符（GTID）等功能。图 6-4 显示了组复制插件架构。

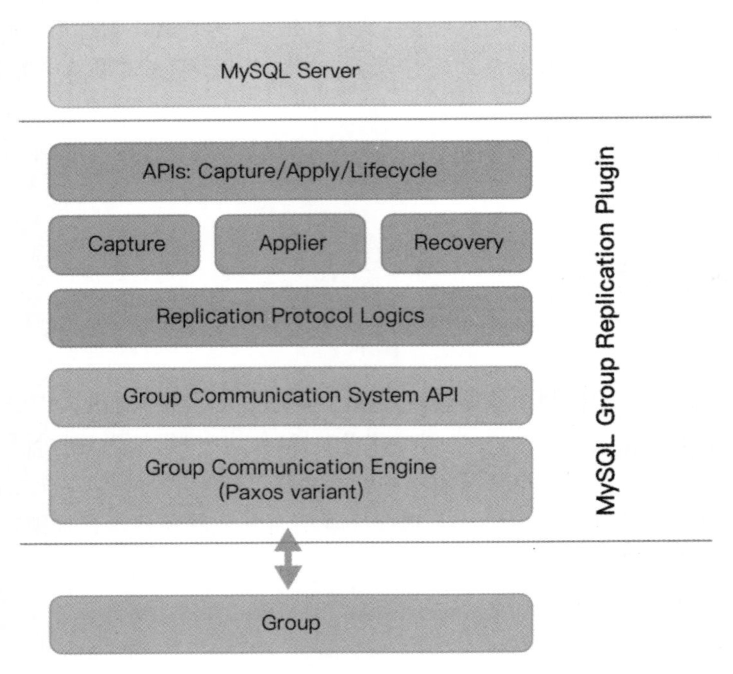

图 6-4 组复制插件架构

组复制插件包含一组捕获、应用和生命周期 API，用于控制插件与 MySQL 服务器的交互方式。

这些接口将 MySQL 服务器核心与组复制插件隔离。服务器向插件通知启动、恢复、准备接收连接、即将提交事务等消息。插件指示服务器执行诸如提交事务、中止正在进行的事务、事务在中继日志中排队等动作。

组复制插件体系结构的下一层是一组组件。捕获组件负责跟踪与正在执行的事务相关的上下文。应用组件负责在数据库上执行远程事务。恢复组件管理分布式恢复，负责选择捐赠者，对故障做出反应，执行追赶程序，使加入该组的服务器获得更新。

堆栈下一层的复制协议模块包含复制协议的特定逻辑。它处理冲突检测，接收事务并将其传播到组。

组复制插件架构的最后两层是组通信系统（Group Communication System，GCS）API 和基于 Paxos 的组通信引擎（XCom）的实现。GCS API 将消息传递层的实现与插件上层分离，组通信引擎处理与复制组成员的通信。

6.1.5　组复制分布式恢复

1. 分布式恢复基础

新成员加入组时会自动执行分布式恢复，它可以概括为服务器从组中获得丢失事务的过程。在分布式恢复期间，加入组的服务器会缓冲组中所需的事务和成员事件。一旦新成员收到了组的所有事务，它就会应用在恢复过程中缓冲的事务。此过程结束时，服务器随之作为在线成员加入组中。

分布式恢复分为两个阶段。第一阶段，选择组中一个在线服务器作为缺失数据的捐赠者。捐赠者负责为新服务器提供加入该组的所有数据，直到它加入组为止。这是通过在捐赠者和新成员之间建立的标准异步复制通道来实现的。复制通道是 MySQL 5.7 中提出的概念，简单讲一个复制通道表示从主库到从库的一条复制路径。通过此通道复制捐赠者的二进制日志，直到新服务器成为组的一部分，并触发视图更改时。新成员在收到捐赠者的二进制日志时应用它们。

新服务器还会缓存在组内交换的每个事务。也就是说，它监听在加入该组之后发生的事务，同时应用来自捐赠者的数据。当第一阶段结束并且关闭捐赠者的复制通道时，开始第二阶段：追赶。在此阶段，新服务器继续执行缓存的事务。排队等待执行的事务数最终达到零时，该成员将声明为在线。

新成员从捐赠者获取二进制日志时，恢复过程可以承受捐赠者故障。在这种情况下，捐赠者在第一阶段期间失败时，新成员将从新捐赠者恢复数据。

2. 视图更改

视图对应当前组成员，当组中有成员加入或离开时，会发生视图更改。任何组成员身份更改都会导致在同一逻辑时间点向所有组成员传达视图更改。视图标识符唯一标识视图，只要视图发生更改，就会生成一个新的视图标识符。

初始时，所有服务器都在线并处理事务，构成一个稳定组，如图 6-5 所示。有些服务器复制的事务可能稍微落后，但最终它们会相同。此时该组充当一个分布式数据库的副本。

每当新成员加入组并因此执行视图更改时，每个联机服务器都会把视图更改日志事件排入队列以备执行。在视图更改之前，服务器上可能有一些属于旧视图的事务排队进行应用，将视图更改事件排在它们之后可确保正确标记何时发生了视图更改。

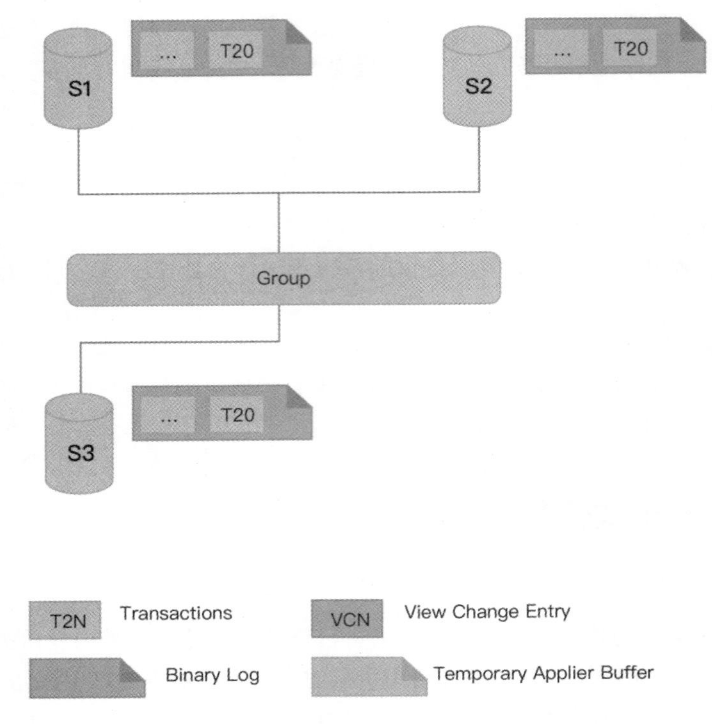

图 6-5 稳定组

同时，新成员从视图的在线服务器列表中选择捐赠者。如图 6-6 所示，成员 S4 加入时生成视图更改条目 VC4，在线成员将此视图更改事件写入二进制日志。

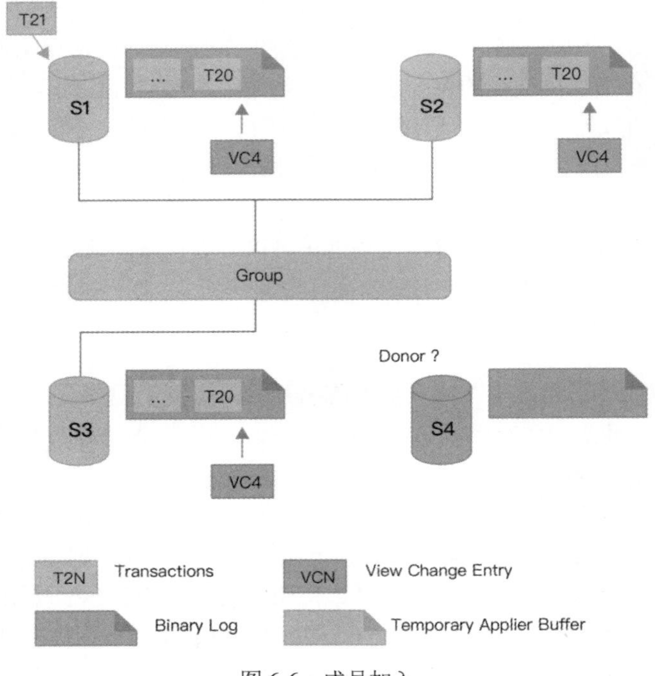

图 6-6 成员加入

一旦新成员选择组中某服务器作为捐赠者，则在两者之间建立异步复制连接并且开始状态转移（第一阶段）。新成员从捐赠者复制，直到视图改变相匹配的视图标识符，如图 6-7 所示。

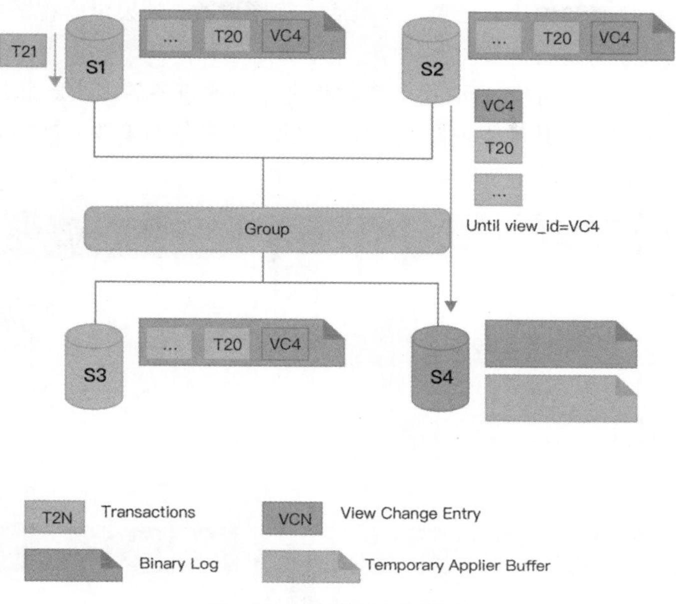

图 6-7　从捐赠者复制数据

新成员知道它应该在哪个视图标识符停止复制。由于视图标识符在相同的逻辑时间点被发送到组中所有成员，避免了复杂的 GTID 集合计算，因为视图 ID 清楚地标记了属于每个组视图的数据。新成员正在从捐赠者复制时，也会缓存来自组的传入事务。最后它停止从捐赠者复制并切换到应用缓存的那些事务，即所谓的追赶，如图 6-8 所示。

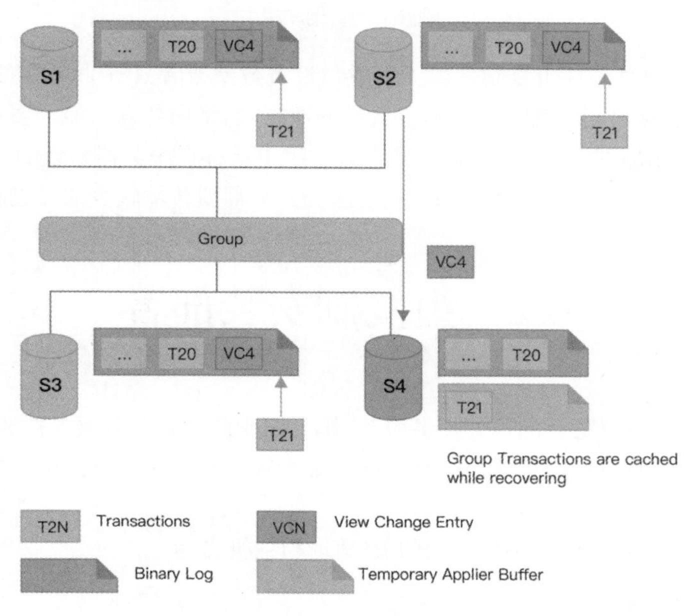

图 6-8　事务排队与追赶

当新成员识别出视图更改日志事件时，终止与捐赠者的连接并开始应用缓存的事务（第二阶段）。视图更改日志事件除了在二进制日志中充当分隔标记，还扮演另一个角色。当新服务器进入组时，它传达所有服务器感知的认证信息，即最后的视图更改。如果没有视图更改事件，新成员将没有必要的信息对后续事务进行冲突检测。

追赶的持续时间是不确定的，它取决于负载和进入组的事务的多少。此过程完全联机，新成员在追赶时不会阻止组中任何其他服务器。当新成员的排队事务为零并且其存储的数据等于其他成员时，状态将更改为联机，如图 6-9 所示。

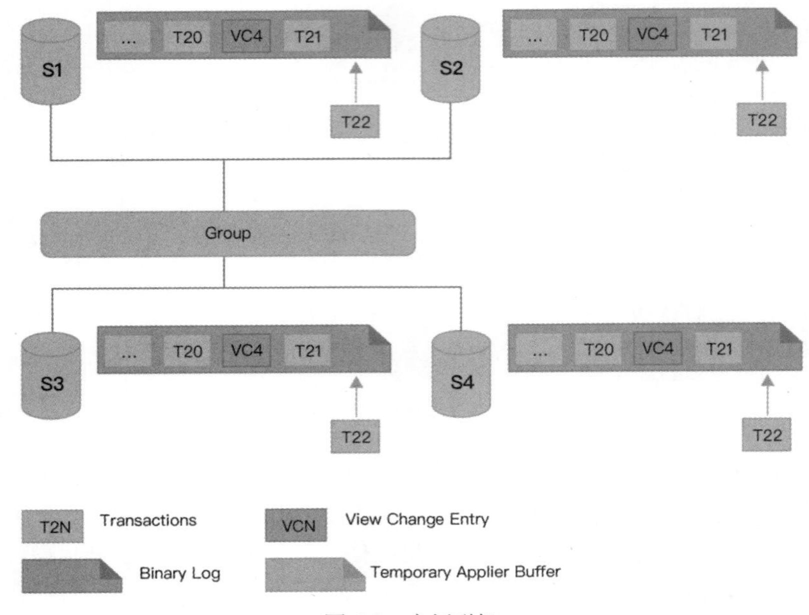

图 6-9　实例联机

分布式恢复基于传统的异步复制，如果新成员没有数据或者只有非常旧的备份数据，那么恢复过程可能很慢。这意味着要在第一阶段传输大量的数据，新增服务器可能需要很长时间才能恢复。因此建议在将服务器添加到组之前，应该为其配置与组中已存在服务器相近的快照。这最小化了第一阶段所需时间并减少了对捐赠服务器的影响，因为它只需要传输较少的二进制日志。

6.2　组复制安装部署

组复制作为 MySQL 服务器的插件提供服务，那么组中的每台服务器就需要安装和配置这个插件。本节将讲述配置具有三台服务器的组复制的详细步骤，安装好三个独立的 MySQL 实例。拓扑结构如图 6-10 所示。

MySQL 版本为 8.0.16，各服务器对应的 IP 地址和主机名为：

- S1：172.16.1.125　hdp2
- S2：172.16.1.126　hdp3

- S3：172.16.1.127　hdp4

所有服务器都要先配置好不同的主机名，并修改/etc/hosts 文件以配置域名解析。

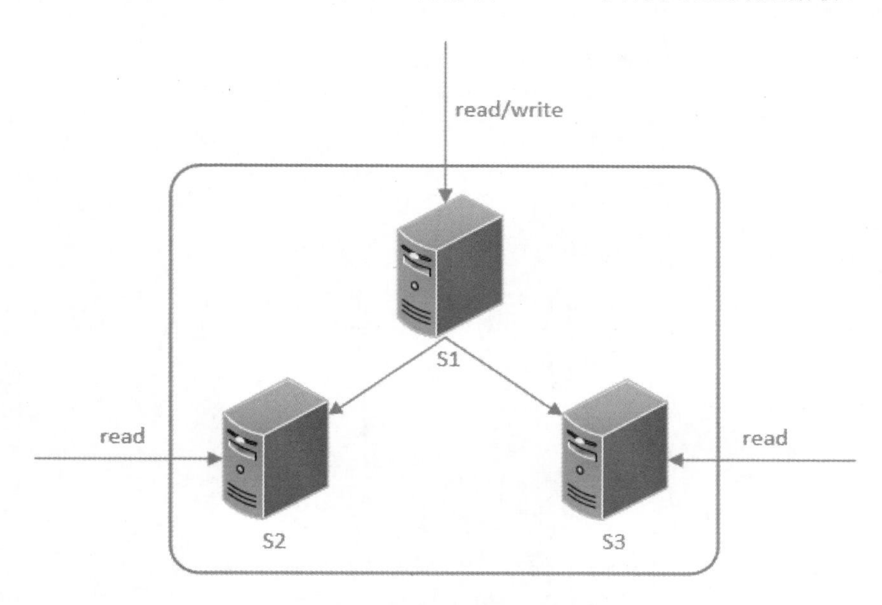

图 6-10　三个实例的组复制

6.2.1　部署单主模式组复制

组复制可以以单主模式或多主模式运行，默认采用单主模式。在单主模式中，只有一个可以读写的服务器，其他服务器为只读。在多主模式中，所有服务器均可读写。本示例配置为单主模式，规划 hdp2 为 PRIMARY，hdp3 和 hdp4 为 SECONDARY。

1. 安装组复制插件

在 hdp2 的 MySQL 实例中安装组复制插件：

```
mysql> install plugin group_replication soname 'group_replication.so';
Query OK, 0 rows affected (0.01 sec)
```

检查组复制插件是否安装成功：

```
mysql> show plugins;
+--------------------+--------+--------------------+---------------------+
---------+
| Name               | Status | Type               | Library              | License |
+--------------------+--------+--------------------+---------------------+-
--------+
| binlog             | ACTIVE | STORAGE ENGINE     | NULL                 | GPL     |
...
| group_replication | ACTIVE | GROUP REPLICATION | group_replication.so |
GPL     |
+--------------------+--------+--------------------+---------------------+-
```

```
--------+
    46 rows in set (0.00 sec)
```

2. 准备配置文件

hdp2 的配置文件/etc/my.cnf 内容如下：

```
[mysqld]
server_id=1125
gtid_mode=ON
enforce-gtid-consistency=true
binlog_checksum=NONE
innodb_buffer_pool_size=4G

disabled_storage_engines="MyISAM,BLACKHOLE,FEDERATED,ARCHIVE,MEMORY"

log_bin=binlog
log_slave_updates=ON
binlog_format=ROW
master_info_repository=TABLE
relay_log_info_repository=TABLE

transaction_write_set_extraction=XXHASH64
group_replication_group_name="aaaaaaaa-aaaa-aaaa-aaaa-aaaaaaaaaaaa"
group_replication_start_on_boot=off
group_replication_local_address= "172.16.1.125:33061"
group_replication_group_seeds= "172.16.1.125:33061,172.16.1.126:
33061,172.16.1.127:33061"
group_replication_bootstrap_group=off
```

主要参数说明：

- 组复制数据必须存储在 InnoDB 事务存储引擎中，使用其他存储引擎可能导致组复制出错。因此设置 disabled_storage_engines 以禁用其他存储引擎。
- 配置 transaction_write_set_extraction 可指示服务器对于每个事务必须收集写集并使用 XXHASH64 哈希算法编码。从 MySQL 8.0.2 开始，此设置是默认设置，因此可以省略此行。
- 配置 group_replication_group_name 可告诉插件它正在加入或创建的组名为 aaaaaaaa-aaaa-aaaa-aaaa-aaaaaaaaaaaa。group_replication_group_name 的值必须是有效的 UUID。在二进制日志中为组复制事件设置 GTID 时，将在内部使用此 UUID。可以使用 SELECT UUID()或 Linux 的 uuidgen 命令生成 UUID。
- 配置 group_replication_start_on_boot 可指示插件在服务器启动时不自动启动。这在设置组复制时很重要，因为它确保可以在手动启动插件之前配置服务器。配置成员后，可以将 group_replication_start_on_boot 设置为 on，以便在服务器引导时自动启动组复制。
- 配置 group_replication_local_address 可告诉插件使用网络地址 172.16.1.125 和端口 33061 与组中的其他成员进行内部通信。组复制将此地址用于涉及组通信引擎（XCom，Paxos 变体）的远程实例的内部成员间的连接。此端口必须不同于客户端连接的端口，并且不得用于客户端应用程序。在运行组复制时，必须为组成员之间的内部通信保留它。

- 配置group_replication_group_seeds以设置组成员的IP和端口，这些成员称为种子成员。组成员建立连接后，组成员身份信息将出现在performance_schema.replication_group_members表中。通常，group_replication_group_seeds列表包含每个组成员的group_replication_local_address的ip:port，但这不是强制性的，可以选择组成员的子集作为种子。启动该组的服务器不使用此选项，因为它是初始服务器，负责引导组。换句话说，引导该组的服务器上的任何现有数据都是用于下一个加入成员的数据。第二个服务器上的任何缺失数据都从引导成员上的数据中复制，然后进行组扩展。加入的第三台服务器可以要求这两台服务器中的任何一台作为捐赠者，数据被同步到新成员，然后该组再次扩展。后续服务器在加入时重复此过程。
- 配置group_replication_bootstrap_group可指示插件是否引导该组。此选项只能在一台服务器实例上使用，通常是第一次引导组时（或者在整个组关闭并重新备份的情况下）。如果多次引导组，例如在多台服务器实例设置了此选项时，可以人工创建一个“脑裂”的情景，存在两个具有相同名称的不同组。

hdp3和hdp4的配置文件中只有server_id和group_replication_local_address两个参数值不同，其他配置参数与hdp2相同：

```
# hdp3:
server_id=1126
group_replication_local_address= "172.16.1.126:33061"

# hdp4:
server_id=1127
group_replication_local_address= "172.16.1.127:33061"
```

3. 重启主库实例

执行下面的命令重启hdp2的MySQL实例：

```
mysqladmin -uroot -p123456 shutdown
mysqld_safe --defaults-file=/etc/my.cnf &
```

4. 启动组复制

在hdp2上执行以下步骤启动组复制。组复制使用异步复制协议实现分布式恢复，在将组成员加入组之前同步数据。分布式恢复过程依赖于名为group_replication_recovery的复制通道，该通道用于将事务从捐赠者转移到加入该组的成员。因此需要设置具有正确权限的复制用户，以便组复制可以建立直接的成员到成员恢复复制通道。具体步骤如下：

步骤01 创建复制用户：

```
create user 'repl'@'%' identified with 'mysql_native_password' by '123456';
grant replication slave on *.* to 'repl'@'%';
```

步骤02 配置用于新成员与捐赠者之间异步复制的复制通道：

```
change master to
      master_user='repl',
      master_password='123456'
  for channel 'group_replication_recovery';
```

步骤 03 启动组复制：

要启动该组，需指示服务器 S1 引导该组，然后启动组复制。此引导程序应仅由单台服务器完成，该服务器启动组只执行一次。

```
set global group_replication_bootstrap_group=on;
start group_replication;
set global group_replication_bootstrap_group=off;
```

步骤 04 确认组复制是否启动成功：

一旦 START GROUP_REPLICATION 语句返回，就启动该组。可以检查组现在是否已创建以及是否包含一个成员：

```
mysql> select * from performance_schema.replication_group_members\G
*************************** 1. row ***************************
  CHANNEL_NAME: group_replication_applier
     MEMBER_ID: 8eed0f5b-6f9b-11e9-94a9-005056a57a4e
   MEMBER_HOST: hdp2
   MEMBER_PORT: 3306
  MEMBER_STATE: ONLINE
   MEMBER_ROLE: PRIMARY
MEMBER_VERSION: 8.0.16
1 row in set (0.00 sec)
```

此表内信息确认组中的成员具有唯一标识符 8eed0f5b-6f9b-11e9-94a9-005056a57a4e，它的状态为 ONLINE 并且在 hdp2 上的端口 3306 上侦听客户端连接。为了证明服务器确实在一个组中并且它能够处理负载，可创建一个表并向其添加一些内容。

```
create database test;
use test;
create table t1(a bigint auto_increment primary key);
```

组复制环境下要求每个表都需要有主键，否则表上的 DML 会报错：

```
ERROR 3098 (HY000): The table does not comply with the requirements by an external plugin.
```

创建一个能长时间执行的存储过程：

```
delimiter //
create procedure p1(a int)
begin
   declare i int default 1;
   while i<=a do
      insert into t1 select null;
      set i=i+1;
end while;
end;
//
delimiter ;
```

```
-- 模拟联机事务
call p1(100000);
```

5. 向组中添加实例

在上一步存储过程运行期间执行下面的步骤，联机向组中添加实例。

步骤 01 从 hdp2 向 hdp3 和 hdp4 联机复制数据。

在 hdp2 上执行以下命令，可以开两个终端并行执行。关于 xtrabackup 的使用细节，参见 1.3.3 小节。

```
# 复制到 hdp3
xtrabackup -uroot -p123456 --socket=/tmp/mysql.sock --no-lock --backup
--compress --stream=xbstream --parallel=4 --target-dir=./ | ssh mysql@172.16.1.126
"xbstream -x -C /usr/local/mysql/data/ --decompress"

# 复制到 hdp4
xtrabackup -uroot -p123456 --socket=/tmp/mysql.sock --no-lock --backup
--compress --stream=xbstream --parallel=4 --target-dir=./ | ssh mysql@172.16.1.127
"xbstream -x -C /usr/local/mysql/data/ --decompress"
```

步骤 02 在 hdp3 和 hdp4 上应用日志。

分别在 hdp3 和 hdp4 上执行以下命令：

```
xtrabackup --prepare --target-dir=/usr/local/mysql/data/
```

步骤 03 启动 hdp3 和 hdp4 的 MySQL 实例。

分别在 hdp3 和 hdp4 上执行以下命令：

```
mysqld_safe --defaults-file=/etc/my.cnf &
```

步骤 04 将 hdp3 和 hdp4 加入到组中。

分别在 hdp3 和 hdp4 上执行以下 SQL 命令：

```
-- 重置 relay log info
reset slave all;
-- 设置复制通道
change master to master_user='repl', master_password='123456' for channel
'group_replication_recovery';
-- 添加到组
start group_replication;
```

此命令返回后查看 performance_schema.replication_group_members 表，MEMBER_STATE 开始时的值为 RECOVERING，表示新增服务器正在追赶主库。当赶上主库时，MEMBER_STATE 值改为 ONLINE，最终显示该组中有三个 ONLINE 状态的服务器。注意，组复制中每个成员执行的事务不是同步的，但最终同步。更确切地说，事务以相同的顺序传递给所有组成员，但是它们的执行不同步，这意味着在接受提交事务之后，每个成员按照自己的进度提交。

```
mysql> select * from performance_schema.replication_group_members;
```

```
    ...
    | group_replication_applier | 5c93a708-a393-11e9-8343-005056a5497f | hdp4 |
3306 | ONLINE | SECONDARY | 8.0.16 |
    | group_replication_applier | 5f045152-a393-11e9-8020-005056a50f77 | hdp3 |
3306 | ONLINE | SECONDARY | 8.0.16 |
    | group_replication_applier | 8eed0f5b-6f9b-11e9-94a9-005056a57a4e | hdp2 |
3306 | ONLINE | PRIMARY  | 8.0.16 |
    ...
    3 rows in set (0.00 sec)
```

performance_schema.replication_group_members 表字段含义如下：

- CHANNEL_NAME：通道名称。组复制插件创建两个复制通道。group_replication_recovery 用于与分布式恢复阶段相关的复制更改。group_replication_applier 用于来自组传入的更改，是应用直接来自组的事务的通道。
- MEMBER_ID：组成员实例的 server_uuid。
- MEMBER_HOST：组成员主机名。如果配置了 report_host 参数，则这里显示 IP 地址。
- MEMBER_ROLE：成员角色，主为 PRIMARY，从为 SECONDARY。
- MEMBER_VERSION：成员数据库实例版本。
- MEMBER_STATE：成员状态，取值和含义如表 6-1 所示。

表 6-1　MEMBER_STATE 取值含义

取值	含义	状态是否在组内同步
ONLINE	表示该成员可正常提供服务	YES
RECOVERING	表示当前成员正在从其他节点恢复数据	YES
OFFLINE	表示组复制插件已经加载，但是该成员不属于任何一个复制组	NO
ERROR	表示成员在恢复阶段出现错误，或从其他节点同步状态中出现错误	NO
UNREACHABLE	成员处于不可达状态，无法与之进行网络通信	NO

从表 6-1 可知，只有 ONLINE 和 RECOVERING 两种状态会在集群中得到同步。这个状态同步是指状态在所有成员上查询均能保持一致。对于 OFFLINE、ERROR 和 UNREABLE：

- 只有在当前 OFFLINE 成员查询 replication_group_members 表才能得到 OFFLINE 状态，在其他成员上查询 replication_group_members 表，则没有该成员的状态，因为 OFFLINE 成员已经不属于这个复制组。
- 只有在当前 ERROR 成员查询 replication_group_members 表才能得到 ERROR 状态，同上面的 OFFLINE，在其他成员上查询也看不到该成员。
- 假设成员 A 与 B 网络通信失败，那么在节点 A 上查询 replication_group_members 表，有可能得到 B 的状态为 UNREACHABLE。

成员状态转移如图 6-11 所示，左边是组成员本身的状态变化，右边为组成员所看到的其他成员的状态变化。

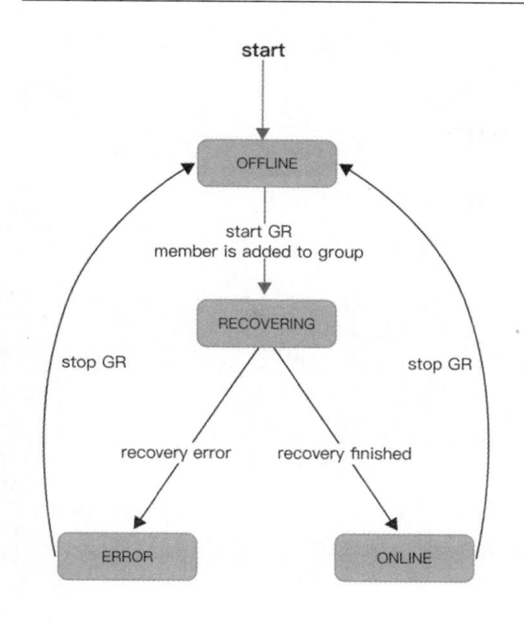

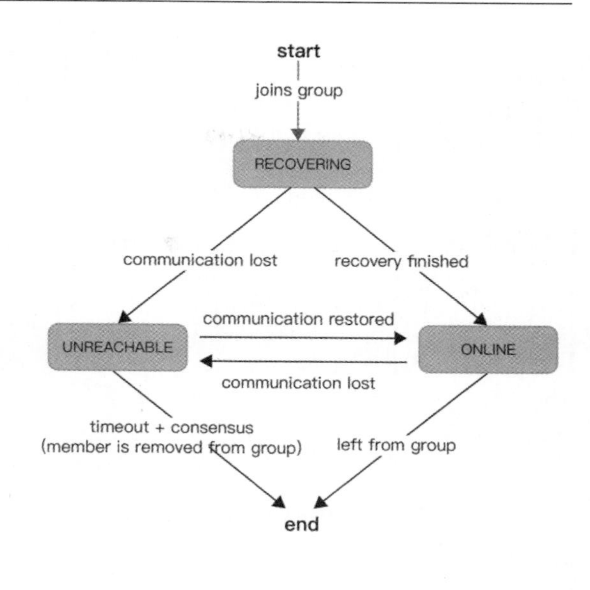

图 6-11　组成员状态转移

当一个成员加入一个复制组，其状态首先变成 RECOVERING，表示当前成员正处于集群恢复阶段。这个阶段下，成员会选择组中一个服务器作为捐赠者（Donor），利用传统的异步复制进行数据恢复。当数据能够成功追平，成员的状态将会变成 ONLINE，这个过程中通过其他成员也可以看到该节点的状态，不管是 RECOVERING 还是最后的 ONLINE。

假如该成员在 RECOVERING 阶段出现了异常，如选择捐赠者进行复制时失败或者在追赶捐赠者数据的过程中失败，那么该成员的状态将会变成 ERROR。注意，这时候在其他成员上查询时，发现该 RECOVERING 节点已经从组里面被移除。

另外，如果一个 ONLINE 成员失去与其他成员的通信（可能因为成员宕机或者网络异常），则该成员在其他成员上面查询到的状态将会是 UNREACHABLE。如果这个 UNREACHABLE 成员在规定的超时时间内没有恢复，那么它将会从组中被移除掉。这个规定的超时时间，取决于组失去这个成员后还能否达到可用状态。如果失去这个成员而组仍可用，那么这个 UNREACHABLE 的超时时间很短，几乎看不到这个状态。但是，如果失去这个成员后组不可用，那么这个成员将会一直处于 UNREACHABLE 状态。

下面以一个例子来验证。kill hdp4 的 MySQL 实例进程，注意是 kill 实例进程而不是正常 shutdown MySQL 实例：

```
ps -ef | grep mysqld | grep -v grep | awk '{print $2}' | xargs kill -9
```

通过其他可用成员查询到，被 kill 掉的实例从组中被移除了：

```
mysql> select * from performance_schema.replication_group_members;
...
| group_replication_applier | 5f045152-a393-11e9-8020-005056a50f77 | hdp3 |
3306 | ONLINE | SECONDARY | 8.0.16 |
```

```
    | group_replication_applier | 8eed0f5b-6f9b-11e9-94a9-005056a57a4e | hdp2 |
3306 | ONLINE | PRIMARY  | 8.0.16 |
    ...
    2 rows in set (0.00 sec)
```

接下来再 kill 掉 hdp3 的 MySQL 实例进程。再次查询 replication_group_members：

```
    mysql> select * from performance_schema.replication_group_members;
    ...
    | group_replication_applier | 5f045152-a393-11e9-8020-005056a50f77
|hdp3|3306| UNREACHABLE | SECONDARY | 8.0.16 |
    | group_replication_applier | 8eed0f5b-6f9b-11e9-94a9-005056a57a4e
|hdp2|3306| ONLINE      | PRIMARY   | 8.0.16      |
    ...
    2 rows in set (0.00 sec)
```

这个时候，UNREACHABLE 状态将一直持续。而且此时，组无法满足 2N + 1 的多数票条件，已经不可用，即使有主成员，它也是不可写的。此时恢复组复制步骤如下：

步骤 01 在 hdp2 重新创建一个新的复制组：

```
    stop group_replication;
    set global group_replication_bootstrap_group=on;
    start group_replication;
    set global group_replication_bootstrap_group=off;
```

步骤 02 启动 hdp3 和 hdp4 的 MySQL 实例：

```
    mysqld_safe --defaults-file=/etc/my.cnf &
```

步骤 03 将 hdp3 和 hdp4 重新加入新复制组：

```
    change master to master_user='repl', master_password='123456' for channel
'group_replication_recovery';
    start group_replication;
```

此时查看二进制日志中的视图更改事件，可以看到创建过两个复制组，每添加一个成员，view_id 的序号加 1：

```
    [mysql@hdp2/usr/local/mysql/data]$mysqlbinlog binlog.000003 | grep view_id
    #190711 12:19:17 server id 1125  end_log_pos 3406788 View_change_log_event:
view_id=15628187590651916:2
    #190711 12:19:17 server id 1125  end_log_pos 6037781 View_change_log_event:
view_id=15628187590651916:3
    #190711 14:59:01 server id 1125  end_log_pos 21750924 View_change_log_event:
view_id=15628283431266410:1
    #190711 14:59:01 server id 1125  end_log_pos 21751315 View_change_log_event:
view_id=15628283431266410:2
    #190711 14:59:01 server id 1125  end_log_pos 21751706 View_change_log_event:
view_id=15628283431266410:3
```

6.2.2　组复制监控

与监控传统主从复制的 show slave status 不同，组复制监控主要依赖以下几个 performance_schema 表：

- performance_schema.replication_group_members
- performance_schema.replication_group_member_stats
- performance_schema.replication_connection_status
- performance_schema.replication_applier_status

performance_schema.replication_group_members 表用于监视作为组成员的不同服务器实例的状态。只要视图更改，就会更新表中的信息。例如，因新成员加入而动态更改组的配置时。服务器通过交换一些元数据以使其自身同步并继续一起协作。信息在作为复制组成员的所有服务器实例之间共享，因此可以从任何成员查询有关所有组成员的信息。此表可用作获取复制组状态的高级视图。

performance_schema.replication_group_member_stats 表提供与认证过程相关的组级信息，以及由复制组的每个成员接收和发起的事务的统计信息。信息在作为复制组成员的所有服务器实例之间共享，因此可以从任何成员查询有关所有组成员的信息。刷新远程成员统计信息的频率由 group_replication_flow_control_period 配置参数中指定的消息周期来控制（默认值为 1 秒），因此可能与本地查询的成员统计信息略有不同。

该表字段对于监控组中连接成员的性能尤为重要。例如，假设组中的一个成员总是在其队列中报告与其他成员相比包含大量事务。这意味着该成员存在延迟，并且无法与该组的其他成员保持同步。根据此信息，可能决定从组中删除该成员，或者延迟处理该组其他成员上的事务，以减少排队事务的数量。此信息还可以帮助决定如何调整组复制插件的流控。关于流控将在 6.3.6 小节进行说明。

performance_schema.replication_connection_status 显示有关组复制的信息，例如已从组接收并在应用程序队列（中继日志）中排队的事务。performance_schema.replication_applier_status 显示与组复制相关的通道和线程的状态。如果有许多不同的工作线程应用事务，那么该表也可用于监控每个工作线程正在执行的操作。

6.2.3　容错示例

本小节以三个成员为例，验证一个 SECONDARY 或 PRIMARY 实例异常宕机对整个组的影响。因为只有三个成员，这两种场景下均能够保证最大票数。无法保证最大票数时，如上面例子中三个成员中的两个异常宕机，则整个组无法正常读写，需要管理员人为介入解决问题。这种情况显然不属于容错的范畴。

1. 一个 SECONDARY 实例异常 shutdown

（1）PRIMARY 上执行长时间运行的事务

```
-- 在 hdp2 上执行
```

```
use test;
truncate table t1;
call p1(100000);
```

（2）在上一步执行期间停止一个 SECONDARY

```
# 停止 hdp4 的 MySQL 实例
ps -ef | grep mysqld | grep -v grep | awk {'print $2'} | xargs kill -9
```

（3）检查剩余组复制成员的状态

在 hdp2 上检查复制组成员的状态：

```
mysql> select * from performance_schema.replication_group_members;
...
| group_replication_applier | 5f045152-a393-11e9-8020-005056a50f77 | hdp3 |
3306 | ONLINE | SECONDARY | 8.0.16 |
| group_replication_applier | 8eed0f5b-6f9b-11e9-94a9-005056a57a4e | hdp2 |
3306 | ONLINE | PRIMARY   | 8.0.16 |
...
2 rows in set (0.00 sec)
```

在 hdp3 检查复制事务和数据：

```
mysql> select * from performance_schema.replication_group_member_stats where
member_id='5f045152-a393-11e9-8020-005056a50f77'\G
*************************** 1. row ***************************
                    CHANNEL_NAME: group_replication_applier
                         VIEW_ID: 15628283431266410:6
                       MEMBER_ID: 5f045152-a393-11e9-8020-005056a50f77
         COUNT_TRANSACTIONS_IN_QUEUE: 0
          COUNT_TRANSACTIONS_CHECKED: 106387
            COUNT_CONFLICTS_DETECTED: 0
      COUNT_TRANSACTIONS_ROWS_VALIDATING: 6385
      TRANSACTIONS_COMMITTED_ALL_MEMBERS:
8eed0f5b-6f9b-11e9-94a9-005056a57a4e:1-2,
aaaaaaaa-aaaa-aaaa-aaaa-aaaaaaaaaaaa:1-200011
          LAST_CONFLICT_FREE_TRANSACTION:
aaaaaaaa-aaaa-aaaa-aaaa-aaaaaaaaaaaa:206397
COUNT_TRANSACTIONS_REMOTE_IN_APPLIER_QUEUE: 0
      COUNT_TRANSACTIONS_REMOTE_APPLIED: 106389
      COUNT_TRANSACTIONS_LOCAL_PROPOSED: 0
      COUNT_TRANSACTIONS_LOCAL_ROLLBACK: 0
1 row in set (0.00 sec)

mysql> select min(a),max(a),count(*) from test.t1;
+--------+--------+----------+
| min(a) | max(a) | count(*) |
+--------+--------+----------+
|      1 |   6385 |     6385 |
+--------+--------+----------+
1 row in set (0.00 sec)
```

可以看到，一个 SECONDARY 实例异常 shutdown，对应用来说只是少了一个只读实例。复制组中的剩余成员状态依然是 ONLINE。PRIMARY 正常读写，剩余 SECONDARY 正常复制，并且没有积压的事务（COUNT_TRANSACTIONS_IN_QUEUE 为 0）。

（4）恢复 shutdown 的实例

启动 hdp4 的 MySQL 实例：

```
mysqld_safe --defaults-file=/etc/my.cnf &
```

在 hdp4 上检查组复制成员的状态：

```
mysql> select * from performance_schema.replication_group_members\G
*************************** 1. row ***************************
  CHANNEL_NAME: group_replication_applier
     MEMBER_ID:
   MEMBER_HOST:
   MEMBER_PORT: NULL
  MEMBER_STATE: OFFLINE
   MEMBER_ROLE:
MEMBER_VERSION:
1 row in set (0.01 sec)
```

此时 hdp4 的状态为 OFFLINE，已经从复制组中被移除。在 hdp4 检查复制事务和数据：

```
mysql> select * from performance_schema.replication_group_member_stats where
member_id='5c93a708-a393-11e9-8343-005056a5497f'\G
Empty set (0.00 sec)

mysql> select min(a),max(a),count(*) from test.t1;
+--------+--------+----------+
| min(a) | max(a) | count(*) |
+--------+--------+----------+
|    1 |   1759 |    1759 |
+--------+--------+----------+
1 row in set (0.02 sec)
```

可以看到，此时 performance_schema.replication_group_member_stats 表中已经没有此成员相关的信息，实例停止时插入了 1759 条数据。

将 hdp4 重新加入复制组：

```
change master to master_user='repl', master_password='123456' for channel
'group_replication_recovery';
start group_replication;
```

再次检查成员状态、复制事务和数据：

```
mysql> select * from performance_schema.replication_group_members;
...
| group_replication_applier | 5c93a708-a393-11e9-8343-005056a5497f
|hdp4|3306| RECOVERING | SECONDARY | 8.0.16 |
| group_replication_applier | 5f045152-a393-11e9-8020-005056a50f77
|hdp3|3306| ONLINE     | SECONDARY | 8.0.16 |
```

```
    | group_replication_applier | 8eed0f5b-6f9b-11e9-94a9-005056a57a4e
|hdp2|3306| ONLINE     | PRIMARY   | 8.0.16 |
    ...
    3 rows in set (0.00 sec)

    mysql> select * from performance_schema.replication_group_member_stats where
member_id='5c93a708-a393-11e9-8343-005056a5497f'\G
    *************************** 1. row ***************************
                      CHANNEL_NAME: group_replication_applier
                           VIEW_ID: 15628283431266410:7
                         MEMBER_ID: 5c93a708-a393-11e9-8343-005056a5497f
       COUNT_TRANSACTIONS_IN_QUEUE: 16608
        COUNT_TRANSACTIONS_CHECKED: 0
          COUNT_CONFLICTS_DETECTED: 0
  COUNT_TRANSACTIONS_ROWS_VALIDATING: 0
  TRANSACTIONS_COMMITTED_ALL_MEMBERS:
8eed0f5b-6f9b-11e9-94a9-005056a57a4e:1-2,
    aaaaaaaa-aaaa-aaaa-aaaa-aaaaaaaaaaaa:1-230667
       LAST_CONFLICT_FREE_TRANSACTION:
    COUNT_TRANSACTIONS_REMOTE_IN_APPLIER_QUEUE: 0
        COUNT_TRANSACTIONS_REMOTE_APPLIED: 0
        COUNT_TRANSACTIONS_LOCAL_PROPOSED: 0
        COUNT_TRANSACTIONS_LOCAL_ROLLBACK: 0
    1 row in set (0.00 sec)

    mysql> select min(a),max(a),count(*) from test.t1;
    +--------+--------+----------+
    | min(a) | max(a) | count(*) |
    +--------+--------+----------+
    |      1 | 38350  |   38350  |
    +--------+--------+----------+
    1 row in set (0.00 sec)
```

hdp4 开始处于 RECOVERING 状态，表明它正在追赶组的复制进度，当赶上后，它的状态将变为 ONLINE。由此可见，一个 SECONDARY 实例异常 shutdown，就是少了一个读成员，基本对复制组没有影响。当它重新加入组时会进行实例和事务的恢复，直至赶上后状态变回 ONLINE。重启 hdp4 MySQL 实例并重新加入组后的整个恢复过程是自动进行的，对用户完全透明。

2. PRIMARY 实例异常 shutdown

（1）PRIMARY 上执行长时间运行的事务

```
-- 在 hdp4 上执行
use test;
truncate table t1;
call p1(100000);
```

（2）在上一步执行期间停止 PRIMARY

```
# 停止 hdp4 的 MySQL 实例
ps -ef | grep mysqld | grep -v grep | awk {'print $2'} | xargs kill -9
```

（3）检查剩余组复制成员的状态

在 hdp2 和 hdp3 上检查复制组成员的状态：

```
mysql> select * from performance_schema.replication_group_members;
...
| group_replication_applier | 5f045152-a393-11e9-8020-005056a50f77 | hdp3 |
3306 | ONLINE | PRIMARY  | 8.0.16 |
| group_replication_applier | 8eed0f5b-6f9b-11e9-94a9-005056a57a4e | hdp2 |
3306 | ONLINE | SECONDARY | 8.0.16 |
...
2 rows in set (0.01 sec)
```

可以看到，复制组中剩余成员的状态依然是 ONLINE，并且自动把其中一个 SECONDARY（本例中为 hdp3）提升为 PRIMARY。在 hdp2 和 hdp3 上检查复制事务和数据：

```
mysql> select * from performance_schema.replication_group_member_stats\G
*************************** 1. row ***************************
                    CHANNEL_NAME: group_replication_applier
                         VIEW_ID: 15628283431266410:10
                       MEMBER_ID: 5f045152-a393-11e9-8020-005056a50f77
       COUNT_TRANSACTIONS_IN_QUEUE: 0
        COUNT_TRANSACTIONS_CHECKED: 1227
          COUNT_CONFLICTS_DETECTED: 0
  COUNT_TRANSACTIONS_ROWS_VALIDATING: 1
  TRANSACTIONS_COMMITTED_ALL_MEMBERS:
8eed0f5b-6f9b-11e9-94a9-005056a57a4e:1-2,
aaaaaaaa-aaaa-aaaa-aaaa-aaaaaaaaaaaa:1-303809
      LAST_CONFLICT_FREE_TRANSACTION:
aaaaaaaa-aaaa-aaaa-aaaa-aaaaaaaaaaaa:303809
COUNT_TRANSACTIONS_REMOTE_IN_APPLIER_QUEUE: 0
   COUNT_TRANSACTIONS_REMOTE_APPLIED: 1227
   COUNT_TRANSACTIONS_LOCAL_PROPOSED: 0
   COUNT_TRANSACTIONS_LOCAL_ROLLBACK: 0
*************************** 2. row ***************************
                    CHANNEL_NAME: group_replication_applier
                         VIEW_ID: 15628283431266410:10
                       MEMBER_ID: 8eed0f5b-6f9b-11e9-94a9-005056a57a4e
       COUNT_TRANSACTIONS_IN_QUEUE: 0
        COUNT_TRANSACTIONS_CHECKED: 2806
          COUNT_CONFLICTS_DETECTED: 0
  COUNT_TRANSACTIONS_ROWS_VALIDATING: 1
  TRANSACTIONS_COMMITTED_ALL_MEMBERS:
8eed0f5b-6f9b-11e9-94a9-005056a57a4e:1-2,
aaaaaaaa-aaaa-aaaa-aaaa-aaaaaaaaaaaa:1-303809
      LAST_CONFLICT_FREE_TRANSACTION:
aaaaaaaa-aaaa-aaaa-aaaa-aaaaaaaaaaaa:303809
COUNT_TRANSACTIONS_REMOTE_IN_APPLIER_QUEUE: 0
   COUNT_TRANSACTIONS_REMOTE_APPLIED: 2808
   COUNT_TRANSACTIONS_LOCAL_PROPOSED: 0
   COUNT_TRANSACTIONS_LOCAL_ROLLBACK: 0
2 rows in set (0.00 sec)
```

```
mysql> select min(a),max(a),count(*) from test.t1;
+--------+--------+----------+
| min(a) | max(a) | count(*) |
+--------+--------+----------+
|      1 |   1226 |     1226 |
+--------+--------+----------+
1 row in set (0.00 sec)
```

（4）恢复 shutdown 的实例

启动 hdp4 的 MySQL 实例：

```
mysqld_safe --defaults-file=/etc/my.cnf &
```

在 hdp4 上检查组复制成员的状态：

```
mysql> select * from performance_schema.replication_group_members\G
*************************** 1. row ***************************
  CHANNEL_NAME: group_replication_applier
     MEMBER_ID:
   MEMBER_HOST:
   MEMBER_PORT: NULL
  MEMBER_STATE: OFFLINE
   MEMBER_ROLE:
MEMBER_VERSION:
1 row in set (0.00 sec)
```

此时 hdp4 的状态为 OFFLINE，已经从复制组中被移除。在 hdp4 检查复制事务和数据：

```
mysql> select * from performance_schema.replication_group_member_stats where
member_id='5c93a708-a393-11e9-8343-005056a5497f'\G
Empty set (0.01 sec)

mysql> select min(a),max(a),count(*) from test.t1;
+--------+--------+----------+
| min(a) | max(a) | count(*) |
+--------+--------+----------+
|      1 |   1225 |     1225 |
+--------+--------+----------+
1 row in set (0.01 sec)
```

可以看到，此时 performance_schema.replication_group_member_stats 表中已经没有此成员相关的信息，实例停止时插入了 1225 条数据。原 PRIMARY 比新的 PRIMARY 还少了一条数据，可见各个实例的提交进度由自己而不是复制组控制。

将 hdp4 重新加入复制组：

```
change master to master_user='repl', master_password='123456' for channel
'group_replication_recovery';
start group_replication;
```

再次检查成员状态、复制事务和数据：

```
mysql> select * from performance_schema.replication_group_members;
```

```
...
| group_replication_applier | 5c93a708-a393-11e9-8343-005056a5497f | hdp4 |
3306 | ONLINE | SECONDARY | 8.0.16 |
| group_replication_applier | 5f045152-a393-11e9-8020-005056a50f77 | hdp3 |
3306 | ONLINE | PRIMARY  | 8.0.16 |
| group_replication_applier | 8eed0f5b-6f9b-11e9-94a9-005056a57a4e | hdp2 |
3306 | ONLINE | SECONDARY | 8.0.16 |
...
3 rows in set (0.00 sec)

mysql> select * from performance_schema.replication_group_member_stats where
member_id='5c93a708-a393-11e9-8343-005056a5497f'\G
*************************** 1. row ***************************
                          CHANNEL_NAME: group_replication_applier
                               VIEW_ID: 15628283431266410:11
                             MEMBER_ID: 5c93a708-a393-11e9-8343-005056a5497f
           COUNT_TRANSACTIONS_IN_QUEUE: 0
            COUNT_TRANSACTIONS_CHECKED: 0
              COUNT_CONFLICTS_DETECTED: 0
    COUNT_TRANSACTIONS_ROWS_VALIDATING: 1
    TRANSACTIONS_COMMITTED_ALL_MEMBERS:
8eed0f5b-6f9b-11e9-94a9-005056a57a4e:1-2,
aaaaaaaa-aaaa-aaaa-aaaa-aaaaaaaaaaaa:1-303809
        LAST_CONFLICT_FREE_TRANSACTION:
COUNT_TRANSACTIONS_REMOTE_IN_APPLIER_QUEUE: 0
         COUNT_TRANSACTIONS_REMOTE_APPLIED: 0
         COUNT_TRANSACTIONS_LOCAL_PROPOSED: 0
         COUNT_TRANSACTIONS_LOCAL_ROLLBACK: 0
1 row in set (0.00 sec)

mysql> select min(a),max(a),count(*) from test.t1;
+--------+--------+----------+
| min(a) | max(a) | count(*) |
+--------+--------+----------+
|    1 |   1226 |     1226 |
+--------+--------+----------+
1 row in set (0.00 sec)
```

hdp4 处于 ONLINE 状态，但角色已经变成了 SECONDARY，数据也已经和其他成员一致。PRIMARY 实例异常 shutdown，组复制会把一个 SECONDARY 提升为新的 PRIMARY。而原来的 PRIMARY 在重新加入组后立即 ONLINE，角色变为 SECONDARY。这一切都是自动进行的，应用所需做的是重新连接新的 PRIMARY 实例，以便继续执行读写事务。

最终得出的结论是，只要保证最大票数的实例可用，组复制中成员的数据恢复、角色交换等容错行为是全自动的，应用只要在必要时修改到 PRIMARY 的连接即可。

6.3 组复制性能

6.3.1 概述

组复制的基本保证是，只有在组中的大多数节点接收到事务，并且就并发事务的相对顺序达成一致之后，才会提交事务。其对事务的基本处理流程为：

（1）在提交之前拦截服务器接收到的客户端事务，并构建事务写入集。该写入集唯一标识了对应事务所修改的数据集合。

（2）将写入集发送给组成员，并就并发事务的应用顺序达成一致。

（3）如果事务与之前的事务不存在冲突则提交，否则回滚该事务。即相互冲突的事务第一个成功提交，其他回滚。

影响组复制性能的组件主要有三个：组通信层、认证和二进制日志应用程序。

1. 组通信层

组复制实现了一个基于 Paxos 协议的组通信层，以允许多个服务器在事务提交顺序上达成一致。发送给组通信层的消息在每个组成员中以相同的顺序接收。Paxos 协议很复杂，但为了理解组复制中实现变体对性能的影响，还是需要了解处理一个典型消息的过程：

（1）从消息源向每个其他组成员发送带有用户事务的消息。

（2）向消息源发送每个接收者的投票信息，当达到多数票后，客户端将消息视为已传递。

（3）将消息源所接收的信息发送给其他所有组成员。

由此可见，组复制性能的关键因素是网络吞吐量，简单说就是可以在网络通道中容纳多少并行消息，以及节点到节点的延迟，即需要多长时间才能与发送方达成一致。Paxos 的组复制实现包括许多优化，如并行处理多条消息，并将多个消息打包成一个消息，以便在可能的时候发送到每个节点。从用户角度来看到的结果是：

- 在现有网络带宽下，发送者将尽可能多的事务发送到其他节点。节点间的实际带宽是网络带宽除以组成员数量。
- 准备好提交之后，每个事务将被延迟至少从发送者到接收者的一个中间网络往返时间。

2. 认证

从组通信层接收到的事务消息会被放入一个队列进行认证。在每个节点中运行的算法用于判断认证结果，以决定如何处理事务。若认证通过，事务被认为在面向客户端的服务器（事务发起方）上完成，然后在本地中继日志中排队，以便在其他组成员上执行后续操作。反之如果认证失败，事务将根据需要被丢弃或回滚。

认证是在每个节点接收到正确排序的事务消息后独立完成的，因此不需要进一步的网络同步活动。认证算法在一个专用线程中运行，该线程遵循由组通信层建立的事务顺序，如果事务是远程的，则将其写入中继日志。因为认证每个事务都需要时间，所以该过程可能成为争用点。影响认证吞吐量的主要因素是中继日志带宽，如果用于本地中继日志的存储子系统，即认证事务排队等待应用程序线程的存储子系统无法跟上写入请求，那么其事务认证队列开始增长，事务应用开始落后于发起事务的成员。从下面的查询可以获得每个组成员认证队列大小的近似值：

```
select member_id, count_transactions_in_queue
from performance_schema.replication_group_member_stats;
```

多主组复制中，由于多个组成员都可以执行写事务，gtid_executed 集合的复杂性可能成为另一个影响认证性能的因素。当多个成员同时向组写入时，认证数据库上执行的大量 gtid_executed 集合可能变得难以处理，因为其中可能出现许多间隙，并且无法按照 GTID 范围进行压缩。那么可以看作是成员节点之间的吞吐量平衡性降低了。

3. 二进制日志应用程序

事务写入中继日志后，它们就可以像异步或半同步复制一样，由复制的二进制日志应用程序执行。然而，组复制的二进制日志应用程序有一个应该注意的细微差别。一旦组成员对事务进行了认证，认证程序会立即将事务更新行标记为已更改，但实际该事务可能会在中继日志中保留一段时间等待应用。在多主模式中，在此期间尝试更改这些行的本地事务都被回滚。

与 MySQL 传统主从异步复制同理，具有足够线程的 MTS（Multi-Threaded Slave）应用程序来处理接收到的工作负载，有助于增加 SECONDARY 的吞吐量。对于异步复制，LOGICAL_CLOCK 调度程序使用二进制日志组提交机制来确定哪些事务可以在从库上并行执行。组提交能提高从库并发度。但是，存储子系统速度太快或客户端并发量太小都会使提交组中的事务变少（一个组提交不能包含来自同一客户端的多个事务），从而降低了效率。尤其是在级联复制中，因为复制线程通常比用户线程少得多，这意味着较低级别从库的并行度更低。

使用 binlog_group_commit_sync_delay 或 binlog_group_commit_sync_no_delay_count 选项可以增加提交组中的事务数量，以抵消存储速度过快的影响。而基于写集（WRITESET）的依赖项跟踪允许更高的并行度，特别是在低并发工作负载下。基于写集的依赖项跟踪可能重新排序来自同一会话的事务，这意味着来自同一会话的两个事务在主从库的提交顺序可能不同。以牺牲一点性能为代价，可以选择强制使用 slave_preserve_commit_order 选项来保持从库上会话历史与主库保持一致。

组复制考虑了用于验证每个成员中事务的写入集，并根据认证顺序和事务更改的数据行构建依赖项。如果两个事务在其写入集中存在相同的行，则它们不能在 SECONDARY 中并行运行，因此后一个更新这些行的事务将依赖于前一个事务。

关于多线程复制的详细讨论，参见 4.2 节。有关 MySQL 官方描述的组复制性能影响因素，参见网址 http://mysqlhighavailability.com/tuning-mysql-group-replication-for-fun-and-profit/。

简单来说，组复制性能受限于组通信层、认证和二进制日志应用程序三个主要组件：

- 缺乏网络带宽会降低在途消息数量，而高的网络往返时间会相应地增加用户事务延迟。
- 缺乏中继日志存储带宽会延迟认证，因此要注意认证队列的增长。
- 二进制日志应用程序瓶颈除了会增加复制延迟外，还可能增加多主使用场景中的事务中止。

对于每个组件，组复制提供了若干选项，适当配置这些选项可以从底层计算资源中获取最佳性能。后面小节中将详细介绍这些选项。

6.3.2 测试规划

在讨论具体选项配置前，希望使用一个统一的测试方法，对比不同配置值对复制性能产生的影响，测量的指标包括 PRIMARY、SECONDARY 的每秒事务数 TPS 和 SECONDARY 的复制延迟时长。测试目的在于对比不同配置情况下组复制的性能，而不是针对测量绝对值进行优化。这里使用的思路是：PRIMARY 加压，同时 SECONDARY 复制。记录加压前后的 GTID，得到需要执行的总事务数。PRIMARY 的执行时间等于加压时间。等待 SECONDARY 应用完所有事务，记录 SECONDARY 的执行时间。用总事务数除以相应执行时间得到 PRIMARY 和 SECONDARY 的 TPS。加压则使用 tpcc-mysql 基准测试工具。

1. 测试环境

单主模式下“一主两从”的组复制基本信息如下：

PRIMARY：172.16.1.125
SECONDARY 1：172.16.1.126
SECONDARY 2：172.16.1.127

MySQL 版本：8.0.16；操作系统：CentOS Linux release 7.2.1511 (Core)；硬件配置为：三台服务器均为 Linux 虚拟机，2.10GHz 的双核双 CPU，8GB 物理内存，100GB 硬盘空间，1000Mbps 网卡。

MySQL 实例的基本配置：

```
[mysqld]
server_id=1125     # 两个 SECONDARY 为 1126、1127
gtid_mode=ON
enforce-gtid-consistency=true
innodb_buffer_pool_size=4G
disabled_storage_engines="MyISAM,BLACKHOLE,FEDERATED,ARCHIVE,MEMORY"

log_bin=binlog
log_slave_updates=ON
binlog_format=ROW
binlog_checksum=NONE
master_info_repository=TABLE
relay_log_info_repository=TABLE

plugin-load=group_replication.so
group_replication_group_name="aaaaaaaa-aaaa-aaaa-aaaa-aaaaaaaaaaaa"
group_replication_start_on_boot=off
group_replication_local_address= "172.16.1.125:33061"  # 两个 SECONDARY 为
"172.16.1.126:33061"、"172.16.1.127:33061"
group_replication_group_seeds=
"172.16.1.125:33061,172.16.1.126:33061,172.16.1.127:33061"
group_replication_bootstrap_group=off
```

为保证不同配置的测试环境相同，每次修改配置后，在开始测试前，重启三个 MySQL 实例，并使用下面的脚本重启组复制：

```
    -- PRIMARY
    change master to master_user='repl', master_password='123456' for channel
'group_replication_recovery';
    set global group_replication_bootstrap_group=on;
    start group_replication;
    set global group_replication_bootstrap_group=off;
    select * from performance_schema.replication_group_members;

    -- SECONDARY
    stop slave;
    reset slave all;
    change master to master_user='repl', master_password='123456' for channel
'group_replication_recovery';
    start group_replication;
```

2. 测试脚本

每次测试只需要执行同一个脚本文件 tpcc_test.sh，内容如下：

```
    # 初始化 tpcc 数据
    mysql -uwxy -p123456 -h172.16.1.125 < tpcc_test.sql

    # 开始 GTID
    read start_gtid < <(mysql -uwxy -p123456 -e "show variables like
'gtid_executed';" --skip-column-names | awk '{print $2}' | sed "s/\\\n//g")

    # 等待 SECONDARY 执行完初始化复制
    mysql -uwxy -p123456 -h172.16.1.126 -e "select
wait_for_executed_gtid_set('$start_gtid');" > /dev/null &
    mysql -uwxy -p123456 -h172.16.1.127 -e "select
wait_for_executed_gtid_set('$start_gtid');" > /dev/null

    # 开始时间
    start_time=`date '+%s'`

    # PRIMARY 执行压测，10 个仓库，32 个并发线程，预热 1 分钟，压测 5 分钟
    tpcc_start -h172.16.1.125 -d tpcc_test -u wxy -p "123456" -w 10 -c 32 -r 60
-l 300 > tpcc_test.log 2>&1

    # 结束 GTID
    read end_gtid < <(mysql -uwxy -p123456 -e "show variables like 'gtid_executed';"
--skip-column-names | awk '{print $2}' | sed "s/\\\n//g")

    # 等待 SECONDARY 执行完复制
    mysql -uwxy -p123456 -h172.16.1.126 -e "select
wait_for_executed_gtid_set('$end_gtid'); select unix_timestamp(now());"
--skip-column-names | awk 'NR>1' > end_time1 &
    mysql -uwxy -p123456 -h172.16.1.127 -e "select
wait_for_executed_gtid_set('$end_gtid'); select unix_timestamp(now());"
--skip-column-names | awk 'NR>1' > end_time2
```

```
# 结束时间
end_time1=`cat end_time1`
end_time2=`cat end_time2`

# 复制执行时长
elapsed1=$(($end_time1 - $start_time))
elapsed2=$(($end_time2 - $start_time))

# 执行的事务数
read start end < <(mysql -uwxy -p123456 -e "select
gtid_subtract('$end_gtid','$start_gtid');" --skip-column-names | awk -F: '{print
$2}' | awk -F- '{print $1,$2}')
trx=$(($end - $start + 1))

# 计算 PRIMARY 和 SECONDARY 的 TPS
Master_TPS=`expr $trx / 360`
Slave1_TPS=`expr $trx / $elapsed1`
Slave2_TPS=`expr $trx / $elapsed2`

# 打印输出
echo "TRX: $trx"
echo "Master TPS: $Master_TPS"
echo "Elapsed1: $elapsed1" "Slave1 TPS: $Slave1_TPS"
echo "Elapsed2: $elapsed2" "Slave2 TPS: $Slave2_TPS"
```

关于 tpcc 工具的安装和 tpcc_test.sql 文件的生成及作用，参见 4.2 节。每条命令前的注释已经明确表明其用途，这里做两点补充说明：

- MySQL 组复制要求表具有主键。tpcc 测试库中的 history 表没有主键，因此需要手工编辑 tpcc_test.sql 文件，在其建表的 create table 语句中增加主键定义。这完全是为了满足组复制的要求，因此只要将所有字段一起定义成联合主键即可，即使出现主键重复的错误，也不会影响测试继续进行。况且这里要得到的只是对比结果，而测试结果的绝对值并无太多参考意义。
- 使用 MySQL 5.7.5 新增的 wait_for_executed_gtid_set 函数，等待从库应用完事务，返回复制结束时间。本测试中采用“一主两从”的组复制拓扑结构，需要并行但分别取得两个从库的结束时间，因此使用 & 符号，将第一条执行 wait_for_executed_gtid_set 的命令在后台执行。这样可以获得两个从库各自的执行时间，而后面的命令会在它们都执行完后才开始执行，而不论它们谁先执行完。

3. 执行默认配置测试

获得默认配置的测试结果，作为后面不同配置的对比基准。测试结果如下：

TRX: 176527
Master TPS: 490
Elapsed1: 708 Slave1 TPS: 249
Elapsed2: 922 Slave2 TPS: 191

6.3.3 消息压缩

当网络带宽成为瓶颈时，消息压缩可以在组通信层提高吞吐量。如图 6-12 所示，压缩发生在组通信系统 API 级别，因此事务负载可以在被发送到组之前压缩，并且在成员接收事务时进行解压缩。压缩使用 LZ4 算法，阈值默认为 1000000 字节，由 group_replication_compression_threshold 参数控制，即只压缩大于该参数指定字节数的事务，设置为 0 将禁用压缩。收到消息后，组成员检查消息是否被压缩。如果需要，该成员再将事务交付给上层之前解压缩该事务。组复制不要求组中的所有服务器都同时启用压缩。

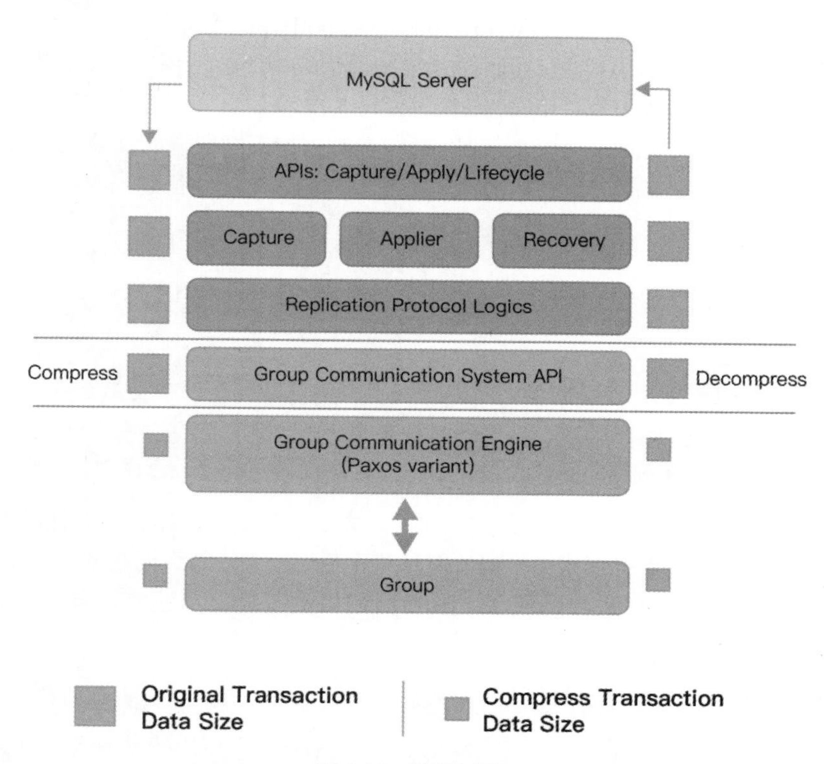

图 6-12　消息压缩

将 group_replication_compression_threshold 设置为 0、100 和 1024 时的测试结果如表 6-2 所示。

表 6-2　不同 group_replication_compression_threshold 值性能对比

group_replication_compression_threshold	TRX	Master TPS	Slave1 Elapsed	Slave1 TPS	Slave2 Elapsed	Slave2 TPS
0	175932	488	727	241	914	192
100	175384	487	715	233	895	195
1024	176601	490	743	237	904	195

禁用压缩、压缩大于 100 字节的事务和压缩大于 1K 字节的事务，与默认压缩大于 1000000 字节的事务相比，在我们的测试环境中，PRIMARY 和 SECONDARY 的 TPS、SECONDARY 复制延

时并无显著差异。结论是此测试环境和负载压力下，网络传输不是瓶颈，因此后面的测试保持 group_replication_compression_threshold 的默认值不变。

6.3.4 组通信线程循环

组通信线程（Group Communication Thread，GCT）在加载组复制插件后循环运行。GCT 从组插件接收消息，处理仲裁和故障检测相关的任务，发送活动消息，并处理服务器和组之间的双向事务传输。当队列中没有消息时，GCT 进入睡眠状态。通过在实际进入睡眠状态之前将等待时间设置为稍长，即进行主动等待，在某些情况下是有益的，因为 GCT 频繁休眠可能浪费可用的 CPU 时间。要强制 GCT 执行主动等待，使用 group_replication_poll_spin_loops 选项。此选项控制 GCT 在进入睡眠状态前尝试接收消息的次数，也可说是控制通信线程的贪婪程度，默认值为 0。在网速很慢的环境中，适当配置该选项可以降低事务延迟。

这里没有针对 group_replication_poll_spin_loops 参数进行不同配置的实验，原因有二：一是在我们的实验环境下网络不是瓶颈，这点从上面的压缩选项测试中已经得到证明；二是即便在慢网环境中，按 MySQL 官方文档的说法，该选项也只是作为一种微调手段。

6.3.5 写入集

在前面测试执行过程中，通过执行下面的查询观察认证队列中的事务数量：

```
select member_id, count_transactions_in_queue
from performance_schema.replication_group_member_stats;
```

发现 count_transactions_in_queue 的值始终很小，大于 0 的时候非常少，而且最大也不过十几。从组复制的流程可知，认证不存在积压情况。但是从库延迟较大，网络无瓶颈，认证不积压，剩下就是针对二进制日志应用程序进行调整了。从前面的组复制性能概述中，自然想到了 MTS 和 WRITESET，于是修改以下配置选项继续测试。

- PRIMARY 启用写入集增加组提交并行度。

```
binlog_transaction_dependency_tracking  = WRITESET
transaction_write_set_extraction       = XXHASH64
```

- SECONDARY 启用 8 线程 MTS。

```
slave_parallel_type = LOGICAL_CLOCK
slave_parallel_workers = 8
slave_preserve_commit_order=1
```

测试结果如下：

TRX: 175459

Master TPS: 487

Elapsed1: 478 Slave1 TPS: 367

Elapsed2: 589 Slave2 TPS: 297

可以看到，总事务数和 PRIMARY 的 TPS 基本不变，但两个 SECONDARY 的 TPS 却增加了 50%以上，吞吐量明显得到提升。即便如此，两个 SECONDARY 的复制延迟还是分别将近 2 分钟和 4 分钟。说明在此压力负载下，SECONDARY 无力追赶上 PRIMARY。如果一定要缩减两者之间的复制延迟时间，就要启用下面介绍的限流措施。

MySQL 官方提供的写入集与多线程下组复制基准测试，参见网址 https://mysqlhighavailability.com/zooming-in-on-group-replication-performance/。

6.3.6 流控

组复制可确保事务仅在组中大多数成员收到事务，并同意所有并发事务之间的相对顺序后提交。如果对组的写入总量不超过组中任何成员的写入容量，则此方法很有效。如果某些成员的写入吞吐量低于其他成员，特别是小于 PRIMARY，那些成员可能会开始落后于 PRIMARY。必须强调的是，MySQL 组复制并不要求成员保持同步，复制延迟只会增加本地中继日志中挂起的事务数。

成员落后于组会带来一些问题，例如：对这些成员的读取可能会使旧数据外化到应用程序；组中其他成员可能必须保存更多复制的上下文，以便能够满足来自慢成员的数据传输请求。组复制协议中存在一种被称为流量控制的机制，可以避免快速成员和慢速成员之间的应用事务存在太大的距离。它试图达到的主要目标是保持成员的已执行事务足够接近。

流量控制依赖于两种基本机制：

- 监控服务器，收集所有组成员的吞吐量和队列大小统计数据，对每个成员可以承受的最大写入量进行有根据的猜测。
- 限制服务器，避免写入超出其可用容量的限制。

1. 探测和统计

监控机制的工作原理是让每个成员部署一组探针，来收集有关其工作队列和吞吐量的信息。成员会定期将该信息传播给组，以便与其他成员共享这些统计数据。此类探针分散在整个插件中，允许建立以下指标：

- 认证队列大小。
- 复制应用程序队列的大小。
- 经认证的事务总数。
- 成员中应用的远程事务总数。
- 本地事务总数。

监控数据每秒与组中其他成员共享一次。一旦成员收到其他成员的统计信息，它将计算在上一个监控周期内认证、应用和本地执行的事务数量。

2. 限流

根据在组中所有服务器上收集的统计数据，限制机制决定成员能够提交新事务的速率。流量控制考虑了两个工作队列：认证队列和二进制日志应用程序队列。因此，用户可以选择在认证或复制应用程序执行流量控制，或者两者都执行。使用以下变量控制相应队列的大小：

- group_replication_flow_control_mode：指定是否启用流控，可取值为 quota 和 disable。默认值为 quota，表示启用流控限额。如果指定为 disable 则禁用流控。
- group_replication_flow_control_certifier_threshold：如果认证队列中的事务数超过该变量值，则触发认证流控。默认值为 25000，值域为 0~2147483647。
- group_replication_flow_control_applier_threshold：如果复制应用队列中的事务数超过该变量值，则触发应用流控。默认值为 25000，值域为 0~2147483647。

如果认证队列或二进制日志应用程序队列大小超过用户定义的阈值，则限制机制将强制执行写入配额。配额将本监控周期的事务数比上一周期减少 10%，以允许触发问题的队列减小其大小。为了避免吞吐量的大幅跳跃，增加吞吐量同样仅允许增长上一个周期的 10%。当前的限制机制不会影响低于配额的事务，但会延迟完成那些超出的事务，直到监控期结束。

从前面的测试中已知，认证队列始终很小，基本不存在积压，复制延迟主要是由 SECONDARY 的复制应用程序缓慢造成的。因此调整 PRIMARY 的 group_replication_flow_control_applier_threshold 变量以对应用队列进行流控，缩小复制延迟。该变量设置为 100、1000 和 10000 时的测试结果如表 6-3 所示。

表 6-3　不同 group_replication_flow_control_applier_threshold 值性能对比

group_replication_flow_control_applier_threshold	TRX	Master TPS	Slave1 Elapsed	Slave1 TPS	Slave2 Elapsed	Slave2 TPS
100	98469	273	361	272	361	272
1000	106442	295	361	294	363	293
10000	162866	452	461	353	556	292

从表 6-3 中看到：

- 当 group_replication_flow_control_applier_threshold=100 时，两个 SECONDARY 基本没有延迟，但整个组的 TPS 只有 273 左右。
- 当 group_replication_flow_control_applier_threshold=1000 时，两个 SECONDARY 的延迟依然很小，整个组的 TPS 为 294，提升了不到 8%。
- 当 group_replication_flow_control_applier_threshold=10000 时，SECONDARY 1 和 SECONDARY 2 分别延迟了 100 秒和 200 秒，PRIMARY、SECONDARY 1 和 SECONDARY 2 的 TPS 分别是 452、353、292。

在本测试环境的压力下，解读表中的结果可得到以下结论：

- 组复制的吞吐量和复制延时无法兼得。
- 要保持复制低延时，组复制的吞吐量取决于最小 TPS 的节点，符合木桶原理。
- group_replication_flow_control_applier_threshold 大于某一阈值时，PRIMARY 与 SECONDARY 间的延迟将变得非常明显。

6.3.7　其他配置

1. 消息分段

当组复制组成员之间发送异常大的消息时，可能会导致某些组成员报告为失败并从组中移除。这是因为组通信引擎使用的单个线程被处理消息占用时间太长，使得某些组成员将接收状态报告为失败。从 MySQL 8.0.16 开始，在默认情况下，大消息会自动拆分为单独发送的消息片段，并由接收方重新组装为原消息。

系统变量 group_replication_communication_max_message_size 指定组复制通信的最大消息之大小，超过该大小的消息将被分段。默认值为 10485760 字节（10MB），允许的最大值与 slave_max_allowed_packet 系统变量的最大值相同，即 1073741824 字节（1GB）。group_replication_communication_max_message_size 的值不能大于 slave_max_allowed_packet 的值，因为复制应用程序的线程无法处理大于 slave_max_allowed_packet 的消息片段。如果要关闭消息分段，可将 group_replication_communication_max_message_size 设置为 0。

当所有组成员收到并重新组装消息的所有片段时，认为消息传递已完成。分段消息包括其消息头中的信息，这些信息在消息传输期间加入，以使成员恢复之前发送的早期消息片段。如果组成员无法恢复消息，则将其从组中移除。为了使复制组使用消息分段，所有组成员必须为 MySQL 8.0.16 或更高版本，并且组使用的复制通信协议版本也必须设置为 8.0.16 或更高版本。

如果复制组因某些成员不支持而无法使用消息分段时，系统变量 group_replication_transaction_size_limit 可用于限制组接收的最大事务。在 MySQL 8.0 中，默认值为 150000000（约 143MB），大于该值的事务将被回滚。

2. XCom 缓存

作为组复制协议的一部分，组复制的通信引擎（XCom，Paxos 变体）包括用于组成员之间交换的消息及其元数据的高速缓存，用于与其他组成员通信超时后重新连接时进行恢复。从 MySQL 8.0.16 开始，可以使用 group_replication_message_cache_size 系统变量设置 XCom 的消息缓存大小，默认值和最小值为 1GB。如果达到缓存大小限制，XCom 将删除已交付的最旧条目。

如果重新连接成员时需要恢复消息，但该消息已从消息高速缓存中删除，则该成员无法重新连接。当使用了 group_replication_member_expel_timeout 系统变量（在 MySQL 8.0.13 中引入）指定可疑成员从组中移除之前的延迟时间，更有可能发生这种情况。组复制通信系统（GCS）通过警告消息告知用户何时从消息缓存中删除了当前无法访问的成员恢复时可能需要的消息。此警告消息记录在所有活动组成员上，表明高速缓存大小可能不足以支持移除成员之前的超时时间。在这种情况下，应参考 group_replication_member_expel_timeout 系统变量指定的时间段内的预期消息量来增加高速缓存大小，以便高速缓存包含成员成功返回所需的所有错过的消息。

可以使用以下语句查询性能模式表 memory_summary_global_by_event_name：

```
select * from performance_schema.memory_summary_global_by_event_name
 where event_name like 'memory/group_rpl/gcs_xcom::xcom_cache'\G
```

这将返回消息缓存的内存使用情况之统计信息，包括当前缓存条目数和缓存的当前大小。如果减小缓存大小的限制，XCom 将删除已确定并交付的最旧条目，直到当前大小低于限制。在此删除过程正在进行时，XCom 可能会暂时超出缓存大小的限制。

3. 故障检测

组复制的故障检测机制旨在识别不再与该组通信的组成员，并在它们可能发生故障时将其移除。通常，所有组成员定期与所有其他组成员交换消息。如果组成员在 5 秒内没有收到来自特定成员的任何消息，会引起对该成员的怀疑。当怀疑超时的时候，就假定被怀疑的成员失败（即出现故障），并将其移出该组。被移除成员从其他成员看到的成员列表中删除，但它不知道自己已被移出该组，因此它将自己视为在线而其他成员无法访问。如果该成员实际上没有失败（例如网络闪断）并且能够恢复与其他成员的通信，则它会收到已被从组中移除的信息。默认情况下，如果怀疑某个成员失败，则会发生以下行为：

（1）当怀疑被创建时，它会立即超时（group_replication_member_expel_timeout 默认值为 0），因此成员只要被怀疑失败就会被立即移除。该成员可能会在超时后存活几秒钟，因为对过期怀疑的检查会定期进行。

（2）如果被驱逐的成员恢复通信并意识到被移除，它不会尝试重新加入该组。

（3）当被驱逐的成员接受其移除时，它会切换到超级只读模式（super_read_only）并等待管理员干预。

这些默认设置优先考虑组能够正确处理请求。但在较慢或不稳定的网络情况下，可能经常需要管理员干预来修复被移除的成员。可使用组复制配置选项永久或临时更改这些行为：

- 可以使用从 MySQL 8.0.13 新增的 group_replication_member_expel_timeout 系统变量，在产生怀疑和移除可疑成员之间留出更多时间，最大为 3600 秒。处于此状态的可疑成员不可访问，但不会从组的成员列表中删除。创建怀疑之前的 5 秒检测周期是不可配置的。
- 可以使用从 MySQL 8.0.16 新增的 group_replication_autorejoin_tries 系统变量指定成员被移除后尝试重新加入该组的次数。默认值为 0 表示成员不尝试重新加入组，并继续执行由 group_replication_exit_state_action 系统变量指定的操作（默认为 READ_ONLY）。每次重试间隔 5 分钟，最多可以进行 2016 次重试。在自动重新加入过程中，被移除成员保持 super_read_only 模式，并在其复制组视图上显示 ERROR 状态。
- 可以使用 group_replication_exit_state_action 系统变量选择未能重新加入（或未尝试）的被移除成员的行为，可选值为 ABORT_SERVER 或 READ_ONLY，默认值为 READ_ONLY。ABORT_SERVER 表示关闭 MySQL 服务器，READ_ONLY 表示切换为 super_read_only 模式。与自动重新加入过程一样，如果成员进入 super_read_only 模式，则读取过时数据的可能性会增加。指示成员自行关闭意味着 MySQL Server 实例不可用，必须重新启动。无论设置了什么退出操作，都需要管理员干预，因为已经耗尽其自动重新加入的尝试（或从未尝试过）并且已被移出组的成员不允许在不重新启动组复制的情况下重新加入。
- 如果希望避免由于不适当的配置产生裂脑情况，可以使用系统变量 group_replication_unreachable_majority_timeout 设置成员在与大多数组成员失去联系后等待的秒数。在此之后，将回滚该成员和少数组中其他成员处理的所有待处理事务，并且该组中的服务器将变为 ERROR 状态，然后执行 group_replication_exit_state_action 指定的退出操作。

6.3.8　主从、半同步、组复制性能对比测试

现在将关注点从组复制性能本身，转移到异步、半同步、组复制三种 MySQL 复制的横向性能对比上。我们最为关心的是不同复制方式对主库 TPS 的影响。测试环境不变，不同的只是更改相应的复制配置。三种复制均为一主两从，具体测试步骤如下：

步骤 01　MySQL 实例基本配置：

```
[mysqld]
server_id=1125      # 两个从库为 1126 和 1127
gtid_mode=ON
enforce-gtid-consistency=true
innodb_buffer_pool_size=4G
disabled_storage_engines="MyISAM,BLACKHOLE,FEDERATED,ARCHIVE,MEMORY"

log_bin=binlog
log_slave_updates=ON
binlog_format=ROW
binlog_checksum=NONE
master_info_repository=TABLE
relay_log_info_repository=TABLE

# 主库启用写集
binlog_transaction_dependency_tracking = WRITESET
transaction_write_set_extraction      = XXHASH64

# 从库使用 8 线程 MTS
slave_parallel_type = LOGICAL_CLOCK
slave_parallel_workers = 8
slave_preserve_commit_order=1
```

步骤 02　启动异步复制：

```
-- 在从库执行
change master to
       master_host = '172.16.1.125',
       master_port = 3306,
       master_user = 'repl',
       master_password = '123456',
       master_auto_position = 1;
start slave;
```

步骤 03　执行异步复制测试：

```
./tpcc_test.sh
```

步骤 04　配置半同步复制：

```
# 主库增加配置：
```

```
plugin-load="rpl_semi_sync_master=semisync_master.so"
rpl_semi_sync_master_enabled=1

# 从库增加配置：
plugin-load="rpl_semi_sync_slave=semisync_slave.so"
rpl_semi_sync_slave_enabled=1
```

然后重启三个 MySQL 实例。

步骤01 执行半同步复制测试：

```
./tpcc_test.sh
```

步骤02 组复制配置：

去掉上一步的半同步配置，并增加组复制配置：

```
plugin-load=group_replication.so
group_replication_group_name="aaaaaaaa-aaaa-aaaa-aaaa-aaaaaaaaaaaa"
group_replication_start_on_boot=off
group_replication_local_address= "172.16.1.125:33061" # 两个 SECONDARY 为
"172.16.1.126:33061"、"172.16.1.127:33061"
group_replication_group_seeds=
"172.16.1.125:33061,172.16.1.126:33061,172.16.1.127:33061"
group_replication_bootstrap_group=off
```

然后重启三个 MySQL 实例：

```
-- PRIMARY 执行
reset master;
reset slave all;
change master to master_user='repl', master_password='123456' for channel
'group_replication_recovery';
set global group_replication_bootstrap_group=on;
start group_replication;
set global group_replication_bootstrap_group=off;

-- SECONDARY 执行
reset master;
stop slave;
reset slave all;
change master to master_user='repl', master_password='123456' for channel
'group_replication_recovery';
start group_replication;
```

步骤03 执行组复制测试：

```
./tpcc_test.sh
```

三种复制测试结果如表 6-4 所示。

表 6-4　异步复制、半同步复制、组复制性能对比

复制方式	TRX	Master TPS	Slave1 Elapsed	Slave1 TPS	Slave2 Elapsed	Slave2 TPS
异步	196808	546	360	546	435	452
半同步	189303	525	361	524	414	457
组复制	166032	461	462	359	561	295

从测试结果可以看到，相同条件下：

- 半同步复制主库的 TPS 是异步复制的 96%，组复制主库的 TPS 是异步复制的 84%。
- 在复制延迟上，从库 1 的异步复制和半同步复制基本无延迟，从库 2 的异步复制延迟还大于半同步复制；而两个从库的组复制延迟都很大（100 秒和 200 秒）。
- 异步复制与半同步复制没有提供缩小复制延迟的机制；组复制可以通过流控机制，减少 SECONDARY 复制延迟，代价是将组复制整体吞吐量拉低到组中吞吐量最差节点的水平。

由于组复制会带来整体吞吐量的下降，大多数情况下，使用传统主从异步复制即可。如果在读写分离场景下要求读取一致性，可以考虑半同步复制。与异步复制相比，半同步复制的吞吐量并没有显著衰减。只有在首要需求为多主可写、自动处理事务冲突、PRIMARY 失败自动转移等情况下，再考虑使用组复制。

MySQL 官方提供的组复制性能基准测试，可参见网址 http://mysqlhighavailability.com/an-overview-of-the-group-replication-performance/。

6.4　组复制要求与限制

目前的 MySQL 组复制对于存储引擎、网络带宽、表设计，以及服务器实例的配置还存在诸多要求与限制，尤其是多主模式，使用时更要格外注意。下面是一些关于 MySQL 组复制的已知问题，在实际应用中使用组复制前，有必要了解它们以帮助做出正确的选择。在“踩坑”前做到未雨绸缪总是有益的。

6.4.1　组复制要求

1. 基础架构

- 数据必须存储在 InnoDB 事务存储引擎中。事务以乐观方式执行，然后在提交时检查冲突。如果存在冲突，为了保持整个组的一致性，将回滚一些事务，因此需要事务存储引擎。此外，InnoDB 还提供了一些附加功能，可以在与组复制一起操作时更好地管理和处理冲突。使用其他存储引擎，包括临时 MEMORY 存储引擎，可能会导致组复制出错。可以通过在组成员上设置 disabled_storage_engines 系统变量来阻止使用其他存储引擎，例如：disabled_storage_engines="MyISAM,BLACKHOLE,FEDERATED,ARCHIVE,MEMORY"。
- 组复制的每个表必须具有主键，或者具有等效的非空唯一键。它们作为表中每一行的唯一标

识符是必需的，这使得系统能够通过准确识别每个事务已修改的行来确定哪些事务存在冲突。

- 网络性能会影响组的性能，网络延迟和网络带宽都会影响组复制性能及稳定性。因此组复制中的 MySQL 服务器实例应该部署在彼此非常接近的集群环境中，使得所有组成员之间始终保持双向通信。如果阻止服务器实例的收发消息，例如通过防火墙限制，则该成员无法在组中运行，并且组成员（包括有问题的成员）可能无法报告受影响的服务器实例的正确成员状态。从 MySQL 8.0.14 开始，可以使用 IPv4 或 IPv6 网络基础结构，或两者的混合，用于远程组复制服务器之间的 TCP 通信。

2. 服务器实例配置

必须在作为组成员的服务器实例上配置以下选项：

- 设置--log-bin [= log_file_name] 激活二进制日志，MySQL 8 中默认启用此选项。与其他 MySQL 复制方式一样，组复制需要复制二进制日志的内容，因此需要打开二进制日志才能运行。
- 设置--log-slave-updates 记录从库更新。服务器需要记录复制应用程序所应用的二进制日志。组中的服务器需要记录它们收到的所有事务并从组中予以应用。这是必不可少的配置，因为分布式恢复要依赖组中成员的二进制日志方能进行。因此，每个事务的副本都需要存放在每台服务器上，即使对于那些未在服务器上启动的事务也是如此。MySQL 8 中默认启用此选项。
- 设置--binlog-format = row 就是将二进制日志设置为行格式。组复制依赖于基于行的复制格式，以在组成员之间一致地传播更改。它依赖于基于行的基础结构来提取必要信息，以检测在组中不同服务器并发执行的事务之间的冲突。
- 设置--binlog-checksum = NONE 关闭二进制日志校验。由于复制事件校验和设计的限制，组复制无法使用它们，因此必须禁用。
- 设置--gtid-mode = ON 可启用全局事务标识符。组复制使用全局事务标识符来准确跟踪在每台服务器实例上已提交的事务，从而能够推断哪些服务器执行的事务可能与其他地方已提交的事务冲突。换句话说，显式的事务标识符是框架的基本部分，以便能够确定哪些事务可能发生冲突。
- 设置--master-info-repository = TABLE 和--relay-log-info-repository = TABLE，将复制信息资料库存储到表中。复制应用程序需要将主库信息和中继日志元数据写入 mysql.slave_master_info 和 mysql.slave_relay_log_info 系统表。这可确保组复制插件具有一致的可复制性和复制元数据的事务管理。从 MySQL 8.0.2 开始，这些选项默认值为 TABLE，而从 MySQL 8.0.3 开始，不推荐使用 FILE 设置。
- 设置--transaction-write-set-extraction = XXHASH64，以便在将数据行记录到二进制日志时，服务器也会收集写入集。写入集基于每行的主键，唯一标识已更改的行，用于检测事务冲突。MySQL 8 中默认启用此选项。
- 推荐设置 slave_parallel_workers 为大于零的值，以启用组成员上的多线程复制应用程序，最多可以指定 1024 个并行复制应用程序线程。设置 slave_preserve_commit_order = 1，确保并行事务的最终提交与原始事务的顺序相同，这是组复制所需的，它依赖于所有组成员以相同顺序接收和应用已提交事务，以保证并行事务的数据一致性。slave_preserved_commit_order = 1 需要设置 slave_parallel_type = LOGICAL_CLOCK，该变量指定用于决定允许哪些事务在从库上并行执行的策略。设置 slave_parallel_workers = 0 会禁用并行复制，并为从库提供单个应用程序线程，而不是协调器线程。使用该设置，slave_parallel_type 和 slave_preserve_commit_order 选项无效并被忽略。

6.4.2　组复制限制

组复制存在以下已知限制：

- 使用 MINIMAL 选项（--upgrade = MINIMAL）的 MySQL 服务器升级后，无法启动组复制，该选项不会升级复制内部所依赖的系统表。
- 认证过程没有考虑间隙锁。除非应用程序中依赖 REPEATABLE READ 语义，否则 MySQL 建议将 READ COMMITTED 隔离级别与组复制一起使用。InnoDB 在 READ COMMITTED 中不使用间隙锁，它将 InnoDB 中的本地冲突检测与组复制执行的分布式冲突检测统一化。
- 认证过程不考虑表锁（LOCK TABLES 和 UNLOCK TABLES）或命名锁（GET_LOCK）。
- 组复制不支持 SERIALIZABLE 隔离级别。将事务隔离级别设置为 SERIALIZABLE 会使组拒绝提交事务。
- 使用多主模式时，不支持针对同一对象但在不同服务器上执行的并发数据定义语句（DDL）和数据操作语句（DML）。
- 多主模式不支持具有多级外键依赖关系的表，特别是具有已定义 CASCADING 外键约束的表。MySQL 建议在多主模式组复制中设置 group_replication_enforce_update_everywhere_checks = ON，以避免未检测到的冲突。
- 当组复制以多主模式运行时，SELECT .. FOR UPDATE 语句可能导致死锁。
- 全局复制过滤器不能在为组复制配置的 MySQL 服务器实例上使用。
- MySQL 服务器自 MySQL 8.0.16 版本起可以支持 TLSv1.3 协议，并且需要使用 OpenSSL 1.1.1 或更高版本编译 MySQL。

如果单个事务太大，以至于在 5 秒钟内无法通过网络在组成员之间复制消息，则可能会怀疑成员失败，然后被移出组。由于内存分配问题，大型事务也可能导致系统速度变慢。要避免这些问题，使用以下缓解措施：

- 尽可能尝试限制事务规模。例如，将与 LOAD DATA 一起使用的文件拆分为较小的块。
- 使用系统变量 group_replication_transaction_size_limit 指定组接收的最大事务的大小。超过此大小的事务将回滚，不会发送到组。在 MySQL 8.0 中，此系统变量默认值为 150000000 字节（大约 143 MB）。
- 从 MySQL 8.0.13 开始，可以使用系统变量 group_replication_member_expel_timeout 来允许在怀疑失败的成员被移出之前有更多的时间。可以在最初的 5 秒检测期后最多延长一个小时。
- 从 MySQL 8.0.16 开始，大型消息会自动分段，这意味着大型消息不会触发引发怀疑的 5 秒检测周期，除非此时存在其他网络问题。为了使复制组使用分段，所有组成员必须处于 MySQL 8.0.16 或更高版本，并且组使用的组复制通信协议版本必须允许分段。如果 MySQL 版本不支持消息分段，可以使用系统变量 group_replication_communication_max_message_size 来调整最大消息的大小，默认值为 10485760 字节（10MB），或通过指定零值来关闭分段。

6.5 小 结

组复制（MySQL Group Replication，MGR）是 MySQL 5.7.17 引入的插件，提供了故障检测、组成员、容错等服务，用以实现多主复制、自动故障转移等集群功能。在实现上，组复制在二进制日志、基于行的日志记录格式、全局事务标识符（GTID）和多线程复制等原有的 MySQL 复制架构基础上，增加了自动的分布式恢复和事务冲突检测。组是动态的，当有服务器加入组时，组会通过从现有服务器获取状态自动更新成员，数据通过 MySQL 异步复制进行传输。如果服务器离开该组，其余服务器会知道它已离开并自动重新配置该组。组复制支持单主模式和多主模式部署方式，默认使用单主模式。就目前来说，组复制限制较多，性能也比异步复制或半同步复制有较大差距，用户应根据自己的实际应用需求来选择使用。

第 7 章

MySQL Router

7.1　MySQL Router 简介

MySQL Router 最早是作为 MySQL-Proxy 的替代方案出现的。MySQL 官方不建议将 MySQL Proxy 用于生产环境，并且已经不提供 MySQL Proxy 的下载。作为一个轻量级中间件，MySQL Router 可在应用程序和后端 MySQL 服务器之间提供透明路由和负载均衡功能，从而有效地提高 MySQL 数据库服务的高可用性与可伸缩性。

MySQL Router 2.0 是其初始版本，可以与 MySQL Fabric 无缝连接，允许 Fabric 存储和管理用于路由的高可用数据库服务器组，使管理 MySQL 服务器组更加简单。但该版本已被弃用，不再支持。MySQL Router 2.1 是为支持 MySQL InnoDB Cluster 而引入，MySQL Router 8.0 则是 2.1 版本上的扩展，其版本号与 MySQL 服务器版本号保持一致，即 Router 2.1.5 作为 Router 8.0.3（版本号同 MySQL Server 8.0.3）发布，2.1.x 分支被 8.0.x 取代。这两个分支完全兼容，当前最新版本为 8.0.17。MySQL 强烈建议使用 Router 8 与 MySQL Server 8 或 5.7 一起使用。

MySQL Router 表现为一个可执行文件，可以与应用程序在同一平台上运行，也可以单独部署。虽然 MySQL Router 是 InnoDB Cluster 的一部分，MySQL 5.6 等版本数据库仍然可以使用 Router 作为它的中间代理层。MySQL Router 的配置文件中包含有关如何执行路由的信息。它与 MySQL 服务器的配置文件类似，也是由多个段组成的，每个段中包含相关配置选项。

7.1.1　透明路由

MySQL Router 通过智能地将客户端路由到 MySQL 服务器来简化应用程序的开发。想象在一个主从复制或组复制环境中，当主库（主服务器）发生问题时执行故障转移。这可能是以手工方式提升一个从库作为新主库，或者是基于 Paxos 协议自动选举一个新主库。无论哪种情况，这里都存在一个问题：如果应用程序直连数据库实例，当发生主从切换时可用的数据库 IP 地址发生了变化，客户端应用程序必须自行修改它的连接配置。这种方案极不现实，因为需要应用程序了解复制的拓扑结构并知道哪个 MySQL 实例是主库，对于应用程序显然有些强人所难，这些处理逻辑本该对应用透明。

对于这种情况，正是 MySQL Router 的用武之地。Router 作为一个流量转发层，位于应用与 MySQL 服务器之间。应用不再直连 MySQL 实例，而是与 Router 相连。根据 Router 的配置，将会把应用程序的读写请求转发给下游的 MySQL 服务器。当下游有多台 MySQL 服务器，无论主库还是从库，Router 可以对读写请求进行负载均衡。当某台 MySQL 服务器失效时，Router 可以将其从 Active 列表中移除，当其在线后再次加入 Active 列表，由此提供了 Failover 特性。

当 MySQL 发生复制拓扑结构变更时（比如增减了从库节点），只需要修改 Router 的配置即可，而无需修改应用的数据库连接，因为应用配置的是 Router 地址而非 MySQL 实例的原始地址，即数据库拓扑结构对应用来说是透明的。

在与 InnoDB Cluster 一起使用时，MySQL Router 充当代理，向应用程序隐藏网络上的多个 MySQL 实例，并将数据请求映射到其中一个集群实例。只要有足够的在线副本并且组件之间的通信完好，客户端就能够连接其中一个实例，保持对外服务的连续性。应用程序要做的只是连接到 Router，其他的交给 Router 处理。MySQL Router 的推荐部署模型是与 InnoDB Cluster 集成。

7.1.2 元数据缓存

如果 MySQL 服务器为 5.7+版本，且构建为 InnoDB Cluster 模式，那么 Router 还能使用元数据缓存机制，感知 MySQL 服务器的主从切换、从库增减等集群拓扑结构的变更，而且基于变更能够实现主从自动切换和服务器列表自动更新。比如主库失效后，集群将会自动选举一个新主库，此时 Router 不需要进行任何调整，可以自动发现此新主库进而继续为应用服务。

MySQL Router 保留在线 MySQL 服务器的缓存列表，以及配置的 InnoDB Cluster 的拓扑结构和状态。Router 启动时，列表从其配置文件加载。当使用--bootstrap 选项引导 Router 时，此列表由 InnoDB Cluster 服务器生成。为了更新缓存，Router 元数据缓存组件与包含元数据的 InnoDB Cluster 服务器之一保持连接，通过从该 MySQL 实例的 performance_schema 库表查询元数据和实时状态信息来实现。每当修改 InnoDB 集群时都会更改集群元数据，并且只要检测到集群状态更改，就会通过 MySQL 服务器的 Group Replication 插件实时更新 performance_schema 库表。

当 Router 检测到连接的 MySQL 服务器关闭时，它会尝试连接到不同的 MySQL 服务器以从新的服务器获取元数据和 InnoDB Cluster 状态。关闭的 MySQL 服务器的应用程序连接会自动关闭。应用程序必须重新连接到 Router，这要求应用程序实现重连机制。Router 将它们重定向到其他在线 MySQL 服务器。

7.1.3 简单重定向

Router 将客户端连接重定向到可用的 MySQL 服务器，这意味着数据包是在未经检查的情况下整体路由的。如果连接失败，应用程序需要重试连接，Router 在尝试连接失败后选择新的 MySQL 服务器，这被称为简单重定向连接路由。也就是说，如果从 Router 到 MySQL 服务器的连接中断，则应用程序会遇到连接失败，但新的连接尝试会触发 Router 查找并连接到另一台 MySQL 服务器。路由连接的服务器和路由策略在配置文件中定义。

使用 Router 不需要特定的库或接口，与不使用 Router 相比，应用程序所感知的唯一区别是如何建立与 MySQL 服务器的连接。因为 Router 在尝试连接时只是单纯重定向，并不读取数据包或执行分析，所以应用程序需要捕获连接错误并重试连接到 Router。

使用 MySQL Router 的工作流程如下：

（1）MySQL 客户端连接到 Router。

（2）Router 检查可用的 MySQL 服务器。

（3）Router 打开一个适用 MySQL 服务器的连接。

（4）Router 在应用程序和 MySQL 服务器之间来回转发数据包。

（5）如果连接的 MySQL 服务器出现故障，Router 将断开连接。当应用程序重试连接到 Router 时，Router 选择另一个可用 MySQL 服务器。

Router 通常是解决比如跨机房部署、流量迁移、异构兼容，或者 MySQL 集群规模性宕机时快速切换等问题。Router 中间件本身不会对请求“拆包”（Unpackage），所以无法在 Router 中间件上实现比如 SQL 审计、限流、分库分表等功能。但是 Router 提供了 C 语言的插件机制，用户可以开发自己的插件来扩展 Router 的额外特性。

7.1.4 部署与使用

考虑到 Router 独立部署可能引入额外的成本、性能下降、连接数上限等问题，部署 MySQL Router 时建议但不强制遵循以下两点：

- 在与应用程序相同的主机上安装并运行 Router。
- 使用配置文件中的 bind_port = 127.0.0.1:<port>将 Router 绑定到 localhost，或者禁用 TCP 连接（--conf-skip-tcp），并将其限制为仅使用 Unix 套接字连接（--conf-use-sockets）。

每当在网络中引入通信组件时，都会产生一定的开销，并且会受到工作负载的严重影响。幸运的是，MySQL Router 的性能影响非常小。官方文档显示，当前版本的简单重定向连接路由，其速度与直连数据库相比仅慢约 1%。

MySQL Router 是一个单独的应用程序，mysqlrouter 程序是 Router 的核心。用户可以自己编写监控 mysqlrouter 运行情况的程序，并在需要时重新启动它。例如，如果目标选项中的服务器列表已用尽，则可以重启 Router 以重试列表中的服务器，或者使用新目标列表重启 Router。

MySQL Router 对应用是透明的，开发与 Router 一起使用的应用程序所增加的工作只是维护 Router 实例。可以在整个网络中运行多个 Router 实例。但是 MySQL 官方并没有提供 Router 集群，即每个 Router 节点均为独立运行，它们之间互不通信，无主从角色，无选举机制。当某个 Router 节点失效，应用层面需要借助 MySQL Connector 的高级特性实现失败转移功能。简而言之，Router 中间件与 Connector 的高级协议互相协作，才能够实现请求在 Router 之间的负载均衡和 Failover 等功能。

Router 本身的 CPU、内存、磁盘消耗都极低，但是要求 Router 节点对网络 I/O 的支撑能力应该较强。考虑到 Router 底层为异步 I/O，如果条件允许，应该构建在较高版本的 Linux 平台下，而且给予合理的 CPU 资源。MySQL Router 在 2.1.4 版本以下，内核基于 select() I/O 模型，存在连接数 500 上限、较大 SQL 请求导致 CPU 过高，以及并发连接过高时 Router 假死等问题，建议升级到 2.1.6+。

Router 对连接的管理是基于"黏性"方式，即应用与 Router 的一个 TCP 连接，将对应一个 Router 与 MySQL 服务器的连接，当应用与 Router 的连接失效时，Router 也将断开其与 MySQL 服务器的连接。只要 Router 上下游网络正常连通，那么 Router 将不会主动断开与应用的连接，也不会切换其与服务器的连接。当应用与 Router 创建一个新连接时，Router 将根据负载均衡算法，选择一台服务器并与其建立连接，此后将唯一绑定，直到此服务器失效时触发重新选择其他服务器。这就引入了一个问题，如果某个连接上发生了"繁重"的 SQL 操作，那么将会导致下游服务器伴随高负载而无法"负载均衡"。

7.1.5 MySQL Router 8.0 的新特性

较以前 2.x 版本，MySQL Router 8.0 主要新增了如下功能：

- 添加了可选的 routing_strategy 配置选项。可用值为 first-available、next-available、round-robin 和 round-robin-with-fallback。在以前版本中，这些策略对应 mode 选项中的调度模式，其中 read-write 对应 first-available，而 read-only 对应 round-robin。它们保留了这些模式的先前行为。
- 添加了--ssl-key 和--ssl-cert 命令行选项，指定客户端证书和私钥以方便客户端身份验证，用于使用 REQUIRE X509 创建 root 账户。
- 添加了 connect_timeout 和 read_timeout 元数据配置文件选项。它们在[DEFAULT]命名空间下定义，并影响元数据服务器连接等内部操作。
- Bootstrap 接受 InnoDB Cluster 的任何成员，并自动查找和重新连接到可写服务器，而以前只接受主库。
- Bootstrap 接受--config 选项并读取[logger]日志级别选项的定义。
- 最大并发客户端连接数从 500 增加到 5000。
- 添加了一个新的 mysqlrouter_plugin_info 实用程序来帮助调试 MySQL Router 插件。

7.2 Router 实现 MySQL 高可用示例

MySQL Router 是个轻量级中间件，核心就一个可执行文件，安装、配置和使用都比较简单。本节用一个例子演示使用 Router 2.1.6 实现 MySQL 高可用、负载均衡与读写分离。

7.2.1 安装和配置

环境：

172.16.1.125：MySQL Router

172.16.1.126：MySQL Replication Master

172.16.1.127：MySQL Replication Slave

我们在 172.16.1.125 上安装和配置 MySQL Router，172.16.1.126 和 172.16.1.127 为本例中要通过 Router 访问的两台 MySQL 服务器地址。在本例中这两台 MySQL 服务器已经配置好主从复制，拓扑结构如图 7-1 所示，7001 和 7002 是路由端口。

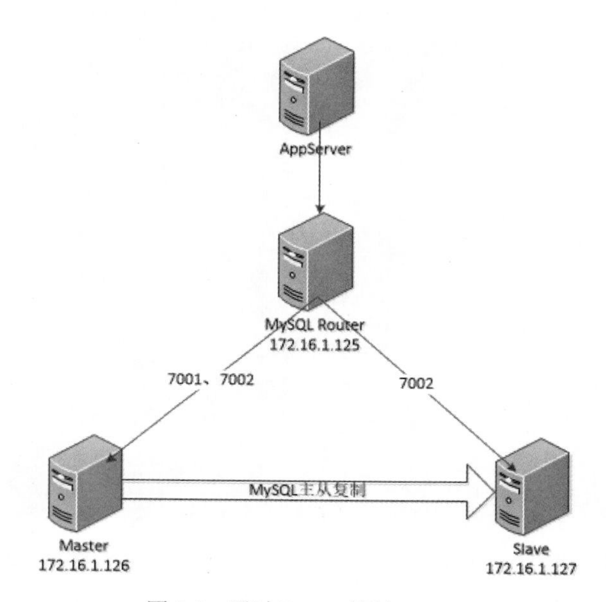

图 7-1 通过 Router 连接 MySQL

1. 安装 MySQL Router 2.1.6

（1）下载二进制安装包

从 https://dev.mysql.com/downloads/router/2.1.html 页面选择下载的安装包，本例为 mysql-router-2.1.6-linux-glibc2.12-x86-64bit.tar.gz。

（2）解压缩

```
tar xzf mysql-router-2.1.6-linux-glibc2.12-x86-64bit.tar.gz
mv mysql-router-2.1.6-linux-glibc2.12-x86-64bit mysql-router-2.1.6
```

（3）添加执行文件路径并使之生效

```
# 在资源文件.bashrc 中添加
echo "export PATH=\$PATH:/home/mysql/mysql-router-2.1.6/bin;" >> ~/.bashrc

# 使资源配置生效
source ~/.bashrc
```

（4）验证安装

```
mysqlrouter --help
```

从 mysqlrouter 联机帮助的输出中，可以看到默认配置文件寻找路径及其顺序，插件路径、日志目录、持久化数据目录和运行时状态目录的默认位置等重要信息。在后面的配置文件和服务启停文件中需要定义这些目录。注意，如果在 mysqlrouter 命令行使用--config 或-c 选项传入用户定义的配置文件，则不会加载默认配置文件。

2. 配置 Router

```
# 复制配置文件
cp /home/mysql/mysql-router-2.1.6/share/doc/mysqlrouter/
sample_mysqlrouter.conf /etc/mysqlrouter.conf
cp /home/mysql/mysql-router-2.1.6/share/doc/mysqlrouter/
sample_mysqlrouter.init /etc/init.d/mysqlrouter

# 修改属主为 mysql
chown mysql:mysql /etc/mysqlrouter.conf
chown mysql:mysql /etc/init.d/mysqlrouter

# 变为可执行
chmod +x /etc/init.d/mysqlrouter

# 系统启动时自动执行
echo "/etc/init.d/mysqlrouter" >> /etc/rc.d/rc.local

# 建立日志目录
mkdir /home/mysql/mysql-router-2.1.6/log
```

配置文件/etc/mysqlrouter.conf 内容如下：

```
[DEFAULT]
# 日志路径
logging_folder = /home/mysql/mysql-router-2.1.6/log
# 插件路径
plugin_folder = /home/mysql/mysql-router-2.1.6/lib/mysqlrouter
# 配置路径
config_folder = /home/mysql/mysql-router-2.1.6/config
# 运行时状态路径
runtime_folder = /home/mysql/mysql-router-2.1.6/run
# 数据文件路径
data_folder = /home/mysql/mysql-router-2.1.6/data

[logger]
# 日志级别
level = INFO

# 以下选项用于路由标识的策略部分
[routing:basic_failover]
# Router 地址
bind_address = 172.16.1.125
# Router 端口
bind_port = 7001
# 读写模式
```

```
mode = read-write
# 目标服务器
destinations = 172.16.1.126:3306,172.16.1.127:3306

[routing:load_balance]
bind_address = 172.16.1.125
bind_port = 7002
mode = read-only
destinations = 172.16.1.126:3306,172.16.1.127:3306
```

MySQL Router 的配置文件比较简单，大部分配置项的含义一目了然。上面的文件中配置了两条路由策略，一个用于故障转移，一个用于负载均衡，绑定端口分别是 7001 和 7002。值得一提的是 mode 参数，该参数的可选值为 read-write 或 read-only，但其实际作用并不是字面含义所示的那样。

对于 read-write 模式，将采用“首个可用”算法，优先使用第一台服务器，当第一台服务器（172.16.1.126:3306）不可达时，将会切换到第二台服务器（172.16.1.127:3306），依次进行。如果都不可达，那么此端口上的请求将会被中断，此端口将不能提供服务，且此时所属的路由策略将不可用。需要注意的是，此算法只遍历一次列表，即逐个验证 destinations 中的服务器，不会循环。一旦所有的服务器依次验证且不可用之后，本条路由策略将不能继续服务，内置状态设定为 aborted（中止），即使此后服务器恢复上线，也不能继续对客户端提供服务，因为它不会与服务器保持心跳检测。对于 Router（路由器）而言，此时会直接拒绝客户端的连接请求，只有重启 Router 节点才能解决。

对于 read-only 模式：将采用“轮询”算法，依次选择服务器新建连接，如果某台服务器不可达，将会重试下一台服务器，如果所有的服务器都不可达，那么此端口上的请求将中断，即读写操作将不可用。同时 Router 将会持续与每台服务器保持心跳探测，当服务器恢复后重新加入 Active 列表，此后那些新建的连接请求将可以分发给此台服务器。

但是比较遗憾的是，Router 不会将已有的连接重新分配给“新加入”列表的服务器，比如 Router 有 2 个服务器（S1, S2），某时刻 S1 不可达，那么在 S1 上黏性的客户端连接也将被断开，新建连接将会全部路由在 S2 上。此后 S1 恢复正常，那么在 S2 上的已有连接将不会迁移到 S1 上，此时 S1 只会接收新的连接，如果没有新连接的请求，那么 S1 将会在一段时间看起来是“不提供服务”的。为了解决此问题，通常要求 Connection Pool（连接池）有管理“连接生命周期”的相关控制，比如一个连接被创建 X 秒后，在返回连接池时应该被主动关闭，这个参数在 tomcat-jdbc-pool 中为“maxAge”。

如果在应用程序中部署方式是“一主多从”，我们完全可以在承接主库请求的 Router 路由策略上也配置为“read-only”模式，那么在此单主库节点失效重启后，可以不需要重启 Router 即可继续服务。因为 Router 不会对 TPC 拆包，所有“read-write”和“read-only”并不会干扰实际的 SQL 执行。严格来说，这两种模式映射两种“路由算法”：“首个可用”和“轮询”。除此之外，再无特殊含义。

对于读和写两种操作，因为 Router 不对请求拆包，所以它无法判断请求的读写类型。我们只能在配置文件中分别为读和写设定不同的配置：使用不同的绑定端口。比如本例“7001”端口接收到的请求都会转发给 172.16.1.126:3306，当它不可用时，都会转发给 172.16.1.127:3306。“7002”

端口接收的请求则会以轮询方式转发给 172.16.1.126:3306 和 172.16.1.127:3306。

MySQL Router 服务启动停止命令文件/etc/init.d/mysqlrouter 内容如下：

```
#! /bin/bash

# Source function library
. /etc/rc.d/init.d/functions

# Source networking configuration
. /etc/sysconfig/network

# add general install path
base_dir=/home/mysql/mysql-router-2.1.6
# fix exec path
exec=${base_dir}/bin/mysqlrouter
prog=mysqlrouter
piddir=${base_dir}/run
pidfile=${piddir}/mysqlrouter.pid
logdir=${base_dir}/log
logfile=$logdir/mysqlrouter.log
lockfile=/var/lock/subsys/$prog

# add conf path
conf=/etc/mysqlrouter.conf

start () {
[ -d $piddir ] || mkdir -p $piddir
chown mysql:mysql $piddir
[ -d $logdir ] || mkdir -p $logdir
chown mysql:mysql $logdir
[ -e $logfile ] || touch $logfile
chown mysql:mysql $logfile
export ROUTER_PID=$pidfile
# add opt -c to resolv mysqlrouter.ini
daemon --user mysql $exec -c $conf >/dev/null 2>&1 &  #
ret=$?
if [ $ret -eq "0" ]; then
action $"Starting $prog: " /bin/true
touch /var/lock/subsys/$prog
else
action $"Starting $prog: " /bin/false
fi
return $ret
}

stop () {
[ -f /var/lock/subsys/$prog ] || return 0
killproc mysqlrouter >/dev/null 2>&1
ret=$?
if [ $ret -eq "0" ]; then
```

```
    rm -f $pidfile
    rm -f /var/lock/subsys/$prog
    action $"Stopping $prog: " /bin/true
    else
    action $"Stopping $prog: " /bin/false
    fi
    }

    restart () {
    stop
    start
    }

    condrestart () {
    [ -e /var/lock/subsys/$prog ] && restart || return 0
    }

    case "$1" in
    start)
    start
    ;;
    stop)
    stop
    ;;
    status)
    status -p "$pidfile" $prog
    ;;
    restart)
    restart
    ;;
    condrestart|try-restart)
    condrestart
    ;;
    reload)
    exit 3
    ;;
    force-reload)
    restart
    ;;
    *)
    echo $"Usage: $0
{start|stop|status|condrestart|try-restart|reload|force-reload}"
    exit 2
    esac

    exit $?
```

当 mysqlrouter 程序意外被 kill 后，可能相关程序运行标识还会存在。这种情况下需要先删除标识文件再启动，也就是脚本中以下两条命令的作用，否则程序会提示有一个实例正在运行而不能启动该服务：

```
    rm -f $pidfile
    rm -f /var/lock/subsys/$prog
```

很多程序需要判断是否当前已经有一个实例在运行，/var/lock/subsys 目录就是让程序判断是否有实例运行的标志。比如说 xinetd，如果目录下存在这个文件，表示 xinetd 已经在运行了，否则就是没有。当然，程序里面还要有相应的判断措施来真正确定是否有实例在运行。通常与该目录配套的还有/var/run 目录，用来存放对应实例的 PID。在编写脚本时，会发现这两个目录结合起来可以很方便地判断出许多服务是否在运行，以及运行的相关信息等等。实际上，判断是否上锁就是判断这个文件，所以文件存在与否也就隐含了是否上锁。但是这个目录的内容并不能表示一定上锁了，因为很多服务在启动脚本里使用 touch 来创建这个加锁文件，在系统结束时该脚本负责清除锁，这本身就不可靠，比如意外失败导致锁文件仍然存在。所以脚本里一般结合 PID 文件，如果有 PID 文件的话，从 PID 文件里得到该实例的 PID，然后用 ps 测试是否存在该 PID，从而判断是否真正有这个实例在运行。更加稳妥的方法是用进程通信，不过这样的话单靠脚本就无法完成了。

3. 启动 Router

用 root 用户执行 mysqlrouter 服务启动命令：

```
[root@hdp2~]#service mysqlrouter start
Starting mysqlrouter (via systemctl):                          [  OK  ]
[root@hdp2~]#
```

查看日志文件，显示两个路由策略的监听器已经启动。

```
    [mysql@hdp2~]$more /home/mysql/mysql-router-2.1.6/log/mysqlrouter.log
    2018-07-18 17:04:11 INFO    [7fa3437fb700] [routing:load_balance] started: 
listening on 172.16.1.125:7002; read-only
    2018-07-18 17:04:11 INFO    [7fa343ffc700] [routing:basic_failover] started: 
listening on 172.16.1.125:7001; read-write
    [mysql@hdp2~]$
```

如果在启动服务时出现类似 log_daemon_msg: command not found 这样的错误，按如下步骤处理：

步骤 01 安装 redhat-lsb-core：

```
yum -y install redhat-lsb-core
```

步骤 02 编辑/usr/lib/lsb/init-functions 文件，在文件最后添加如下内容：

```
log_daemon_msg () {
    # Dummy function to be replaced by LSB library.
    echo $@
}
log_progress_msg() {
   echo $@
}
log_end_msg () {
    # Dummy function to be replaced by LSB library.
    if test "$1" != "0"; then
```

```
        echo "Error with $DESCRIPTION: $NAME"
    fi
    return $1
}
```

7.2.2 自动故障转移

当 172.16.1.126 可用时，对于 7001 端口的请求，会全部发送到 172.16.1.126；对 7002 端口的请求，会以轮询方式发送给 172.16.1.126 和 172.16.1.127。

```
C:\>mysql -utest -p123456 -h172.16.1.125 -P7001 -N -s -e "show variables like
'server_id'"
mysql: [Warning] Using a password on the command line interface can be insecure.
server_id       126

C:\>mysql -utest -p123456 -h172.16.1.125 -P7001 -N -s -e "show variables like
'server_id'"
mysql: [Warning] Using a password on the command line interface can be insecure.
server_id       126

C:\>mysql -utest -p123456 -h172.16.1.125 -P7002 -N -s -e "show variables like
'server_id'"
mysql: [Warning] Using a password on the command line interface can be insecure.
server_id       126

C:\>mysql -utest -p123456 -h172.16.1.125 -P7002 -N -s -e "show variables like
'server_id'"
mysql: [Warning] Using a password on the command line interface can be insecure.
server_id       127
```

当 172.16.1.126 不可用时，对于 7001 端口的请求，会自动切换到 172.16.1.127；而对于 7002 端口的请求，会全部转移到 172.16.1.127。

首先杀掉 172.16.1.126 的 mysqld 进程：

```
pkill -9 mysqld
```

然后查看路由的目标服务器：

```
C:\>mysql -utest -p123456 -h172.16.1.125 -P7001 -N -s -e "show variables like
'server_id'"
mysql: [Warning] Using a password on the command line interface can be insecure.
server_id       127

C:\>mysql -utest -p123456 -h172.16.1.125 -P7001 -N -s -e "show variables like
'server_id'"
mysql: [Warning] Using a password on the command line interface can be insecure.
server_id       127

C:\>mysql -utest -p123456 -h172.16.1.125 -P7002 -N -s -e "show variables like
'server_id'"
```

```
    mysql: [Warning] Using a password on the command line interface can be insecure.
    server_id       127

    C:\>mysql -utest -p123456 -h172.16.1.125 -P7002 -N -s -e "show variables like
'server_id'"
    mysql: [Warning] Using a password on the command line interface can be insecure.
    server_id       127
```

当 172.16.1.126 再次可用时（假设复制已经重新搭建，主从角色已经互换，172.16.1.127 为主，172.16.1.126 为从），对于 7001 端口的请求，还是会路由到 172.16.1.127，而不会自动转到 172.16.1.126。而对于 7002 端口的请求，会自动继续以轮询方式发送给 172.16.1.126 和 172.16.1.127 这两台服务器。

启动 172.16.1.126 的 MySQL 服务：

```
    service mysql start
```

查看路由的目标服务器：

```
    C:\>mysql -utest -p123456 -h172.16.1.125 -P7001 -N -s -e "show variables like
'server_id'"
    mysql: [Warning] Using a password on the command line interface can be insecure.
    server_id       127

    C:\>mysql -utest -p123456 -h172.16.1.125 -P7001 -N -s -e "show variables like
'server_id'"
    mysql: [Warning] Using a password on the command line interface can be insecure.
    server_id       127

    C:\>mysql -utest -p123456 -h172.16.1.125 -P7002 -N -s -e "show variables like
'server_id'"
    mysql: [Warning] Using a password on the command line interface can be insecure.
    server_id       126

    C:\>mysql -utest -p123456 -h172.16.1.125 -P7002 -N -s -e "show variables like
'server_id'"
    mysql: [Warning] Using a password on the command line interface can be insecure.
    server_id       127
```

此时重启 mysqlrouter 服务，回到初始状态，对于 7001 端口的请求，只会路由到 172.16.1.126。对 7002 端口的请求，路由策略不变，会以轮询方式发送给 172.16.1.126 和 172.16.1.127。

在 172.16.1.125 上重启 mysqlrouter 服务：

```
    service mysqlrouter restart
```

查看路由的目标服务器：

```
    C:\>mysql -utest -p123456 -h172.16.1.125 -P7001 -N -s -e "show variables like
'server_id'"
    mysql: [Warning] Using a password on the command line interface can be insecure.
    server_id       126
```

```
    C:\>mysql -utest -p123456 -h172.16.1.125 -P7001 -N -s -e "show variables like
'server_id'"
    mysql: [Warning] Using a password on the command line interface can be insecure.
    server_id       126

    C:\>mysql -utest -p123456 -h172.16.1.125 -P7002 -N -s -e "show variables like
'server_id'"
    mysql: [Warning] Using a password on the command line interface can be insecure.
    server_id       126

    C:\>mysql -utest -p123456 -h172.16.1.125 -P7002 -N -s -e "show variables like
'server_id'"
    mysql: [Warning] Using a password on the command line interface can be insecure.
    server_id       127
```

注意，此时 172.16.1.126 在复制中的角色依然是从（slave），但只有它接受读写请求，实际上是以 172.16.1.126 作为复制的主（master），这次 Router 的重启已经破坏了复制的数据一致性，因此这种情况下需要重新手工搭建复制互换角色。

```
    C:\>mysql -utest -p123456 -h172.16.1.125 -P7001 -e "use test; create table t1
 (a int); insert into t1 values (1);"
    mysql: [Warning] Using a password on the command line interface can be insecure.

    C:\>mysql -utest -p123456 -h172.16.1.125 -P7002 -N -s -e "show variables like
'server_id';select * from test.t1;"
    mysql: [Warning] Using a password on the command line interface can be insecure.
    server_id       126
    1

    C:\>mysql -utest -p123456 -h172.16.1.125 -P7002 -N -s -e "show variables like
'server_id';select * from test.t1;"
    mysql: [Warning] Using a password on the command line interface can be insecure.
    server_id       127
    ERROR 1146 (42S02) at line 1: Table 'test.t1' doesn't exist

    # 验证 read-only 模式下的写请求：
    C:\>mysql -utest -p123456 -h172.16.1.125 -P7002 -e "insert into test.t1 values
(2);"
    mysql: [Warning] Using a password on the command line interface can be insecure.

    C:\>mysql -utest -p123456 -h172.16.1.125 -P7002 -N -s -e "show variables like
'server_id';select * from test.t1;"
    mysql: [Warning] Using a password on the command line interface can be insecure.
    server_id       126
    1
    2
```

7.2.3 负载均衡

从上面的实验可以看出，在“一主一从”的配置中，只要将读请求发送到 7002 端口，请求就会被以轮询方式发送到两台 MySQL 服务器，从而达到读负载均衡的目的。对于读写负载均衡，则需要配置双主复制，然后将两台 MySQL 服务器都放到 read-only 下，例如两台 MySQL 服务器互为主从的拓扑结构，只需要配置如下一条路由策略即可。

```
[routing:load_balance]
bind_address = 172.16.1.125
bind_port = 7001
mode = read-only
destinations = 172.16.1.126:3306,172.16.1.127:3306
```

设置的 read-only（只读）模式，就是指出路由方式为“轮询”。正如上面测试看到的，两台服务器会以轮询方式进行读写，也就实现了最简单读写负载均衡。

7.2.4 读写分离

从上面的实验可以看出，在“一主一从”的配置中，只要将写请求发送到 7001 端口，读请求发送到 7002 端口，就可实现读写分离。正常情况下，主服务器接收写请求，主和从服务器接收读请求。如果主服务器宕机，所有读写请求都切换到从服务器上。

7.2.5 多实例

一个 Router 实例可以同时路由多组毫无关系的主从复制拓扑结构的 MySQL 实例。从应用角度只是看到多个独立的 MySQL 实例，而屏蔽了每个 MySQL 实例后面的复制拓扑结构。下面是一个多实例的示例。因为相关配置和效果与单一主从类似，所以下面只给出了实验步骤和结果，不再做重复性的说明。

1. 环境

172.16.1.125：3306 和 3307 这两个端口对应两个 MySQL 主库实例。

172.16.1.126：3306 和 3307 端口分别对应 172.16.1.125 上的 3306 和 3307 两个主库的从库实例。

MySQL Router 绑定 VIP 172.16.1.100。

2. Router 配置

配置文件/etc/mysqlrouter.conf 内容如下：

```
[DEFAULT]
logging_folder = /home/mysql/mysql-router-2.1.6/log
plugin_folder = /home/mysql/mysql-router-2.1.6/lib/mysqlrouter
config_folder = /home/mysql/mysql-router-2.1.6/config
runtime_folder = /home/mysql/mysql-router-2.1.6/run
data_folder = /home/mysql/mysql-router-2.1.6/data
```

```
[logger]
level = INFO

[routing:db1_write]
bind_address = 172.16.1.100
bind_port = 33060
mode = read-write
destinations = 172.16.1.125:3306,172.16.1.126:3306

[routing:db1_read]
bind_address = 172.16.1.100
bind_port = 33061
mode = read-only
destinations = 172.16.1.126:3306

[routing:db2_write]
bind_address = 172.16.1.100
bind_port = 33070
mode = read-write
destinations = 172.16.1.125:3307,172.16.1.126:3307

[routing:db2_read]
bind_address = 172.16.1.100
bind_port = 33071
mode = read-only
destinations = 172.16.1.126:3307
```

3. 功能测试

（1）在 172.16.1.125 上绑定 VIP：

```
/sbin/ifconfig ens32:1 172.16.1.100
```

（2）客户端用 VIP 访问数据库：

```
C:\>mysql -uroot -p123456 -P33060 -h172.16.1.100 -N -s -e "show variables like 'server_id'; show variables like 'port'"
mysql: [Warning] Using a password on the command line interface can be insecure.
server_id       125
port    3306

C:\>mysql -uroot -p123456 -P33061 -h172.16.1.100 -N -s -e "show variables like 'server_id'; show variables like 'port'"
mysql: [Warning] Using a password on the command line interface can be insecure.
server_id       126
port    3306

C:\>mysql -uroot -p123456 -P33070 -h172.16.1.100 -N -s -e "show variables like 'server_id'; show variables like 'port'"
mysql: [Warning] Using a password on the command line interface can be insecure.
server_id       125
```

```
port    3307

C:\>mysql -uroot -p123456 -P33071 -h172.16.1.100 -N -s -e "show variables like
'server_id'; show variables like 'port'"
mysql: [Warning] Using a password on the command line interface can be insecure.
server_id       126
port    3307
```

（3）停止两个主库（即主服务器）：

```
mysqladmin -uroot -p123456 -P3306 -h127.0.0.1 shutdown
mysqladmin -uroot -p123456 -P3307 -h127.0.0.1 shutdown
```

（4）客户端用 VIP 访问数据库：

```
C:\>mysql -uroot -p123456 -P33060 -h172.16.1.100 -N -s -e "show variables like
'server_id'; show variables like 'port'"
mysql: [Warning] Using a password on the command line interface can be insecure.
server_id       126
port    3306

C:\>mysql -uroot -p123456 -P33061 -h172.16.1.100 -N -s -e "show variables like
'server_id'; show variables like 'port'"
mysql: [Warning] Using a password on the command line interface can be insecure.
server_id       126
port    3306

C:\>mysql -uroot -p123456 -P33070 -h172.16.1.100 -N -s -e "show variables like
'server_id'; show variables like 'port'"
mysql: [Warning] Using a password on the command line interface can be insecure.
server_id       126
port    3307

C:\>mysql -uroot -p123456 -P33071 -h172.16.1.100 -N -s -e "show variables like
'server_id'; show variables like 'port'"
mysql: [Warning] Using a password on the command line interface can be insecure.
server_id       126
port    3307

C:\WINDOWS\system32>
```

（5）重新启动两台主库：

```
mysqld_safe --defaults-file=/home/mysql/mysql-5.6.14/my.cnf &
mysqld_safe --defaults-file=/home/mysql/mysql-5.6.14/my_2.cnf &
```

（6）重启 MySQL Router：

```
service mysqlrouter restart
```

（7）客户端用 VIP 访问数据库：

```
C:\>mysql -uroot -p123456 -P33060 -h172.16.1.100 -N -s -e "show variables like
'server_id'; show variables like 'port'"
```

```
mysql: [Warning] Using a password on the command line interface can be insecure.
server_id       125
port    3306

C:\>mysql -uroot -p123456 -P33061 -h172.16.1.100 -N -s -e "show variables like
'server_id'; show variables like 'port'"
mysql: [Warning] Using a password on the command line interface can be insecure.
server_id       126
port    3306
C:\>mysql -uroot -p123456 -P33070 -h172.16.1.100 -N -s -e "show variables like
'server_id'; show variables like 'port'"
mysql: [Warning] Using a password on the command line interface can be insecure.
server_id       125
port    3307

C:\>mysql -uroot -p123456 -P33071 -h172.16.1.100 -N -s -e "show variables like
'server_id'; show variables like 'port'"
mysql: [Warning] Using a password on the command line interface can be insecure.
server_id       126
port    3307
```

7.3 Router 8 安装和配置

MySQL Router 8 版本较之前的 Router 2 做了一些改进，主要体现在启动方式、路由策略和配置选项等方面。

7.3.1 安装

Router 8 的安装过程依赖于所使用的操作系统和安装介质，二进制包的安装通常非常简单，而源代码包则需要先编译再安装。例如在 Linux 上安装最新的 MySQL Router 二进制包，只需要用 MySQL 用户执行一条解压命令就完成了：

```
tar -Jxvf mysql-router-8.0.17-linux-glibc2.12-x86_64.tar.xz
```

解压后生成如下目录：

```
[mysql@hdp1~]$ll mysql-router-8.0.17-linux-glibc2.12-x86_64
total 108
drwxrwxr-x 2 mysql mysql    126 Sep  2 15:26 bin
drwxrwxr-x 3 mysql mysql   4096 Sep  2 15:26 lib
-rw-r--r-- 1 mysql mysql 101805 Jun 25 18:23 LICENSE.router
drwxrwxr-x 3 mysql mysql     17 Sep  2 15:22 man
-rw-r--r-- 1 mysql mysql    700 Jun 25 18:23 README.router
drwxrwxr-x 3 mysql mysql     16 Sep  2 15:22 share
```

bin 目录下存放的是可执行文件，可将该目录添加到PATH环境变量中以方便路由程序的执行，例如：

```
    export PATH=.:/sbin:/bin:/usr/sbin:/usr/bin:/usr/X11R6/bin:/home/mysql
/mysql-5.6.14/bin:/home/mysql/mysql-router-8.0.17-linux-glibc2.12-x86_64/bin;
```

然后就可以执行 mysqlrouter 命令，如检查版本或显示帮助信息：

```
    [mysql@hdp1~]$mysqlrouter --version
    MySQL Router  Ver 8.0.17 for linux-glibc2.12 on x86_64 (MySQL Community - GPL)

    [mysql@hdp1~]$mysqlrouter --help
    MySQL Router  Ver 8.0.17 for linux-glibc2.12 on x86_64 (MySQL Community - GPL)
    Copyright (c) 2015, 2019, Oracle and/or its affiliates. All rights reserved.

    Oracle is a registered trademark of Oracle Corporation and/or its
    affiliates. Other names may be trademarks of their respective
    owners.

    Configuration read from the following files in the given order (enclosed
    in parentheses means not available for reading):
      (/home/mysql/mysql-router-8.0.17-linux-glibc2.12-x86_64/bin/.././mysqlro
uter.conf)
      (/home/mysql/mysql-router-8.0.17-linux-glibc2.12-x86_64/bin/.././mysqlro
uter.ini)
      (/home/mysql/.mysqlrouter.conf)
      (/home/mysql/.mysqlrouter.ini)
    ...
```

各种安装包的下载地址为 https://dev.mysql.com/downloads/router/。其他情况的安装可参考 https://dev.mysql.com/doc/mysql-router/8.0/en/mysql-router-installation.html。

7.3.2 启动

MySQL Router 8 启动时需要读取其配置文件获取连接路由信息。Router 插件执行基于连接的路由，这意味着它将数据包转发到服务器而不检查它们。这是一种提供高吞吐量的简单方法。一个基本的连接路由设置如下所示：

```
    [logger]
    level = INFO

    [routing:secondary]
    bind_address = localhost
    bind_port = 7001
    destinations = 172.16.1.125:3306,172.16.1.126:3306,172.16.1.127:3306
    routing_strategy = round-robin

    [routing:primary]
    bind_address = localhost
    bind_port = 7002
    destinations = 172.16.1.125:3306,172.16.1.126:3306,172.16.1.127:3306
```

```
    routing_strategy = first-available
```

这里设置了两个路由策略：通过本地 7001 端口，循环连接到 172.16.1.125:3306、172.16.1.126:3306 和 172.16.1.127:3306 这三个 MySQL 实例，由 round-robin 路由策略所定义；通过本地 7002 端口，对同样的三个 MySQL 实例设置首个可用策略。首个可用策略使用目标列表中的第一台可用服务器，即当 172.16.1.125:3306 可用时，所有 7002 端口的连接都转发到它，否则就转发到 172.16.1.126:3306，以此类推。Router 不会检查数据包，也不会根据分配的策略或模式限制连接，因此应用程序可以据此确定将读和写请求发送到不同的服务器。本例中可将读请求发送到本地 7001 端口，将读负载均衡到三台服务器。同时将写请求发送到 7002，这样只写一台服务器，从而实现了读写分离。

将以上配置保存到/home/mysql/.mysqlrouter.conf 文件，并在后台启动 MySQL Router：

```
    [mysql@hdp1~]$mysqlrouter -c /home/mysql/.mysqlrouter.conf &
```

查看进程确认 Router 已经启动：

```
    [mysql@hdp1~]$ps -ef | grep router
    mysql  327410 326543  0 16:31 pts/0  00:00:00 mysqlrouter -c
/home/mysql/.mysqlrouter.conf
    ...
```

默认的日志文件中显示如下信息：

```
    [mysql@hdp1~]$more ~/mysql-router-8.0.17-linux-glibc2.12-x86_64
/mysqlrouter.log
    2019-09-02 16:42:36 routing INFO [7f151540b700] [routing:primary] started:
listening on localhost:7002, routing strategy = first-available
    2019-09-02 16:42:36 routing INFO [7f1514c0a700] [routing:secondary] started:
listening on localhost:7001, routing strategy = round-robin
    [mysql@hdp1~]$
```

查看监听状态：

```
    [mysql@hdp1~]$netstat -tnlp
    (Not all processes could be identified, non-owned process info
     will not be shown, you would have to be root to see it all.)
    Active Internet connections (only servers)
    Proto Recv-Q Send-Q Local Address   Foreign
Address         State    PID/Program name
    tcp      0    0 0.0.0.0:22     0.0.0.0:*          LISTEN   -
    tcp      0    0 127.0.0.1:199   0.0.0.0:*          LISTEN   -
    tcp6     0    0 :::80         :::*             LISTEN   -
    tcp6     0    0 :::22         :::*             LISTEN   -
    tcp6     0    0 ::1:7001       :::*             LISTEN   329852/mysqlrouter
    tcp6     0    0 ::1:7002       :::*             LISTEN   329852/mysqlrouter[
mysql@hdp1~]$
```

现在，Router 正在侦听端口 7001 和 7002，并将请求发送到适当的 MySQL 实例：

```
    [mysql@hdp1~]$mysql -uwxy -p123456 -P7001 --protocol=TCP -N -s -e"
select @@hostname"
```

```
    Warning: Using a password on the command line interface can be insecure.
    hdp2
    [mysql@hdp1~]$mysql -uwxy -p123456 -P7001 --protocol=TCP -N -s -e"select
@@hostname"
    Warning: Using a password on the command line interface can be insecure.
    hdp3
    [mysql@hdp1~]$mysql -uwxy -p123456 -P7001 --protocol=TCP -N -s -e"select
@@hostname"
    Warning: Using a password on the command line interface can be insecure.
    hdp4
    [mysql@hdp1~]$mysql -uwxy -p123456 -P7001 --protocol=TCP -N -s -e"select
@@hostname"
    Warning: Using a password on the command line interface can be insecure.
    hdp2
    [mysql@hdp1~]$mysql -uwxy -p123456 -P7002 --protocol=TCP -N -s -e"select
@@hostname"
    Warning: Using a password on the command line interface can be insecure.
    hdp2
    [mysql@hdp1~]$mysql -uwxy -p123456 -P7002 --protocol=TCP -N -s -e"select
@@hostname"
    Warning: Using a password on the command line interface can be insecure.
    hdp2
    [mysql@hdp1~]$mysql -uwxy -p123456 -P7002 --protocol=TCP -N -s -e"select
@@hostname"
    Warning: Using a password on the command line interface can be insecure.
    hdp2
    [mysql@hdp1~]$
```

由上可见，发送到本地 7001 端口的请求，被循环转发到三台服务器，而发送到本地 7002 端口的请求，全部被转发到 172.16.1.125:3306。

routing_strategy 是 MySQL Router 的核心选项，从 8.0.4 版本开始引入，当前有效值为 first-available、next-available、round-robin 和 round-robin-with-fallback。顾名思义，该选项实际控制路由策略，即客户端请求最终连接到哪台 MySQL 服务器实例。

- round-robin：每个新连接都以循环方式连接到下一台可用的服务器，以实现负载平衡。
- round-robin-with-fallback：用于 InnoDB Cluster。每台新的连接都以循环方式连接到下一台可用的 SECONDARY 服务器。如果 SECONDARY 服务器不可用，则以循环方式使用 PRIMARY 服务器。
- first-available：新连接从目标列表路由到第一台可用服务器。如果失败，则使用下一台可用的服务器，如此循环，直到所有服务器都不可用为止。
- next-available：与 first-available 类似，新连接从目标列表路由到第一台可用服务器。与 first-available 不同的是，如果一台服务器被标记为不可访问，那么它将被中止使用，并且不会再次用作目标服务器。重启 Router 后，所有被中止使用的服务器将再次可选。此策略向后兼容 MySQL Router 2.x 中 mode 设置为 read-write 的行为。

相对于以前版本的 mode 选项，routing_strategy 选项更为灵活。不能同时设置 routing_strategy 和 mode，静态路由的设置只能选择其中之一。对于 InnoDB Cluster 而言，该设置是可选的，而默认使用 round-robin 策略。

本例中 7002 端口的路由策略为 first-available，下面依次停止 172.16.1.125:3306、172.16.1.126:3306 和 172.16.1.127:3306，观察 7002 端口的访问情况。

```
    # 停止 172.16.1.125:3306
    [mysql@hdp2~]$mysqladmin -uwxy -p123456 -h172.16.1.125 -P3306 shutdown
    mysqladmin: [Warning] Using a password on the command line interface can be insecure.
    [mysql@hdp2~]$

    # 访问 7002
    [mysql@hdp1~]$mysql -uwxy -p123456 -P7002 --protocol=TCP -N -s -e"select @@hostname"
    Warning: Using a password on the command line interface can be insecure.
    hdp3

    # 停止 172.16.1.126:3306
    [mysql@hdp3~]$mysqladmin -uwxy -p123456 -h172.16.1.126 -P3306 shutdown
    mysqladmin: [Warning] Using a password on the command line interface can be insecure.
    [mysql@hdp3~]$

    # 访问 7002
    [mysql@hdp1~]$mysql -uwxy -p123456 -P7002 --protocol=TCP -N -s -e"select @@hostname"
    Warning: Using a password on the command line interface can be insecure.
    hdp4
    [mysql@hdp1~]$

    # 停止 172.16.1.127:3306
    [mysql@hdp4~]$mysqladmin -uwxy -p123456 -h172.16.1.127 -P3306 shutdown
    mysqladmin: [Warning] Using a password on the command line interface can be insecure.
    [mysql@hdp4~]$

    # 启动 172.16.1.125:3306
    [mysql@hdp2~]$mysqld_safe --defaults-file=/etc/my.cnf &

    # 访问 7002
    [mysql@hdp1~]$mysql -uwxy -p123456 -P7002 --protocol=TCP -N -s -e"select @@hostname"
    Warning: Using a password on the command line interface can be insecure.
    hdp2
    [mysql@hdp1~]$
```

first-available 总是循环查找目标列表中第一台可用服务器并连接它。将 7002 的策略改为 next-available，把上面的步骤再做一遍，会看到最后的查询失败。即使此时 172.16.1.125:3306 可用也不会再次连接它，除非重启 Router。

```
    # 最后的查询连接失败
```

```
[mysql@hdp1~]$mysql -uwxy -p123456 -P7002 --protocol=TCP -N -s -e"select
@@hostname"
Warning: Using a password on the command line interface can be insecure.
ERROR 2003 (HY000): Can't connect to remote MySQL server for client connected
to 'localhost:7002'

# 重启路由
[mysql@hdp1~]$kill `ps -ef | grep router | grep -v grep | awk '{print $2}'`
[mysql@hdp1~]$mysqlrouter -c ~/.mysqlrouter.conf &

# 再次查询
[mysql@hdp1~]$mysql -uwxy -p123456 -P7002 --protocol=TCP -N -s -e"select
@@hostname"
Warning: Using a password on the command line interface can be insecure.
hdp2
[mysql@hdp1~]$
```

可以在网络上的单台或多台主机上运行多个 Router 实例，而无需将它隔离到单台服务器上。这是因为 Router 对任何特定服务器或主机都不具有亲和性。要停止 Router，只需用 kill 或 killall 命令直接杀掉相关进程。Router 只起到一个转发流量的作用，它实在太“轻”了，以至于都没有一个单独的停止命令。

7.3.3 配置

MySQL Router 启动时可以从配置文件、扩展配置文件和命令行选项这三个地方获取配置信息。本节说明 MySQL Router 配置文件的位置、配置文件语法和配置选项。

1. 配置文件位置

（1）默认配置文件

MySQL Router 在启动时扫描默认配置文件，并可选择在运行时从命令行加载用户定义的配置文件。在默认情况下，MySQL Router 扫描依赖于操作系统的特定位置以查找配置文件。执行 mysqlrouter --help 可以查看系统上默认配置文件的位置：

```
[mysql@hdp1~]$mysqlrouter --help | more
MySQL Router  Ver 8.0.17 for linux-glibc2.12 on x86_64 (MySQL Community - GPL)
Copyright (c) 2015, 2019, Oracle and/or its affiliates. All rights reserved.
...
Configuration read from the following files in the given order (enclosed
in parentheses means not available for reading):
  (/home/mysql/mysql-router-8.0.17-linux-glibc2.12-x86_64/bin/.././mysqlro
uter.conf)
  (/home/mysql/mysql-router-8.0.17-linux-glibc2.12-x86_64/bin/.././mysqlro
uter.ini)
  /home/mysql/.mysqlrouter.conf
  (/home/mysql/.mysqlrouter.ini)
...
```

从帮助信息的开始部分可以看到MySQL Router查找的默认配置文件和查找顺序。括号中的文件表示不存在，这里只有/home/mysql/.mysqlrouter.conf一个可用的默认配置文件。注意，如果使用-c或--config选项传入用户定义的配置文件，则不会加载默认配置文件。

（2）用户定义的和扩展的配置文件

用户定义的配置文件由-c或--config命令行选项指定，扩展的配置文件由-a或--extra-config命令行选项来指定。从默认配置文件或使用--config选项指定的文件读取配置后，再读取扩展的配置文件：

```
shell> mysqlrouter --config /custom/path/to/router.conf --extra-config
/another/config.conf
```

可以传入多个扩展配置选项，并按照输入的顺序加载文件：

```
shell> mysqlrouter --config b.conf --extra-config a.conf --extra-config c.conf
```

在上面的示例中，首先加载b.conf，然后按顺序加载a.conf和c.conf。此外，由于使用了--config，因此不会加载默认配置文件。每个加载的配置文件中的配置选项设置都会覆盖掉先前读取的配置文件中相同的选项设置。

2. 配置文件语法

MySQL Router配置文件格式类似于MySQL服务器的配置文件，符合传统INI文件格式，其中包含多个段，每个段中包含多个选项。选项中的目录支持正斜杠和反斜杠，反斜杠无条件复制而不是转义字符。配置文件可以包含注释行，注释行以井号（#）或分号（;）开头，并继续到行尾。要注意的是，注释行只支持整行注释，不支持行中和行尾注释。

（1）段

每个配置文件都包含一个配置段列表，其中每个段包含一系列配置选项，每个配置选项都有名称和值，形式如下：

```
[section name]
option = value
option = value
option = value

[section name:optional section key]
option = value
option = value
option = value
```

在中括号中是段名和可选的段键，段名与段键之间用冒号（:）分隔。段名和段键的组合对于配置是唯一的，中括号内不能有首尾空格。段名与段键是由一个或多个字母、数字或下画线（_）所组成的序列，不允许使用其他字符。

段类似于命名空间，例如user选项，其含义取决于其关联的段。[DEFAULT]段中的user指的是运行MySQL Router的系统用户，该用户也受--user命令行选项所控制。[metadata_cache]段中定义的user，指的是访问MySQL服务器元数据的MySQL用户。

下面列举几个主要的段：

- [DEFAULT]：DEFAULT 是一个特殊的段，用于选项的默认值，在该段中查找其他段中没出现的选项，通常用于设置扩展配置文件所在目录、日志所在目录、MySQL Router 运行时的目录（如 pid 文件）等。DEFAULT 段不接受段键。
- [logger]：logger 段只有一个选项，设置日志的记录级别。
- [routing:NAME]：该段是 MySQL Router 主要的配置部分，设置不同的路由实例，其中 NAME 是自定义的段键名。

（2）选项

在段名之后，可以有多个形式如下的选项行：

```
name = value
```

处理之前将删除选项名称或选项值上的任何首尾空格。选项名称不区分字母大小写。选项值支持在大括号中引用选项名称，作为变量引用值。变量引用值用于检索选项值，它不是在读取配置文件时完成的。如果变量未定义，则不进行替换，并按字面读取选项值。参考下面的示例配置文件：

```
[DEFAULT]
prefix = /usr/

[sample]
bin = {prefix}bin/{name}
lib = {prefix}lib/{name}
name = magic
directory = /data/{mysqlrouterdata}
```

这里 bin 的值是“/usr/bin/magic”，lib 的值是“/usr/lib/magic”。由于未定义名为“{mysqlrouterdata}”的变量，因此 directory 的值是“/data/{mysqlrouterdata}”。

MySQL Router 定义了配置文件可用的如下预定义变量。变量使用大括号，例如{program}作为 Router 可执行程序预定义变量。

- program：程序名称，通常是 mysqlrouter。
- origin：二进制文件所在的目录。
- logging_folder：日志文件所在的目录。
- plugin_folder：插件文件所在的目录。
- runtime_folder：运行时数据文件所在的目录。
- config_folder：配置文件所在的目录。

3. 配置文件示例

启动时，MySQL Router 会读取配置文件列表，这些配置文件一起构成 Router 的配置。至少需要一个配置文件，否则启动 Router 时会报出“Error: No valid configuration file available”的错误。下面是一个名为 mycluster 的 MySQL InnoDB cluster 的基本连接路由示例，读写流量发送到 6446 端口，只读访问则使用 6447 端口。

```
[DEFAULT]
logging_folder=/opt/routers/myrouter/log
runtime_folder=/opt/routers/myrouter/run
```

```
data_folder=/opt/routers/myrouter/data
connect_timeout=30
read_timeout=30

[logger]
level = INFO

[metadata_cache:mycluster]
router_id=5
bootstrap_server_addresses=mysql://localhost:3310,mysql://localhost:3320,m
ysql://localhost:3330
user=mysql_router
metadata_cluster=mycluster
ttl=5

[routing:mycluster_default_rw]
bind_address=0.0.0.0
bind_port=6446
destinations=metadata-cache://mycluster/default?role=PRIMARY
routing_strategy=round-robin

[routing:mycluster_default_ro]
bind_address=0.0.0.0
bind_port=6447
destinations=metadata-cache://mycluster/default?role=SECONDARY
routing_strategy=round-robin
```

该配置中定义了五个段：

- [DEFAULT]中定义了数据文件目录、日志文件目录、运行时文件目录的位置，以及连接和读取 MySQL 元数据服务器的超时时间为 30 秒。
- [logger]中定义日志级别为 info，可选值有 info、debug、warning、error 和 fatal，不区分字母大小写。
- [metadata_cache:mycluster]中定义 MySQL 服务器元数据访问信息：router_id 是整型的路由 ID。bootstrap_server_addresses 是逗号分隔的 InnoDB cluster 中的 MySQL 实例。user 定义访问元数据的用户。metadata_cluster 定义 InnoDB cluster 名称。ttl 定义缓存时间为 5 秒。
- [routing:mycluster_default_rw]和[routing:mycluster_default_ro]分别定义读写路由和只读路由。bind_address 指定路由使用本机 IP 地址。配置文件中的 bind_address 选项只接受一个地址，但是可以使用 bind_addres = 0.0.0.0 绑定到 localhost 上的所有地址。bind_port 指定客户端连接 Router 使用的端口。destinations 引用由 metadata-cache 定义的目标地址，以利用 InnoDB Cluster 的元数据缓存来动态配置主机信息。如果不使用 InnoDB Cluster，destinations 可以指定为逗号分隔的 IP:port 地址，但 bind_address 中的地址不能出现在 destinations 列表中。routing_strategy 定义路由策略，本例中读写和只读均为循环访问三个节点。

配置选项的完整说明，参见：https://dev.mysql.com/doc/mysql-router/8.0/en/mysql-router-conf-options.html。

7.4 小 结

MySQL Router 是官方提供的一个轻量级中间件，它提供透明路由、元数据缓存和简单重定向功能。它可以实现 MySQL 实例级的高可用、负载均衡与读写分离，并且对应用透明。Router 是一个单独的程序，核心是 mysqlrouter 可执行文件。MySQL 建议的部署方式是 Router 与 InnoDB Cluster 集成。出于性能考虑，最好将 Router 与应用程序客户端安装在同一台主机上。总的来说，MySQL Router 从原理、安装、配置到部署使用都是如此轻而易举，可满足简单高可用应用场景的需求。下一章将介绍另一个 MySQL 官方的中间件产品——MySQL Fabric。

第 8 章

MySQL Fabric

8.1 MySQL Fabric 简介

Oracle 在 2014 年 5 月推出了一个被寄予厚望的产品 MySQL Fabric，其大致功能从名字便有迹可循。Fabric 是“织物”的意思，这意味着它是用来“织”起一片 MySQL 数据库。MySQL Fabric 是一套数据库服务器场（Database Server Farm）的架构管理系统。

8.1.1 MySQL Fabric 的用途

当表的数据量达到一定程度，如大于 1TB 时，无论在查询、更新的效率上，或者是备份、更改表结构所需要的时间上，都会造成很大的问题。通常在这种情况下需要进行数据分片，将大表分散到多台数据库服务器中。然而单纯地分片又会使每台数据库服务器都有可能成为单个故障点，一台挂掉就会影响整个系统的操作。另一方面，应用程序也会因为每个查询都要依据其查询条件分别指向不同的数据库而变得更加复杂。再者，当数据分片的拓扑结构改变时，例如增加一个数据库节点，会使应用端的所有程序都必须做出相应地修改，从而导致维护变得极为困难。以上种种性能或维护上的问题正是 MySQL Fabric 要解决的。

Fabric 架构为 MySQL 提供了高可用（HA）和基于数据分片（Sharding）的横向扩展能力。在实际部署中，可以单独启用高可用或数据分片，也可以结合使用这两个特性。MySQL Fabric 能组织多个 MySQL 数据库实例，使应用系统能够将大表分散到多个数据库，即数据分片。在同一个分片内又可以含有多个数据库，并且由 Fabric 自动挑选一个适合的实例作为主库，其他实例配置成从库，来做主从复制。在主库挂掉时，从各个从库中挑选一个提升为主库。之后，其他从库转向新主库复制数据。这里说的“自动”是指这些过程由 MySQL Fabric 在后台完成，不需要用户手动更改配置。MySQL Fabric 是符合 GPL 规范的开源软件，可以自由使用和修改它。

8.1.2 MySQL Fabric 架构

如图 8-1 所示，MySQL Fabric 由管理节点、数据库服务器场和连接器三部分构成。

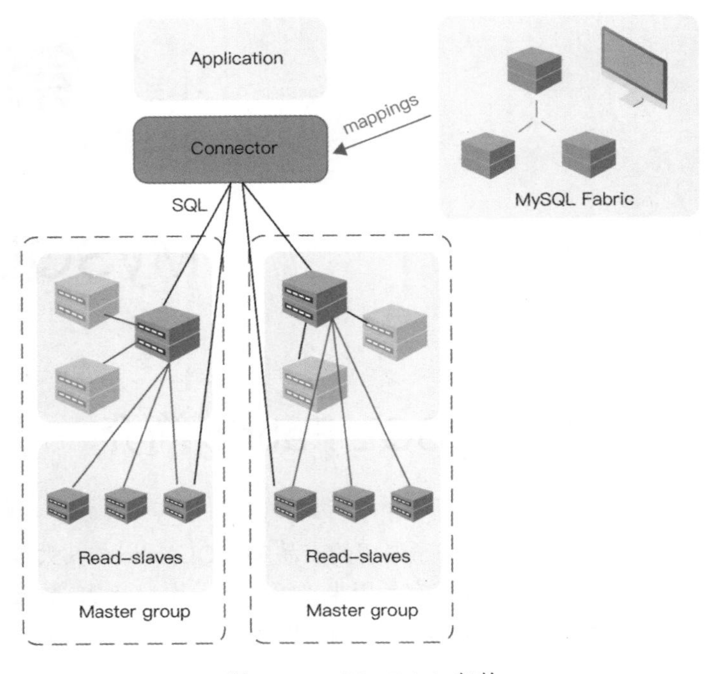

图 8-1 MySQL Fabric 架构

1. MySQL Fabric 管理节点

MySQL Fabric 管理节点在软件层面是一个 Python 程序，包括一个特殊的库实现所有功能，通常安装在独立的服务器上，如图 8-1 中右上所示的部分。它是 Fabric 架构的核心，主要功能是管理整个数据库服务器场。安装管理节点时使用的配置文件（默认是/etc/mysql/fabric.cnf）中，需要指定一个 MySQL 实例的 IP、端口和连接账号等信息。在 Fabric 中该实例被称为状态存储（State Store）。

Fabric 提供了一个名为 mysqlfabric 的命令行工具与其交互，该工具提供了一套命令完成创建和管理服务器组、定义和维护数据分片等操作。Fabric 在执行 mysqlfabric manage setup 命令进行初始化时，会在状态存储 MySQL 实例上创建一个数据库（通常数据库名称为 fabric），用来存储 Fabric 状态、服务器场复制拓扑结构和配置的相关信息，如哪些服务器组由哪些数据库实例构成，各服务器组中的主从数据库实例分别是哪些等等。Fabric 管理节点在启动时会对服务器场中各数据库下达建立主从复制的命令。正常运行时会定期 ping 各组的主服务器，当发现主库没有正常运行时，它会启动故障转移程序，在该服务器组的从库中找一个合适的服务器提升为主库。其他从库则转向新的主数据库继续复制数据。

2. 数据库服务器场（Database Server Farm）

数据库服务器场由被 Fabric 管理的多台 MySQL 服务器构成，是整个架构中的工作引擎。在传统的数据库应用中这是单一的 MySQL 数据库实例，Fabric 则是以多个数据库实例支持大数据量表（TB 级以上）和高可用性数据库的需求。这些数据库实例分成若干高可用组（HA Group），每组包含一台以上的 MySQL 数据库服务器，图 8-1 中每个虚线部分的方框代表一个高可用组。

如果高可用组中有多个 MySQL 实例，那么用 mysqlfabric group promote 命令会使 Fabric 挑选其中一个提升为主库（Master），其他实例则成为从库（Slave），从库复制主库的变化，完成设定为同一个高可用组内的主从（异步或半同步）复制。以后，Fabric 会定期监视这个主数据库。当主库宕掉之后，Fabric 会从高可用组内挑选一个从库提升为新主库，其他的数据库会转向这个新主库继续复制。另一方面，Fabric 会指示应用端的连接器对这些主从数据库进行读写分离。当应用程序对数据库执行读写兼有的操作时，连接器会将该指令提交给主数据库。如果应用程序对数据库进行只读操作，并且 MySQL 实例的 read_only 参数设置为“ON”，则所有的查询均轮流传送到这些数据库实例中。借助读写分离，应用系统的数据处理能力得以增加。此外，如前所述，Fabric 还能处理需要拆分到多个数据库服务器的表（Sharding Table），每一个高可用组都可能存放分片表的部分数据。应用端的连接器会将对分片表的指令依照 Fabric 管理节点的设定发送到不同的高可用组，这样可使数据库的容量随着高可用组的数量增加而增长，从而达到横向扩展的目的。同时，对非分片表所下达的指令和所有的 DDL 语句会由连接器送到全局高可用组（Global Group），全局高可用组的主库被 Fabric 设置为其他高可用组的主库。所有存储分片表的高可用组的主库复制全局高可用组的变化，这样其他高可用组都有一份非分片表的数据，从而使得 SQL 语句中对分片表与非分片表的连接（join）操作变得更加简单。

3. 连接器（Connector）

图 8-1 中间部分的“Connector”即为 MySQL Fabric 的连接器。Fabric 的主要特点是把路由功能合并到各应用端的连接器中，以避免单一代理的单点故障和性能瓶颈。应用系统在运行时，每个 SQL 指令都会经由连接器发送到数据库服务器。Fabric 所搭配的连接器是一个扩展的 MySQL 连接器版本，使用 XML-RPC 协议访问 Fabric。和一般的单机版 MySQL 连接器相比，Fabric 感知连接器多了一些能处理数据库服务器场的功能。Fabric 连接器能在建立数据库连接时，以 XML-RPC 协议检查 Fabric 管理节点中数据库服务器场的配置，然后该连接下的查询可依据 Fabric 的指示送到适合的数据库实例。如此一来，常见的数据库分片方案中可能造成性能瓶颈的代理部分放到了应用的连接器中，从而解决了这个问题。

Fabric 连接器支持的语言有 Java、Python、PHP，即 Connector/J、Connector/Python 和 Connector/PHP 都是 Fabric-aware（感知 Fabric 的）。以 Java 为例，JDBC driver 必须是 Connector/J 5.1.30 以后的版本。使用 Fabric 的 Java 程序和一般访问单机 MySQL 的 Java 程序差不多，只是在建立数据库连接对象（Database Connection Object）时数据库连接 URL 不是指向数据库，而是指向 MySQL Fabric 管理节点。当查询的表是全局表（不做表分片 table shard）或 SQL 语句为 DDL（例如建表或改表结构）时，建立连接对象时要加上"fabricServerGroup="参数，之后通过这个连接对象所下的 SQL 指令会送到全局组的主数据库，再由该数据库实例复制到其他的高可用组中。如果 SQL 命令所要操作的表是分片表，则建立连接对象时要加上"fabricShardTable="参数，之后通过这个连接对象所下的 SQL 命令会根据 Fabric 所设定的分片原则送到各分片的高可用组。这样一来，应用程序对这些分片表下达 SQL 指令时，不需要在 SQL 中判断要送到哪个数据库，完全由连接器在建立数据库连接时根据从 Fabric 所查到的服务器场的配置信息（哪个数据库实例属于哪个高可用组，各分片表的拆分规则等）来决定。而且这个配置在建立连接后就缓存在连接器所在的应用端。这样，在每次下达 SQL 指令时就不需要重复查询 Fabric 管理节点，而根据存放在应用端的分表配置直接送到正确的数据库，同时应用程序的效率不会因为进行了表的拆分而有任何的降低。

8.2 高可用性

8.2.1 概述

MySQL Fabric 在 MySQL 复制上增加了一个管理和监控层，它和一组 Fabric-aware 连接器一起，把写和一致性读操作路由到当前的主数据库。Fabric 有一个 HA 组（HA Group）的概念，它是由两台或两台以上的 MySQL 服务器组成的服务器场。在任一时间点，一个 HA 组中只有一个主库，其他的都是从库。HA 组的作用是确保该组中的数据总是可访问的。MySQL 复制通过把数据拷贝多份来提供数据的安全性，同时 Fabric 管理节点使用 GTID（全局事务标识）的事务集合来检查和维护 MySQL 服务器之间的数据一致性。Fabric 提供两个高可用特性：

- 故障检测与提升：Fabric 进程监控 HA 组中的主数据库服务器。当主库宕机，该进程会在 HA 组中选择一个从库，把它提升为主库，而 HA 组中其他的从库将接收新主库的数据修改。当连接器观察到主库出现问题时会提醒 Fabric，而 Fabric 进程会使用该信息作为产生服务器场中相关状态决策的一部分。
- 路由数据库请求：当 Fabric 提升了一个新主库，它会修改 State Store 中存储的服务器状态，并且提示连接器用更新的路由信息刷新其缓存。通过这种方式，应用不需要感知复制拓扑结构的改变，也不需要知道写操作指向了不同的目的地址。

8.2.2 安装与配置

1. 安装准备与规划

安装和使用 MySQL Fabric 需要满足以下前置条件：

- 所有 MySQL 服务器版本大于等于 5.6.10，需要支持 GTID。
- Python 版本大于等于 2.6。
- Connector/Python 版本大于等于 1.2.1，Connector/J 版本大于等于 5.1.27。
- 一个 MySQL 实例用于状态存储（State Store），该实例不是 Fabric HA 组的成员，但其版本需要跟其他在 HA 中的 MySQL 实例版本保持一致。
- 应用使用 Fabric，必须有一个 Fabric-aware 连接器安装在运行应用的系统上。

示例环境为两个 CentOS release 6.4 虚拟机，安装了 Python 2.6，并且关闭了 iptables 和 selinux。主机说明如表 8-1 所示。

表 8-1　Fabric HA 示例主机信息

主机名	IP	说明
fab_connector	192.168.56.101	安装 Fabric 和 MySQL 服务器，建立一个 MySQL 实例用作状态存储，使用默认的 3306 端口，存储 HA 组的状态和配置信息
fab_group1	192.168.56.102	安装 MySQL 服务器，建立三个 MySQL 数据库实例，端口分别是 3326、3327 和 3328，组成一个 HA 组

因为只是出于实验目的，所以这里使用 root 用户安装 MySQL 和 Fabric，使用的软件包分别是 mysql-5.7.10-linux-glibc2.5-x86_64.tar 和 mysql-utilities-1.5.6.tar.gz（Fabric 被打包到了 MySQL Utilities 中）。部署规划如图 8-2 所示。

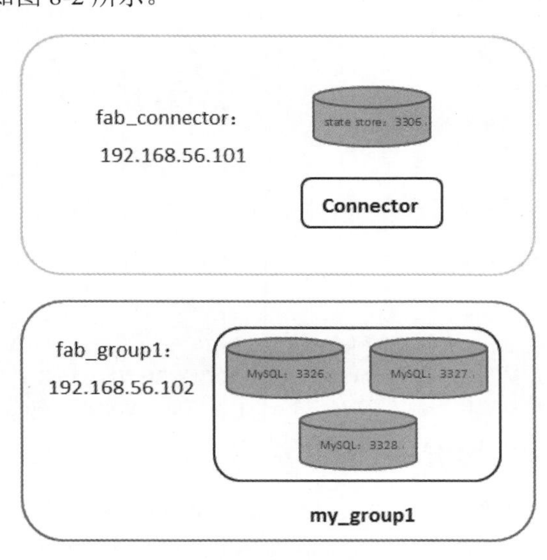

图 8-2　Fabric HA 示例部署规划

2. 安装 MySQL 服务器

在两台主机上安装 MySQL 服务器：

```
cd /root
tar xvf mysql-5.7.10-linux-glibc2.5-x86_64.tar
tar zxvf mysql-5.7.10-linux-glibc2.5-x86_64.tar.gz
ln -s mysql-5.6.13-linux-glibc2.5-x86_64 mysql
groupadd mysql
useradd -r -g mysql mysql
chown -R mysql .
```

3. 安装 Fabric

在 fab_connector 主机上安装 Fabric：

```
cd /root
tar zxvf mysql-utilities-1.5.6.tar.gz
cd mysql-utilities-1.5.6
sudo python setup.py install
```

4. 启动用于状态存储的 MySQL 实例

在 fab_connector 上配置并启动 MySQL 实例，配置文件/etc/my_fabric.cnf 内容如下：

```
[mysqld]
basedir=/root/mysql
datadir=/var/lib/mysql
socket=/var/lib/mysql/mysql.sock
binlog-format=ROW
log-slave-updates=true
gtid-mode=on
enforce-gtid-consistency=true
master-info-repository=TABLE
relay-log-info-repository=TABLE
sync-master-info=1
port=3306
report-host=fab_connector
report-port=3306
server-id=1
log-bin=fab-bin.log
```

启动 MySQL 实例并建立 Fabric 使用的 MySQL 用户，该用户对数据库 fabric 有完全权限：

```
mysqld --defaults-file=/etc/my_fabric.cnf --initialize
mysqld --defaults-file=/etc/my_fabric.cnf --user=mysql &
mysql -h 127.0.0.1 -P3306 -u root -p -e "CREATE USER 'fabric'@'localhost'
IDENTIFIED BY 'secret';GRANT ALL ON fabric.* TO 'fabric'@'localhost'";
```

5. 启动 HA 组的 MySQL 实例

HA 组的三个 MySQL 实例初始配置文件分别为 my_group1_1_init.cnf、my_group1_2_init.cnf 和 my_group1_3_init.cnf，实例配置文件分别为/etc/my_group1_1.cnf、/etc/my_group1_2.cnf 和/etc/my_group1_3.cnf。my_group1_1_init.cnf 和/etc/my_group1_1.cnf 文件的内容如下：

```
# /etc/my_group1_1_init.cnf
[mysqld]
basedir=/root/mysql
datadir=/var/lib/group1_1
port=3326
socket=/var/lib/group1_1/mysql.sock

# /etc/my_group1_1.cnf
[mysqld]
basedir=/root/mysql
datadir=/var/lib/group1_1
port=3326
socket=/var/lib/group1_1/mysql.sock
binlog-format=ROW
log-slave-updates=true
gtid-mode=on
enforce-gtid-consistency=true
master-info-repository=TABLE
```

```
relay-log-info-repository=TABLE
sync-master-info=1
report-host=fab_group1
report-port=3326
server-id=11
log-bin=fab1a-bin.log
log_error_verbosity=1
```

其他两个实例的相关配置文件内容与实例 1 的类似，只是数据目录、socket、server-id 和端口不同。

执行如下操作启动实例 1：

```
# 初始化实例并记录临时密码
mysqld --defaults-file=/etc/my_group1_1_init.cnf --initialize

# 启动实例
chown -R mysql /var/lib/group1_1
mysqld --defaults-file=/etc/my_group1_1_init.cnf --user=mysql &

# 修改初始密码，创建使用 Fabric 的 MySQL 用户
mysql -h 127.0.0.1 -P3328 -u root -p
alter user user() identified by 'new_password';
create user 'fabric'@'%' identified by 'secret';
grant all on *.* to 'fabric'@'%';

# 重启实例
mysqladmin -u root --protocol=tcp -h127.0.0.1 -P3326 -p shutdown
mysqld --defaults-file=/etc/my_group1_1.cnf --user=mysql &
```

其他两个 MySQL 实例执行同样的操作。注意，Mysql Fabric 使用相同的用户名密码连接所有它所管理的数据库服务器场中的 MySQL 实例。本例中 Fabric 连接三个 MySQL 实例使用的用户名都是 fabric，密码都是 secret，但这些用户名和密码可以与连接状态存储（State Store）实例所使用的用户名和密码不同。

6. 启动 Fabric

在 fab_connector 主机上配置并启动 Fabric，配置文件/etc/mysql/fabric.cfg 的内容如下：

```
[DEFAULT]
sysconfdir = /etc
logdir = /var/log

[statistics]
prune_time = 3600

[logging]
url = file:///var/log/fabric.log
level = INFO

[storage]
auth_plugin = mysql_native_password
```

```
database = fabric
user = fabric
address = localhost:3306
connection_delay = 1
connection_timeout = 6
password = secret
connection_attempts = 6

[failure_tracking]
notification_interval = 60
notification_clients = 50
detection_timeout = 1
detection_interval = 6
notifications = 300
detections = 3
failover_interval = 0
prune_time = 3600

[servers]
restore_user = fabric
unreachable_timeout = 5
backup_password = secret
backup_user = fabric
user = fabric
restore_password = secret
password = secret

[connector]
ttl = 1

[protocol.xmlrpc]
disable_authentication = no
realm = MySQL Fabric
threads = 5
user = admin
address = 192.168.16.119:32274
password = secret

[executor]
executors = 5

[sharding]
mysqldump_program = /root/mysql/bin/mysqldump
mysqlclient_program = /root/mysql/bin/mysql

[protocol.mysql]
disable_authentication = no
user = admin
address = 192.168.16.119:32275
password = secret
```

下面说明Fabric配置文件中每个段及其选项的具体含义。

（1）default段

如果请求的选项没有在命令行指定，或没有在配置文件中找到，那么Fabric将查看default段的信息。

- sysconfdir：配置文件所在的目录。
- logdir：日志文件所存储的位置，绝对路径，由fabric守护进程创建。

（2）statistics段

本段设置统计信息的相关选项。

- prune_time：删除大于1小时的内部统计信息。

（3）logging段

设置Fabric日志信息记录到哪里，如果不是启动为后台进程，则将日志输出到标准输出设备。

- url：存储日志的文件，可为绝对或相对路径。如果是相对路径，则将参照default段中logdir参数指定的目录。
- level：日志级别，支持DEBUG、INFO、WARNING、ERROR和CRITICAL。

（4）storage段

配置状态存储的相关选项。

- auth_plugin：设置使用的认证插件。
- database：存储Fabric表的数据库。
- user：连接到状态存储实例的用户名。
- address：指定状态存储的MySQL实例地址和端口。
- connection_delay：连续尝试创建连接之间的延迟时间，默认为1秒。
- connection_timeout：中断请求之前等待的最大秒数。
- password：认证密码，也能设置为空密码。
- connection_attempts：创建连接的最大尝试次数。

（5）failure_tracking段

连接器和其他外部实体会报告错误，Fabric保持跟踪服务器的健康状态和采取相应的行为，例如提升一个新的主库。如果一个非主库服务器不稳定，就简单地将它标记为faulty（有缺陷）。

- notification_interval：评估服务器是否错误的秒数，不考虑比这更早的时间。
- notification_clients：多少不同源的报告错误。
- detection_timeout：错误检查程序尝试连接到一个组中服务器的超时时间，单位是秒。
- detection_interval：连续检查之间的间隔秒数。
- notifications：多少次报告错误后，将服务器标记为不可用。
- detections：错误检查监控一个组，需要连续尝试访问当前库3次（默认值），如果都错误后，才能提升新主库。
- failover_interval：为了避免整个系统不可用，自上次提升间隔多少秒后才能再次选取新主库。

- prune_time：在错误日志中保留错误信息多长时间，超过此秒数的错误信息将被删除。

（6）servers 段

本段用于 Fabric 连接所管理的 MySQL 服务器的信息。

- restore_user：恢复 MySQL 服务器时使用的用户名。
- unreachable_timeout：用于检查连接故障服务器或服务器场中新服务器的超时时间，默认为 5 秒。
- backup_password：备份 MySQL 服务器时使用的密码。
- backup_user：备份 MySQL 服务器时使用的用户名。
- user：连接 MySQL 服务器使用的用户名。
- restore_password：恢复 MySQL 服务器时使用的密码。
- password：连接 MySQL 服务器使用的密码。

（7）connector 段

Fabric-aware 连接器连接到 Fabric，获取组、分片和服务器的信息，把结果缓存到本地以提高性能。本段包含传递给连接器的配置参数。

- ttl：缓存生存秒数，决定多长时间，连接器认为从 Fabric 获取的信息是有效的。在 TTL（Time To Live）过期后，从 MySQL Fabric 管理节点重新加载它们。

（8）protocol.xmlrpc 段

本段包含有关客户端使用 XML-RPC 协议连接到 MySQL Fabric 节点的配置参数。

- disable_authentication：命令行请求是否禁用身份验证，默认要认证。
- realm：XML-RPC 服务器在身份验证时的标识，如 RFC 2617 中所定义。
- threads：XML-RPC 会话的并发线程数，决定 Fabric 能接受多少并发请求。
- user：客户端用来连接到 XML-RPC 服务器的用户。
- address：XML-RPC 服务器的主机和端口，在客户端连接到 MySQL Fabric 节点时使用。端口号通常是 32274，主机通常是 localhost。这里的 192.168.16.119 为虚拟机桥接网卡的 IP 地址，用于宿主机访问虚拟机。
- password：客户端连接到服务器时使用的密码。如果没有提供，那么客户端命令行将要求输入密码。

（9）executor 段

本段包含执行器的配置参数。执行器按顺序执行，保证了请求不会冲突。接收到的请求映射到可以直接执行或通过执行器调度的过程。通过执行器调度的过程在执行器生成的线程上下文中处理。通常，读操作由 XML-RPC 会话线程立即执行，写操作由执行器调度和执行。

- executors：执行器处理请求使用的线程数。

（10）sharding 段

Fabric 使用 mysqldump 和 mysql 客户端程序执行数据迁移和分片，本段指定这些客户端程序的路径。

- mysqldump_program：mysqldump 客户端程序所在的目录。
- mysqlclient_program：mysql 客户端程序所在的目录。

（11）protocol.mysql 段

本段包含客户端如何使用 MySQL 客户端/服务器协议连接到 Fabric 管理节点的信息。

- disable_authentication：禁用身份验证。
- user：客户端用来连接到 Fabric 管理节点的用户。
- address：Fabric 管理节点的主机和端口，端口号通常是 32275，主机通常是 localhost。
- password：客户端连接到 Fabric 管理节点时使用的密码。如果未提供，客户端将在命令行中请求密码。

在用于 state store 的 MySQL 实例中建立数据库表，存储 Fabric 的配置和状态：

```
mysqlfabric manage setup
```

命令执行后会创建的库表如下：

```
mysql> show tables from fabric;
+-------------------+
| Tables_in_fabric  |
+-------------------+
| checkpoints       |
| error_log         |
| group_replication |
| group_view        |
| groups            |
| log               |
| permissions       |
| proc_view         |
| role_permissions  |
| roles             |
| servers           |
| shard_maps        |
| shard_ranges      |
| shard_tables      |
| shards            |
| user_roles        |
| users             |
+-------------------+
```

使用下面的命令启动 Fabric 管理节点，并以后台守护进程的方式来执行：

```
mysqlfabric manage start --daemonize
```

检查 fabric 进程是否运行：

```
mysqlfabric manage ping
```

7. 建立 HA 组

在 fab_connector 主机上建立 HA 组 my_group1，并在其中添加三个 MySQL 实例。执行这步前要确认组中所有 MySQL 实例的 server-uuid 都不相同，否则将服务器添加到组中时会报错。本例中可检查 fab_group1 主机上的 /var/lib/group1_1/auto.cnf 、 /var/lib/group1_2/auto.cnf 和 /var/lib/group1_3/auto.cnf 文件的内容并加以确认，然后执行下面的操作。

```
# 创建 HA 组
mysqlfabric group create my_group1

# 向组中添加 MySQL 实例
mysqlfabric group add my_group1 192.168.56.102:3326
mysqlfabric group add my_group1 192.168.56.102:3327
mysqlfabric group add my_group1 192.168.56.102:3328

# 自动在 my_group1 中选出一个实例提升为主库
mysqlfabric group promote my_group1

# 查看 my_group1 中的实例
mysqlfabric group lookup_servers my_group1

# 默认 Fabric 并不会执行故障时的主从自动切换，需要激活该功能
mysqlfabric group activate my_group1
```

至此，具有 HA 功能的 MySQL Fabric 搭建完成。

8.2.3 HA 功能测试

1. 没有活动会话，正常关闭实例

初始实例的状态：3328 为主库、3326 和 3327 为从库：

```
[root@fab_connector ~]# mysqlfabric group lookup_servers my_group1
Fabric UUID:  5ca1ab1e-a007-feed-f00d-cab3fe13249e
Time-To-Live: 1

                         server_uuid                address    status       mode weight
------------------------------------ ------------------- --------- ---------- ------
e488a44d-aa02-11e5-876a-080027a5c938 192.168.56.102:3326 SECONDARY READ_ONLY    1.0
ee359882-aa02-11e5-89aa-080027a5c938 192.168.56.102:3327 SECONDARY READ_ONLY    1.0
f6bea0b0-aa02-11e5-89c7-080027a5c938 192.168.56.102:3328 PRIMARY   READ_WRITE   1.0
```

三个实例中的 GTID 为：

3326：e488a44d-aa02-11e5-876a-080027a5c938:1-3113

3327：ee359882-aa02-11e5-89aa-080027a5c938:1-3892

3328：f6bea0b0-aa02-11e5-89c7-080027a5c938:1-21101

正常关闭 3328 实例：

```
mysqladmin -u root --protocol=tcp -h127.0.0.1 -P3328 -p shutdown
```

自动故障转移后，3328 的状态变为 FAULTY，3327 提升为主库，3326 为从库：

```
[root@fab_connector ~]# mysqlfabric group lookup_servers my_group1
Fabric UUID:  5ca1ab1e-a007-feed-f00d-cab3fe13249e
Time-To-Live: 1

                         server_uuid         address    status       mode weight
------------------------------------ --------- ---------- ------
e488a44d-aa02-11e5-876a-080027a5c938 192.168.56.102:3326 SECONDARY READ_ONLY    1.0
ee359882-aa02-11e5-89aa-080027a5c938 192.168.56.102:3327 PRIMARY    READ_WRITE   1.0
f6bea0b0-aa02-11e5-89c7-080027a5c938 192.168.56.102:3328 FAULTY    READ_WRITE    1.0
```

在新主库上执行一些操作：

```
use test;
create table t1(a int);
insert into t1 values(1);
commit;
```

由于执行了两个事务，3327 的 GTID 由 3892 变成了 3894，两个实例中的 GTID 如下：

3326：e488a44d-aa02-11e5-876a-080027a5c938:1-3113

3327：ee359882-aa02-11e5-89aa-080027a5c938:1-3894

3328：f6bea0b0-aa02-11e5-89c7-080027a5c938:1-21101

再把 3328 加回到组中：

```
[root@fab_connector ~]# mysqlfabric group remove my_group1 192.168.56.102:3328
[root@fab_group1 ~]# mysqld --defaults-file=/etc/my_group1_3.cnf --user=mysql &
[root@fab_connector ~]# mysqlfabric group add my_group1 192.168.56.102:3328
```

加回 3328 实例后，三个实例的状态是 3327 为主库，3326 和 3328 为从库：

```
[root@fab_connector ~]# mysqlfabric group lookup_servers my_group1
Fabric UUID:  5ca1ab1e-a007-feed-f00d-cab3fe13249e
Time-To-Live: 1

                         server_uuid         address    status       mode   weight
------------------------------------ --------- ---------- ------
e488a44d-aa02-11e5-876a-080027a5c938 192.168.56.102:3326 SECONDARY  READ_ONLY    1.0
ee359882-aa02-11e5-89aa-080027a5c938 192.168.56.102:3327   PRIMARY READ_WRITE   1.0
f6bea0b0-aa02-11e5-89c7-080027a5c938 192.168.56.102:3328 SECONDARY  READ_ONLY    1.0
```

三个实例中的 GTID 如下：

3326：e488a44d-aa02-11e5-876a-080027a5c938:1-3113

3327：ee359882-aa02-11e5-89aa-080027a5c938:1-3894

3328：f6bea0b0-aa02-11e5-89c7-080027a5c938:1-21101

在 3328 上查询 t1，数据已经自动复制：

```
mysql> select * from test.t1;
+------+
| a    |
+------+
```

```
|    1 |
+------+
1 row in set (0.00 sec)
```

在“没有活动会话且正常关闭实例”的情况下，可以进行自动故障转移（Auto Failover）。当实例重新加回到组中时，下线期间的事务会自动复制以保持数据一致性。这种场景意义不大，除了有计划的停机，一般不会在生产系统中出现。

2. 没有活动会话，异常关闭实例

用 kill -9 强杀主库实例进程，结果和上一种情况类似，可以进行自动故障转移，当实例重新加回到组中，下线期间的事务会自动复制以保持数据一致性。这种场景意义也不大，在繁忙的生产系统中出现的概率几乎没有。

3. 有活动会话，正常关闭实例

初始实例的状态：3328 为主库、3326 和 3327 为从库：

```
[root@fab_connector ~]# mysqlfabric group lookup_servers my_group1
Fabric UUID:  5ca1ab1e-a007-feed-f00d-cab3fe13249e
Time-To-Live: 1

                         server_uuid          address     status       mode weight
------------------------------------ ------------------- --------- ---------- ------
e488a44d-aa02-11e5-876a-080027a5c938 192.168.56.102:3326 SECONDARY  READ_ONLY    1.0
ee359882-aa02-11e5-89aa-080027a5c938 192.168.56.102:3327 SECONDARY  READ_ONLY    1.0
f6bea0b0-aa02-11e5-89c7-080027a5c938 192.168.56.102:3328   PRIMARY READ_WRITE    1.0
```

三个实例中的 GTID 如下：

3326：e488a44d-aa02-11e5-876a-080027a5c938:1-3113

3327：ee359882-aa02-11e5-89aa-080027a5c938:1-3896

3328：f6bea0b0-aa02-11e5-89c7-080027a5c938:1-23809

运行应用程序或存储过程，持续向 test.chat_message 表里添加数据。在执行过程中正常关闭 3328 实例：

```
    [root@fab_group1 ~]# mysqladmin -u root --protocol=tcp -h127.0.0.1 -P3328
-p shutdown
```

此时三个实例的状态如下，3328 的状态为有缺陷（FAULTY），3326 和 3327 的状态还是从（SECONDARY），并没有进行自动故障转移：

```
[root@fab_connector ~]# mysqlfabric group lookup_servers my_group1
Fabric UUID:  5ca1ab1e-a007-feed-f00d-cab3fe13249e
Time-To-Live: 1

                         server_uuid          address    status       mode  weight
------------------------------------ ---------------- -------- -------- ------
e488a44d-aa02-11e5-876a-080027a5c938 192.168.56.102:3326 SECONDARY  READ_ONLY    1.0
ee359882-aa02-11e5-89aa-080027a5c938 192.168.56.102:3327 SECONDARY  READ_ONLY    1.0
f6bea0b0-aa02-11e5-89c7-080027a5c938 192.168.56.102:3328    FAULTY READ_WRITE    1.0
```

3326 上三个实例的 GTID 为：

3326：e488a44d-aa02-11e5-876a-080027a5c938:1-3113
3327：ee359882-aa02-11e5-89aa-080027a5c938:1-3896
3328：f6bea0b0-aa02-11e5-89c7-080027a5c938:1-24647

3327 上三个实例的 GTID 为：

3326：e488a44d-aa02-11e5-876a-080027a5c938:1-3113
3327：ee359882-aa02-11e5-89aa-080027a5c938:1-3896
3328：f6bea0b0-aa02-11e5-89c7-080027a5c938:1-24166

3326 上由复制执行的事务多于 3327（24647 大于 24166），分别在两个实例上查询记录数也可以证明这一点。3326 上 test.chat_message 表的记录数是 838（24647-23809），3327 上是 357（24166-23809）。

再看一下两个从库的状态，3326 的状态如下：

```
mysql> show slave status\G
*************************** 1. row ***************************
           Slave_IO_State: Reconnecting after a failed master event read
              Master_Host: 192.168.56.102
              Master_User: fabric
              Master_Port: 3328
            Connect_Retry: 60
          Master_Log_File: fab1a-bin.000007
      Read_Master_Log_Pos: 434518
           Relay_Log_File: fab_group1-relay-bin.000002
            Relay_Log_Pos: 434611
    Relay_Master_Log_File: fab1a-bin.000007
         Slave_IO_Running: Connecting
        Slave_SQL_Running: Yes
...
      Exec_Master_Log_Pos: 434518
          Relay_Log_Space: 434823
...
    Seconds_Behind_Master: NULL
Master_SSL_Verify_Server_Cert: No
                Last_IO_Errno: 2003
                Last_IO_Error: error reconnecting to master
'fabric@192.168.56.102:3328' - retry-time: 60  retries: 6
...
           Retrieved_Gtid_Set:
f6bea0b0-aa02-11e5-89c7-080027a5c938:23809-24647
            Executed_Gtid_Set: e488a44d-aa02-11e5-876a-080027a5c938:1-3113,
ee359882-aa02-11e5-89aa-080027a5c938:1-3896,
f6bea0b0-aa02-11e5-89c7-080027a5c938:1-24647
                Auto_Position: 1
         Replicate_Rewrite_DB:
                 Channel_Name:
```

```
        Master_TLS_Version:
1 row in set (0.00 sec)
```

3327 的状态如下：

```
mysql> show slave status\G
*************************** 1. row ***************************
               Slave_IO_State:
                  Master_Host: 192.168.56.102
                  Master_User: fabric
                  Master_Port: 3328
                Connect_Retry: 60
              Master_Log_File: fab1a-bin.000007
          Read_Master_Log_Pos: 185847
               Relay_Log_File: fab_group1-relay-bin.000003
                Relay_Log_Pos: 185493
        Relay_Master_Log_File: fab1a-bin.000007
             Slave_IO_Running: No
            Slave_SQL_Running: Yes
...
          Exec_Master_Log_Pos: 185360
              Relay_Log_Space: 224784
...
        Seconds_Behind_Master: 0
Master_SSL_Verify_Server_Cert: No
                Last_IO_Errno: 0
                Last_IO_Error:
...
           Retrieved_Gtid_Set:
f6bea0b0-aa02-11e5-89c7-080027a5c938:23735-24166
            Executed_Gtid_Set: e488a44d-aa02-11e5-876a-080027a5c938:1-3113,
ee359882-aa02-11e5-89aa-080027a5c938:1-3896,
f6bea0b0-aa02-11e5-89c7-080027a5c938:1-24166
                Auto_Position: 1
         Replicate_Rewrite_DB:
                 Channel_Name:
           Master_TLS_Version:
1 row in set (0.00 sec)
```

从以上状态信息可以看到，3326 的 I/O 线程一直在尝试连接主库，而 3327 已经停止 I/O 线程。这种情况下 Fabric 并没有进行自动故障转移。此时只能进行人为干预：

（1）停止两个从库：在 3326 和 3327 两个实例上执行 stop slave。

（2）提升一个从库成为主库：在 fabric 上执行 mysqlfabric group promote my_group1。

此时三个实例的状态：3328 的状态为有缺陷（FAULTY），3327 的状态是主（PRIMARY），3326 的状态是从（SECONDARY）：

```
[root@fab_connector ~]# mysqlfabric group lookup_servers my_group1
Fabric UUID:  5ca1ab1e-a007-feed-f00d-cab3fe13249e
Time-To-Live: 1
```

```
                server_uuid                  address      status    mode    weight
------------------------------------ ------------------ --------- ---------- ------
e488a44d-aa02-11e5-876a-080027a5c938 192.168.56.102:3326 SECONDARY READ_ONLY    1.0
ee359882-aa02-11e5-89aa-080027a5c938 192.168.56.102:3327   PRIMARY READ_WRITE   1.0
f6bea0b0-aa02-11e5-89c7-080027a5c938 192.168.56.102:3328    FAULTY READ_WRITE   1.0
```

3326 和 3327 上三个实例的 GTID 如下：

3326：e488a44d-aa02-11e5-876a-080027a5c938:1-3113

3327：ee359882-aa02-11e5-89aa-080027a5c938:1-3896

3328：f6bea0b0-aa02-11e5-89c7-080027a5c938:1-24647

可见在 Fabric 把 3327 提升为主时，已经把落后于 3326 的事务补齐，这时分别在两个实例上查询 test.chat_message 表的记录数都是 838。

和前面的情况一样，此时如果再把 3328 加回到组中，它将成为 3327 的从库，并自动复制下线期间的数据变化。由实验得出在“有活动会话且正常关闭实例”情况下的结论：

- 不能进行自动故障转移，需要人为干预。
- 当失败的实例重新加回到组中，下线期间的事务会自动复制以保持数据一致性。
- 手工提升可能会造成数据丢失。

这种场景应该不是常见的情况，即便要人为停止主库，常规操作也会先停掉复制。

1. 有活动会话，异常关闭实例

用 kill -9 强杀主库实例进程，结果和上一种情况类似。不能进行自动故障转移，需要人为干预。当实例重新加回到组中，下线期间的事务会自动复制以保持数据一致性。这种场景是最常见的异常情况，在繁忙的生产系统中主库出现问题，基本上肯定要人为处理了。

2. 启动半同步复制后，有活动会话，异常关闭实例

（1）启动 MySQL 5.7 的半同步复制

分别在 3326、3327 和 3328 三个实例上执行下面的操作以启动半同步复制：

① 在 cnf 配置文件的[mysqld]中添加插件目录：

```
plugin_dir=/root/mysql-5.7.10-linux-glibc2.5-x86_64/lib/plugin
```

② 重启实例后安装半同步复制插件：

```
install plugin rpl_semi_sync_master soname 'semisync_master.so';
install plugin rpl_semi_sync_slave soname 'semisync_slave.so';
```

③ 在 cnf 配置文件的[mysqld]段中添加半同步复制配置：

```
rpl_semi_sync_slave_enabled=1
rpl_semi_sync_master_enabled = 1
rpl_semi_sync_master_timeout = 1000
```

④ 再次重启实例。

（2）测试 Fabric 自动故障转移

测试步骤与第 3 种情况一模一样，结果却迥然不同。在启用半同步复制的情况下，得到以下结果：

- 可以进行自动故障转移。
- 当实例重新加回到组中时，下线期间的事务会自动复制以保持数据一致性。
- 没有数据丢失。

MySQL Fabric 的自动故障转移（Auto Failover）是以主从数据一致性为前提的，如果主从数据不一致，则不会进行自动故障转移，这时需要人为干预，把是否需要手工切换的选择权交给用户。这样的设计是合理的，若要在生产系统中要使用自动故障转移，则一定要启用半同步复制。当然，这又引入了快速网络、快速存储、多线程复制等加快复制速度的优化问题了，可参考 4.2 节中有关复制性能的内容。

8.3 数据分片

当单台 MySQL 服务器（或 HA 组）的写性能达到极限时，可以使用 Fabric 把数据分布到多个 MySQL 服务器组。注意这里说的组可以是单一服务器，也可以是 HA 组。管理员通过建立一个分片映射来定义数据如何在多台服务器中分片。一个分片映射作用于一个或多个表，由管理员指定每个表上的哪些列作为分片键，MySQL Fabric 使用分片键计算一个表的特定行应该存放于哪个分片上。当多个表使用相同的映射和分片键时，这些表上包含相同分片列值的数据行将存放于同一个分片中。单一事务可以访问一个分片中的所有数据。Fabric 提供两种用分片键来计算分片号的方法：

- HASH：在分片键上执行一个哈希函数生成分片号。如果作为分片键的列只有很少的重复值，那么哈希函数的结果会平均分布在多个分片上。
- RANGE：管理员显式地定义分片键的取值范围和分片之间的映射关系。这可以尽可能让用户控制数据分片，并确定哪一行被分配到哪一个分片。

应用程序访问分片的数据库时，需要设置一个连接属性来指定分片键。Fabric 连接器会应用正确的范围或哈希映射，并将事务路由到正确的分片。当需要更多的分片时，MySQL Fabric 可以把现有的一个分片分成两个，同时修改状态存储和连接器中缓存的路由数据。类似地，一个分片可以从一个 HA 组迁移到另一个。

注意，单一的事务或查询只能访问一个单一的分片，所以基于对数据的理解和应用的访问模式来选择一个分片键是非常重要的。由于有不能交叉分片查询的限制，将某些小表的全部数据存储到每一个组中可能会更好。这些全局表被写入到“全局组”，表中数据的任何改变都会自动复制到所有其他非全局组中。全局组中模式（表结构等）的改变也会复制到其他非全局组中以保证一致性。为了得到映射，在没有可“自然选择”的分片键时可能就需要修改数据库的表结构。

在图 8-3 所示的 Fabric Sharding 部署中共有四个 MySQL 实例，其中端口号为 3306 的实例是用来存储 Fabric 状态的状态存储（State Store），另外三个实例各为一个数据分片，端口号 3326

的实例是一个全局组。与该示例相关的 Fabric Sharding 的配置、使用与应用程序代码，可参考 https://blog.csdn.net/wzy0623/article/details/50441091。

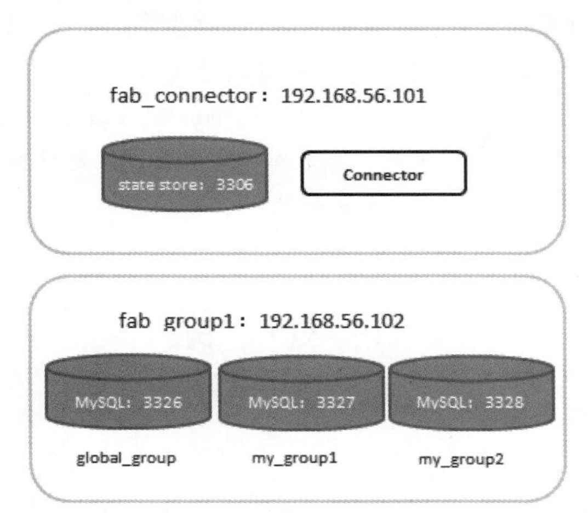

图 8-3　Fabric Sharding 部署示例

也可以将 Fabric 的高可用性（HA）与数据分片（Sharding）结合在一起使用。图 8-4 所示的是一个具有高可用和数据分片特性的 MySQL Fabric 架构，图中共有 10 个 MySQL 实例，其中一个存储 Fabric 状态，另外 9 个是工作节点。每行的三个实例是一个 HA 组，第一行是全局组，每列的三个实例是一个数据分片。该部署的示例可参考：https://blog.csdn.net/wzy0623/article/details/50447833。

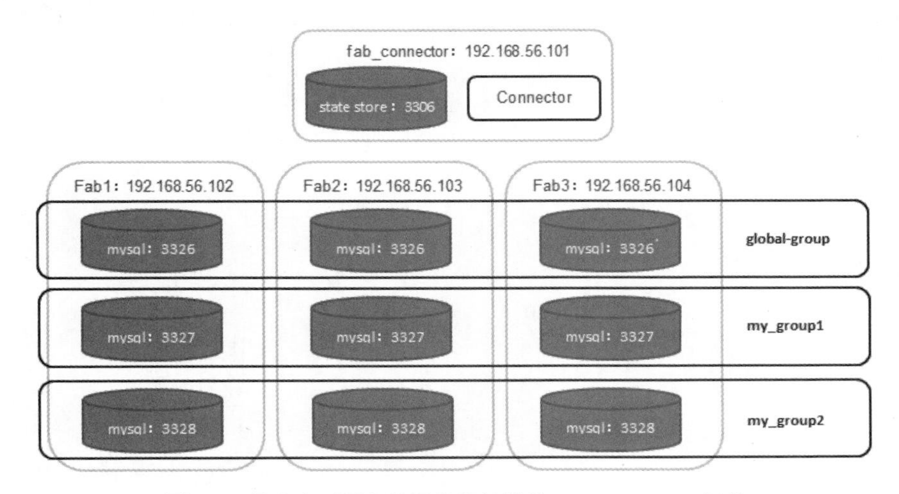

图 8-4　具有高可用和数据分片特性的 MySQL Fabric 架构

8.4　MySQL Fabric 限制

MySQL Fabric 有如下限制：

- 分片对应用并不完全透明。应用不需要知道数据行存储在哪台服务器，也不需要关心数据存储的位置是否改变，但它在访问数据库时需要提供分片键。
- 所有的事务和查询被限制在分片表的单一分片和非分片表中，不支持多个分片的连接查询。
- 连接器执行路由功能，避免了基于代理的解决方案带来的额外延迟，但同时意味着需要 Fabric-aware 的扩展连接器，当前支持 PHP、Python 和 Java 三种语言。
- MySQL Fabric 本身并没有失败冗余，失败时需要重启。但是当 MySQL Fabric 进程失效时，连接器仍然可以使用本地缓存继续路由操作。

8.5 小 结

MySQL Fabric 提供了原生的数据库服务器高可用和数据分片功能。这个产品起初的愿景很好：基于 MySQL 原有的异步或半同步复制和 GTID，实现了自动主从故障转移（Failover）；将代理功能迁移到应用端连接器（Connector），既能够对应用程序提供透明的路由功能，又有效地缓解了传统代理（Proxy）的单点故障和性能瓶颈，可谓一举两得。

但是，MySQL Fabric 发展得似乎并不理想，HA 功能逐渐被组复制和 InnoDB Cluster 所取代。分片（Sharding）设计缺陷也比较明显，例如对应用不够透明，需要在程序中指定分片键，不支持跨分片查询等硬伤，因而早已停止了进一步的开发。现在 MySQL 官方已经完全删除了 Fabric 下载和文档的相关链接。为了保持讨论 MySQL 高可用性方案的完整和系统，我们还是在本章用示例详细说明了 Fabric 的安装和配置以及实际操作。从下一章开始，我们将介绍 4 个常用的 MySQL 高可用第三方中间件的解决方案。

第 9 章 MMM

9.1 MMM 简介

MMM 的全称是 Master-Master Replication Manager for MySQL，是一套支持 MySQL 主主复制、故障转移、日常管理和配置监控的 Perl 脚本程序套件。

9.1.1 基本功能

MMM 主要用来监控和管理 MySQL Master-Master 主主复制，可以说是 MySQL 主主复制管理器。正如 4.1.2 小节所描述的，MySQL 推荐的是主动-被动模式下的主主复制，也就是指两个（或多个）主库的配置相同，但在同一时刻只有其中一个主库可以接受写请求。MMM 所支持的就是这种拓扑结构。虽然名字叫做 Master-Master Replication，但是它在任何时刻只有一个节点可以被写入，另外的节点提供部分读服务，以加速在主主切换时备选主库的预热过程。MMM 包含 writer 和 reader 两类角色，分别对应读写节点和只读节点。

MMM 这套脚本程序一方面实现了自动故障转移的功能，另一方面，在基于标准的主从复制配置中，MMM 内部附加的工具脚本也可以实现对任意数量的从库进行读负载均衡。可以用 MMM 为一组基于复制的 MySQL 服务器启动 VIP。MMM 提供了自动和手动两种方式，移除一组 MySQL 服务器中复制延迟较高的服务器上的 VIP。除此之外，MMM 套件中还有实现数据备份、节点之间数据同步功能的脚本。

由于 MMM 无法完全保证数据一致性，因此它适用于对数据的一致性要求不是很高，但是又想最大程度地保证业务可用性的场景。MySQL 的标准主从复制本身并没有提供复制故障转移（Replication Failover）的解决方案，通过 MMM 就可以实现服务器的自动故障转移，从而实现 MySQL 数据库的高可用。对于那些对数据一致性要求极高的业务，不建议采用 MMM 这种高可用架构。

MMM 的特点归纳如下：

- 稳定和成熟的开源产品，经过了时间的考验。相对应地，由于发布时间较早，因此不支持GTID、多线程复制等 MySQL 功能。
- 依托 MySQL 原生的复制技术，只是使用脚本程序来控制，所以在原理上比较容易理解，而且管理能够更智能化。
- 安装、配置和使用比较简单。
- 除了提供基本的 MySQL 数据库高可用性，自动故障转移功能，还有其他工具套件。
- 由于架构里只有一个写入点，因此扩展性有限。
- 监控节点本身是单故障点，但可以通过与 Keepalived 等工具一起使用，保证它自身的高可用。
- 整体性能主要取决于最差 I/O 的主机，因此各 MySQL 服务器节点软硬件配置应该相同。

9.1.2 工作原理

一个典型的 MMM 架构如图 9-1 所示，其中包含一个监控节点（mmm-monitor）、两个 MySQL 主节点（Master）和若干 MySQL 从节点（Slave）。监控节点上运行 MMM 监控程序，每个 MySQL 服务器节点上运行一个 MMM 代理程序（mmm-agent）。监控节点会提供多个虚拟 IP（VIP），包括一个可写 VIP、多个可读 VIP。通过监控程序的设置，这些 VIP 会绑定在可用 MySQL 上，应用程序通过连接不同的 VIP 将读请求和写请求分离。

作为架构核心的监控节点负责监控所有 MySQL 实例。当一台 MySQL 主库宕机时，监控程序会将写 VIP 迁移至另一个 MySQL 主库。除了管理双主节点，MMM 也会管理从节点，在某个从库出现宕机、复制延迟或复制错误时，监控程序会移除该节点的读 VIP，直到它恢复正常。

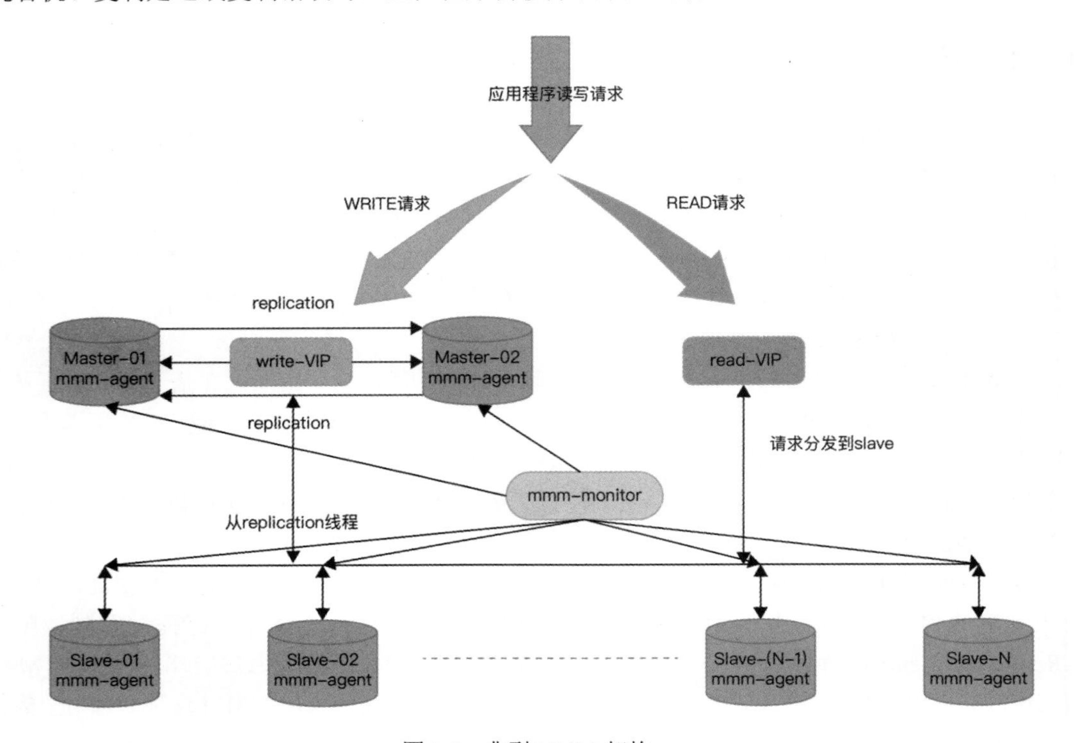

图 9-1 典型 MMM 架构

在整个监管过程中，需要在 MySQL 中添加相关授权用户，以便让 MySQL 可以接受监控节点的维护。授权的用户包括一个 mmm_monitor 用户和一个 mmm_agent 用户，如果想使用 MMM 的备份工具，则还要添加一个 mmm_tools 用户。

MMM 的主要功能通过以下三个脚本来提供：

- mmm_mond：监控守护进程，执行所有监控工作并做出有关角色切换的所有决定。此脚本需要在监控节点上运行，一般和它所管理的 MySQL 服务器分开部署。
- mmm_agent：运行在每台 MySQL 服务器上的代理进程，完成监控的探针工作和执行简单的远端服务设置。此脚本需要在被监管的 MySQL 服务器节点上运行。
- mmm_control：一个提供管理 mmm_mond 进程命令的简单脚本。

9.1.3 典型用例

1. 双主节点

双主节点架构如图 9-2 所示。在双节点主-主设置中，MMM 使用 5 个 IP：每个节点的单个永久 IP，2 个读 VIP（只读）和 1 个写 VIP（更新）。3 个 VIP 在节点之间迁移，具体取决于节点可用性。通常在没有复制延迟时，活动主库有读、写 2 个 VIP，备用主库有 1 个读 VIP。如果发生故障，读写操作都会迁移到另一个工作节点。

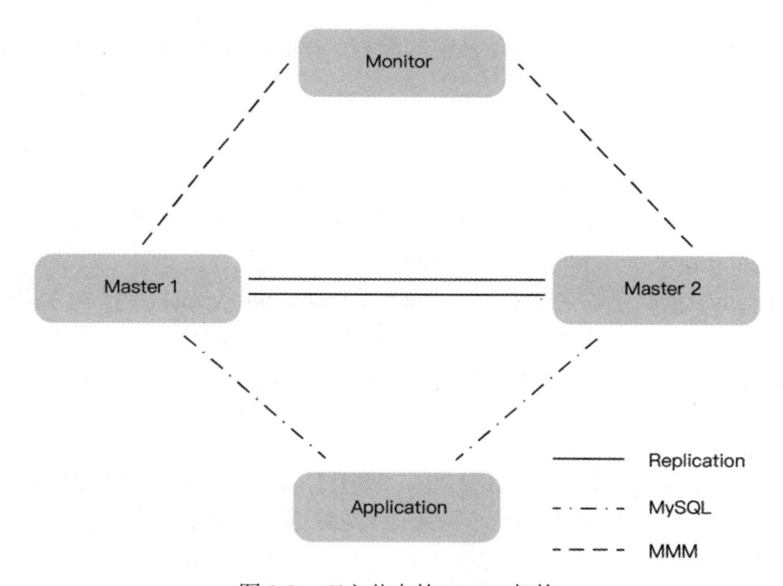

图 9-2 双主节点的 MMM 架构

2. 双主+一个/多个从节点

“双主一（多）从”的架构如图 9-3 所示。

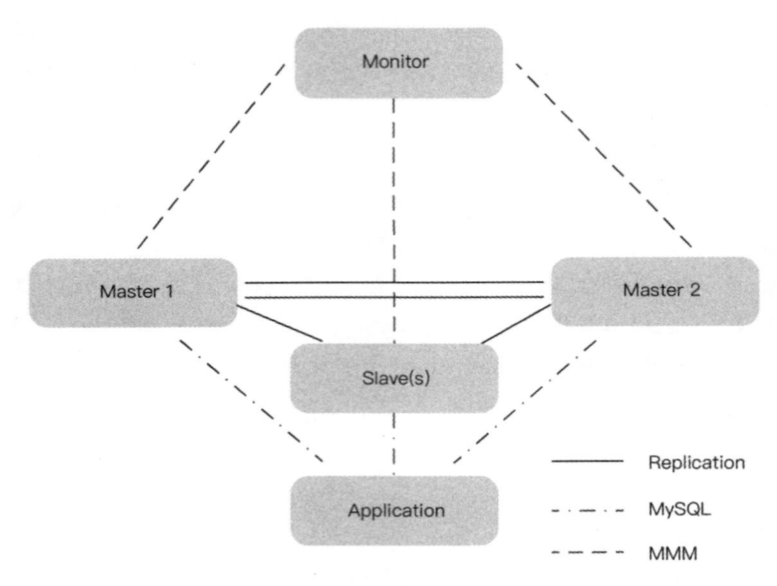

图 9-3 “双主一（多）从”的 MMM 架构

9.1.4 切换流程

以图 9-3 的架构为例，描述一下故障转移的流程。现在假设 Master1 宕机，MMM 将执行以下步骤：

步骤 01 Monitor 检测到 Master1 连接失败。

步骤 02 Monitor 发送 set_offline 指令到 Master1 的 Agent。

步骤 03 Master1 Agent 如果存活，下线写 VIP，尝试把 Master1 设置为 read_only=1。

步骤 04 Monitor 发送 set_online 指令到 Master2。

步骤 05 Master2 Agent 接收到指令，执行 select master_pos_wait()等待同步完毕。

步骤 06 Master2 Agent 上线写 VIP，把 Master2 节点设为 read_only=0。

步骤 07 Monitor 发送更改同步对象的指令到各个从（Slave）节点的 Agent。

步骤 08 各个从节点（Slave）从新主节点（Master）同步数据。

从整个流程可以看到，如果主节点出现故障，MMM 会自动故障转移（也称为自动故障切换），不需要人工干预。同时我们也可以看到一个问题，就是数据库挂掉后，只是进行了故障转移或切换，而不会主动补齐丢失的数据，所以 MMM 会有数据不一致性的风险。

9.1.5 系统需求

对于管理 n 台 MySQL 服务器的 MMM 设置，有以下需求：

- n + 1 台主机：每台 MySQL 服务器对应一台主机；MMM 监控节点对应一台主机。
- 2 * n + 1 个 IP 地址：每台主机占用一个固定 IP +读角色的一个 VIP（虚拟 IP），一个写入 VIP

（虚拟 IP）。

- monitor user：MMM 监控程序使用的、具有 REPLICATION CLIENT 权限的 MySQL 用户。
- agent user：MMM 代理进程使用的、具有 SUPER 和 REPLICATION CLIENT 权限的 MySQL 用户。
- replication user：用于复制的、具有 REPLICATION SLAVE 权限的 MySQL 用户。
- tools user：使用 MMM 工具所需的具有 SUPER、REPLICATION CLIENT 和 RELOAD 权限的 MySQL 用户。

监控主机需要安装以下支持包：

（1）perl
（2）fping（如果想以非 root 用户身份运行 mmm_mond）
（3）Perl 模块：

- Algorithm::Diff
- Class:Singleton
- DBI and DBD::mysql
- File::Basename
- File::stat
- File::Temp
- Log::Dispatch
- Log::Log4perl
- Mail::Send
- Net::Ping
- Proc::Daemon
- Thread::Queue
- Time::HiRes

对于 MySQL 节点，初始应该在所有 MySQL 服务器的配置中设置 read_only=1，MMM 将使用 active_master_role 在活动主机上将其更改为 read_only=0。MySQL 节点主机需要安装以下支持包：

（1）perl
（2）iproute
（3）send_arp（solaris）
（4）Perl 模块：

- Algorithm::Diff
- DBI and DBD::mysql
- File::Basename
- File::stat
- Log::Dispatch
- Log::Log4perl
- Mail::Send

- Net::ARP（Linux）
- Proc::Daemon
- Time::HiRes

如果要使用 MMM 工具（mmm_backup、mmm_restore、mmm_clone 等），则必须将 LVM 用于 MySQL 数据库和日志所在的分区，并且需要对回滚段空间进行自由物理扩展，参见网址 https://www.percona.com/blog/2008/06/09/estimating-undo-space-needed-for-lvm-snapshot/。

MMM 工具还需要以下 perl 模块：

- Path::Class
- Data::Dumper

9.2 实现 MySQL 主主复制的高可用

下面用一个示例详细说明 MMM 的安装和配置，并测试其功能。

9.2.1 基本环境

示例使用的操作系统版本为 CentOS Linux release 7.2.1511 (Core)，MySQL 数据库服务器的版本是 5.6.14，主机和 VIP 信息分别如表 9-1 和表 9-2 所示。

表 9-1 MMM 示例主机信息

角色	IP	主机名	网卡名
db1	172.16.1.125	hdp2	ens32
db2	172.16.1.126	hdp3	ens32
monitor	172.16.1.127	hdp4	-

表 9-2 MMM 示例 VIP 信息

VIP	角色	描述
172.16.1.100	write	应用程序连接该 VIP 对主库进行写请求
172.16.1.210	read	应用程序连接该 VIP 进行读请求
172.16.1.211	read	应用程序连接该 VIP 进行读请求

示例架构如图 9-4 所示。

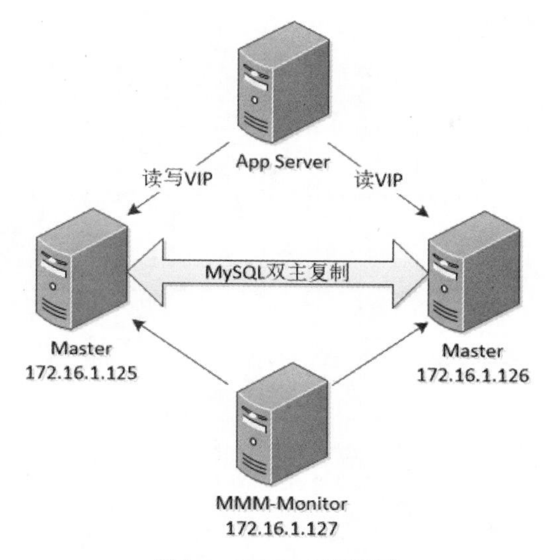

图 9-4 MMM 示例架构

9.2.2 安装和配置

1. 配置 MySQL 双主复制

双主复制本质上就是将两个主从复制组合在一起，因此分别在两个 MySQL 实例上配置主从复制即可。只要将主库信息指向对方，并注意自增选项的配置以避免自增列复制冲突。

（1）配置文件

172.16.1.125 的 MySQL 配置文件内容如下：

```
server-id=125                        # 保证两台 MySQL 配置不同值
log-bin=mysql-bin                    # 启用二进制日志
auto_increment_increment=2           # 自增步长为 2，一般有 n 台主 MySQL 就设置为 n
auto_increment_offset=1              # 自增初始值，第 n 台主 MySQL 设置为 n
```

172.16.1.126 的 MySQL 配置文件内容如下：

```
server-id=126
log-bin=mysql-bin
auto_increment_increment=2
auto_increment_offset=2
```

配置完成后重启两个 MySQL 实例。

（2）创建复制使用的 MySQL 用户

在两个 MySQL 实例中创建复制使用的用户：

```
create user 'repl'@'%' identified by '123456';
grant replication slave on *.* to 'repl'@'%';
```

（3）设置复制主库

分别在两个 MySQL 实例上查看二进制日志名和位置：

```
    mysql> show master status;
    +------------------+----------+--------------+------------------+---------
----------+
    | File             | Position | Binlog_Do_DB | Binlog_Ignore_DB |
Executed_Gtid_Set |
    +------------------+----------+--------------+------------------+---------
----------+
    | mysql-bin.000001 |      120 |              |                  |
    +------------------+----------+--------------+------------------+---------
----------+
    1 row in set (0.00 sec)
```

用上面命令的输出指定主库信息：

```
-- 125 的 MySQL 实例执行
change master to
      master_host='172.16.1.126',
      master_user='repl',
      master_password='123456',
      master_log_file='mysql-bin.000001',
      master_log_pos=120;

-- 126 的 MySQL 实例执行
change master to
      master_host='172.16.1.125',
      master_user='repl',
      master_password='123456',
      master_log_file='mysql-bin.000001',
      master_log_pos=120;
```

（4）启动复制

分别在两个 MySQL 实例上启动复制：

```
start slave;
```

2. 安装 MMM

在三台主机执行下面的 yum 命令来安装 MMM 软件包：

```
yum -y install mysql-mmm-*
```

3. 建立 MySQL 数据库用户

在 db1 和 db2 中建立 mmm_agent 和 mmm_monitor 用户：

```
    grant super,replication client,process on *.* to 'mmm_agent'@'%' identified
by '123456';
    grant replication client on *.* to 'mmm_monitor'@'%' identified by '123456';
```

4. 配置 MMM

（1）通用配置

编辑 db1 上的/etc/mysql-mmm/mmm_common.conf 文件，内容如下：

```
active_master_role      writer
```

```
<host default>
    cluster_interface       ens32
    pid_path                /var/run/mmm_agentd.pid
    bin_path                /usr/libexec/mysql-mmm/
    replication_user        repl
    replication_password    123456
    agent_user              mmm_agent
    agent_password          123456
</host>

<host db1>
    ip      172.16.1.125
    mode    master
    peer    db2
</host>

<host db2>
    ip      172.16.1.126
    mode    master
    peer    db1
</host>

<role writer>
    hosts    db1, db2
    ips     172.16.1.100
    mode    exclusive
</role>

<role reader>
    hosts    db1, db2
    ips     172.16.1.210, 172.16.1.211
    mode    balanced
</role>
```

主要配置项说明：

- active_master_role：活动主机角色名称，agent 与 monitor 使用。
- replication_user：用于复制的用户。
- agent_user：mmm-agent 用户。
- host 段中的 mode：标明是否为主库或者备选主库，或者从库。
- role 段中的 mode：exclusive 为独占模式，同一时刻只能有一个主库。balanced 对应的 ips 可能包含多个 IP，读负载将在这些主机之间平衡。
- <role write>中 hosts：表示当前主库和备选主库的真实主机 IP 或者主机名，ips 为对外提供的 VIP 地址。
- <role reader>中 hosts：代表从库真实的 IP 或主机名，ips 代表从库的 VIP 地址。

将该文件复制到其他所有节点（db2、Monitor）：

```
scp /etc/mysql-mmm/mmm_common.conf 172.16.1.126:/etc/mysql-mmm/
scp /etc/mysql-mmm/mmm_common.conf 172.16.1.127:/etc/mysql-mmm/
```

（2）agent 配置

db1 的/etc/mysql-mmm/mmm_agent.conf 文件内容为：

```
include mmm_common.conf
this db1
```

db2 的/etc/mysql-mmm/mmm_agent.conf 文件内容为：

```
include mmm_common.conf
this db2
```

（3）monitor 配置

monitor 上的配置文件/etc/mysql-mmm/mmm_mon.conf 内容为：

```
include mmm_common.conf

<monitor>
    ip                  172.16.1.127
    pid_path            /var/run/mmm_mond.pid
    bin_path            /usr/libexec/mysql-mmm
    status_path         /var/lib/mysql-mmm/mmm_mond.status
    ping_ips            172.16.1.125,172.16.1.126
    auto_set_online     60
</monitor>

<host default>
    monitor_user        mmm_monitor
    monitor_password    123456
</host>

debug 0
```

auto_set_online 表示将节点状态从 AWAITING_RECOVERY 切换到 ONLINE 之前等待的秒数，0 表示已禁用。

9.2.3 功能测试

1. 启动 MMM

（1）在 db1 和 db2 上启动 agent

```
/etc/init.d/mysql-mmm-agent start
/etc/init.d/mysql-mmm-agent start
```

（2）在 monitor 上启动监控

```
/etc/init.d/mysql-mmm-monitor start
```

（3）检查MMM启动后的状态

MMM启动成功后，在monitor上执行mmm_control show和mmm_control checks命令，结果如下：

```
[root@hdp4~]#mmm_control show
  db1(172.16.1.125) master/ONLINE. Roles: reader(172.16.1.210)
  db2(172.16.1.126) master/ONLINE. Roles: reader(172.16.1.211),
writer(172.16.1.100)

[root@hdp4~]#mmm_control checks
db2  ping        [last change: 2018/08/02 08:57:38]  OK
db2  mysql       [last change: 2018/08/02 08:57:38]  OK
db2  rep_threads  [last change: 2018/08/02 08:57:38]  OK
db2  rep_backlog  [last change: 2018/08/02 08:57:38]  OK: Backlog is null
db1  ping        [last change: 2018/08/02 08:57:38]  OK
db1  mysql       [last change: 2018/08/02 08:57:38]  OK
db1  rep_threads  [last change: 2018/08/02 08:57:38]  OK
db1  rep_backlog  [last change: 2018/08/02 08:57:38]  OK: Backlog is null
```

可以看到，db2上分配了读、写两个VIP，角色为主库。db1分配了读VIP，角色为备用主库。读、写VIP是自动绑定的，因此这里与图9-4中所示的VIP分配有所不同。两个MySQL实例与双主复制状态正常。

2. 测试切换

（1）停止db1上的MySQL服务

```
service mysql stop
```

查看状态，db1上的VIP reader(172.16.1.210) 自动迁移到db2上：

```
[root@hdp4~]#mmm_control show
  db1(172.16.1.125) master/HARD_OFFLINE. Roles:
  db2(172.16.1.126) master/ONLINE. Roles: reader(172.16.1.210),
reader(172.16.1.211), writer(172.16.1.100)
[root@hdp4~]#
```

（2）启动db1上的MySQL服务

```
service mysql start
```

一分钟之后，状态恢复：

```
[root@hdp4~]#mmm_control show
  db1(172.16.1.125) master/ONLINE. Roles: reader(172.16.1.210)
  db2(172.16.1.126) master/ONLINE. Roles: reader(172.16.1.211),
writer(172.16.1.100)
[root@hdp4~]#
```

（3）停止db2上的MySQL服务

```
service mysql stop
```

db2上负责读的VIP（172.16.1.211）以及负责写的VIP（172.16.1.100）会自动迁移到db1上：

```
[root@hdp4~]#mmm_control show
  db1(172.16.1.125) master/ONLINE. Roles: reader(172.16.1.210),
reader(172.16.1.211), writer(172.16.1.100)
  db2(172.16.1.126) master/HARD_OFFLINE. Roles:
[root@hdp4~]#
```

（4）启动 db2 上的 MySQL 服务

```
service mysql start
```

一分钟之后，db1 上负责读的 VIP（172.16.1.210）自动迁移到 db2 上，但是负责写的 VIP（172.16.1.100）仍在 db1 上：

```
[root@hdp4~]#mmm_control show
  db1(172.16.1.125) master/ONLINE. Roles: reader(172.16.1.211),
writer(172.16.1.100)
  db2(172.16.1.126) master/ONLINE. Roles: reader(172.16.1.210)
[root@hdp4~]#
```

（5）只读节点上停止复制

在 db2 上停止复制：

```
mysql> stop slave;
```

查看状态，db2 上的 VIP（172.16.1.210）会自动迁移到 db1 上：

```
[root@hdp4~]#mmm_control show
  db1(172.16.1.125) master/ONLINE. Roles: reader(172.16.1.210),
reader(172.16.1.211), writer(172.16.1.100)
  db2(172.16.1.126) master/REPLICATION_FAIL. Roles:
[root@hdp4~]#
```

（6）只读节点上启动复制

在 db2 上启动复制：

```
mysql> start slave;
```

状态恢复：

```
[root@hdp4~]#mmm_control show
  db1(172.16.1.125) master/ONLINE. Roles: reader(172.16.1.211),
writer(172.16.1.100)
  db2(172.16.1.126) master/ONLINE. Roles: reader(172.16.1.210)
[root@hdp4~]#
```

（7）读写节点上停止复制

在 db1 上停止复制：

```
mysql> stop slave;
```

查看状态无任何变化。理论上也应该是对现有的环境无任何影响：

```
[root@hdp4~]#mmm_control show
  db1(172.16.1.125) master/ONLINE. Roles: reader(172.16.1.211),
writer(172.16.1.100)
```

```
    db2(172.16.1.126) master/ONLINE. Roles: reader(172.16.1.210)
[root@hdp4~]#
```

（8）停止 MMM 监控主机上的 monitor 服务

```
/etc/init.d/mysql-mmm-monitor stop
```

VIP 都还在之前的节点上：

```
[root@hdp2~]#ip a | grep ens32
2: ens32: <BROADCAST,MULTICAST,UP,LOWER_UP> mtu 1500 qdisc pfifo_fast state
UP qlen 1000
    inet 172.16.1.125/24 brd 172.16.1.255 scope global ens32
    inet 172.16.1.100/32 scope global ens32
    inet 172.16.1.211/32 scope global ens32

[mysql@hdp3~]$ip a | grep ens32
2: ens32: <BROADCAST,MULTICAST,UP,LOWER_UP> mtu 1500 qdisc pfifo_fast state
UP qlen 1000
    inet 172.16.1.126/24 brd 172.16.1.255 scope global ens32
    inet 172.16.1.210/32 scope global ens32
```

（9）启动 MMM 监控服务

```
/etc/init.d/mysql-mmm-monitor start
```

对 db1 和 db2 上的相关服务无影响：

```
[root@hdp4~]#mmm_control show
  db1(172.16.1.125) master/ONLINE. Roles: reader(172.16.1.211),
writer(172.16.1.100)
  db2(172.16.1.126) master/ONLINE. Roles: reader(172.16.1.210)
[root@hdp4~]#
```

（10）查看监控日志

以上的角色切换过程都在监控日志中记录：

```
[root@hdp4~]#tail -f /var/log/mysql-mmm/mmm_mond.log
...
2018/08/02 09:07:46 FATAL State of host 'db1' changed from ONLINE to HARD_OFFLINE
(ping: OK, mysql: not OK)
2018/08/02 09:10:53 FATAL State of host 'db1' changed from HARD_OFFLINE to
AWAITING_RECOVERY
2018/08/02 09:11:54 FATAL State of host 'db1' changed from AWAITING_RECOVERY
to ONLINE because of auto_set_online(60 seconds). It was in state AWAITING_RECOVERY
for 61 seconds
2018/08/02 09:14:06 FATAL State of host 'db2' changed from ONLINE to HARD_OFFLINE
(ping: OK, mysql: not OK)
2018/08/02 09:16:22 FATAL State of host 'db2' changed from HARD_OFFLINE to
AWAITING_RECOVERY
2018/08/02 09:17:24 FATAL State of host 'db2' changed from AWAITING_RECOVERY
to ONLINE because of auto_set_online(60 seconds). It was in state AWAITING_RECOVERY
for 62 seconds
```

```
    2018/08/02 09:20:02 FATAL State of host 'db2' changed from ONLINE to
REPLICATION_FAIL
    2018/08/02 09:22:14 FATAL State of host 'db2' changed from REPLICATION_FAIL
to ONLINE
```

9.3 小 结

MMM 是一套 Perl 脚本套件，通过 VIP 的自动迁移实现 MySQL 多主复制的高可用性。它基于 MySQL 原生的复制技术，只是使用脚本程序来控制复制角色切换过程，所以工作机制上比较容易理解。再者由于功能单一，又是脚本实现，使得 MMM 的安装、配置、使用、监控都较为简单。MMM 的核心是一个 mmm-monitor 进程，它完成大部分的功能，通常与被监控的 MySQL 服务器分开部署。每台 MySQL 服务器上需要运行一个代理进程 mmm-agent。除这两种程序外，MMM 套件中还提供了工具脚本程序来实现数据备份、节点之间数据同步等功能。需要注意的是，MMM 并不能严格确保多个 MySQL 实例的数据一致，因此不适用于对数据一致性要求极高的场景。下一章将介绍与 MMM 类似的另一个中间件产品——MHA。

第10章 MHA

10.1 MHA简介

MHA（Master High Availability）在MySQL高可用方面是一个相对成熟的解决方案。与上一章介绍MMM的一样，MHA也是用Perl语言开发的。不同的是，MHA更多关注的是MySQL主从复制中的主库，当主库崩溃时，能够快速地在从库中找到最佳候选者，并将其提升为新主库。另一个与MMM的显著区别是，当MySQL主服务器宕机时，MHA会尝试从主服务器尽可能多地保存二进制日志，从而最大程度保证不丢失事务，以达到真正意义上的高可用。

10.1.1 基本功能

1. 主服务器监控和自动故障转移（自动切换）

MHA监控复制架构中的主服务器，一旦检测到主服务器故障，就会自动进行故障转移。即使有些从服务器没有收到最新的relay log也不要紧，MHA自动从最新的从服务器上识别差异的relay log，并把这些日志应用到其他从服务器上，因此所有的从服务器保持数据一致。MHA通常在十几秒内完成故障转移：9~12秒可以检测出主服务器故障，7~10秒内关闭故障的主服务器以避免出现“脑裂”问题，几秒钟内应用差异的relay log到新的主服务器上，整个过程可以在10~30秒内完成。还可以设置优先级指定其中的一台从服务器作为主服务器的候选者。由于MHA在从服务器之间修复一致性，因此不会发生因为数据不一致所导致的复制失败问题。

2. 交互式主服务器故障转移（手工切换）

可以只使用MHA的故障转移功能，而不用于监控主服务器。当主服务器故障时，人工调用MHA来进行故障转移。

3. 非交互式主服务器故障转移

不监控主服务器，但自动实现故障转移。这种特征适用于已经使用其他软件来监控主服务器状态，比如用 heartbeat 来检测主服务器故障和 VIP 地址接管。可以使用 MHA 来实现故障转移，并将从服务器提升为主服务器。

4. 在线切换主服务器

在许多情况下，需要将现有的主服务器迁移到另外一台服务器上，比如主服务器更换硬件，内核更新，MySQL 升级等常规维护操作。维护主服务器会引起性能下降，导致停机时间或无法写入数据等问题。另外，阻塞或杀掉当前运行的会话可能导致主从库之间数据不一致的问题发生。MHA 提供快速切换和优雅的阻塞写入，整个切换过程只需要 0.5~2 秒的时间，这段时间内数据是无法写入的。在很多情况下，0.5~2 秒的阻塞写入是可以接受的，因此切换主服务器通常不需要计划分配维护的时间窗口。

10.1.2 切换流程

当主服务器出现故障时，MHA 可以自动将最新数据的从服务器提升为新的主服务器，然后将所有其他的从服务器重新指向新的主服务器，整个故障转移过程对应用程序完全透明。具体切换流程如下：

（1）把宕机的主服务器的二进制日志保存下来。
（2）识别含有最新 relay-log 的从服务器。
（3）在上一步找到的从服务器上，把差异的中继日志（relay log）应用于其他从服务器。
（4）将第 1 步保存下来的二进制日志恢复到第 2 步识别出的从服务器上。
（5）将第 2 步识别出的从服务器提升为新的主服务器。
（6）将其他从服务器重新指向新提升的主服务器，并开启主从复制。

10.1.3 部署架构

MHA 软件由两部分组成：MHA Manager（管理节点）和 MHA Node（数据节点）。MHA Manager 可以单独部署在一台独立的机器上管理多个 master-slave 集群，也可以部署在一台从节点上，但建议不要将 Manager 安装在主服务器上，以防止主机层面出现问题。MHA Node 运行在每台 MySQL 服务器上，MHA Manager 会定时（默认为 3 秒）探测集群中的主服务器，当其出现问题时自动执行故障转移。图 10-1 所示为一个典型的 MHA 部署架构。

在 MHA 自动故障切换过程中，MHA 试图从宕机的主服务器上保存二进制日志，最大程度的保证数据不丢失，但这并不总是可行的。例如，如果主服务器硬件故障或无法通过 ssh 访问，MHA 没法保存二进制日志，只进行故障转移而丢失了最新的数据。使用 MySQL 5.5 以后的半同步复制，可以大大降低数据丢失的风险。MHA 可以与半同步复制结合使用。

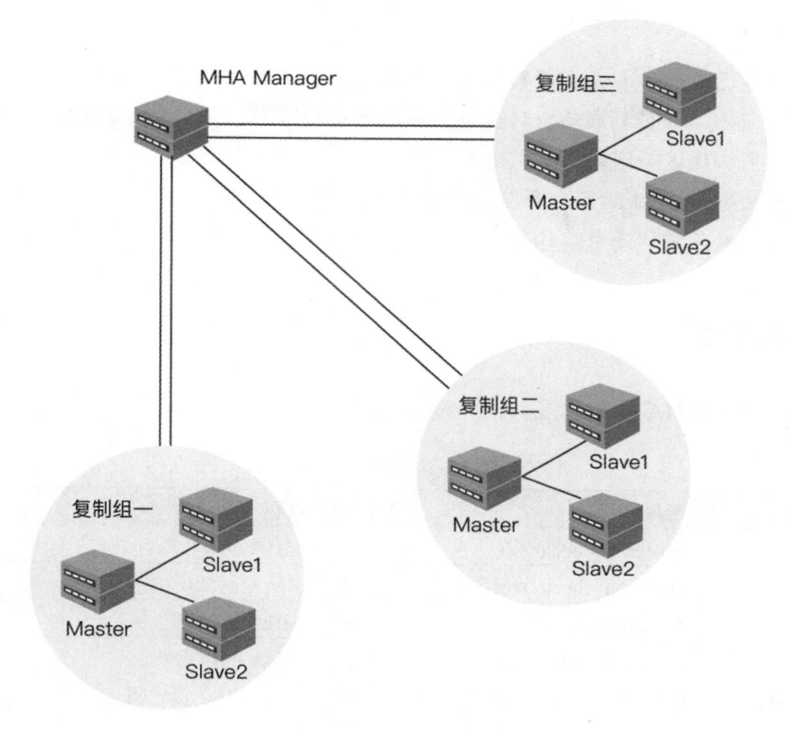

图 10-1　MHA 部署架构

MHA 主要支持“一主多从”架构。要搭建 MHA，要求一个复制集群中至少有“一主两从”三台 MySQL 数据库服务器，，即一台充当主库，另一台充当备用主库，第三台充当从库。因为至少需要三台服务器，出于机器成本和简化架构的考虑，有些公司在该基础上进行了改造，以支持“一主一从”。从代码层面看，MHA 就是一套 Perl 脚本，有一定技术实力的用户将 MHA 改成支持“一主一从”也并非难事。

10.1.4　主要特点

MHA 具有以下优点：

- 自动故障转移速度快，只需 10~30 秒。
- 加强了数据安全性，切换过程不易出现数据丢失。
- 不需要对当前 MySQL 环境做重大修改。
- 仅一台 MHA manager 就可监控多个主从复制结构，无需再额外增加服务器。
- 可工作在半同步复制和异步复制。
- 支持 GTID。
- 当监控 MySQL 状态时，仅需要每隔 N 秒向主服务器发送 ping 包（默认为 3 秒），所以对性能无影响。可以理解为 MHA 的性能和简单的主从复制框架性能一样。
- 只要复制支持的存储引擎，MHA 都支持，不局限于 InnoDB。

MHA 的缺点为：

- 需要用户自己编写脚本或利用第三方工具来实现虚拟 IP 的配置，增加了使用复杂度。
- 与 MMM 不同，MHA 启动后只会对主服务器进行监控，不监控复制链路，也就无法发现复制中断、主从延迟等问题。
- 需要基于 SSH 免认证配置，存在一定的安全隐患。
- 没有提供从服务器的读负载均衡功能。

10.1.5 工具介绍

表 10-1 中包含了 MHA 的工具及其简单功能说明。

表 10-1 MHA 工具

所属组件	工具名称	功能说明
MHA Manager	masterha_check_ssh	检查 MHA 的 ssh-key
	masterha_check_repl	检查主从复制情况
	masterha_manager	启动 MHA
	masterha_check_status	检测 MHA 的运行状态
	masterha_master_monitor	检测主服务器是否宕机
	masterha_master_switch	手动故障转移
	masterha_conf_host	手动添加服务器信息
	masterha_secondary_check	建立到远程服务器的 TCP 连接
	masterha_stop	停止 MHA
MHA Node	save_binary_logs	保存和应用主服务器的二进制日志
	apply_diff_relay_logs	识别差异的中继日志事件并将差异事件应用于其他从服务器。
	filter_mysqlbinlog	去除二进制日志中不必要的 ROLLBACK 事件（MHA 已不再使用这个工具）
	purge_relay_logs	清除中继日志（不会阻塞 SQL 线程）

10.2 安装和配置

可从 https://github.com/yoshinorim/mha4mysql-manager/wiki/Downloads 下载 MHA 安装包。虽然都是 Perl 程序，但 MHA 的安装和配置比 MMM 复杂一些，毕竟 MHA 的功能相对来说更完备，尤其体现在故障转移后尽量保证不丢失数据。下面用一个示例详细说明 MHA 的安装和配置过程，并测试其主要功能。示例使用的基本环境为：

- 操作系统版本：CentOS Linux release 7.2.1511 (Core)
- MySQL 服务器版本：5.6.14
- VIP：172.16.1.100

- 主机信息：见表 10-2。

表 10-2 MHA 示例主机信息

角色	IP	主机名	网卡名	server_id	功能
Monitor Host	172.16.1.124	hdp1	-	-	监控复制组
Master	172.16.1.127	hdp4	ens160	127	响应写请求
Candidate Master	172.16.1.126	hdp3	ens32	126	响应读请求
Slave	172.16.1.125	hdp2	ens32	125	响应读请求

示例架构如图 10-2 所示。

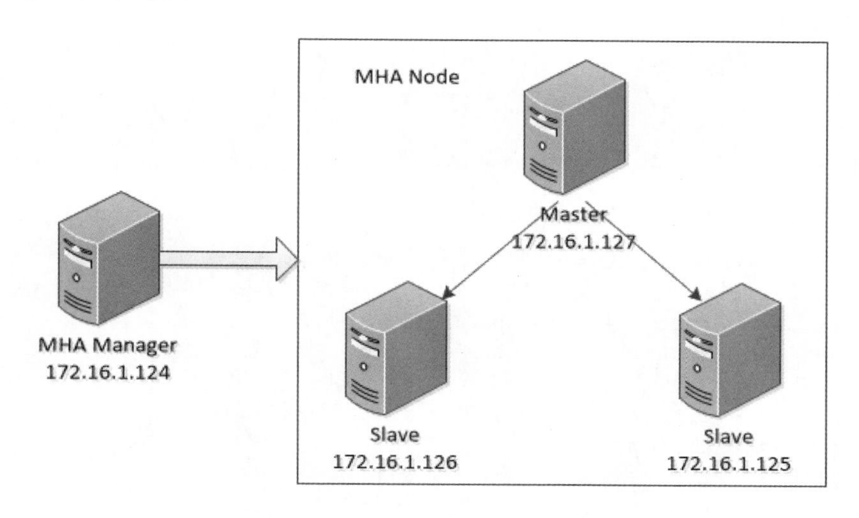

图 10-2 MHA 示例架构

10.2.1 配置主从复制

首先需要对三台 MySQL 服务器配置“一主两从”的标准异步复制。MySQL 主从复制的配置较为简单。如果是全新搭建的复制，只要打开主库的 binlog，然后将从库切换为主库（change master），即指定到 file 和 pos，再启动从库（start slave）即可。如果是为已经存在且正在使用的数据库搭建从库，通常有两种方式：一种是用 mysqldump 的 master-data 参数记录 master 的 file 和 pos，但可能会卡库；另一种比较好的方法是用 innobackupex 联机搭建从库，具体过程可参考 1.3.3 小节。

10.2.2 安装 Perl 依赖模块

MHA 是用 Perl 语言编写的，运行需要依赖某些 Perl 模块。以 root 用户在所有 4 个节点执行下面的操作来安装依赖模块：

```
    # 安装一个 epel 源
    wget -O /etc/yum.repos.d/epel-7.repo
http://mirrors.aliyun.com/repo/epel-7.repo
```

```
# 用 yum 安装依赖包
yum install perl-DBD-MySQL perl-Config-Tiny perl-Log-Dispatch
perl-Parallel-ForkManager perl-Time-HiRes -y
```

10.2.3 配置 SSH 免密登录

MHA Manager 节点通过 ssh 连接 MySQL 服务器，MHA Node 节点通过 scp 发送最新的 relay log 到其他从服务器上。为了使这些过程自动化，使用 SSH 公钥验证密码。

在 hdp1 172.16.1.124（Monitor）上以 root 用户执行下面的命令，配置 MHA 管理节点对 3 个 MySQL 节点免密登录：

```
ssh-keygen -t rsa
ssh-copy-id -i /root/.ssh/id_rsa.pub root@172.16.1.125
ssh-copy-id -i /root/.ssh/id_rsa.pub root@172.16.1.126
ssh-copy-id -i /root/.ssh/id_rsa.pub root@172.16.1.127
```

然后配置 3 个 MySQL 节点之间都可以免密登录：

```
# 在 hdp4 172.16.1.127（Master）上以 root 用户执行如下命令
ssh-keygen -t rsa
ssh-copy-id -i /root/.ssh/id_rsa.pub root@172.16.1.125
ssh-copy-id -i /root/.ssh/id_rsa.pub root@172.16.1.126

# 在 hdp3 172.16.1.126（slave1）上以 root 用户执行如下命令
ssh-keygen -t rsa
ssh-copy-id -i /root/.ssh/id_rsa.pub root@172.16.1.125
ssh-copy-id -i /root/.ssh/id_rsa.pub root@172.16.1.127

# 在 hdp2 172.16.1.125（slave2）上以 root 用户执行如下命令
ssh-keygen -t rsa
ssh-copy-id -i /root/.ssh/id_rsa.pub root@172.16.1.126
ssh-copy-id -i /root/.ssh/id_rsa.pub root@172.16.1.127
```

10.2.4 安装 MHA Node

在 hdp2、hdp3 和 hdp4 上以 root 用户执行下面的命令：

```
rpm -ivh mha4mysql-node-0.56-0.el6.noarch.rpm
```

安装完成后，在/usr/bin/目录中有如下 MHA 相关文件：

```
apply_diff_relay_logs
filter_mysqlbinlog
purge_relay_logs
save_binary_logs
```

这些脚本工具通常由 MHA Manager 的脚本触发，无须人为操作。

10.2.5 安装 MHA Manager

在 hdp1 上以 root 用户执行下面的命令：

```
rpm -ivh mha4mysql-manager-0.56-0.el6.noarch.rpm
```

安装完成后，在/usr/bin/目录下有如下 MHA 相关文件：

```
masterha_check_repl
masterha_check_ssh
masterha_check_status
masterha_conf_host
masterha_manager
masterha_master_monitor
masterha_master_switch
masterha_secondary_check
masterha_stop
apply_diff_relay_logs
filter_mysqlbinlog
purge_relay_logs
save_binary_logs
```

10.2.6 配置 MHA

在 hdp1 上以 root 用户执行下面步骤的操作。

步骤 01 建立配置文件目录：

```
mkdir -p /etc/masterha
```

步骤 02 创建如下内容的配置文件/etc/masterha/app1.cnf：

```
[server default]
manager_log=/var/log/masterha/app1/manager.log
manager_workdir=/var/log/masterha/app1.log
master_binlog_dir=/data
master_ip_failover_script=/usr/bin/master_ip_failover
master_ip_online_change_script=/usr/bin/master_ip_online_change
password=123456
ping_interval=1
remote_workdir=/tmp
repl_password=123456
repl_user=repl
secondary_check_script=/usr/bin/masterha_secondary_check -s hdp4 -s hdp3
--user=root --master_host=hdp4 --master_ip=172.16.1.127 --master_port=3306
shutdown_script=""
ssh_user=root
user=root
```

```
[server1]
hostname=172.16.1.127
port=3306

[server2]
candidate_master=1
check_repl_delay=0
hostname=172.16.1.126
port=3306

[server3]
hostname=172.16.1.125
port=3306
```

server default 段是 manager 的一些基本配置参数，server1、server2 和 server3 段分别对应复制中的主服务器、第一台从服务器和第二台从服务器。该配置文件的语法要求严格，变量值后不能有多余的空格。主要配置项说明如下：

- manager_log：设置 manager 的日志文件。
- manager_workdir：设置 manager 的工作目录。
- master_binlog_dir：设置主库（master）保存 binlog 的位置，以便 MHA 可以找到主库的日志，这里就指定为 MySQL 的数据目录。
- master_ip_failover_script：设置自动故障转移时候的切换脚本。
- master_ip_online_change_script：设置手动故障转移时候的切换脚本。
- password：设置 MySQL 中 root 用户的密码。
- ping_interval：设置监控主库时发送 ping 包的时间间隔，默认是 3 秒，尝试三次没有回应的时候自动进行故障转移。
- remote_workdir：设置远端 MySQL 在发生故障转移时的 binlog 保存位置。
- repl_password：设置复制用户的密码。
- repl_user：设置复制使用的用户名。
- secondary_check_script：一旦 MHA 到 hdp4 的监控之间出现问题，MHA Manager 将会尝试从 hdp3 登录到 hdp4。
- shutdown_script：设置故障发生后关闭故障主机的脚本。该脚本的主要作用是关闭主机以防止发生“脑裂”，这里没有使用。
- ssh_user：设置 ssh 的登录用户名。
- user：设置监控用户为 root。
- candidate_master：设置为候选主库（master）。设置该参数以后，发生主从切换时会将此从库提升为主库，即使它不是集群中事件最新的从库（slave）。
- check_repl_delay：默认情况下，如果一个从库（slave）落后主库（master）100MB 的 relay logs 的话，MHA 将不会选择该从库作为一个新的主库，因为对于这个从库的恢复需要花费很长时间。通过设置 check_repl_delay=0，MHA 触发主从切换时，在选择一个新主库的时候将会忽略复制延迟。该参数对于设置了 candidate_master=1 的主机非常有用，因为这个候选者在切换过程中一定是新的主库。

步骤03 建立软链接：

```
ln -s /home/mysql/mysql-5.6.14/bin/mysqlbinlog /usr/bin/mysqlbinlog
ln -s /home/mysql/mysql-5.6.14/bin/mysql /usr/bin/mysql
```

步骤04 设置复制中从服务器的relay_log_purge参数：

在hdp3和hdp2上以mysql用户执行：

```
mysql -uroot -p123456 -e "set global relay_log_purge=0"
```

注意，MHA在发生主从切换的过程中，从库的恢复过程依赖于relay log，所以这里要将relay log的自动清除设置为OFF，即采用手动清除relay log的方式。在默认情况下，从服务器上的中继日志会在SQL线程执行完毕后被自动删除。但是在MHA环境中，这些中继日志在恢复其他从服务器时可能会被用到，因此需要禁用中继日志的自动删除功能。定期清除中继日志需要考虑到复制延时的问题。在ext3文件系统下，删除大文件需要一定的时间，可能会导致严重的复制延迟。为了避免发生这种情况，需要暂时为中继日志创建硬链接，因为在Linux系统中通过硬链接删除大文件速度会很快。在MySQL数据库中删除大表时，通常也采用建立硬链接的方式，本书最后一章包含一个相关的优化案例。

10.2.7 创建相关脚本

创建相关脚本的操作步骤如下：

步骤01 创建定期清理relay log脚本。

在hdp3和hdp2两台从服务器上建立/root/purge_relay_log.sh文件，内容如下：

```
#!/bin/bash
. /home/mysql/.bashrc

user=root
passwd=123456
port=3306
log_dir='/data'
work_dir='/data'
purge='/usr/bin/purge_relay_logs'

if [ ! -d $log_dir ]
then
   mkdir $log_dir -p
fi

$purge --user=$user --password=$passwd --disable_relay_log_purge --port=$port
--workdir=$work_dir >> $log_dir/purge_relay_logs.log 2>&1
```

purge_relay_logs参数说明：

- user mysql：MySQL用户名。

- password mysql：MySQL 用户密码。
- port：MySQL 端口号。
- workdir：指定创建 relay log 硬链接的位置，默认是/var/tmp。由于在系统不同分区创建硬链接文件会失败，故需要指定硬链接的具体位置。在成功执行脚本后，硬链接的中继日志文件被删除。
- disable_relay_log_purge：在默认情况下，如果 relay_log_purge=1，脚本会什么都不清理，自动退出。通过设置这个参数，当 relay_log_purge=1 的情况下会将 relay_log_purge 设置为 0，清理 relay log 之后，最后将参数设置为 OFF。

改脚本文件模式为可执行：

```
chmod 755 purge_relay_log.sh
```

手工执行/root/purge_relay_log.sh，在控制台输出：

```
2018-07-31 12:45:20: purge_relay_logs script started.
 Found relay_log.info: /data/relay-log.info
 Opening /data/hdp2-relay-bin.000001 ..
 Opening /data/hdp2-relay-bin.000002 ..
 Executing SET GLOBAL relay_log_purge=1; FLUSH LOGS; sleeping a few seconds
so that SQL thread can delete older relay log
 files (if it keeps up); SET GLOBAL relay_log_purge=0; .. ok.
2018-07-31 12:45:23: All relay log purging operations succeeded.
```

将脚本文件添加到 crontab 中定期自动执行：

```
0 4 * * * /bin/bash /root/purge_relay_log.sh
```

步骤 02 创建自动故障转移脚本。

在 hdp1 上创建/usr/bin/master_ip_failover 文件，内容如下：

```
#!/usr/bin/env perl
use strict;
use warnings FATAL => 'all';
use Getopt::Long;

my (
    $command, $ssh_user, $orig_master_host, $orig_master_ip,
    $orig_master_port, $new_master_host, $new_master_ip, $new_master_port
);

my $vip = '172.16.1.100';  # Virtual IP
my $key = "1";
my $ssh_start_vip = "/sbin/ifconfig ens32:$key $vip";
my $ssh_stop_vip = "/sbin/ifconfig ens160:$key down";

GetOptions(
    'command=s'          => \$command,
    'ssh_user=s'         => \$ssh_user,
    'orig_master_host=s' => \$orig_master_host,
```

```
        'orig_master_ip=s'   => \$orig_master_ip,
        'orig_master_port=i' => \$orig_master_port,
        'new_master_host=s'  => \$new_master_host,
        'new_master_ip=s'    => \$new_master_ip,
        'new_master_port=i'  => \$new_master_port,
    );

    exit &main();

    sub main {
        print "\n\nIN SCRIPT TEST====$ssh_stop_vip==$ssh_start_vip===\n\n";

        if ( $command eq "stop" || $command eq "stopssh" ) {
            my $exit_code = 1;
            eval {
                print "Disabling the VIP on old master: $orig_master_host \n";
                &stop_vip();
                $exit_code = 0;
            };
            if ($@) {
                warn "Got Error: $@\n";
                exit $exit_code;
            }
            exit $exit_code;
        }
        elsif ( $command eq "start" ) {
            my $exit_code = 10;
            eval {
                print "Enabling the VIP - $vip on the new master - $new_master_host
\n";
                &start_vip();
                $exit_code = 0;
            };
            if ($@) {
                warn $@;
                exit $exit_code;
            }
            exit $exit_code;
        }
        elsif ( $command eq "status" ) {
            print "Checking the Status of the script.. OK \n";
            `ssh $ssh_user\@$orig_master_host \" $ssh_start_vip \"`;
            exit 0;
        }
        else {
            &usage();
            exit 1;
        }
    }
```

```
    sub start_vip() {
        `ssh $ssh_user\@$new_master_host \" $ssh_start_vip \"`;
    }

    sub stop_vip() {
        `ssh $ssh_user\@$orig_master_host \" $ssh_stop_vip \"`;
    }

    sub usage {
        print
        "Usage: master_ip_failover --command=start|stop|stopssh|status
--orig_master_host=host --orig_master_ip=ip --orig_master_port=port
--new_master_host=host --new_master_ip=ip --new_master_port=port\n";
    }
```

注意脚本中 VIP 漂移的部分。

步骤 03 创建手动故障转移脚本。

在 hdp1 上创建/usr/bin/master_ip_online_change 文件，内容如下：

```
    #!/usr/bin/env perl

    use strict;
    use warnings FATAL => 'all';

    use Getopt::Long;
    use MHA::DBHelper;
    use MHA::NodeUtil;
    use Time::HiRes qw(sleep gettimeofday tv_interval);
    use Data::Dumper;

    my $_tstart;
    my $_running_interval = 0.1;
    my (
     $command, $orig_master_host, $orig_master_ip,
     $orig_master_port, $orig_master_user,
     $new_master_host, $new_master_ip, $new_master_port,
     $new_master_user,
    );

    my $vip = '172.16.1.100';  # Virtual IP
    my $key = "1";
    my $ssh_start_vip = "/sbin/ifconfig ens32:$key $vip";
    my $ssh_stop_vip = "/sbin/ifconfig ens160:$key down";
    my $ssh_user = "root";
    my $new_master_password = "123456";
    my $orig_master_password = "123456";

    GetOptions(
```

```
  'command=s'              =>\$command,
  'orig_master_host=s'     =>\$orig_master_host,
  'orig_master_ip=s'       =>\$orig_master_ip,
  'orig_master_port=i'     =>\$orig_master_port,
  'orig_master_user=s'     =>\$orig_master_user,
  'new_master_host=s'      =>\$new_master_host,
  'new_master_ip=s'        =>\$new_master_ip,
  'new_master_port=i'      =>\$new_master_port,
  'new_master_user=s'      =>\$new_master_user,
);

exit &main();

sub current_time_us {
  my ($sec, $microsec ) = gettimeofday();
  my$curdate = localtime($sec);
 return $curdate . " " . sprintf( "%06d", $microsec);
}

sub sleep_until {
  my$elapsed = tv_interval($_tstart);
  if ($_running_interval > $elapsed ) {
   sleep( $_running_interval - $elapsed );
  }
}

sub get_threads_util {
  my$dbh                    = shift;
  my$my_connection_id       = shift;
  my$running_time_threshold = shift;
  my$type                   = shift;
  $running_time_threshold = 0 unless ($running_time_threshold);
  $type                   = 0 unless($type);
  my@threads;

  my$sth = $dbh->prepare("SHOW PROCESSLIST");
  $sth->execute();

 while ( my $ref = $sth->fetchrow_hashref() ) {
   my$id         = $ref->{Id};
   my$user       = $ref->{User};
   my$host       = $ref->{Host};
   my$command    = $ref->{Command};
   my$state      = $ref->{State};
   my$query_time = $ref->{Time};
   my$info       = $ref->{Info};
   $info =~ s/^\s*(.*?)\s*$/$1/ if defined($info);
   next if ( $my_connection_id == $id );
   next if ( defined($query_time) && $query_time <$running_time_threshold );
   next if ( defined($command)   && $command eq "Binlog Dump" );
```

```
    next if ( defined($user)      && $user eq "system user" );
    next
      if ( defined($command)
      && $command eq "Sleep"
      && defined($query_time)
      && $query_time >= 1 );

     if( $type >= 1 ) {
      next if ( defined($command) && $command eq "Sleep" );
      next if ( defined($command) && $command eq "Connect" );
     }

     if( $type >= 2 ) {
      next if ( defined($info) && $info =~ m/^select/i );
      next if ( defined($info) && $info =~ m/^show/i );
     }

    push @threads, $ref;
   }
  return @threads;
 }

 sub main {
   if ($command eq "stop" ) {
     my$exit_code = 1;
    eval {
      my $new_master_handler = new MHA::DBHelper();

      $new_master_handler->connect( $new_master_ip, $new_master_port,
        $new_master_user, $new_master_password, 1 );
      print current_time_us() . " Set read_only on the new master..";
      $new_master_handler->enable_read_only();
      if ( $new_master_handler->is_read_only() ) {
        print "ok.\n";
      }
      else {
        die "Failed!\n";
      }
      $new_master_handler->disconnect();

      my $orig_master_handler = new MHA::DBHelper();
      $orig_master_handler->connect( $orig_master_ip, $orig_master_port,
        $orig_master_user, $orig_master_password, 1 );

      my $time_until_read_only = 15;
      $_tstart = [gettimeofday];
      my @threads = get_threads_util( $orig_master_handler->{dbh},
        $orig_master_handler->{connection_id} );
      while ( $time_until_read_only > 0 && $#threads >= 0 ) {
        if ( $time_until_read_only % 5 == 0 ) {
```

```
        printf
  "%s Waiting all running %d threads aredisconnected.. (max %d milliseconds)\n",
          current_time_us(), $#threads + 1, $time_until_read_only * 100;
        if ( $#threads < 5 ) {
          print Data::Dumper->new( [$_] )->Indent(0)->Terse(1)->Dump ."\n"
            foreach (@threads);
        }
      }
      sleep_until();
      $_tstart = [gettimeofday];
      $time_until_read_only--;
      @threads = get_threads_util( $orig_master_handler->{dbh},
        $orig_master_handler->{connection_id} );
    }

    print current_time_us() . " Set read_only=1 on the orig master..";
    $orig_master_handler->enable_read_only();
    if ( $orig_master_handler->is_read_only() ) {
      print "ok.\n";
    }
    else {
      die "Failed!\n";
    }

    my $time_until_kill_threads = 5;
    @threads = get_threads_util( $orig_master_handler->{dbh},
      $orig_master_handler->{connection_id} );
    while ( $time_until_kill_threads > 0 && $#threads >= 0 ) {
      if ( $time_until_kill_threads % 5 == 0 ) {
        printf
  "%s Waiting all running %d queries aredisconnected.. (max %d milliseconds)\n",
          current_time_us(), $#threads + 1, $time_until_kill_threads * 100;
        if ( $#threads < 5 ) {
          print Data::Dumper->new( [$_] )->Indent(0)->Terse(1)->Dump ."\n"
            foreach (@threads);
        }
      }
      sleep_until();
      $_tstart = [gettimeofday];
      $time_until_kill_threads--;
      @threads = get_threads_util( $orig_master_handler->{dbh},
        $orig_master_handler->{connection_id} );
    }

    print "Disabling the VIPon old master: $orig_master_host \n";
    &stop_vip();

    print current_time_us() . " Killing all application threads..\n";
    $orig_master_handler->kill_threads(@threads) if ( $#threads >= 0);
    print current_time_us() . " done.\n";
```

```
        $orig_master_handler->disconnect();

        $exit_code = 0;
       };
       if($@) {
        warn "Got Error: $@\n";
        exit $exit_code;
       }
     exit $exit_code;
    }
  elsif ( $command eq "start" ) {
       my$exit_code = 10;
       eval{
        my $new_master_handler = new MHA::DBHelper();

        $new_master_handler->connect( $new_master_ip, $new_master_port,
          $new_master_user, $new_master_password, 1 );

        print current_time_us() . " Set read_only=0 on the newmaster.\n";
        $new_master_handler->disable_read_only();

        $new_master_handler->disconnect();

        print "Enabling the VIP -$vip on the new master - $new_master_host \n";
        &start_vip();
        $exit_code = 0;
       };
       if($@) {
        warn "Got Error: $@\n";
        exit $exit_code;
       }
     exit $exit_code;
    }
  elsif ( $command eq "status" ) {
     exit 0;
    }
    else{
     &usage();
     exit 1;
    }
}

sub start_vip() {
   `ssh $ssh_user\@$new_master_host \" $ssh_start_vip \"`;
}

sub stop_vip() {
   `ssh $ssh_user\@$orig_master_host \" $ssh_stop_vip \"`;
}
```

```
    sub usage {
     print
    "Usage: master_ip_online_change --command=start|stop|status--orig_master_
host=host --orig_master_ip=ip --orig_master_port=port--new_master_host=host
--new_master_ip=ip --new_master_port=port\n";
      die;
    }
```

10.2.8 检查 MHA 配置

1. 检查 SSH 配置

在 hdp1 上以 root 用户执行如下操作：

```
    [root@hdp1~]#masterha_check_ssh --conf=/etc/masterha/app1.cnf
    Tue Jul 31 12:50:22 2018 - [warning] Global configuration file
/etc/masterha_default.cnf not found. Skipping.
    Tue Jul 31 12:50:22 2018 - [info] Reading application default configuration
from /etc/masterha/app1.cnf..
    Tue Jul 31 12:50:22 2018 - [info] Reading server configuration from
/etc/masterha/app1.cnf..
    Tue Jul 31 12:50:22 2018 - [info] Starting SSH connection tests..

    ...

    Tue Jul 31 12:50:25 2018 - [info] All SSH connection tests passed successfully.
    [root@hdp1~]#
```

输出中显示所有 SSH 连接均通过测试。

2. 检查整个复制环境状况

在 hdp1 上以 root 用户操作如下：

```
    [root@hdp1~]#masterha_check_repl --conf=/etc/masterha/app1.cnf
    Tue Jul 31 12:52:19 2018 - [warning] Global configuration file
/etc/masterha_default.cnf not found. Skipping.
    Tue Jul 31 12:52:19 2018 - [info] Reading application default configuration
from /etc/masterha/app1.cnf..
    Tue Jul 31 12:52:19 2018 - [info] Reading server configuration from
/etc/masterha/app1.cnf..

    ...

    Checking the Status of the script.. OK
    SIOCSIFADDR: No such device
    ens32:1: ERROR while getting interface flags: No such device
    Tue Jul 31 12:52:25 2018 - [info]  OK.
    Tue Jul 31 12:52:25 2018 - [warning] shutdown_script is not defined.
    Tue Jul 31 12:52:25 2018 - [info] Got exit code 0 (Not master dead).
```

```
MySQL Replication Health is OK.
```

输出中只有几个警告信息，没有错误，复制显示正常。

3. 检查 MHA Manager 的状态

在 hdp1 上以 root 用户执行如下操作：

```
[root@hdp1~]#masterha_check_status --conf=/etc/masterha/app1.cnf
app1 is stopped(2:NOT_RUNNING).
[root@hdp1~]#
```

显示"NOT_RUNNING"代表 MHA 监控没有开启。执行下面的命令后台启动 MHA：

```
    mkdir -p /var/log/masterha/app1/
    nohup masterha_manager --conf=/etc/masterha/app1.cnf
--remove_dead_master_conf --ignore_last_failover < /dev/null >
/var/log/masterha/app1/manager.log 2>&1 &
```

启动参数说明：

- remove_dead_master_conf：该参数代表当发生主从切换后，原主服务器的 IP 将会从配置文件中移除。
- manager_log：日志存放位置。
- ignore_last_failover：在默认情况下，如果 MHA 检测到连续发生宕机，且两次宕机间隔不足 8 小时的话，则不会进行故障转移。之所以这样限制是为了避免 ping-pong 效应。该参数代表忽略上次 MHA 触发主从切换产生的文件。在默认情况下，MHA 发生主从切换后会在日志目录，也就是上面设置的/data 中产生 app1.failover.complete 文件，下次再次主从切换的时候如果发现该目录下存在该文件将不允许触发主从切换，除非在第一次主从切换后手工删除该文件。为了方便测试，这里设置为--ignore_last_failover。

再次检查 MHA Manager 的状态：

```
[root@hdp1~]#masterha_check_status --conf=/etc/masterha/app1.cnf
app1 (pid:298237) is running(0:PING_OK), master:172.16.1.127
[root@hdp1~]#
```

可以看到正在执行监控，主服务器是 172.16.1.127。

4. 查看启动日志

在 hdp1 上以 root 用户操作如下：

```
    [root@hdp1~]#cat /var/log/masterha/app1/manager.log
    ...
    Checking the Status of the script.. OK
    ...
    Tue Jul 31 12:57:06 2018 - [info] Starting ping health check on
172.16.1.127(172.16.1.127:3306)..
    Tue Jul 31 12:57:06 2018 - [info] Ping(SELECT) succeeded, waiting until MySQL
doesn't respond..
    [root@hdp1~]#
```

10.3 功能测试

10.3.1 初始绑定 VIP

在 hdp4 172.16.1.127（master）上以 root 用户执行：

```
/sbin/ifconfig ens160:1 172.16.1.100/24
```

查看 VIP：

```
[root@hdp4~]#ip a
1: lo: <LOOPBACK,UP,LOWER_UP> mtu 65536 qdisc noqueue state UNKNOWN
    link/loopback 00:00:00:00:00:00 brd 00:00:00:00:00:00
    inet 127.0.0.1/8 scope host lo
       valid_lft forever preferred_lft forever
    inet6 ::1/128 scope host
       valid_lft forever preferred_lft forever
2: ens160: <BROADCAST,MULTICAST,UP,LOWER_UP> mtu 1500 qdisc mq state UP qlen
1000
    link/ether 00:50:56:a5:49:7f brd ff:ff:ff:ff:ff:ff
    inet 172.16.1.127/24 brd 172.16.1.255 scope global ens160
       valid_lft forever preferred_lft forever
    inet 172.16.1.100/16 brd 172.16.255.255 scope global ens160:1
       valid_lft forever preferred_lft forever
    inet6 fe80::250:56ff:fea5:497f/64 scope link
       valid_lft forever preferred_lft forever
[root@hdp4~]#
```

10.3.2 测试自动切换

1. 模拟主从延时

在 slave1 库（172.16.1.126）上停掉 slave I/O 线程：

```
mysql -uroot -p123456 -e "stop slave io_thread;"
```

2. 主服务器生成数据

在主库上（172.16.1.127）安装 sysbench，执行 sysbench 生成数据，在 sbtest 库下生成 sbtest 表，共 10W 记录：

```
# 用 root 用户安装 sysbench
yum install sysbench -y

# 以 mysql 用户建立 sbtest 数据库
mysql -uroot -p123456 -e "create database sbtest;"
```

```
    # 以mysql用户执行sysbench生成数据
    sysbench /usr/share/sysbench/tests/include/oltp_legacy/oltp.lua
--mysql-host=127.0.0.1 --mysql-port=3306 --mysql-user=root
--mysql-password=123456 --oltp-test-mode=complex --oltp-tables-count=10
--oltp-table-size=10000 --threads=10 --time=120 --report-interval=10
--db-driver=mysql prepare
```

3. 停止主服务器的 MySQL 服务

以 root 用户停止主服务器上的 MySQL 服务：

```
    service mysql stop
```

4. 验证 VIP 漂移

在 hdp3 上以 root 用户执行如下操作：

```
    [root@hdp3~]#ip a
    1: lo: <LOOPBACK,UP,LOWER_UP> mtu 65536 qdisc noqueue state UNKNOWN
        link/loopback 00:00:00:00:00:00 brd 00:00:00:00:00:00
        inet 127.0.0.1/8 scope host lo
           valid_lft forever preferred_lft forever
        inet6 ::1/128 scope host
           valid_lft forever preferred_lft forever
    2: ens32: <BROADCAST,MULTICAST,UP,LOWER_UP> mtu 1500 qdisc pfifo_fast state
UP qlen 1000
        link/ether 00:50:56:a5:0f:77 brd ff:ff:ff:ff:ff:ff
        inet 172.16.1.126/24 brd 172.16.1.255 scope global ens32
           valid_lft forever preferred_lft forever
        inet 172.16.1.100/16 brd 172.16.255.255 scope global ens32:1
           valid_lft forever preferred_lft forever
        inet6 fe80::250:56ff:fea5:f77/64 scope link
           valid_lft forever preferred_lft forever
    [root@hdp3~]#
```

在 hdp4 上以 root 用户操作：

```
    [root@hdp4~]#ip a
    1: lo: <LOOPBACK,UP,LOWER_UP> mtu 65536 qdisc noqueue state UNKNOWN
        link/loopback 00:00:00:00:00:00 brd 00:00:00:00:00:00
        inet 127.0.0.1/8 scope host lo
           valid_lft forever preferred_lft forever
        inet6 ::1/128 scope host
           valid_lft forever preferred_lft forever
    2: ens160: <BROADCAST,MULTICAST,UP,LOWER_UP> mtu 1500 qdisc mq state UP qlen
1000
        link/ether 00:50:56:a5:49:7f brd ff:ff:ff:ff:ff:ff
        inet 172.16.1.127/24 brd 172.16.1.255 scope global ens160
           valid_lft forever preferred_lft forever
        inet6 fe80::250:56ff:fea5:497f/64 scope link
           valid_lft forever preferred_lft forever
    [root@hdp4~]#
```

可以看到 VIP 已经从 hdp4（master）漂移到了 hdp3（slave1）。

5. 客户端用 VIP 访问数据库

```
C:\WINDOWS\system32>mysql -uroot -p123456 -h172.16.1.100 -e "use sbtest; show tables; select count(*) from sbtest1; select count(*) from sbtest10;"
mysql: [Warning] Using a password on the command line interface can be insecure.
+-------------------+
| Tables_in_sbtest  |
+-------------------+
| sbtest1           |
| sbtest10          |
| sbtest2           |
| sbtest3           |
| sbtest4           |
| sbtest5           |
| sbtest6           |
| sbtest7           |
| sbtest8           |
| sbtest9           |
+-------------------+
+----------+
| count(*) |
+----------+
|   10000  |
+----------+
+----------+
| count(*) |
+----------+
|   10000  |
+----------+

C:\WINDOWS\system32>
```

在还没有创建 sbtest 库的时候，172.16.1.126 就停了 slave I/O 线程。在新的主库 172.16.1.126 上查看数据，可以看到落后的数据也同步过来了，数据没有丢失。

6. 查看复制的主从切换

```
C:\WINDOWS\system32>mysql -uroot -p123456 -h172.16.1.125 -e "show slave status\G"
mysql: [Warning] Using a password on the command line interface can be insecure.
*************************** 1. row ***************************
               Slave_IO_State: Waiting for master to send event
                  Master_Host: 172.16.1.126
                  Master_User: repl
                  Master_Port: 3306
                Connect_Retry: 60s
              Master_Log_File: mysql-bin.000001
          Read_Master_Log_Pos: 19093607
               Relay_Log_File: hdp2-relay-bin.000002
                Relay_Log_Pos: 283
```

```
            Relay_Master_Log_File: mysql-bin.000001
              Slave_IO_Running: Yes
              Slave_SQL_Running: Yes
    ...
              Master_Server_Id: 126
                   Master_UUID: fadd5b7d-7d9f-11e8-90b4-13ccc7802b56
              Master_Info_File: /data/master.info
                     SQL_Delay: 0
           SQL_Remaining_Delay: NULL
       Slave_SQL_Running_State: Slave has read all relay log; waiting for the
slave I/O thread to update it
    ...

    C:\WINDOWS\system32>mysql -uroot -p123456 -h172.16.1.126 -e "show slave
status\G"
    mysql: [Warning] Using a password on the command line interface can be insecure.

    C:\WINDOWS\system32>
```

可以看到，172.16.1.126 成为新的主库，而 172.16.1.125 也指向了这个新的主库。

7. 检查 MHA Manager 状态

在 hdp1 上以 root 用户执行如下操作：

```
    [root@hdp1~]#masterha_check_status --conf=/etc/masterha/app1.cnf
    app1 is stopped(2:NOT_RUNNING).
    [1]+  Done    nohup masterha_manager --conf=/etc/masterha/app1.cnf
--remove_dead_master_conf --ignore_last_failover < /dev/null >
/var/log/masterha/app1/manager.log 2>&1
    [root@hdp1~]#
```

发现在执行了一次自动故障转移后，MHA Manager 进程停止了。官网上对这种情况的解释如下：

Running MHA Manager from daemontools

Currently MHA Manager process does not run as a daemon. If failover completed successfully or the master process was killed by accident, the manager stops working. To run as a daemon, daemontool. or any external daemon program can be used. Here is an example to run from daemontools.

1. Install daemontools

```
* For RedHat
manager_host# yum install daemontools
```

2. Create run file under /service/masterha(*app_name*)/run

```
manager_host# mkdir /service/masterha_app1
manager_host# cat /service/masterha_app1/run
#!/bin/sh
exec masterha_manager --conf=/etc/app1.cnf --wait_on_monitor_error=60 --wait_on_failover_error=60 >> /var/log/masterha/ap
manager_host# chmod 755 /service/masterha_app1/run
```

You can stop/restart monitoring by daemontool commands.

```
## stopping monitoring
manager_host# svc -d /service/masterha_app1

## starting monitoring
manager_host# svc -u /service/masterha_app1
```

意思是安装一个进程工具，通过该工具结合脚本来管理进程。

10.3.3　测试手工切换

先执行以下操作还原环境。

还原数据库复制：

```
-- 在 hdp4、hdp3 和 hdp2 上重置主库 master 和从库 slave
stop slave;
drop database sbtest;
reset master;
reset slave all;

-- 在 hdp3 和 hdp2 上重新指向 hdp4 为主库 master
change master to
master_host='172.16.1.127',
master_port=3306,
master_user='repl',
master_password='123456',
master_log_file='mysql-bin.000001',
master_log_pos=120;

start slave;
show slave status\G
```

还原 VIP 绑定：

```
# 在 hdp3 上用 root 用户执行
/sbin/ifconfig ens32:1 down

# 在 hdp4 上用 root 用户执行
/sbin/ifconfig ens160:1 172.16.1.100
```

还原配置文件：编辑 hdp1 上的/etc/masterha/app1.cnf，将[server1]段添加回去。

启动 MHA Manage：

```
# 在 hdp1 上以 root 用户执行
nohup masterha_manager --conf=/etc/masterha/app1.cnf
--remove_dead_master_conf --ignore_last_failover < /dev/null >
/var/log/masterha/app1/manager.log 2>&1 &
```

至此环境还原完毕，可以开始测试手工主从切换。当主服务器故障时，人工手动调用 MHA 来进行故障转移操作，步骤如下：

步骤 01　停止 MHA Manage。

在 hdp1 上以 root 用户操作如下：

```
masterha_stop --conf=/etc/masterha/app1.cnf
```

步骤 02　关闭主库。

在 hdp4 上以 root 用户操作如下：

```
service mysql stop
```

步骤 03 执行手工主从切换。

在 hdp1 上以 root 用户操作如下：

```
masterha_master_switch --master_state=dead --conf=/etc/masterha/app1.cnf
--dead_master_host=172.16.1.127 --dead_master_port=3306 --new_master_host
=172.16.1.126 --new_master_port=3306 --ignore_last_failover
```

步骤 04 验证 VIP 漂移：

```
[root@hdp3~]#ip a
1: lo: <LOOPBACK,UP,LOWER_UP> mtu 65536 qdisc noqueue state UNKNOWN
    link/loopback 00:00:00:00:00:00 brd 00:00:00:00:00:00
    inet 127.0.0.1/8 scope host lo
       valid_lft forever preferred_lft forever
    inet6 ::1/128 scope host
       valid_lft forever preferred_lft forever
2: ens32: <BROADCAST,MULTICAST,UP,LOWER_UP> mtu 1500 qdisc pfifo_fast state
UP qlen 1000
    link/ether 00:50:56:a5:0f:77 brd ff:ff:ff:ff:ff:ff
    inet 172.16.1.126/24 brd 172.16.1.255 scope global ens32
       valid_lft forever preferred_lft forever
    inet 172.16.1.100/16 brd 172.16.255.255 scope global ens32:1
       valid_lft forever preferred_lft forever
    inet6 fe80::250:56ff:fea5:f77/64 scope link
       valid_lft forever preferred_lft forever
[root@hdp3~]#
```

步骤 05 验证复制关系：

```
C:\WINDOWS\system32>mysql -uroot -p123456 -h172.16.1.125 -e "show slave
status\G"
mysql: [Warning] Using a password on the command line interface can be insecure.
*************************** 1. row ***************************
            Slave_IO_State: Waiting for master to send event
               Master_Host: 172.16.1.126
               Master_User: repl
               Master_Port: 3306
             Connect_Retry: 60
           Master_Log_File: mysql-bin.000001
       Read_Master_Log_Pos: 120
            Relay_Log_File: hdp2-relay-bin.000002
             Relay_Log_Pos: 283
     Relay_Master_Log_File: mysql-bin.000001
          Slave_IO_Running: Yes
         Slave_SQL_Running: Yes
...
          Master_Server_Id: 126
```

```
                Master_UUID: fadd5b7d-7d9f-11e8-90b4-13ccc7802b56
            Master_Info_File: /data/master.info
                   SQL_Delay: 0
         SQL_Remaining_Delay: NULL
     Slave_SQL_Running_State: Slave has read all relay log; waiting for the
slave I/O thread to update it
    ...

    C:\WINDOWS\system32>mysql -uroot -p123456 -h172.16.1.126 -e "show slave
status\G"
    mysql: [Warning] Using a password on the command line interface can be insecure.

    C:\WINDOWS\system32>
```

步骤 06 验证客户端 VIP 访问：

```
    C:\WINDOWS\system32>mysql -uroot -p123456 -h172.16.1.100 -e "show variables
like 'server_id';"
    mysql: [Warning] Using a password on the command line interface can be insecure.
    +---------------+-------+
    | Variable_name | Value |
    +---------------+-------+
    | server_id     | 126   |
    +---------------+-------+

    C:\WINDOWS\system32>
```

10.3.4 测试在线主从切换

MHA 在线主从切换的大致过程如下：

（1）检测复制设置和确定当前主服务器。

（2）确定新的主服务器。

（3）阻塞到当前主服务器的写入。

（4）等待所有从服务器赶上复制。

（5）授予写入到新的主服务器。

（6）重新设置从服务器。

注意，在线主从切换的时候应用程序需要考虑以下两个问题：

- 自动识别主库和从库的问题（主库的机器可能会切换），如果采用了 VIP 的方式，基本可以解决这个问题。
- 负载均衡问题。可以定义大概的读写比例，以及每台机器可承担的负载比例。当有机器离开集群时，需要考虑这个问题。

为保证数据完全一致性，并在最快的时间内完成主从切换，MHA 的在线主从切换必须满足以下条件才会切换成功，否则切换将失败。

- 所有从库的I/O线程和SQL线程都在运行。
- 所有的show slave status的输出中Seconds_Behind_Master参数小于或者等于running_updates_limit秒。如果在切换命令中不指定running_updates_limit参数，那么在默认情况下running_updates_limit为1秒。
- 在主库端，通过show processlist输出，没有一个更新花费的时间大于running_updates_limit秒。

在测试前，先按照上面测试手工主从切换前的步骤进行环境还原（在线主从切换不用修改/etc/masterha/app1.cnf配置文件），然后按以下步骤测试在线主从切换。

步骤01 停止MHA Manage。

在hdp1上以root用户操作如下：

```
masterha_stop --conf=/etc/masterha/app1.cnf
```

步骤02 执行在线主从切换命令。

在hdp1上以root用户执行如下操作：

```
masterha_master_switch --conf=/etc/masterha/app1.cnf --master_state=alive
--new_master_host=172.16.1.126 --new_master_port=3306
--orig_master_is_new_slave --running_updates_limit=10000
```

步骤03 验证复制关系。

在hdp2、hdp3和hdp4查看从库状态（slave status）：

```
C:\WINDOWS\system32>mysql -uroot -p123456 -h172.16.1.125 -e "show slave
status\G"
mysql: [Warning] Using a password on the command line interface can be insecure.
*************************** 1. row ***************************
             Slave_IO_State: Waiting for master to send event
                Master_Host: 172.16.1.126
                Master_User: repl
                Master_Port: 3306
              Connect_Retry: 60
            Master_Log_File: mysql-bin.000001
        Read_Master_Log_Pos: 120
             Relay_Log_File: hdp2-relay-bin.000002
              Relay_Log_Pos: 283
      Relay_Master_Log_File: mysql-bin.000001
           Slave_IO_Running: Yes
          Slave_SQL_Running: Yes
...
           Master_Server_Id: 126
                Master_UUID: fadd5b7d-7d9f-11e8-90b4-13ccc7802b56
           Master_Info_File: /data/master.info
                  SQL_Delay: 0
        SQL_Remaining_Delay: NULL
    Slave_SQL_Running_State: Slave has read all relay log; waiting for the
slave I/O thread to update it
```

```
...

C:\WINDOWS\system32>mysql -uroot -p123456 -h172.16.1.126 -e "show slave
status\G"
mysql: [Warning] Using a password on the command line interface can be insecure.

C:\WINDOWS\system32>mysql -uroot -p123456 -h172.16.1.127 -e "show slave
status\G"
mysql: [Warning] Using a password on the command line interface can be insecure.
*************************** 1. row ***************************
               Slave_IO_State: Waiting for master to send event
                  Master_Host: 172.16.1.126
                  Master_User: repl
                  Master_Port: 3306
                Connect_Retry: 60
              Master_Log_File: mysql-bin.000001
          Read_Master_Log_Pos: 120
               Relay_Log_File: hdp4-relay-bin.000002
                Relay_Log_Pos: 283
        Relay_Master_Log_File: mysql-bin.000001
             Slave_IO_Running: Yes
            Slave_SQL_Running: Yes
...
             Master_Server_Id: 126
                  Master_UUID: fadd5b7d-7d9f-11e8-90b4-13ccc7802b56
             Master_Info_File: /data/master.info
                    SQL_Delay: 0
          SQL_Remaining_Delay: NULL
      Slave_SQL_Running_State: Slave has read all relay log; waiting for the
slave I/O thread to update it
...

C:\WINDOWS\system32>
```

可以看到hdp3成为新的主库，而hdp2和hdp4成为指向新的主库的从库。

步骤04 验证VIP自动漂移：

```
[root@hdp3~]#ip a
1: lo: <LOOPBACK,UP,LOWER_UP> mtu 65536 qdisc noqueue state UNKNOWN
    link/loopback 00:00:00:00:00:00 brd 00:00:00:00:00:00
    inet 127.0.0.1/8 scope host lo
       valid_lft forever preferred_lft forever
    inet6 ::1/128 scope host
       valid_lft forever preferred_lft forever
2: ens32: <BROADCAST,MULTICAST,UP,LOWER_UP> mtu 1500 qdisc pfifo_fast state
UP qlen 1000
    link/ether 00:50:56:a5:0f:77 brd ff:ff:ff:ff:ff:ff
    inet 172.16.1.126/24 brd 172.16.1.255 scope global ens32
       valid_lft forever preferred_lft forever
    inet 172.16.1.100/16 brd 172.16.255.255 scope global ens32:1
```

```
        valid_lft forever preferred_lft forever
    inet6 fe80::250:56ff:fea5:f77/64 scope link
        valid_lft forever preferred_lft forever
[root@hdp3~]#
```

步骤 05 验证客户端通过 VIP 访问数据库：

```
C:\WINDOWS\system32>mysql -uroot -p123456 -h172.16.1.100 -e "show variables like 'server_id'"
mysql: [Warning] Using a password on the command line interface can be insecure.
+---------------+-------+
| Variable_name | Value |
+---------------+-------+
| server_id     | 126   |
+---------------+-------+

C:\WINDOWS\system32>
```

10.3.5 修复宕机的主库

通常情况下，自动主从切换以后，原主库可能已经废弃掉，待原主库的主机修复后，如果数据完整，可能想把原来主库重新作为新主库的从库。这时我们可以借助当时自动切换时刻的 MHA 日志来完成对原主库的角色改变。下面是提取相关日志的命令：

```
grep -i "All other slaves should start" /var/log/masterha/app1/manager.log
```

可以看到类似下面的信息：

```
All other slaves should start replication from here. Statement should be: CHANGE MASTER TO MASTER_HOST='172.16.1.126', MASTER_PORT=3306, MASTER_LOG_FILE='mysql-bin.000005', MASTER_LOG_POS=120, MASTER_USER='repl', MASTER_PASSWORD='123456';
```

意思是说，如果主库的主机修复好了，可以在修复好后的主库上执行 CHANGE MASTER 操作，作为新的从库。

10.4 小 结

MHA 是用 Perl 语言实现的一个成熟的 MySQL 高可用方案。它主要的功能是监控 MySQL 主从复制中的主服务器，当主服务器出现问题时，自动提升一个最新的从服务器成为新的主服务器。集群中的其他从服务器将从新主服务器上重新进行数据复制。MHA 的主要特点是最大限度地保证故障转移过程中不丢失数据，这是通过保存并应用原主服务器的二进制日志来实现的。本章通过一个示例详细说明了 MHA 的安装和配置以及功能测试。但是，MHA 并不监控主从复制链路，因此无法感知复制本身的问题，也没有实现从服务器的读负载均衡。下一章将介绍一个更为通用的高可用中间件架构——Keepalived + LVS。

第 11 章

Keepalived + LVS

前面介绍的 4 种高可用中间件都是只针对 MySQL 的解决方案。Router 和 Fabric 自不必说，本来就是 MySQL 自己的产品，MMM 和 MHA 本质上还是利用 MySQL 原生的复制功能，只是编写 Perl 脚本加以监控并在必要时实施自动故障转移。本章将了解一个更为通用的产品组合——Keepalived 和 LVS，使用它们能够很好地满足多种应用的高可用与负载均衡的需求。Keepalived 提供健康检查和故障转移以及提高系统可用性，LVS 提供负载均衡。

11.1 Keepalived 简介

Keepalived 是 Linux 下一个轻量级高可用解决方案，部署和使用都比较简单，所有设置只需要一个配置文件即可完成。Keepalived 起初是为 LVS 设计的，专门用来监控集群系统中各个服务节点的状态。它根据 TCP/IP 参考模型的第三、四和五层的交换机制检测每个服务节点的状态，如果某个服务器节点出现异常，或者出现工作故障，Keepalived 能检测到，并将出现故障的服务器节点从集群系统中剔除。这些工作全部自动完成，不需要人工干涉。后来 Keepalived 又加入了 VRRP 功能，以实现网络不间断稳定运行。因此，Keepalived 一方面具有服务器状态检测和故障隔离功能，另外一方面也具有高可用集群功能。

11.1.1 VRRP 协议

在现实的网络环境中，两台需要通信的主机（End-Host）在大多数情况下并没有直接物理连接。对于这样的情况，它们之间的路由怎么选择？通常有两种方法解决如何选定到达目的主机的下一跳路由的问题：

- 使用动态路由协议，如 RIP、OSPF 等。

- 配置静态路由。

很明显，在主机上配置动态路由，因为管理、维护成本以及是否支持等诸多问题有些不切实际，那么配置静态路由就变得很流行了，这种方式一直沿用至今。但是，路由器或者说默认网关（Default Gateway）却经常成为单故障点。就算配置了多个静态路由，却因为必须重启网络才能生效而变得不实用。

VRRP（Virtual Router Redundancy Protocol，虚拟路由冗余协议）的目的就是为了解决静态路由的单点故障问题。它通过一种竞选（Election）协议动态地将路由任务交给虚拟路由器中的某台 VRRP 路由器。这里有两个关键名词：VRRP 路由器和虚拟路由器。

- VRRP 路由器：简单来说，就是一个实现了 VRRP 协议的物理路由器。一个 VRRP 路由器可以位于多个虚拟路由器中。
- 虚拟路由器：所谓虚拟，是指并不是实际存在的，不是物理的路由器而是逻辑的路由器。虚拟路由器通常由多个 VRRP 路由器通过某种方式组成，就好像将这些物理路由器都放到一个池里面去，整个池对外看起来就像是一个路由器，但其实内部有多个路由器。虚拟路由器的标识被称为 VRID。

在一个 VRRP 虚拟路由中，有多个物理的 VRRP 路由器，但是这多个物理路由并不同时工作，而是由一个称为 Master 的主路由器负责路由工作的，其他的都是称为 Backup 的备份路由器。主路由器并非一成不变，VRRP 协议让每个 VRRP 路由器参与竞选，最终获胜的就称为主路由器。主路由器有一些特权，比如拥有虚拟路由器的 IP 地址，等等。拥有特权的主路由器要负责转发给网关地址的包和响应 ARP 请求。

VRRP 通过竞选协议来实现虚拟路由器的功能，所有的协议报文都是通过 IP 多播（Multicast）包的形式发送，多播地址为 224.0.0.18。虚拟路由器由 VRID（范围 0~255）和一组 IP 地址组成，对外表现为一个通用的 MAC 地址：00-00-5E-00-01-{VRID}。因此，在一个虚拟路由器中，不管谁是主路由器，对外都是相同的 MAC 和 IP，该 IP 即为通常所说的 VIP（虚拟 IP）。客户端主机并不需要因为主路由器的改变而修改自己的路由配置，对它们来说，这种主-备（Master-Backup）路由器的切换是完全透明的。

在一个虚拟路由器中，只有作为主路由器的 VRRP 路由器会一直发送 VRRP 广告包（VRRP Advertisement Message），备份路由器不会抢占主路由器，除非它的优先级更高。当主路由器不可用时，备份路由器收不到广告包，多个备份路由器中优先级最高的那个就会抢占为主路由器。这种抢占的速度非常之快（<1 秒），以保证服务的连续性。出于安全考虑，发送 VRRP 报文的路由器利用认证密钥（Authentication Key）和 MD5 算法对 VRRP 报文进行加密。

11.1.2 Keepalived 架构设计

Keepalived 运行在 LVS 之上，主要功能是实现集群节点的故障隔离及负载均衡器间的故障转移，以提高系统的可用性。Keepalived 工作在 TCP/IP 参考模型的第三层网络层、第四层传输层和第五层应用层。

在网络层 Keepalived 最常采用的工作方式，是通过 ICMP 协议向服务器集群中的每个节点发送 ICMP 数据包，这有点类似于 ping 的功能。如果某个节点没有返回响应数据包，则认为该节点发生了故障。Keepalived 将报告这个节点失效（即故障），并从服务器集群中剔除这个故障节点。

传输层提供了两个众所周知的协议：传输控制协议 TCP 和用户数据报协议 UDP。IP 地址和端口代表 TCP 的一个连接端。要获得 TCP 服务，需要在发送机的一个端口和接收机的一个端口上建立连接。Keepalived 在传输层里利用了 TCP 协议的端口连接和扫描技术来判断集群节点的端口是否正常。一旦在传输层探测到端口没有数据响应和数据返回，Keepalived 就认为这些端口出现异常，然后强制将异常端口所对应的节点从服务器集群中剔除。

Keepalived 在应用层的运行方式更加全面和复杂，用户可以自定义 Keepalived 的工作方式。例如，可以通过编写程序或者脚本来运行 Keepalived，而 Keepalived 将根据用户设定的参数检测各种应用程序或者服务是否正常。如果 Keepalived 的检测结果和用户设定的不一致，Keepalived 将把对应的服务器从服务器集群中剔除。

Keepalived 架构如图 11-1 所示，它可以实现对集群节点的状态检测，而 IPVS 可以实现负载均衡功能。因此，Keepalived 借助于第三方模块 IPVS 就可以很方便地搭建一套负载均衡系统。如果因为 Keepalived 可以和 IPVS 一起很好地工作，就认为它只是一个负载均衡软件，那么这种理解就是一个误解。Keepalived 中的 IPVS 模块是可配置的，在编译时可以通过参数打开或关闭负载均衡功能。NetLINK 模块主要用于实现一些高级路由框架和一些相关参数的网络功能，完成用户空间层 Netlink Reflector 模块发来的各种网络请求。

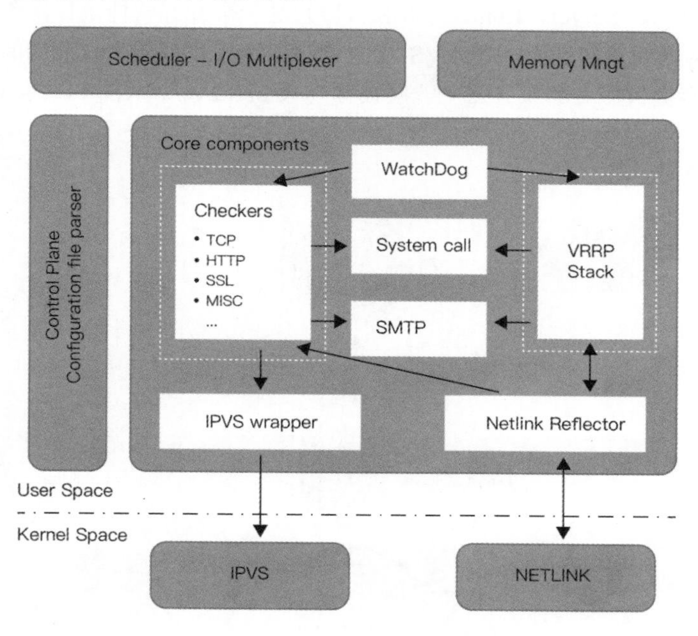

图 11-1 Keepalived 架构

从图 11-1 中我们可以看到用户空间层建立在内核空间层之上，主要包含 4 个部分：

- Scheduler I/O Multiplexer 是一个 I/O 复用分发调度器，负责调度 Keepalived 所有内部的任务请求。
- Memory Mngt 是一个内存管理框架，提供了访问内存的一些通用方法。

- Control Plane 是 Keepalived 的控制版面，可以实现对配置文件的编译和解析。
- Core components 是核心组件，主要包括了以下 5 个部分:
  - ➢ Watchdog: 是计算机可靠领域中极为简单又非常有效的检测工具，Keepalived 正是通过它监控 Checkers 和 VRRP 进程。
  - ➢ Checkers: 提供 Keepalived 最基础也是最主要的功能，实现对服务器运行状态的检测和故障隔离。
  - ➢ VRRP Stack: 这是 Keepalived 后来引入的 VRRP 模块，可以实现高可用集群中的故障转移功能。
  - ➢ IPVS wrapper: 该模块可以将设置好的 IPVS 规则发送到内核空间并且提供给 IPVS 模块，最终实现 IPVS 模块的负载均衡功能。
  - ➢ Netlink Reflector: 用来实现高可用集群故障转移时 VIP 的设置和切换，其所有请求最后都发送到内核空间层的 NETLINK 模块来完成。

11.2 LVS 简介

LVS（Linux Virtual Server，Linux 虚拟服务器）是一个开源软件，作用于网络层，可以实现 Linux 平台下多服务器的负载均衡。在 LVS 构建的服务器集群系统中，前端的负载均衡层被称为 Director Server，后端提供服务的服务器集群层被称为 Real Server。从图 11-2 可以大致了解 LVS 的基本架构。

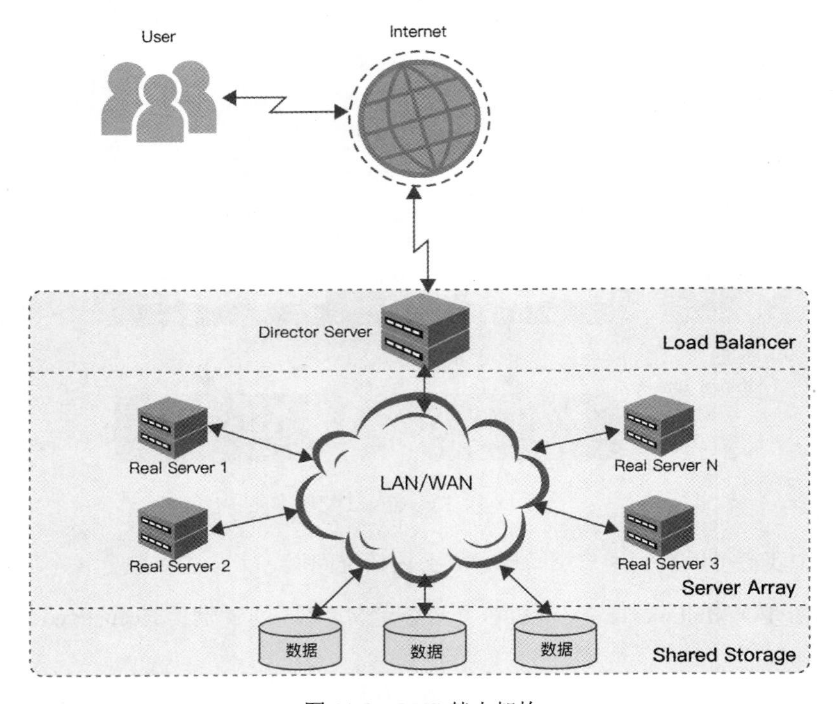

图 11-2 LVS 基本架构

LVS 由 ipvs 和 ipvsadm 两部分组成。ipvs 是工作在内核空间 netfilter 的 input 链上的框架，通过用户空间工具进行管理，其中运行的是真正实现调度的代码。ipvs 会监听 input 链上的请求，一旦请求的是集群服务，ipvs 钩子函数会将请求拉出并进行报文修改，强制转发到 postrouting 进行处理，它的工作流程如图 11-3 所示。ipvsadm 负责为 ipvs 内核框架编写规则，是管理配置内核中 ipvs 程序用于用户空间的工具。LVS 实现了三种负载均衡技术（VS/NAT、VS/TUN 和 VS/DR）和八种调度算法（rr、wrr、lc、wlc、lblc、lblcr、dh 和 sh），它的官网（http://www.linux-vs.org/zh/index.html）上有对相关技术和算法的详细介绍。

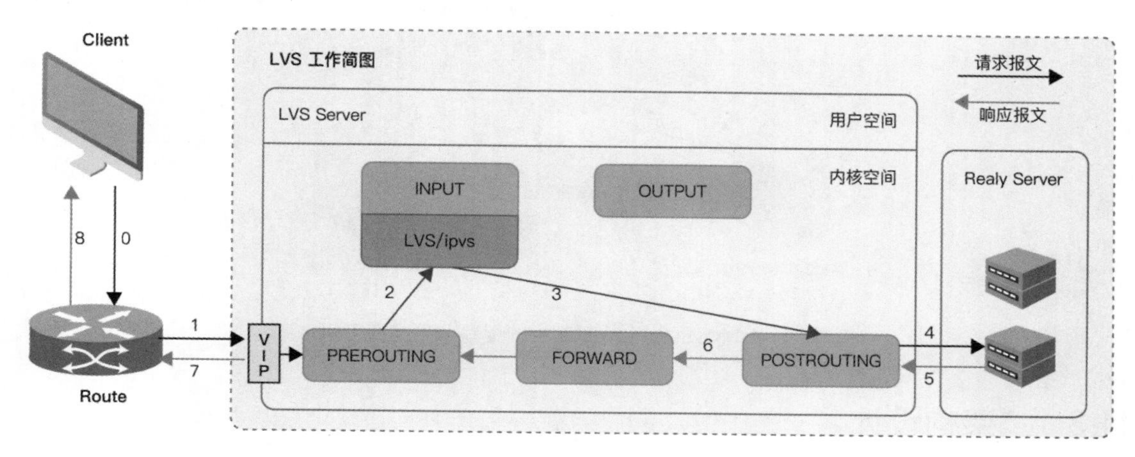

图 11-3 LVS 工作流程

在客户端看来，LVS 就是一个真实的应用服务器。客户端向 LVS 发送请求信息，LVS 接收数据报文至内核空间，工作在 input 链上的 ipvs 模块会判断用户请求的是不是定义的后端服务器。如果是，数据报文传送到 input 链上时，input 链会强行将数据报文转发给 postrouting，postrouting 将数据报文传送给后端真实服务器（Real Server）。LVS 的特点在于超强的分流功能，但它只负责调度流量的去向，不实现在业务层的分流负载。

LVS 可以独立使用，但更普遍的做法是与 Keepalived 结合在一起进行部署。Keepalived 中的 LVS 配置包括虚拟服务器组（Virtual Server Group）和虚拟服务器（Virtual Server），这些配置会传递给 ipvsadm 作为参数。采用这样的架构以后，现有系统的伸缩就非常容易，当后端添加或者减少真实服务器后，只需要更改 Keepalived 配置文件中的 LVS 部分即可，实现了无缝的配置变更。

下面我们用两个示例来详细说明 Keepalived 和 LVS 在 MySQL 高可用架构中的应用。示例环境的操作系统版本为 Linux release 7.2.1511，MySQL 服务器版本为 5.6.14。

11.3 实现 MySQL 主从复制中的读写分离及高可用

第一个示例是使用 Keepalived 和 LVS，实现 MySQL 主从复制中的读写分离及自动故障转移，具体架构如图 11-4 所示。

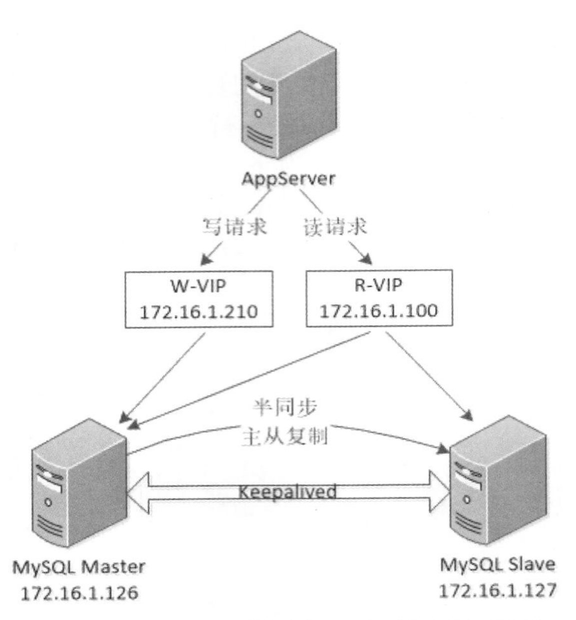

图 11-4　MySQL 主从复制中的读写分离及高可用

11.3.1　架构设计

图 11-4 中，使用两台主机实现 MySQL 主从复制并实现读写分离，用于提高查询性能。采用 MySQL 5.6.x 的半同步复制实现主从数据同步。使用 Keepalived 来监控 MySQL，并提供读写 VIP 漂移。Keepalived 在这里主要用作真实服务器的健康状态检查以及负载均衡主机和备份主机之间故障转移的实现。任何一台主机故障都不会影响对外提供的服务（读写 VIP 可以漂移），保持 MySQL 数据库服务的高可用性。

Keepalived 基于 VRRP 协议来实现高可用解决方案，可以避免单点故障。在这个解决方案中，通常至少有两台服务器运行 Keepalived，一台为主服务器（Master）；另一台为备份服务器（Backup），但对外表现为一个或一组 VIP。主服务器会发送特定消息给备份服务器，当备份服务器收不到该消息时，则认为主服务器出现了故障，备份服务器会接管 VIP 继续对外提供服务，从而保证了 MySQL 数据库的高可用。LVS 在本例的作用是提供读负载均衡。整体架构的设计原理和异常处理可描述如下：

（1）真实服务器 A 和 B，配置为 MySQL 半同步复制的主库和从库。

（2）通过 Keepalived 启用两个虚拟 IP：W-VIP/R-VIP，前者负责响应写入，后者负责响应读取，实现读写分离。

（3）服务器 A 和 B 都存在时，LVS 通过 W-VIP 将写请求转发至主机 A，通过 R-VIP 将读请求转发给 A 和 B，实现读负载均衡。

（4）当主机 A 异常时，B 接管服务，W-VIP/R-VIP 此时都漂移到主机 B 上。

（5）当主机 B 异常时，R-VIP 会将 B 剔除，其他不变。

11.3.2 安装和配置

示例环境各主机的 IP 地址及其角色如下：

- 172.16.1.126：Keepalived + LVS Master + MySQL Semisync-Replication Master
- 172.16.1.127：Keepalived + LVS Backup + MySQL Semisync-Replication Slave
- 172.16.1.100：R-VIP
- 172.16.1.210：W-VIP

步骤 01 安装和配置 MySQL 半同步复制。

按角色安装和配置 MySQL 半同步复制，172.16.1.126 为主库，172.16.1.127 为从库。具体安装和配置过程参见 2.5 节。

步骤 02 下载安装 LVS。

在 172.16.1.126 和 172.16.1.127 上以 root 用户执行以下命令：

```
yum -y install ipvsadm
```

步骤 03 下载安装 Keepalived。

在 172.16.1.126 和 172.16.1.127 上以 root 用户执行以下命令：

```
wget -q http://www.keepalived.org/software/keepalived-1.2.13.tar.gz
tar -zxvf keepalived-1.2.13.tar.gz
cd keepalived-1.2.13
./configure && make && make install
cp /usr/local/etc/rc.d/init.d/keepalived /etc/rc.d/init.d/
cp /usr/local/etc/sysconfig/keepalived /etc/sysconfig/
mkdir /etc/keepalived
cp /usr/local/etc/keepalived/keepalived.conf /etc/keepalived/
cp /usr/local/sbin/keepalived /usr/sbin/
chkconfig --add keepalived
chkconfig --level 345 keepalived on
```

步骤 04 iptables 配置。

在本示例中，由于要实现 MySQL 读写主机 VIP 的漂移，因此 Keepalived、LVS 和 MySQL 需要部署到相同的主机上。当 LVS 架构中的 Director 服务器与 Real Server 是同一台机器时，会出现两台 Director 无限循环转发请求的情况，具体分析如下：

（1）director1 收到目的地址为 VIP 的数据包。

（2）director1 经过 ipvs 对数据包做负载均衡，部分包转发给 realserver2（director2）。

（3）director2 收到的转发包，目的地址也是 VIP，再经过 ipvs 做负载均衡，部分包转发给 realserver1（director1）。

解决该问题的办法是，Director服务器使用iptables对数据包做标记，LVS对fwm转发而不是直接转发TCP。在每台Director服务器上，标记自己发送的数据包，ipvs转发这些被标记的包。亦即其他Director服务器发送的数据包就不再进行ipvs转发了，而是直接交给上层监听程序。

在172.16.1.126（主库）上执行：

```
iptables -t mangle -I PREROUTING -d 172.16.1.100 -p tcp -m tcp --dport 3306 -m mac ! --mac-source 00:50:56:a5:49:7f -j MARK --set-mark 0x1
```

在172.16.1.127（从库）上执行：

```
iptables -t mangle -I PREROUTING -d 172.16.1.100 -p tcp -m tcp --dport 3306 -m mac ! --mac-source 00:50:56:a5:0f:77 -j MARK --set-mark 0x2
```

其中172.16.1.100是读VIP，3306是VPORT，00:50:56:a5:49:7f是172.16.1.127主机网卡的MAC地址，00:50:56:a5:0f:77是172.16.1.126主机网卡的MAC地址。

步骤05 配置Keepalived。

172.16.1.126初始为Keepalived的主服务器，其上的/etc/keepalived/keepalived.conf配置文件内容如下：

```
global_defs {
   router_id LVS_DEVEL
}

vrrp_sync_group VG1 {
group {
VI_1
}
}

vrrp_instance VI_1 {
    state BACKUP
    interface ens32
    virtual_router_id 51
    priority 100
    notify_master "/home/mysql/remove_slave.sh"
    advert_int 1
    nopreempt
    authentication {
        auth_type PASS
        auth_pass 1234
    }
    virtual_ipaddress {
        172.16.1.100
172.16.1.210
    }
}

# 写VIP虚拟服务器，只配置本地机器
virtual_server 172.16.1.210 3306 {
```

```
# 定义虚拟服务器，地址与上面的 virtual_ipaddress 相同
    delay_loop 3                    # 健康检查时间间隔，3 秒
    lb_algo rr                      # 负载均衡调度算法：rr|wrr|lc|wlc|sh|dh|lblc
    lb_kind DR                      # 负载均衡转发规则：NAT|DR|TUN
    # persistence_timeout 5         # 会话保持时间 5 秒，建议开启动态服务
    protocol TCP                    # 转发协议一般有 TCP 和 UDP 两种

    real_server 172.16.1.126 3306 {
        weight 1                    # 权重越大，负载分就越大，0 表示失效
        notify_down /home/mysql/mysql_down.sh
        TCP_CHECK {
            connect_timeout 3
            nb_get_retry 3
            delay_before_retry 3
            connect_port 3306
        }
    }
}

# 读 VIP 虚拟服务器，配置 fwmark 转发
virtual_server fwmark 1 {
    delay_loop 3
    lb_algo rr
    lb_kind DR
    # persistence_timeout 5
    protocol TCP

    real_server 172.16.1.126 3306 {
        weight 1
        notify_down /home/mysql/mysql_down.sh
        TCP_CHECK {
            connect_timeout 3
            nb_get_retry 3
            delay_before_retry 3
            connect_port 3306
        }
    }
    real_server 172.16.1.127 3306 {
        weight 1
        TCP_CHECK {
            connect_timeout 3
            nb_get_retry 3
            delay_before_retry 3
            connect_port 3306
        }
    }
}
```

R-VIP 绑定 172.16.1.126、172.16.1.127，W-VIP 绑定 172.16.1.126。

172.16.1.127 初始为 Keepalived 的备份服务器，其上的/etc/keepalived/keepalived.conf 配置文件

内容如下：

```
global_defs {
  router_id LVS_DEVEL
}

vrrp_sync_group VG1 {
group {
VI_1
}
}

vrrp_instance VI_1 {
    state BACKUP
    interface ens160
    virtual_router_id 51
    priority 90
    notify_master "/home/mysql/remove_slave.sh"
    advert_int 1
    nopreempt
    authentication {
        auth_type PASS
        auth_pass 1234
    }
    virtual_ipaddress {
        172.16.1.100
172.16.1.210
    }
}

# 写 VIP 虚拟服务器，只配置本地机器
virtual_server 172.16.1.210 3306 {
# 定义虚拟服务器，地址与上面的 virtual_ipaddress 相同
    delay_loop 3              # 健康检查时间间隔，3 秒
    lb_algo rr                # 负载均衡调度算法：rr|wrr|lc|wlc|sh|dh|lblc
    lb_kind DR                # 负载均衡转发规则：NAT|DR|TUN
    # persistence_timeout 5 # 会话保持时间 5 秒，建议开启动态服务
    protocol TCP              # 转发协议一般有 TCP 和 UDP 两种

     real_server 172.16.1.127 3306 {
        weight 1              # 权重越大，负载分就越大，0 表示失效
         notify_down /home/mysql/mysql_down.sh
        TCP_CHECK {
            connect_timeout 3
            nb_get_retry 3
            delay_before_retry 3
            connect_port 3306
        }
    }
}
```

```
# 读VIP虚拟服务器，配置fwmark转发
virtual_server fwmark 2 {
    delay_loop 3
    lb_algo rr
    lb_kind DR
    # persistence_timeout 5
    protocol TCP

    real_server 172.16.1.126 3306 {
        weight 1
        TCP_CHECK {
            connect_timeout 3
            nb_get_retry 3
            delay_before_retry 3
            connect_port 3306
        }
    }
    real_server 172.16.1.127 3306 {
        weight 1
        notify_down /home/mysql/mysql_down.sh
        TCP_CHECK {
            connect_timeout 3
            nb_get_retry 3
            delay_before_retry 3
            connect_port 3306
        }
    }
}
```

主服务器（Master）与备份服务器（Backup）的Keepalived配置文件中有以下4点不同：

- priority设置不同，master为100，backup为90。
- 写VIP 172.16.1.210绑定的是本机地址。
- /home/mysql/mysql_down.sh脚本在本机地址下定义。
- 根据iptables的配置，主服务器的fwmark标记为1，备份服务器的fwmark标记为2。

Keepalived配置文件是以块形式组织的，每个块都在{}包围的范围内，以符号#和!开头的行都是注释行。

global_defs为全局定义，对整个Keepalived起作用，而不管是否使用LVS。其中的router_id是指运行Keepalived机器的一个标识号。

vrrp_sync_group配置VRRP同步组。不使用Sync Group的话，如果机器有两个网段，一个内网和一个外网，每个网段开启一个VRRP实例，假设VRRP配置为检查内网，那么当外网出现问题时，VRRPD认为自己仍然健康，就不会触发主服务器和备份服务器的切换。Sync Group用于解决该问题，可以把两个实例都放进同一个Sync Group，这样的话，Sync Group里面任何一个实例出现问题都会发生切换。其中的group设置同一组中的VRRP实例名，本示例中只有一个实例VI_1。

vrrp_instance配置VRRP实例，VRRP实例表示启用了VRRP协议。可以在每块网卡上开启一

个实例。配置块中说明了 VRRP 的一些特性，比如主备角色、VRID 等，还定义 vrrp_sync_group 里面每个组的漂移 IP。其中：

- state：指定实例的初始状态。在两个路由都启动后，马上会发生竞争，高优先级的路由器会竞选为主路由器，所以这里的 state 并不表示这台主机就一直是备份路由器。
- interface：实例绑定的网卡。
- virtual_router_id：VRID 标记，值为 0~255，这里使用默认的 51。
- priority：高优先级的竞选为主服务器。这里 MySQL 主从库两台主机的优先级分别设置为 100 和 90，因此当 Keepalived 启动后，MySQL 主库会被选为主服务器。
- notify_master：指定当切换到主服务器时执行的脚本。
- advert_int：检查间隔，这里设置为默认的 1 秒。
- nopreempt：设置为不抢占，注意这个配置只能设置在 state 为 BACKUP 的主机上。当主服务器出现问题后，备份服务器会竞选为新的主服务器，那么当之前的主服务器重新上线后，是继续成为主服务器还是变成备份服务器呢？默认设置是抢占，那么之前的主服务器重新启动后会继续抢占成为主服务器。这样的频繁切换对于业务是不能容忍的，我们希望原主服务器重新上线后成为备份服务器，所以要设置为不抢占。又因为 nopreempt 配置只能用在 state 为 BACKUP 的主机上，因此主服务器的 state 也得设置为 BACKUP，也就是说 172.16.1.126 和 172.16.1.127 都要将 state 设置为 BACKUP。通过在两台备份服务器上面设置不同的优先级（priority），让它们一启动就竞选，高优先级的 172.16.1.126 成为最初的主服务器。
- authentication：设置认证类型和认证密码。auth_type 为认证类型，支持 PASS 和 AH 两种认证类型，通常使用 PASS 类型。auth_pass 为明文认证密码，同一 VRRP 实例的主服务器与备份服务器使用相同的密码才能正常通信。
- virtual_ipaddress：指定漂移地址（VIP），也就是切换到主服务器时，这些 IP 会被添加，切换到备份服务器时，这些 IP 会被删除。因此每台服务器上可以不绑定任何 VIP，而把它们都放到 virtual_ipaddress 里面，可以有多个。Keepalived 会自动调用 ip addr 进行绑定。

virtual_server 块设置一个虚拟服务器，指定它的虚拟 IP 和虚拟端口。其中：

- delay_loop：指定服务轮询的时间间隔，单位是秒。
- lb_algo：指定 LVS 的调度算法，本例指定为 rr，即轮询。
- lb_kind：指定 LVS 转发模式，本例设置为 DR，这也是大多数生产环境的配置。
- persistence_timeout：指定以秒为单位的会话保持时间，这里做了注释，意为不保持，目的是为了方便查看后面测试 rr 负载均衡算法的效果。生产环境建议启用该参数。
- protocol：指定转发协议使用 TCP 还是 UDP。
- real_server：每台真实服务器都需要一个 real_server 配置项，指定真实服务器的 IP 和端口。本例指定两个 MySQL 数据库服务器的地址和端口。weight 指定 realserver 权重，默认为 1，而 0 为失效。TCP_CHECK 指定健康检查方式为 TCP，其中：connect_timeout 定义连接超时时间，单位是秒；nb_get_retry 指定重连次数；delay_before_retry 指定重连的时间间隔，单位是秒；connect_port：指定健康检查的端口。

当 LVS 检测到主库宕机时，就执行/home/mysql/mysql_down.sh 脚本文件，停止主库上的

Keepalived 服务，从而使得 VIP 漂移到从库。该文件内容如下：

```
#!/bin/bash
/etc/init.d/keepalived stop
```

当主库宕机时，172.16.1.127 切换成为主服务器时执行/home/mysql/remove_slave.sh 脚本文件，停止复制，重置 MySQL 实例的主从状态，关闭 read_only，将原来主从复制中的从库变为主库。该文件的内容如下：

```
#!/bin/bash
. /home/mysql/.bashrc

user=root
password=123456
log=/home/mysql/remove_slave.log

echo "`date`" >> $log
mysql -u$user -p$password -e "set global read_only=OFF;stop slave;reset
master;reset slave all;" >> $log
/bin/sed -i 's#read-only#\#read-only#' /home/mysql/mysql-5.6.14/my.cnf
```

步骤 01 编写 realserver 的网络配置脚本。

在 172.16.1.126 和 172.16.1.127 上建立/etc/init.d/realserver 文件，内容如下：

```
#!/bin/sh
RVIP=172.16.1.100
WVIP=172.16.1.210
. /etc/rc.d/init.d/functions

case "$1" in
# 禁用本地的 ARP 请求、绑定本地回环地址
start)
    /sbin/ifconfig lo down
    /sbin/ifconfig lo up
    echo "1" >/proc/sys/net/ipv4/conf/lo/arp_ignore
    echo "2" >/proc/sys/net/ipv4/conf/lo/arp_announce
    echo "1" >/proc/sys/net/ipv4/conf/all/arp_ignore
    echo "2" >/proc/sys/net/ipv4/conf/all/arp_announce
    /sbin/sysctl -p >/dev/null 2>&1

    # 在回环地址上绑定 VIP，设置掩码，与 Director 服务器上自身的 IP 保持通信
    /sbin/ifconfig lo:0 $RVIP netmask 255.255.255.255 up
    /sbin/ifconfig lo:1 $WVIP netmask 255.255.255.255 up

    /sbin/route add -host $RVIP dev lo:0
    /sbin/route add -host $WVIP dev lo:1

    echo "LVS-DR real server starts successfully.\n"
    ;;
stop)
```

```
    /sbin/ifconfig lo:0 down
    /sbin/ifconfig lo:1 down
    /sbin/route del $RVIP >/dev/null 2>&1
    /sbin/route del $WVIP >/dev/null 2>&1

    echo "1" >/proc/sys/net/ipv4/conf/lo/arp_ignore
    echo "2" >/proc/sys/net/ipv4/conf/lo/arp_announce
    echo "1" >/proc/sys/net/ipv4/conf/all/arp_ignore
    echo "2" >/proc/sys/net/ipv4/conf/all/arp_announce
echo "LVS-DR real server stopped.\n"
    ;;
status)
    isLoOn=`/sbin/ifconfig lo:0 | grep "$RVIP"`
    isRoOn=`/bin/netstat -rn | grep "$RVIP"`
    if [ "$isLoON" == "" -a "$isRoOn" == "" ]; then
       echo "LVS-DR real server has run yet."
    else
       echo "LVS-DR real server is running."
    fi
    exit 3
    ;;
*)
    echo "Usage: $0 {start|stop|status}"
    exit 1
esac
exit 0
```

执行下面的命令将该脚本加入开机自启动：

```
chmod +x /etc/init.d/realserver
echo "/etc/init.d/realserver" >> /etc/rc.d/rc.local
```

步骤02 启动realserver和Keepalived。

在172.16.1.126和172.16.1.127上执行：

```
service realserver start
/etc/init.d/
```

命令执行后，172.16.1.126的IP地址如下：

```
[root@hdp3~]#ip a
1: lo: <LOOPBACK,UP,LOWER_UP> mtu 65536 qdisc noqueue state UNKNOWN
    link/loopback 00:00:00:00:00:00 brd 00:00:00:00:00:00
    inet 127.0.0.1/8 scope host lo
      valid lft forever preferred_lft forever
    inet 172.16.1.100/32 scope global lo:0
      valid_lft forever preferred_lft forever
    inet 172.16.1.210/32 scope global lo:1
      valid_lft forever preferred_lft forever
    inet6 ::1/128 scope host
      valid_lft forever preferred_lft forever
```

```
    2: ens32: <BROADCAST,MULTICAST,UP,LOWER_UP> mtu 1500 qdisc pfifo_fast state
UP qlen 1000
        link/ether 00:50:56:a5:of:77 brd ff:ff:ff:ff:ff:ff
        inet 172.16.1.126/24 brd 172.16.1.255 scope global ens32
           valid_lft forever preferred_lft forever
        inet 172.16.1.100/32 scope global ens32
           valid_lft forever preferred_lft forever
        inet 172.16.1.210/32 scope global ens32
           valid_lft forever preferred_lft forever
        inet6 fe80::250:56ff:fea5:f77/64 scope link
           valid_lft forever preferred_lft forever
    [root@hdp3~]#
```

172.16.1.127 的 IP 地址如下：

```
    [root@hdp4~]#ip a
    1: lo: <LOOPBACK,UP,LOWER_UP> mtu 65536 qdisc noqueue state UNKNOWN
        link/loopback 00:00:00:00:00:00 brd 00:00:00:00:00:00
        inet 127.0.0.1/8 scope host lo
           valid lft forever preferred_lft forever
        inet 172.16.1.100/32 scope global lo:0
           valid_lft forever preferred_lft forever
        inet 172.16.1.210/32 scope global lo:1
           valid_lft forever preferred_lft forever
        inet6 ::1/128 scope host
           valid_lft forever preferred_lft forever
    2: ens160: <BROADCAST,MULTICAST,UP,LOWER_UP> mtu 1500 qdisc pfifo_fast state
UP qlen 1000
        link/ether 00:50:56:a5:49:7f brd ff:ff:ff:ff:ff:ff
        inet 172.16.1.127/24 brd 172.16.1.255 scope global ens160
           valid_lft forever preferred_lft forever
        inet6 fe80::250:56ff:fea5:497f/64 scope link
           valid_lft forever preferred_lft forever
    [root@hdp4~]#
```

可以看到 172.16.1.126 和 172.16.1.127 两个主机的回环地址都绑定了读、写两个 VIP，而只有 172.16.1.126 的本地 IP 绑定了读、写两个 VIP。此时从 172.16.1.126 查看 LVS 集群状态如下：

```
    [root@hdp3~]#ipvsadm -Ln
    IP Virtual Server version 1.2.1 (size=4096)
    Prot LocalAddress:Port Scheduler Flags
      -> RemoteAddress:Port           Forward Weight ActiveConn InActConn
    TCP  172.16.1.210:3306 rr
      -> 172.16.1.126:3306            Route   1      0          0
    FWM  1 rr
      -> 172.16.1.126:3306            Route   1      0          0
      -> 172.16.1.127:3306            Route   1      0          0
    [root@hdp3~]#
```

11.3.3 功能测试

步骤01 验证通过 172.16.1.100 连接的读负载均衡转发策略。

MySQL 客户端读 VIP 连接数据库，并查看所连接 MySQL 实例的 server_id。可以看到，每次执行依次连接到 172.16.1.126 和 172.16.1.127 的 MySQL，证明是轮询策略产生的结果。

```
C:\>mysql -utest -p123456 -h172.16.1.100 -N -s -e "show variables like
'server_id'"
mysql: [Warning] Using a password on the command line interface can be insecure.
server_id       126

C:\>mysql -utest -p123456 -h172.16.1.100 -N -s -e "show variables like
'server_id'"
mysql: [Warning] Using a password on the command line interface can be insecure.
server_id       127

C:\>mysql -utest -p123456 -h172.16.1.100 -N -s -e "show variables like
'server_id'"
mysql: [Warning] Using a password on the command line interface can be insecure.
server_id       126

C:\>mysql -utest -p123456 -h172.16.1.100 -N -s -e "show variables like
'server_id'"
mysql: [Warning] Using a password on the command line interface can be insecure.
server_id       127
```

步骤02 验证通过 172.16.1.210 连接的读负载均衡转发策略。

通过写 VIP 只能连接到 172.16.1.126。

```
C:\>mysql -utest -p123456 -h172.16.1.210 -N -s -e "show variables like
'server_id'"
mysql: [Warning] Using a password on the command line interface can be insecure.
server_id       126

C:\>mysql -utest -p123456 -h172.16.1.210 -N -s -e "show variables like
'server_id'"
mysql: [Warning] Using a password on the command line interface can be insecure.
server_id       126
```

步骤03 模拟从库的 mysqld crash。

在 172.16.1.127 上执行以下命令：

```
pkill -9 mysqld
```

步骤04 再次使用两个 VIP 连接，查看数据库服务是否正常。

可以看到应用不受影响，但两个 VIP 都连接到 IP 地址为 172.16.1.126 的一台 MySQL 服务器上。

```
    C:\>mysql -utest -p123456 -h172.16.1.100 -N -s -e "show variables like
'server_id'"
    mysql: [Warning] Using a password on the command line interface can be insecure.
    server_id       126

    C:\>mysql -utest -p123456 -h172.16.1.210 -N -s -e "show variables like
'server_id'"
    mysql: [Warning] Using a password on the command line interface can be insecure.
    server_id       126
```

此时从 172.16.1.127 查看 LVS 集群状态如下，可以看到读取地址只有 172.16.1.126：

```
    [root@hdp4~]#ipvsadm -Ln
    IP Virtual Server version 1.2.1 (size=4096)
    Prot LocalAddress:Port Scheduler Flags
      -> RemoteAddress:Port           Forward Weight ActiveConn InActConn
    TCP  172.16.1.210:3306 rr
      -> 172.16.1.126:3306            Route   1      0          0
    FWM  2 rr
      -> 172.16.1.126:3306            Route   1      0          0
    [root@hdp4~]#
```

步骤 05 重新启动从库的 MySQL 服务。

在 172.16.1.127 上执行以下命令：

```
    service mysql start
```

步骤 06 再次验证通过 172.16.1.100 连接的读负载均衡转发策略。

```
    C:\>mysql -utest -p123456 -h172.16.1.100 -N -s -e "show variables like
'server_id'"
    mysql: [Warning] Using a password on the command line interface can be insecure.
    server_id       127

    C:\>mysql -utest -p123456 -h172.16.1.100 -N -s -e "show variables like
'server_id'"
    mysql: [Warning] Using a password on the command line interface can be insecure.
    server_id       126

    C:\>mysql -utest -p123456 -h172.16.1.100 -N -s -e "show variables like
'server_id'"
    mysql: [Warning] Using a password on the command line interface can be insecure.
    server_id       127

    C:\>mysql -utest -p123456 -h172.16.1.100 -N -s -e "show variables like
'server_id'"
    mysql: [Warning] Using a password on the command line interface can be insecure.
    server_id       126
```

此时从 172.16.1.127 查看 LVS 集群状态，可以看到读取地址 172.16.1.127 被自动加回：

```
    [root@hdp4~]#ipvsadm -Ln
```

```
IP Virtual Server version 1.2.1 (size=4096)
Prot LocalAddress:Port Scheduler Flags
  -> RemoteAddress:Port           Forward Weight ActiveConn InActConn
TCP  172.16.1.210:3306 rr
  -> 172.16.1.126:3306            Route   1      0          0
FWM  2 rr
  -> 172.16.1.126:3306            Route   1      0          0
  -> 172.16.1.127:3306            Route   1      0          0
[root@hdp4~]#
```

步骤07 模拟主库的 mysqld 失效。

在 172.16.1.126 上执行以下命令：

```
pkill -9 mysqld
```

步骤08 再次验证通过 VIP 连接的读负载均衡转发策略。

```
C:\>mysql -utest -p123456 -h172.16.1.100 -N -s -e "show variables like
'server_id'"
mysql: [Warning] Using a password on the command line interface can be insecure.
server_id       127

C:\>mysql -utest -p123456 -h172.16.1.210 -N -s -e "show variables like
'server_id'"
mysql: [Warning] Using a password on the command line interface can be insecure.
server_id       127
```

此时查看 172.16.1.127 上绑定的 IP 如下：

```
[root@hdp4~]#ip a
1: lo: <LOOPBACK,UP,LOWER_UP> mtu 65536 qdisc noqueue state UNKNOWN
    link/loopback 00:00:00:00:00:00 brd 00:00:00:00:00:00
    inet 127.0.0.1/8 scope host lo
       valid_lft forever preferred_lft forever
    inet 172.16.1.100/32 scope global lo:0
       valid_lft forever preferred_lft forever
    inet 172.16.1.210/32 scope global lo:1
       valid_lft forever preferred_lft forever
    inet6 ::1/128 scope host
       valid_lft forever preferred_lft forever
2: ens160: <BROADCAST,MULTICAST,UP,LOWER_UP> mtu 1500 qdisc pfifo_fast state
UP qlen 1000
    link/ether 00:50:56:a5:49:7f brd ff:ff:ff:ff:ff:ff
    inet 172.16.1.126/24 brd 172.16.1.255 scope global ens160
       valid_lft forever preferred_lft forever
    inet 172.16.1.100/32 scope global ens160
       valid_lft forever preferred_lft forever
    inet 172.16.1.210/32 scope global ens160
       valid_lft forever preferred_lft forever
    inet6 fe80::250:56ff:fea5:497f/64 scope link
       valid_lft forever preferred_lft forever
```

```
[root@hdp4~]#
```

可以看到读、写两个 VIP 已经漂移到 172.16.1.127 上。

步骤 09 验证切换后的新主库可读写：

```
C:\>mysql -utest -p123456 -h172.16.1.210
...
mysql> select * from test.t1;
+-------+
| a     |
+-------+
|     1 |
|     2 |
|     3 |
+-------+
3 rows in set (0.00 sec)

mysql> delete from test.t1;
Query OK, 3 rows affected (10.00 sec)

mysql> commit;
Query OK, 0 rows affected (0.00 sec)
```

11.4 实现 MySQL 双主复制的读写负载均衡及高可用

11.4.1 架构设计

上一个示例用读、写两个 VIP 实现对 MySQL 数据库的读写分离。同时利用 Keepalived 监控两个 MySQL 实例默认的 3306 端口，复制中的从库出问题就直接剔除，若主库出问题则触发自动故障转移，从而实现数据库服务高可用。同时我们还指出，当 LVS 架构中的 Director 服务器和真实服务器（Real Server）工作在同一台机器上时，需要注意 SYN_RECV 问题，即会出现两台 Director 服务器无限循环转发请求的情况。

在本示例中，我们利用 Keepalived 的 IPVS 功能，调用 LVS 实现 MySQL 双主复制的读写负载均衡，同时保证负载均衡器和 MySQL 的高可用性。示例架构如图 11-5 所示。这次将 LVS 的 Director 服务器和真实服务器部署在不同的主机上，只使用一个 VIP。Keepalived 监控的是 LVS 的主服务器和备份服务器，即 Director 服务器，保证 LVS 的高可用，继而利用 LVS 的 VIP 和轮询算法，间接实现 MySQL 服务器，即真实服务器（Real Server）的高可用。

本示例之所以要使用 MySQL 双主复制而不是主从复制，是因为方案中并没有涉及读写分离，而是在两个等价的 MySQL 服务器之间实现读写负载均衡。本例中除了简单配置外，没有任何自定义的脚本。Keepalived 利用 VRRP 实现了 LVS 的 HA，避免了 LVS 服务器的单点故障，出现问题时可以自动切换到正常的节点。LVS 服务器起到了负载均衡的作用，将用户请求分发到多个真实服务器。同时，一台真实服务器故障并不会影响整个集群，因为 LVS 会检测真实服务器的状态，并据此自动添加或删除集群中真实服务器的服务。如本例的配置，需要考虑单台 MySQL 服务器的负载最好不要超过 50%，否则一旦某台 MySQL 服务器故障，可能出现另一台正常 MySQL 不堪重负的情况。

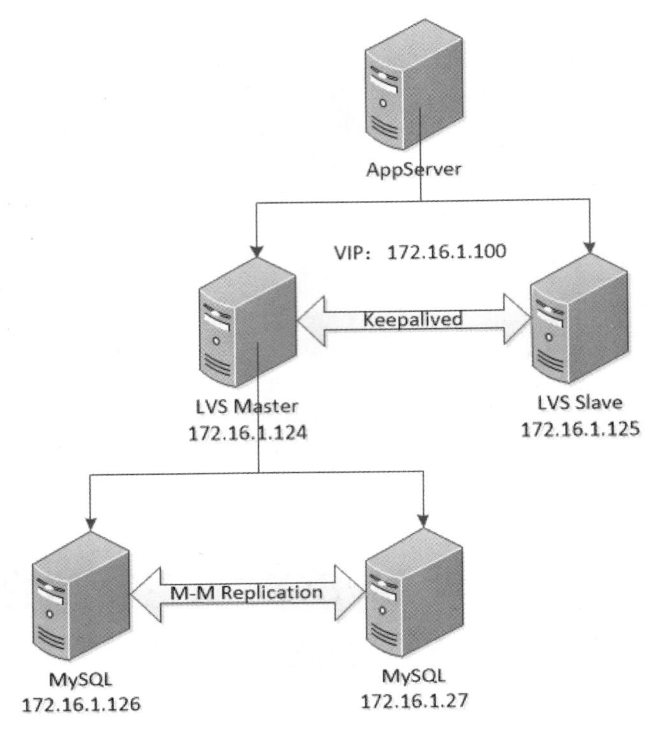

图 11-5 MySQL 双主复制的读写负载均衡及高可用

11.4.2 安装和配置

示例环境各主机的 IP 地址及其角色如下：

- 172.16.1.124：Keepalived + LVS Master
- 172.16.1.125：Keepalived + LVS Slave
- 172.16.1.126：MySQL Replication Master
- 172.16.1.127：MySQL Replication Master
- 172.16.1.100：VIP

步骤 01 安装和配置 MySQL 双主复制。

按角色安装和配置 MySQL 双主复制。具体安装和配置过程参见 9.2.2 小节。与主从复制相比，双主复制需要注意以下三个参数的设置：

- log_slave_updates：要设置为 true，将复制事件写入本机 binlog。一台服务器既做主库又做从库时，此选项必须开启。

- auto_increment_offset 和 auto_increment_increment：为避免自增列冲突，需要设置这两个参数，例如可以配置如下：

```
# masterA 自增长 ID
auto_increment_offset = 1      #奇数 ID
auto_increment_increment = 2
# masterB 自增加 ID
auto_increment_offset = 2      # 偶数 ID
auto_increment_increment = 2
```

步骤 02　下载安装 LVS。

参见 11.3.2 小节的示例。

步骤 03　下载安装 Keepalived。

参见 11.3.2 小节的示例。

步骤 04　配置 Keepalived。

172.16.1.124 初始为 Keepalived 的主服务器，其中的/etc/keepalived/keepalived.conf 配置文件内容如下：

```
global_defs {
  router_id LVS_DEVEL
}

vrrp_sync_group VG1 {
group {
VI_1
}
}

vrrp_instance VI_1 {
    state BACKUP
    interface ens32
    virtual_router_id 51
    priority 100
    advert_int 1
    nopreempt
    authentication {
        auth_type PASS
        auth_pass 1234
    }
    virtual_ipaddress {
        172.16.1.100
    }
}
```

```
virtual_server 172.16.1.100 3306 {
# 定义虚拟服务器，地址与上面的 virtual_ipaddress 相同
    delay_loop 3                    # 健康检查时间间隔，3 秒
    lb_algo rr                      # 负载均衡调度算法：rr|wrr|lc|wlc|sh|dh|lblc
    lb_kind DR                      # 负载均衡转发规则：NAT|DR|TUN
    # persistence_timeout 5         # 会话保持时间 5 秒，建议开启动态服务
    protocol TCP                    # 转发协议一般有 TCP 和 UDP 两种

    # 后端真实服务器，有几台就设置几个
    real_server 172.16.1.126 3306 {
        weight 1                    # 权重越大,负载分就越大，0 表示失效
        TCP_CHECK {
            connect_timeout 3
            nb_get_retry 3
            delay_before_retry 3
            connect_port 3306
        }
    }
    real_server 172.16.1.127 3306 {
        weight 1
        TCP_CHECK {
            connect_timeout 3
            nb_get_retry 3
            delay_before_retry 3
            connect_port 3306
        }
    }
}
```

172.16.1.125 初始为 Keepalived 的从服务器，其上的/etc/keepalived/keepalived.conf 配置文件的内容与主服务器配置文件只有优先级（priority）设置不同，主服务器的优先级为 100，从服务器的优先级为 90，其他全都一样。各选项说明参见 11.3.2 小节的示例。

步骤 05 编写真实服务器的网络配置脚本。

在 172.16.1.126 和 172.16.1.127 上建立/etc/init.d/realserver 文件，内容如下：

```
#!/bin/sh
VIP=172.16.1.100
. /etc/rc.d/init.d/functions

case "$1" in
# 禁用本地的 ARP 请求、绑定本地回环地址
start)
    /sbin/ifconfig lo down
    /sbin/ifconfig lo up
    echo "1" >/proc/sys/net/ipv4/conf/lo/arp_ignore
    echo "2" >/proc/sys/net/ipv4/conf/lo/arp_announce
    echo "1" >/proc/sys/net/ipv4/conf/all/arp_ignore
    echo "2" >/proc/sys/net/ipv4/conf/all/arp_announce
/sbin/sysctl -p >/dev/null 2>&1
```

```
# 在回环地址上绑定 VIP，设置掩码，与 Director 服务器上自身的 IP 保持通信
    /sbin/ifconfig lo:0 $VIP netmask 255.255.255.255 up
    /sbin/route add -host $VIP dev lo:0
    echo "LVS-DR real server starts successfully.\n"
    ;;
stop)
    /sbin/ifconfig lo:0 down
    /sbin/route del $VIP >/dev/null 2>&1
    echo "1" >/proc/sys/net/ipv4/conf/lo/arp_ignore
    echo "2" >/proc/sys/net/ipv4/conf/lo/arp_announce
    echo "1" >/proc/sys/net/ipv4/conf/all/arp_ignore
    echo "2" >/proc/sys/net/ipv4/conf/all/arp_announce
echo "LVS-DR real server stopped.\n"
    ;;
status)
    isLoOn=`/sbin/ifconfig lo:0 | grep "$VIP"`
    isRoOn=`/bin/netstat -rn | grep "$VIP"`
    if [ "$isLoON" == "" -a "$isRoOn" == "" ]; then
        echo "LVS-DR real server has run yet."
    else
        echo "LVS-DR real server is running."
    fi
    exit 3
    ;;
*)
    echo "Usage: $0 {start|stop|status}"
    exit 1
esac
exit 0
```

执行下面的命令将该脚本加入开机自启动：

```
chmod +x /etc/init.d/realserver
echo "/etc/init.d/realserver" >> /etc/rc.d/rc.local
```

步骤 06 启动真实服务器。

执行以下命令启动 realserver：

```
service realserver start
```

命令执行后 172.16.1.126 上的 IP 地址如下：

```
[root@hdp3~]#ip a
1: lo: <LOOPBACK,UP,LOWER_UP> mtu 65536 qdisc noqueue state UNKNOWN
   link/loopback 00:00:00:00:00:00 brd 00:00:00:00:00:00
   inet 127.0.0.1/8 scope host lo
      valid lft forever preferred_lft forever
   inet 172.16.1.100/32 scope global lo:0
      valid_lft forever preferred_lft forever
   inet6 ::1/128 scope host
```

```
        valid_lft forever preferred_lft forever
    2: ens32: <BROADCAST,MULTICAST,UP,LOWER_UP> mtu 1500 qdisc pfifo_fast state
UP qlen 1000
       link/ether 00:50:56:a5:0f:77 brd ff:ff:ff:ff:ff:ff
       inet 172.16.1.126/24 brd 172.16.1.255 scope global ens32
         valid_lft forever preferred_lft forever
       inet6 fe80::250:56ff:fea5:f77/64 scope link
         valid_lft forever preferred_lft forever
    [root@hdp3~]#
```

172.16.1.127 上的 IP 地址如下：

```
    [root@hdp4~]#ip a
    1: lo: <LOOPBACK,UP,LOWER_UP> mtu 65536 qdisc noqueue state UNKNOWN
       link/loopback 00:00:00:00:00:00 brd 00:00:00:00:00:00
       inet 127.0.0.1/8 scope host lo
         valid lft forever preferred_lft forever
       inet 172.16.1.100/32 scope global lo:0
         valid_lft forever preferred_lft forever
       inet6 ::1/128 scope host
         valid_lft forever preferred_lft forever
    2: ens160: <BROADCAST,MULTICAST,UP,LOWER_UP> mtu 1500 qdisc pfifo_fast state
UP qlen 1000
       link/ether 00:50:56:a5:49:7f brd ff:ff:ff:ff:ff:ff
       inet 172.16.1.127/24 brd 172.16.1.255 scope global ens160
         valid_lft forever preferred_lft forever
       inet6 fe80::250:56ff:fea5:497f/64 scope link
         valid_lft forever preferred_lft forever
    [root@hdp4~]#
```

步骤 07 启动 Keepalived。

在 172.16.1.124 和 172.16.1.125 上执行以下命令：

```
    /etc/init.d/keepalived start
```

命令执行后主服务器上的 VIP 如下所示：

```
    [root@hdp1~/keepalived-1.2.13]#ip a
    1: lo: <LOOPBACK,UP,LOWER_UP> mtu 65536 qdisc noqueue state UNKNOWN
       link/loopback 00:00:00:00:00:00 brd 00:00:00:00:00:00
       inet 127.0.0.1/8 scope host lo
         valid lft forever preferred_lft forever
       inet6 ::1/128 scope host
         valid_lft forever preferred_lft forever
    2: ens160: <BROADCAST,MULTICAST,UP,LOWER_UP> mtu 1500 qdisc pfifo_fast state
UP qlen 1000
       link/ether 00:50:56:a5:13:84 brd ff:ff:ff:ff:ff:ff
       inet 172.16.1.124/24 brd 172.16.1.255 scope global ens160
         valid_lft forever preferred_lft forever
       inet 172.16.1.100/32 scope global ens160
         valid_lft forever preferred_lft forever
```

```
    inet6 fe80::250:56ff:fea5:1384/64 scope link
       valid_lft forever preferred_lft forever
[root@hdp1~/keepalived-1.2.13]#
```

从服务器上的 VIP 如下所示：

```
[root@hdp2~]#ip a
1: lo: <LOOPBACK,UP,LOWER_UP> mtu 65536 qdisc noqueue state UNKNOWN
    link/loopback 00:00:00:00:00:00 brd 00:00:00:00:00:00
    inet 127.0.0.1/8 scope host lo
       valid lft forever preferred_lft forever
    inet6 ::1/128 scope host
       valid_lft forever preferred_lft forever
2: ens32: <BROADCAST,MULTICAST,UP,LOWER_UP> mtu 1500 qdisc pfifo_fast state
UP qlen 1000
    link/ether 00:50:56:a5:7a:4e brd ff:ff:ff:ff:ff:ff
    inet 172.16.1.125/24 brd 172.16.1.255 scope global ens32
       valid_lft forever preferred_lft forever
    inet6 fe80::250:56ff:fea5:7a4e/64 scope link
       valid_lft forever preferred_lft forever
[root@hdp2~]#
```

可以看到 VIP 已经成功绑定到 172.16.1.124。

此时查看 LVS 集群状态，可以看到集群下有两个 realserver（真实服务器），以及调度算法，权重等信息。ActiveConn 代表当前 realserver 的活跃连接数。

```
[root@hdp1~]#ipvsadm -Ln
IP Virtual Server version 1.2.1 (size=4096)
Prot LocalAddress:Port Scheduler Flags
  -> RemoteAddress:Port           Forward Weight ActiveConn InActConn
TCP  172.16.1.100:3306 rr
  -> 172.16.1.126:3306            Route   1      0          0
  -> 172.16.1.127:3306            Route   1      0          0
[root@hdp1~]#
```

11.4.3 功能测试

步骤 01 验证 LVS 负载均衡转发策略。

客户端使用 VIP 连接数据库，并查看所连接 MySQL 实例的 server_id。可以看到，每次执行依次连接到 172.16.1.126 和 172.16.1.127 的 MySQL，证明是轮询策略产生的结果。

```
C:\>mysql -uwxy -p123456 -h172.16.1.100 -N -s -e "show variables like
'server_id'"
mysql: [Warning] Using a password on the command line interface can be insecure.
server_id      126

C:\>mysql -uwxy -p123456 -h172.16.1.100 -N -s -e "show variables like
'server_id'"
mysql: [Warning] Using a password on the command line interface can be insecure.
```

```
    server_id       127

    C:\>mysql -uwxy -p123456 -h172.16.1.100 -N -s -e "show variables like
'server_id'"
    mysql: [Warning] Using a password on the command line interface can be insecure.
    server_id       126

    C:\>mysql -uwxy -p123456 -h172.16.1.100 -N -s -e "show variables like
'server_id'"
    mysql: [Warning] Using a password on the command line interface can be insecure.
    server_id       127
```

步骤02 模拟 LVS 的主服务器失效。

在 172.16.1.124 上执行以下命令停止主服务器上的 Keepalived 服务：

```
    /etc/init.d/keepalived stop
```

再次查看 172.16.1.124 绑定的 VIP 如下：

```
    [root@hdp1~]#ip a
    1: lo: <LOOPBACK,UP,LOWER_UP> mtu 65536 qdisc noqueue state UNKNOWN
        link/loopback 00:00:00:00:00:00 brd 00:00:00:00:00:00
        inet 127.0.0.1/8 scope host lo
           valid_lft forever preferred_lft forever
        inet6 ::1/128 scope host
           valid_lft forever preferred_lft forever
    2: ens160: <BROADCAST,MULTICAST,UP,LOWER_UP> mtu 1500 qdisc pfifo_fast state
UP qlen 1000
        link/ether 00:50:56:a5:13:84 brd ff:ff:ff:ff:ff:ff
        inet 172.16.1.124/24 brd 172.16.1.255 scope global ens160
           valid_lft forever preferred_lft forever
        inet6 fe80::250:56ff:fea5:1384/64 scope link
           valid_lft forever preferred_lft forever
    [root@hdp1~]#
```

172.16.1.125 绑定的 VIP 如下：

```
    [root@hdp2~]#ip a
    1: lo: <LOOPBACK,UP,LOWER_UP> mtu 65536 qdisc noqueue state UNKNOWN
        link/loopback 00:00:00:00:00:00 brd 00:00:00:00:00:00
        inet 127.0.0.1/8 scope host lo
           valid_lft forever preferred_lft forever
        inet6 ::1/128 scope host
           valid_lft forever preferred_lft forever
    2: ens32: <BROADCAST,MULTICAST,UP,LOWER_UP> mtu 1500 qdisc pfifo_fast state
UP qlen 1000
        link/ether 00:50:56:a5:7a:4e brd ff:ff:ff:ff:ff:ff
        inet 172.16.1.125/24 brd 172.16.1.255 scope global ens32
           valid_lft forever preferred_lft forever
        inet 172.16.1.100/32 scope global ens32
           valid_lft forever preferred_lft forever
```

```
    inet6 fe80::250:56ff:fea5:7a4e/64 scope link
       valid_lft forever preferred_lft forever
[root@hdp2~]#
```

可以看到 VIP 已经漂移到 172.16.1.125 上，它成为了新的主服务器。 此时连接 MySQL、负载均衡都不受影响。

```
C:\>mysql -uwxy -p123456 -h172.16.1.100 -N -s -e "show variables like 'server_id'"
mysql: [Warning] Using a password on the command line interface can be insecure.
server_id       127

C:\>mysql -uwxy -p123456 -h172.16.1.100 -N -s -e "show variables like 'server_id'"
mysql: [Warning] Using a password on the command line interface can be insecure.
server_id       126

C:\>mysql -uwxy -p123456 -h172.16.1.100 -N -s -e "show variables like 'server_id'"
mysql: [Warning] Using a password on the command line interface can be insecure.
server_id       127

C:\>mysql -uwxy -p123456 -h172.16.1.100 -N -s -e "show variables like 'server_id'"
mysql: [Warning] Using a password on the command line interface can be insecure.
server_id       126
```

若再次启动 172.16.1.124 上 Keepalived 服务，可以看到它已经变为从服务器，而且并不会去抢占成为主服务器，这是由 nopreempt 参数决定的。

步骤 03 模拟 mysqld 失效。

在 172.16.1.126 上执行以下命令：

```
pkill -9 mysqld
```

LVS 检测到了 172.16.1.126 上的 MySQL 服务器宕机，集群自动剔除故障节点。此时集群中只有一个 realserver 的地址，即 172.16.1.127:3306。

```
[root@hdp1~]#ipvsadm -Ln
IP Virtual Server version 1.2.1 (size=4096)
Prot LocalAddress:Port Scheduler Flags
  -> RemoteAddress:Port           Forward Weight ActiveConn InActConn
TCP  172.16.1.100:3306 rr
  -> 172.16.1.127:3306            Route   1      0          0
[root@hdp1~]#
```

连接 MySQL 时可以看到应用不受影响，但只连接到一台 MySQL 服务器。

```
C:\>mysql -uwxy -p123456 -h172.16.1.100 -N -s -e "show variables like 'server_id'"
mysql: [Warning] Using a password on the command line interface can be insecure.
```

```
server_id       127
```

重新启动172.16.1.126上的MySQL后，LVS自动将已修复节点加入集群：

```
[root@hdp1~]#ipvsadm -Ln
IP Virtual Server version 1.2.1 (size=4096)
Prot LocalAddress:Port Scheduler Flags
  -> RemoteAddress:Port           Forward Weight ActiveConn InActConn
TCP  172.16.1.100:3306 rr
  -> 172.16.1.126:3306            Route   1      0          0
  -> 172.16.1.127:3306            Route   1      0          0
[root@hdp1~]#
```

11.5 小 结

Keepalived + LVS组合是常用的高可用负载均衡解决方案。对MySQL而言，可以有几种不同的部署方式：只使用Keepalived保证MySQL主从高可用；使用Keepalived保证LVS高可用，而LVS提供MySQL的读写负载均衡和高可用；将Director服务器（LVS）和真实服务器（MySQL）配置为同一主机，以提供数据库读写分离和高可用性等。后两种我们各用了一个示例进行了详细说明。下一章将介绍另一种通用的高可用架构——Heartbeat + HAProxy。

第12章 Heartbeat + HAProxy

上一章我们使用 Keepalived 和 LVS 实现了 MySQL 高可用与负载均衡，本章将介绍如何利用 Heartbeat 和 HAProxy 来完成相同的功能。

12.1 Heartbeat 简介

Heartbeat 是一个专业的、功能完善的高可用软件，提供了心跳检查、资源接管、检测集群中的服务、在集群节点转移 VIP 所有者等基本功能。Heartbeat 的功能与 Keepalived 有很多相似之处，但对于实际的业务应用还是有区别的，例如 Keepalived 主要是控制 IP 的漂移，而 Heartbeat 不但可以控制 IP 漂移，更擅长对资源服务的操控。相对于 Keepalived，Heartbeat 功能更强大，配套工具更全，适用于大型集群的管理。同时，Heartbeat 的部署和使用相对麻烦一些，尤其是 Heartbeat 2.1.4 后拆分成三个子项目，安装、配置和维护都比较复杂。通常建议 LVS 高可用使用 Keepalived，而业务高可用使用 Heartbeat。

12.1.1 Heartbeat 组成模块

Heartbeat 有 1.x、2.x 和 3.x 三个大版本，1.x 和 2.x 的组成结构比较简单，所有模块都集中在 heartbeat 中。到了 3.x 版本后，整个 Heartbeat 项目进行了拆分，每个子项目分别进行开发。Heartbeat 1.x 与 2.x 包含以下组件：

- heartbeat：节点间通信检测模块。
- ha-logd：集群事件日志服务。
- CCM（Consensus Cluster Membership）：集群成员一致性管理模块。
- LRM （Local Resource Manager）：本地资源管理模块。

- Stonith Daemon：节点监控进程。
- CRM（Cluster Resource Management）：集群资源管理模块。
- Cluster Policy Engine：集群策略引擎。
- Cluster Transition Engine：集群转移引擎，也叫策略执行引擎。

Heartbeat 1.x 与 Heartbeat 2.x 的主要区别在于，1.x 中的集群资源管理器是 haresource，2.x 为向后兼容保留了 haresource，但同时又新增了一个功能更强大的 CRM 资源管理器。图 12-1 显示了 Heartbeat 2.0 的内部组成模块。

CCM 用于管理集群节点成员，同时管理成员之间的关系和节点间资源的分配。heartbeat 模块负责检测节点的运行状态，以决定节点是否失效。ha-logd 模块用于记录集群中所有模块和服务的运行信息。

LRM 负责本地资源的启动、停止和监控，一般由 LRM 守护进程 lrmd 和节点监控进程 Stonith Daemon 组成。lrmd 守护进程负责节点间的通信，Stonith Daemon 是一个 Fence 设备，主要用于监控节点状态。当一个节点出现问题时处于正常状态的节点会通过 Fence 设备将其重启或关机以释放 IP、磁盘等资源，始终保持资源被一个节点拥有，防止资源争用。

CRM 用于处理节点和资源之间的依赖关系。同时管理节点对资源的使用，一般由 CRM 守护进程 crmd、集群策略引擎和集群转移引擎三个部分组成。集群策略引擎（Cluster Policy Engine）具体实施这些管理和依赖，集群转移引擎（Cluster Transition Engine）监控 CRM 模块的状态，当一个节点出现故障时，负责协调另一个节点上的进程进行合理的资源接管。

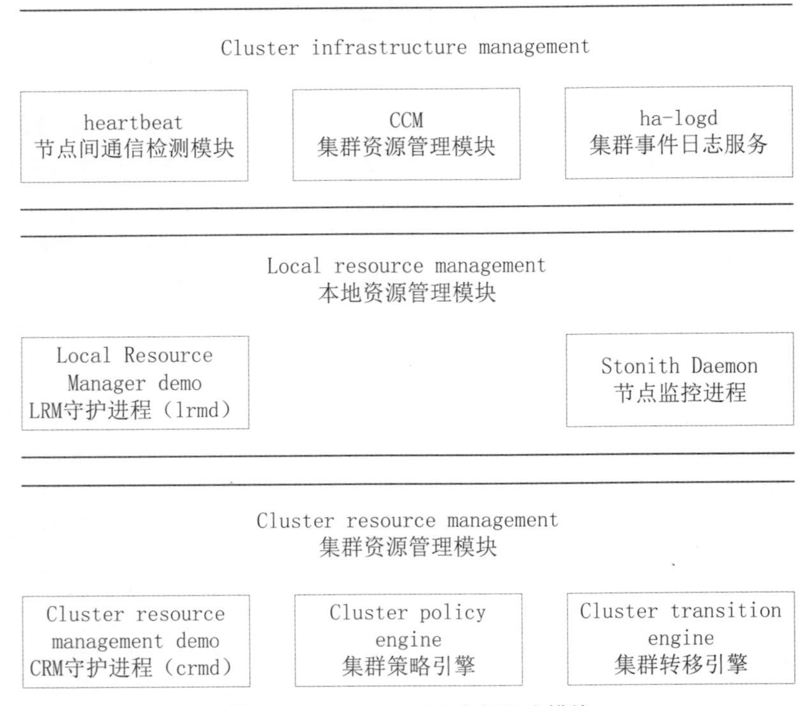

图 12-1 Heartbeat 2.0 内部组成模块

Heartbeat 3.x 按模块把原来 2.x 拆分为 heartbeat、cluster-glue 和 resource-agent 三部分。Heartbeat

部分是整个集群的基础，负责维护集群各节点的信息以及它们之间的通信。cluster-glue 部分相当于一个中间层，可以将 heartbeat 和 CRM 联系起来，主要包含 LRM 和 STONITH 两部分。resource-agent 部分是一个脚本集合，这些脚本将被 LRM 调用从而实现各种资源的启动、停止和监控。

原来的 GUI 工具 CRM 也被拆分，由另一个独立项目 Pacemaker 负责。Pacemaker 提供了多种用户接口：

- crm shell 基于字符的管理方式。
- 一个使用 Ajax Web 配置方式的 Web Konsole 窗口。
- hb_gui 图形配置工具，这也是 2.x 的默认 GUI 配置工具。

Heartbeat 包含了 ipfail、Stonith 和 Ldirector 几个插件，监控资源和应用程序是否运行正常。ipfail 主要用于检测网络故障，并做出合理的反应。为了实现这个功能，ipfail 使用 ping 节点或者 ping 节点组来检测网络连接是否出现故障，从而及时实施故障转移措施。Stonith 插件可以在一个没有响应的节点恢复后，合理接管集群服务资源，防止数据冲突，保证共享存储环境中的数据完整性。Ldirector 是一个监控集群服务节点运行状态的插件，常用于负载均衡集群中。Ldirector 如果监控到集群节点中某个服务出现故障，就屏蔽此节点的对外连接功能，同时将后续请求转移到正常的节点提供服务。

12.1.2 Heartbeat 工作原理

通过修改 Heartbeat 的配置文件，可以指定一台 Heartbeat 服务器作为主服务器（Master），另一台自动成为备份服务器（Backup）。在备份服务器上面配置 Heartbeat 守护程序，监听来自主服务器的心跳信息。如果备份服务器在规定时间内无法监听到来自主服务器的心跳，那么就启动故障转移程序，并取得主服务器上相关资源服务的所有权，接替主服务器继续不间断地提供服务，从而达到资源以及服务高可用的目的。

以上描述的是主备模式，Heartbeat 还支持主主模式，即两台服务器互为“主备”，它们之间会相互发送报文来告诉对方自己当前的状态。如果在指定的时间内未收到对方发送的心跳报文，那么一方就会认为对方失效或者宕机。这时运行正常的主机就会启动自身的资源接管模块来接管运行在对方主机上的资源或者服务，继续为用户提供服务。在通常情况下，这种方式可以较好地实现在一台主机出现故障后，业务仍能够不间断地持续运行。注意，所谓的业务不间断，在故障转移期间还是需要切换时间的，如停止数据库及存储服务等。Heartbeat 的主备模式高可用的切换时间一般在 5~20 秒左右。

另外，和 Keepalived 一样，Heartbeat 高可用是服务器级别的，不是服务级别的。它可以监测以下系统问题：

- 服务器宕机
- Heartbeat 软件故障
- 心跳连接线故障

服务故障不会导致主备切换（类似之前所说的主从切换），但可以通过服务故障触发停掉 Heartbeat 服务，从而主机实现自动切换。至少需要两台 Heartbeat 主机才可以实现高可用服务。两台 Heartbeat 服务器之间可以通过以下方式进行通信：

- 串口直连，是首选的连接方式，缺点是主机间距离不能太远。
- 一根以太网线直连两块网卡，这是常用的连接方式。
- 以太网线通过交换机等网络设备连接。

Heartbeat 通过 IP 地址接管以及通过 ARP 广播进行故障转移。在主服务器故障时，备用节点接管资源后，会立即强制更新所有客户端本地的 ARP（Address Resolution Protocol，地址解析协议）表，即清除客户端本地缓存的故障服务器的 VIP 和 MAC 地址解析记录，确保客户端与新主服务器的对话。

图 12-2 为 Heartbeat 集群的一般拓扑结构示意图。在实际应用中，由于节点数目、网络结构和磁盘类型配置的差异，拓扑结构可能会有所不同。

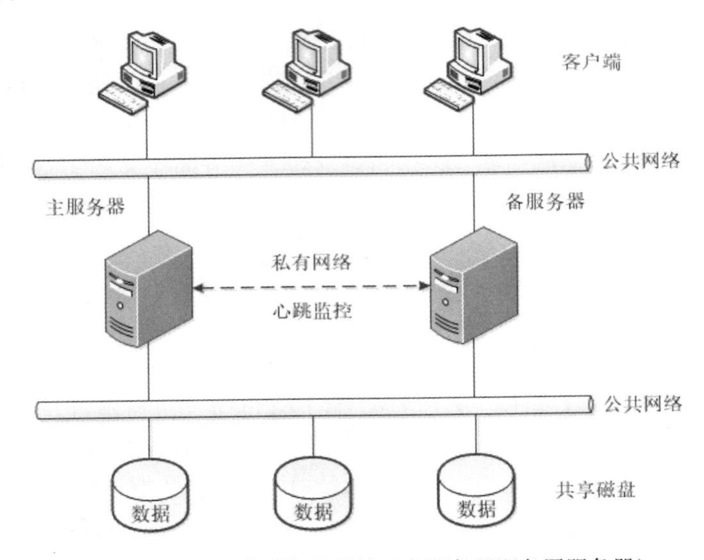

图 12-2　Heartbeat 集群拓扑结构（备服务器即备用服务器）

12.2　HAProxy 简介

HAProxy 是一个开源的高性能反向代理，或者说是负载均衡软件，支持双机热备、虚拟主机、基于 TCP 和 HTTP 应用代理等功能。相对于 LVS 来说，HAProxy 的配置较为简单，而且拥有很好的对服务器节点的健康检查功能（相当于 Keepalived 的健康检查）。当代理的后端服务器出现故障时，HAProxy 会自动将该故障服务器剔除。在服务器故障恢复后，HAProxy 还会自动重新把服务器主机加回来。

HAProxy 实现了一种事件驱动的单一进程模型，此模型支持非常大的并发连接数。根据官方测试数据，HAProxy 单位时间处理的最大请求数为 20000 个，可以同时维护 40000~50000 个并发连接，最大的数据处理能力为 10Gbps。多进程或多线程模型受内存限制、系统调度器限制以及无处不在的锁限制，很少能处理数千的并发连接。事件驱动模型因为在有更多的资源和更好的时间管理的用户空间（User-Space）中可以完成所有任务，所以没有这些问题。此模型的弊端是，在多核系统上通常扩展性较差，这就是为什么必须对其进行优化以使每个 CPU 时间片（Cycle，也称为

CPU 周期）可以完成更多的工作。

HAProxy 支持四层和七层两种的代理模式。在四层模式下，HAProxy 仅在客户端和服务器之间转发双向流量。七层模式下，HAProxy 会分析协议，并且能通过允许、拒绝、交换、增加、修改或者在请求（Request）或回应（Response）里指定内容来控制协议，这种操作要基于特定规则。新版 HAProxy 引入了前端（Frontend）和后端（Backend）功能，前端根据任意 HTTP 请求报头的内容进行 ACL 规则匹配，然后把请求定向到相关的后端（等待前端请求的服务器组）。通过前端和后端可以更容易地实现 HAProxy 的七层代理功能。

HAProxy 特别适用于那些负载很大的 Web 站点，这些站点通常需要会话保持或进行七层处理。HAProxy 的运行模式使得它可以简单安全地整合进当前的架构中，同时可以保护 Web 服务器不被暴露到外网上。

下面我们用两个示例来详细说明 Heartbeat 和 HAProxy 在 MySQL 高可用架构中的应用，它们与上一章的两个示例功能一样。操作系统版本和 MySQL 服务器版本也与上一章的示例相同，分别为 Linux release 7.2.1511 和 MySQL5.6.14。利用 Heartbeat 和 HAProxy 可以对后端的 MySQL 服务进行故障转移和负载均衡，不过在后端的 MySQL 从服务器的数量超过 10 台时性能不如 LVS，因此推荐 Keepalived + LVS。在用于生产环境时，还是要根据实际需求进行测试，以确定最佳的 MySQL 高可用方案。

12.3 实现 MySQL 主从复制的读写分离及高可用

第一个示例是实现 MySQL 主从复制中的读写分离及自动故障转移，具体架构如图 12-3 所示。

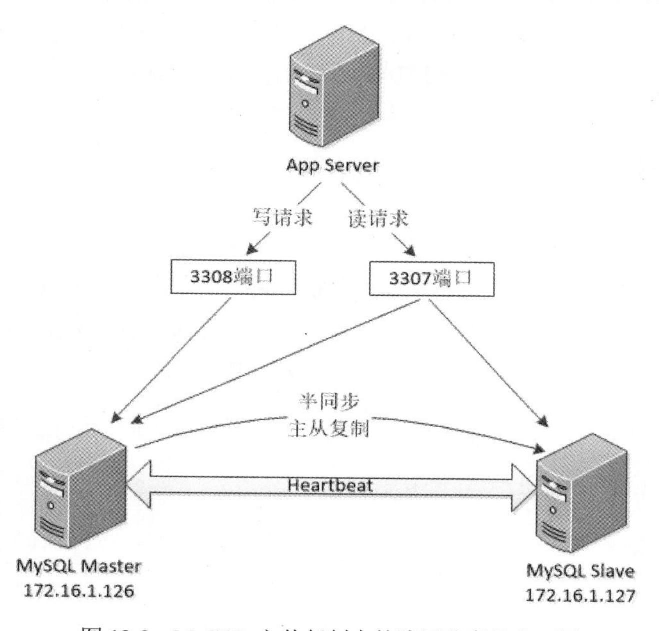

图 12-3 MySQL 主从复制中的读写分离及高可用

12.3.1 架构设计

图 12-3 中，使用两台主机实现 MySQL 主从复制，实现读写分离，用于提高查询性能。采用 MySQL 5.6.x 的半同步实现数据复制和同步。Heartbeat 在这里主要用作主机健康状态检查以及实现 HAProxy 两台主机之间的自动故障转移。任何一台主机宕机都不会影响对外提供的服务（VIP 可以漂移），保持 MySQL 数据库服务的高可用性。

Heartbeat 是使用心跳进行通信和通过选举方式来实现的高可用解决方案，这种方案可以避免单点故障。在这个解决方案中，通常至少有两台服务器运行 Heartbeat，一台为主服务器（Master）；另一台为备份服务器（Backup，也称为从服务器），但对外表现为一个或一组 VIP（虚拟 IP）。主服务器会发送特定消息给备份服务器，当备份服务器收不到该消息时，则认为主服务器出现了故障，备份服务器会接管 VIP，继续提供服务，从而保证了高可用性。HAProxy 在本例的作用是提供读负载均衡。整体架构的设计原理和异常处理可描述如下：

（1）主机 A 和 B，分别配置为 MySQL 半同步复制的主库和从库。A 和 B 都启动 Heartbeat 服务，但只有 A 启动 HAProxy 服务。

（2）通过 Heartbeat 启用一个 VIP，实现 A 和 B 的自动切换。

（3）在 HAProxy 中配置两对前端/后端（Frontend/Backend），使用不同的端口，一个端口用于响应写请求；另一个端口用于响应读请求，实现读写分离。注意，当前端与后端是同一台物理主机时，就像本例的情况，前端不能绑定与后端相同的端口，否则会报错。

（4）初始时，主机 A 和 B 都存在。A 上的 HAProxy 通过写端口将写请求转发至主机 A，通过读端口将读请求转发给主机 A 和 B，实现读负载均衡。

（5）当主机 A 异常时，主机 B 接管服务。此时将完成三项工作：VIP 漂移到主机 B 上；重置主机 B 的 MySQL 角色，使之切换为主库；启动主机 B 上的 HAProxy，继续接收应用的请求。

（6）当主机 B 异常时，主机 A 上的 HAProxy 会将主机 B 剔除，其他不变。

12.3.2 安装和配置

示例环境各主机的 IP 地址及其角色如下：

- 172.16.1.126：Heartbeat + HAProxy + MySQL Semisync-Replication Master。
- 172.16.1.127：Heartbeat + HAProxy + MySQL Semisync-Replication Slave。
- 172.16.1.100：VIP，3307 为读访问端口，3308 为写访问端口。

与上一章的 LVS 示例不同，这里只使用了一个 VIP，通过不同的端口区分读写请求。

步骤 01 安装和配置 MySQL 半同步复制。

按角色安装和配置 MySQL 半同步复制，172.16.1.126 为主库，172.16.1.127 为从库。具体安装和配置过程参见 2.5 节。

步骤 02 安装和配置 HAProxy。

以 root 用户在 172.16.1.126 和 172.16.1.127 两台主机上执行以下操作：

（1）从以下地址下载 HAProxy 源码

https://www.haproxy.org/download/1.8/src/haproxy-1.8.12.tar.gz

（2）创建 HAProxy 运行账户和组

```
groupadd haproxy # 添加 haproxy 组
useradd -g haproxy haproxy -s /bin/false # 创建账户 haproxy 并加入到 haproxy 组
```

（3）解压

```
tar zxvf haproxy-1.8.12.tar.gz
```

（4）编译与安装

```
cd haproxy-1.8.12
make TARGET=linux3100 CPU=x86_64 PREFIX=/usr/local/haproxy
make install PREFIX=/usr/local/haproxy
```

参数说明：

- TARGET：使用 uname -r 查看内核，如 2.6.18-371.el5，此时该参数就为 linux26；kernel 等于 2.6.28，则 TARGET=linux2628。示例环境的内核版本为 3.10.0-327.el7.x86_64，因此 TARGET=linux3100。
- CPU：使用 uname -r 查看系统信息，如为 x86_64 GNU/Linux，此时该参数就为 x86_64。
- PREFIX：指定 HAProxy 安装的路径。

（5）创建目录与文件

```
mkdir -p /usr/local/haproxy/conf # 创建配置文件目录
mkdir -p /etc/haproxy  # 创建配置文件目录
touch /usr/local/haproxy/conf/haproxy.cfg # 创建配置文件
ln -s /usr/local/haproxy/conf/haproxy.cfg /etc/haproxy/haproxy.cfg # 添加配置文件软链接
cp -r /root/haproxy-1.8.12/examples/errorfiles /usr/local/haproxy/errorfiles
# 拷贝错误页面
ln -s /usr/local/haproxy/errorfiles /etc/haproxy/errorfiles    # 添加软链接
mkdir -p /usr/local/haproxy/log                  # 创建日志文件目录
touch /usr/local/haproxy/log/haproxy.log    # 创建日志文件
ln -s /usr/local/haproxy/log/haproxy.log /var/log/haproxy.log # 添加软链接
cp /root/haproxy-1.8.12/examples/haproxy.init /etc/rc.d/init.d/haproxy
# 拷贝开机启动文件
chmod +x /etc/rc.d/init.d/haproxy                # 添加脚本执行权限
chkconfig haproxy on                             # 设置开机启动
ln -s /usr/local/haproxy/sbin/haproxy /usr/sbin # 添加软链接
```

（6）配置 HAProxy

172.16.1.126 上的/usr/local/haproxy/conf/haproxy.cfg 配置文件内容如下：

```
global
    log       127.0.0.1 local2              # 日志定义级别
```

```
    chroot      /usr/local/haproxy          # 当前工作目录
    pidfile     /var/run/haproxy.pid        # 进程 id 文件
    maxconn     4000                        # 最大连接数
    user       haproxy                      # 运行该程序的用户
    group      haproxy                      # 运行该程序的组
    daemon                                  # 以后台方式运行
    stats socket /usr/local/haproxy/stats       # socket 文件所在目录

defaults
    mode      tcp             # haproxy 运行模式（http | tcp | health）
    log       global          # 采用全局定义的日志
    option     dontlognull    # 不记录健康检查的日志信息
    option     redispatch     # 服务器挂掉后，将其上的请求重新分发到其他健康的服务器
    retries    3              # 三次连接失败则服务器不可用
    timeout http-request   10s  # http 请求超时
    timeout queue        1m     # 排队超时
    timeout connect       10s   # 连接超时
    timeout client       1m     # 客户端超时
    timeout server       1m     # 服务器超时
    timeout http-keep-alive 10s     # keep-alive 连接超时
    timeout check        10s        # 心跳检测
    maxconn             600         # 最大连接数

listen stats                        # 配置 HAProxy 状态页（用来查看的页面）
    mode http
    bind :8888
    stats enable
    stats hide-version                          # 隐藏 HAProxy 版本号
    stats uri      /haproxyadmin?stats          # 状态页 uri
    stats realm    Haproxy\ Statistics          # 输入账户和密码时的提示文字
    stats auth     admin:admin                  # 用户名:密码

frontend  read
     bind *:3307
    # 监听前端端口，表示任何 IP 访问 3307 端口都会将数据轮番转发到 MySQL 服务器群组中
    default_backend  mysql_read                 # 后端服务器组名

backend mysql_read
    balance        roundrobin                   # 使用轮询方式调度
    server mysql1 172.16.1.126:3306 check port 3306 maxconn 300
    server mysql2 172.16.1.127:3306 check port 3306 maxconn 300

frontend  write
    bind *:3308
    # 监听前端端口，表示任何 IP 访问 3308 端口都会将数据轮番转发到 MySQL 服务器群组中
    default_backend  mysql_write                # 后端服务器组名

backend mysql_write
    server mysql1 172.16.1.126:3306 check port 3306 maxconn 300
```

HAProxy 配置文件中分成以下五个部分：

- global：设置全局配置参数，属于进程的配置，通常是和操作系统相关。
- defaults：配置默认参数，可以被 frontend、backend、listen 段继承使用。
- frontend：接收请求的前端虚拟节点，可以根据规则直接指定具体使用后端的 backend。
- backend：后端服务集群的配置，一个 backend 对应一个或多个实体服务器。
- listen：frontend 和 backend 的组合体，关联这两部分。

如前所述，这里配置了两对 frontend 和 backend。read 绑定 3307 端口来接收读请求，其对应的 backend 为 mysql_read，其中定义了两台 MySQL 服务器，使用轮询策略实现读负载均衡。write 绑定 3308 端口来接收写请求，它对应的 backend 为 mysql_write，其中只定义了 MySQL 主库，即由它来接收写请求。

将该配置文件最后一行的 172.16.1.126 替换为 172.16.1.127，就是 172.16.1.127 上 /usr/local/haproxy/conf/haproxy.cfg 配置文件的内容。当服务切换到备份服务器上时，接收 MySQL 读写请求的只有 172.16.1.127 一台主机。

（7）启动日志

编辑/etc/rsyslog.conf 文件，去掉以下两行的注释，并在其后添加一行：

```
# Provides TCP syslog reception
$ModLoad imtcp                        # 去掉注释
$InputTCPServerRun 514                # 去掉注释
local2.*    /var/log/haproxy.log      # 添加此行
```

步骤 03 编译安装 Heartbeat。

对 Heartbeat 的源代码进行编译和安装，需要依次安装 Cluster Glue、Resource Agents 和 Heartbeat 三个软件，并且三个软件要安装在相同的目录下。在 172.16.1.126 和 172.16.1.127 两台机器上以 root 用户执行以下操作。

（1）下载软件包

分别从以下地址下载 Heartbeat 3.0.6、Cluster Glue 1.0.12 和 Resource Agents 3.9.6：

http://hg.linux-ha.org/heartbeat-STABLE_3_0/archive/958e11be8686.tar.bz2

http://hg.linux-ha.org/glue/archive/0a7add1d9996.tar.bz2

https://github.com/ClusterLabs/resource-agents/archive/v3.9.6.tar.gz

（2）初始配置

- 禁用防火墙

在/etc/selinux/config 文件中设置 SELINUX=disabled，并执行以下命令：

```
systemctl disable firewalld
```

- 设置主机名

编辑/etc/hosts 文件，它的内容如下：

```
127.0.0.1   localhost localhost.localdomain localhost4
localhost4.localdomain4
::1         localhost localhost.localdomain localhost6 localhost6.localdomain6

172.16.1.126 hdp3
172.16.1.127 hdp4
```

设置主机名：

```
# 在172.16.1.126上
hostnamectl set-hostname hdp3
# 在172.16.1.127上
hostnamectl set-hostname hdp4
```

- 安装依赖包

```
yum -y install glib2-devel libtool-ltdl-devl net-snmp-devel bzip2-devel
ncurses-devel openssl-devel libtool libxml2 libxml2-devel gettext bison flex
zlib-devel mailx which libxslt docbook-dtds docbook-style-xsl PyXML shadow-utils
opensp autoconf automake bzip2 e2fsprogs-devel libxslt-devel libtool-ltdl-devel
make asciidoc libuuid-devel
```

- 建立用户

```
groupadd haclient
useradd -g haclient -M -s /sbin/nologin hacluster
```

（3）安装 Cluster Glue 1.0.12

```
bunzip2 0a7add1d9996.tar.bz2
tar -xf 0a7add1d9996.tar
cd Reusable-Cluster-Components-glue--0a7add1d9996
./autogen.sh
./configure --prefix=/usr/local/heartbeat && make && make install
```

（4）安装 Resource Agents 3.9.6

```
bunzip2 0a7add1d9996.tar.bz2
tar -xf 0a7add1d9996.tar
cd Reusable-Cluster-Components-glue--0a7add1d9996
./autogen.sh
./configure --prefix=/usr/local/heartbeat && make && make install
```

（5）安装 Heartbeat 3.0.6

```
tar -zxf resource-agents-3.9.6.tar.gz
cd resource-agents-3.9.6
./autogen.sh
export CFLAGS="$CFLAGS -I/usr/local/heartbeat/include
-L/usr/local/heartbeat/lib"
./configure --prefix=/usr/local/heartbeat/
ln -s  /usr/local/heartbeat/lib/* /lib/
ln -s  /usr/local/heartbeat/lib/* /lib64/
make && make install
```

如果在编译过程中出现以下错误：

```
    /usr/local/heartbeat/include/heartbeat/glue_config.h:105:0: error:
"HA_HBCONF_DIR" redefined [-Werror]
```

说明在配置文件中，宏 HA_HBCONF_DIR 被定义了多次。删除掉/usr/local/heartbeat/include/heartbeat/glue_config.h 文件中最后一行定义宏的代码，然后再进行编译即可通过：

```
    define HA_HBCONF_DIR "/usr/local/heartbeat/etc/ha.d/"
```

步骤04 配置 Heartbeat。

（1）创建配置文件

Heartbeat 需要 ha.cf、authkeys 和 haresources 三个配置文件，文件所在目录为/usr/local/heartbeat/etc/ha.d/。

```
    cd /usr/local/heartbeat/share/doc/heartbeat
    cp -a ha.cf authkeys haresources /usr/local/heartbeat/etc/ha.d/
    cd /usr/local/heartbeat/etc/ha.d
    chmod 600 etc/ha.d/authkeys
    ln -svf /usr/local/heartbeat/lib64/heartbeat/plugins/RAExec/*
/usr/local/heartbeat/lib/heartbeat/plugins/RAExec/
    ln -svf /usr/local/heartbeat/lib64/heartbeat/plugins/*
/usr/local/heartbeat/lib/heartbeat/plugins/
```

（2）编辑配置文件 ha.cf

172.16.1.126 上的/usr/local/heartbeat/etc/ha.d/ha.cf 文件内容如下：

```
    # 保存调试信息文件
    debugfile /var/log/ha-debug

    # 日志文件
    logfile /var/log/ha-log

    # 表示使用系统日志
    logfacility local0

    # 心跳的时间间隔，单位为秒
    keepalive 1

    # 超出该时间间隔未收到对方节点的心跳，则判定对方死亡
    deadtime 30

    # 超出该时间间隔未收到对方节点的心跳，则发出警告记录到日志中
    warntime 10

    # 在某些系统上，系统启动或重启之后需要经过一段时间网络才能正常工作
    # 该选项用于解决这种情况产生的时间间隔，取值至少为 deadtime 的 2 倍
    initdead 120

    # 设置广播通信使用的端口，694 为默认使用的端口号
```

```
udpport 694

# 传播心跳广播的网卡信息
bcast ens32

# 设置对方机器心跳检测的 IP 地址
ucast ens32 172.16.1.127

# 在该选项设为 on 的情况下，一旦主节点恢复运行，则自动获取资源并取代备份节点
auto_failback off

# 配置主备节点信息
node hdp3
node hdp4

# 如果 ping 不通该地址，就认为当前断网，需要转移 VIP
ping 172.16.1.254

# 指定与 heartbeat 一同启动和关闭的进程，该进程被自动监视，遇到故障则重新启动
# 最常用的进程是 ipfail，该进程用于检测和处理网络故障，
# 需要配合 ping 语句指定的 ping node 来检测网络连接。若系统是 64bit，则要注意该文件的路径
respawn hacluster /usr/local/heartbeat/libexec/heartbeat/ipfail

# 指定用户和组
apiauth ipfail gid=haclient uid=hacluster
```

172.16.1.127 上的/usr/local/heartbeat/etc/ha.d/ha.cf 文件除了以下两个配置项的信息外，其他和172.16.1.126 的相同：

```
# 传播心跳的广播网卡信息
bcast ens160

# 设置对方机器心跳检测的网卡和 IP 地址
ucast ens160 172.16.1.126
```

（3）编辑认证文件 authkeys

172.16.1.126 与 172.16.1.127 上的/usr/local/heartbeat/etc/ha.d/authkeys 文件相同，内容如下：

```
auth 2
2 sha1 HI!
```

（4）编辑资源配置文件 haresources

所有 Heartbeat 节点上的/usr/local/heartbeat/etc/ha.d/haresources 文件必须完全一致，本例中只有一行：

```
hdp3 172.16.1.100 mysql
```

hdp3 是主节点，172.16.1.100 是 VIP，mysql 是 haresources 文件所在目录 resource.d 目录下的一个自定义脚本文件名，即/usr/local/heartbeat/etc/ha.d/resource.d/mysql 文件，其内容为如下两行：

```
/home/mysql/remove_slave.sh
/etc/init.d/haproxy restart
```

当 Heartbeat 主机获得资源时，将自动把 MySQL 主从复制中的从库置为主库，这是通过调用 /home/mysql/remove_slave.sh 文件完成的。之后重启 HAProxy 服务，接收应用请求。

步骤05 创建相关脚本文件。

以下两个脚本文件在 172.16.1.126 与 172.16.1.127 上相同。

（1）remove_slave.sh

/home/mysql/remove_slave.sh 脚本的作用是重置 MySQL 复制中的角色，将从库切换为主库，文件内容如下：

```
#!/bin/bash
. /home/mysql/.bashrc

user=root
password=123456
log=/home/mysql/remove_slave.log

echo "'date'" >> $log

rm -rf /tmp/kill.sql
mysql -u$user -p$password -e "select * into outfile '/tmp/kill.sql' from (select
concat('kill ',id,';') from information_schema.processlist where command='sleep'
union all select 'set global read_only=OFF;'
union all select 'stop slave;'
union all select 'reset slave all;') t;"

mysql -u$user -p$password < /tmp/kill.sql >> $log

/bin/sed -i 's#read-only#\#read-only#' /home/mysql/mysql-5.6.14/my.cnf
```

（2）mysql_check.sh

/home/mysql/mysql_check.sh 脚本用于检查本机 MySQL 的服务器状态，文件内容如下：

```
#!/bin/bash
. /home/mysql/.bashrc

count=1

while true
do

mysql -uroot -p123456 -S /data/mysql.sock -e "show status;" > /dev/null 2>&1
i=$?
ps aux | grep mysqld | grep -v grep > /dev/null 2>&1
j=$?
if [ $i = 0 ] && [ $j = 0 ]
then
   sleep 3
else
```

```
        if [ $i = 1 ] && [ $j = 0 ]
        then
            sleep 3
        else
            if [ $count -gt 5 ]
            then
                break
            fi
        let count++
        continue
        fi
    fi

done
/etc/init.d/heartbeat stop
```

mysql_check.sh 脚本每三秒检测一次 MySQL 状态，如果本机 MySQL 不可用，则关闭 Heartbeat 服务，使 VIP 漂移到对方主机。先将该文件放到 crontab 中自动执行，待启动后再将其从 crontab 中去掉即可。

步骤 06 启动 Heartbeat 和 HAProxy。

（1）启动两个主机的 Heartbeat 服务

在我们这个场景中，一定要注意两个主机 Heartbeat 服务的启动顺序，要先启动 172.16.1.127，再启动 172.16.1.126。如果反过来先启动 172.16.1.126，则再启动 172.16.1.127 时，会执行本地的 mysql 脚本（当 172.16.1.126 获得 VIP 资源时）。这时将会调用 remove_slave.sh，重置 172.16.1.127 在 MySQL 主从复制中从库（slave）的角色，以前的复制就失效了，那就必须重建复制。先后在 172.16.1.127 和 172.16.1.126 上执行以下命令：

```
chkconfig heartbeat on
systemctl start heartbeat
```

（2）在 172.16.1.126 上启动 HAProxy

```
systemctl start haproxy
```

启动完成后，可以看到 VIP 绑定在 172.16.1.126 上：

```
[root@hdp3~]#ip a
1: lo: <LOOPBACK,UP,LOWER_UP> mtu 65536 qdisc noqueue state UNKNOWN
    link/loopback 00:00:00:00:00:00 brd 00:00:00:00:00:00
    inet 127.0.0.1/8 scope host lo
       valid_lft forever preferred_lft forever
    inet6 ::1/128 scope host
       valid_lft forever preferred_lft forever
2: ens32: <BROADCAST,MULTICAST,UP,LOWER_UP> mtu 1500 qdisc pfifo_fast state
UP qlen 1000
    link/ether 00:50:56:a5:0f:77 brd ff:ff:ff:ff:ff:ff
    inet 172.16.1.126/24 brd 172.16.1.255 scope global ens32
       valid_lft forever preferred_lft forever
```

```
    inet 172.16.1.100/24 brd 172.16.1.255 scope global secondary ens32:1
       valid_lft forever preferred_lft forever
    inet6 fe80::250:56ff:fea5:f77/64 scope link
       valid_lft forever preferred_lft forever
[root@hdp3~]#
```

在 172.16.1.126 上有 heartbeat、haproxy 和 mysql_check 相关进程：

```
[root@hdp3~]#ps -ef | grep heartbeat | grep -v grep
root     608426      1  0 Jul26 ?        00:01:55 heartbeat: master control
process
root     608429 608426  0 Jul26 ?        00:00:03 heartbeat: FIFO reader
root     608430 608426  0 Jul26 ?        00:00:10 heartbeat: write: bcast ens32
root     608431 608426  0 Jul26 ?        00:00:10 heartbeat: read: bcast ens32
root     608432 608426  0 Jul26 ?        00:00:08 heartbeat: write: ucast ens32
root     608433 608426  0 Jul26 ?        00:00:14 heartbeat: read: ucast ens32
root     608434 608426  0 Jul26 ?        00:00:17 heartbeat: write: ping
172.16.1.254
root     608435 608426  0 Jul26 ?        00:00:07 heartbeat: read: ping
172.16.1.254
haclust+   608455 608426  0 Jul26 ?        00:00:04
/usr/local/heartbeat/libexec/heartbeat/ipfail
[root@hdp3~]#ps -ef | grep haproxy | grep -v grep
haproxy   608854      1  0 Jul26 ?        00:00:19 /usr/sbin/haproxy -D -f
/etc/haproxy/haproxy.cfg -p /var/run/haproxy.pid
[root@hdp3~]#ps -ef | grep mysql_check | grep -v grep
root     575823 575822  0 Jul26 ?        00:01:24 /bin/bash
/home/mysql/mysql_check.sh
[root@hdp3~]#
```

在 172.16.1.127 上有 heartbeat 和 mysql_check 相关进程：

```
[root@hdp4/usr/local/haproxy/conf]#ps -ef | grep heartbeat | grep -v grep
root     612316      1  0 Jul26 ?        00:01:37 heartbeat: master control
process
root     612319 612316  0 Jul26 ?        00:00:03 heartbeat: FIFO reader
root     612320 612316  0 Jul26 ?        00:00:12 heartbeat: write: bcast ens160
root     612321 612316  0 Jul26 ?        00:00:07 heartbeat: read: bcast ens160
root     612322 612316  0 Jul26 ?        00:00:10 heartbeat: write: ucast ens160
root     612323 612316  0 Jul26 ?        00:00:13 heartbeat: read: ucast ens160
root     612324 612316  0 Jul26 ?        00:00:16 heartbeat: write: ping
172.16.1.254
root     612325 612316  0 Jul26 ?        00:00:06 heartbeat: read: ping
172.16.1.254
haclust+   612475 612316  0 Jul26 ?        00:00:04
/usr/local/heartbeat/libexec/heartbeat/ipfail
[root@hdp4/usr/local/haproxy/conf]#ps -ef | grep haproxy | grep -v grep
[root@hdp4/usr/local/haproxy/conf]#ps -ef | grep mysql_check | grep -v grep
root     290127 290126  0 Jul25 ?        00:04:21 /bin/bash
/home/mysql/mysql_check.sh
[root@hdp4/usr/local/haproxy/conf]#
```

至此，示例的安装和配置工作已经完成，下面进行功能测试。

12.3.3 功能测试

步骤 01 验证 3307 端口的读负载均衡转发策略。

```
C:\>mysql -utest -p123456 -P3307 -h172.16.1.100 -N -s -e "show variables like 'server_id'"
mysql: [Warning] Using a password on the command line interface can be insecure.
server_id       126

C:\>mysql -utest -p123456 -P3307 -h172.16.1.100 -N -s -e "show variables like 'server_id'"
mysql: [Warning] Using a password on the command line interface can be insecure.
server_id       127

C:\>mysql -utest -p123456 -P3307 -h172.16.1.100 -N -s -e "show variables like 'server_id'"
mysql: [Warning] Using a password on the command line interface can be insecure.
server_id       126

C:\>mysql -utest -p123456 -P3307 -h172.16.1.100 -N -s -e "show variables like 'server_id'"
mysql: [Warning] Using a password on the command line interface can be insecure.
server_id       127
```

每次请求轮流连接至两个主从 MySQL 实例。

步骤 02 验证 3308 端口的读负载均衡转发策略。

```
C:\>mysql -utest -p123456 -P3308 -h172.16.1.100 -N -s -e "show variables like 'server_id'"
mysql: [Warning] Using a password on the command line interface can be insecure.
server_id       126

C:\>mysql -utest -p123456 -P3308 -h172.16.1.100 -N -s -e "show variables like 'server_id'"
mysql: [Warning] Using a password on the command line interface can be insecure.
server_id       126
```

所有请求都只连接 ID 为 126 服务器上的一个 MySQL 实例，服务器 ID 为 126 其实就是 IP 地址为 172.16.1.126，后同。

步骤 03 模拟从库的 mysqld crash。

在 172.16.1.127 上执行以下命令：

```
pkill -9 mysqld
```

步骤 04 再次使用两个端口连接，数据库服务正常。

```
C:\>mysql -utest -p123456 -P3307 -h172.16.1.100 -N -s -e "show variables like 'server_id'"
mysql: [Warning] Using a password on the command line interface can be insecure.
server_id       126

C:\>mysql -utest -p123456 -P3308 -h172.16.1.100 -N -s -e "show variables like 'server_id'"
mysql: [Warning] Using a password on the command line interface can be insecure.
server_id       126
```

步骤 05 重新启动从库的 MySQL 服务。

在 172.16.1.127 上执行以下命令：

```
service mysql start
```

步骤 06 再次验证 3307 端口的读负载均衡转发策略。

```
C:\>mysql -utest -p123456 -P3307 -h172.16.1.100 -N -s -e "show variables like 'server_id'"
mysql: [Warning] Using a password on the command line interface can be insecure.
server_id       126

C:\>mysql -utest -p123456 -P3307 -h172.16.1.100 -N -s -e "show variables like 'server_id'"
mysql: [Warning] Using a password on the command line interface can be insecure.
server_id       127

C:\>mysql -utest -p123456 -P3307 -h172.16.1.100 -N -s -e "show variables like 'server_id'"
mysql: [Warning] Using a password on the command line interface can be insecure.
server_id       126

C:\>mysql -utest -p123456 -P3307 -h172.16.1.100 -N -s -e "show variables like 'server_id'"
mysql: [Warning] Using a password on the command line interface can be insecure.
server_id       127
```

ID 为 127 服务器上的 MySQL 实例被自动添加回集群中，读负载均衡随之恢复。

步骤 07 模拟主库的 mysqld 失效。

在 172.16.1.126 上执行以下命令：

```
pkill -9 mysqld
```

步骤 08 再次验证两个端口的读负载均衡转发策略。

```
C:\>mysql -utest -p123456 -P3307 -h172.16.1.100 -N -s -e "show variables like 'server_id'"
mysql: [Warning] Using a password on the command line interface can be insecure.
server_id       127
```

```
    C:\>mysql -utest -p123456 -P3308 -h172.16.1.100 -N -s -e "show variables like
'server_id'"
    mysql: [Warning] Using a password on the command line interface can be insecure.
    server_id       127
```

主库宕机导致发生自动故障转移，但读写应用都不受影响。

此时查看 172.16.1.127 上绑定的 IP 如下：

```
    [root@hdp4/usr/local/haproxy/conf]#ip a
    1: lo: <LOOPBACK,UP,LOWER_UP> mtu 65536 qdisc noqueue state UNKNOWN
        link/loopback 00:00:00:00:00:00 brd 00:00:00:00:00:00
        inet 127.0.0.1/8 scope host lo
           valid_lft forever preferred_lft forever
        inet6 ::1/128 scope host
           valid_lft forever preferred_lft forever
    2: ens160: <BROADCAST,MULTICAST,UP,LOWER_UP> mtu 1500 qdisc mq state UP qlen
1000
        link/ether 00:50:56:a5:49:7f brd ff:ff:ff:ff:ff:ff
        inet 172.16.1.127/24 brd 172.16.1.255 scope global ens160
           valid_lft forever preferred_lft forever
        inet 172.16.1.100/24 brd 172.16.1.255 scope global secondary ens160:0
           valid_lft forever preferred_lft forever
        inet6 fe80::250:56ff:fea5:497f/64 scope link
           valid_lft forever preferred_lft forever
    [root@hdp4/usr/local/haproxy/conf]#
```

可以看到 VIP 已经漂移到 172.16.1.127 上。

步骤 09 验证切换后的新主库可读写。

```
    C:\>mysql -utest -p123456 -P3308 -h172.16.1.100 -e "use test; create table t1
(a int); insert into t1 values (1),(2),(3); commit; select * from t1;"
    mysql: [Warning] Using a password on the command line interface can be insecure.
    +-------+
    | a     |
    +-------+
    |    1 |
    |    2 |
    |    3 |
    +-------+
```

查看 172.16.1.127 上 MySQL 的从库状态（slave status），已经为空：

```
    [mysql@hdp4~]$mysql -u root -p123456 -e "show slave status\G"
    Warning: Using a password on the command line interface can be insecure.
    [mysql@hdp4~]$
```

MySQL 配置文件的 read-only 选项也已经被注释掉：

```
    [mysql@hdp4~]$more /home/mysql/mysql-5.6.14/my.cnf | grep read-only
    #read-only
    [mysql@hdp4~]$
```

12.4 实现MySQL双主复制的读写负载均衡及高可用

12.4.1 架构设计

上一个示例用一个 VIP 的两个端口 3307 和 3308 分别响应 MySQL 读写请求。利用 HAProxy 轮询算法将读请求轮流转发至 MySQL 主从库两个实例，使得读负载均衡，写请求只转发至主库，实现读写分离。初始时 HAProxy 只在 Heartbeat 主服务器上启动。当定期执行的自定义检查脚本 mysql_check 发现主服务器的 MySQL 出现问题时，就停止其上的 Heartbeat 服务，从而造成心跳丢失。此时 Heartbeat 从库接管 VIP 资源，同时调用 remove_slave 将 MySQL 从库提升为主库，并启动备份服务器上的 HAProxy 服务，接收应用请求。整个故障转移过程对应用是透明的，实现了 MySQL 数据库服务的高可用。

在本示例中，我们利用 Heartbeat 和 HAProxy，实现了 MySQL 双主复制的读写负载均衡与高可用，整体架构如图 12-4 所示。

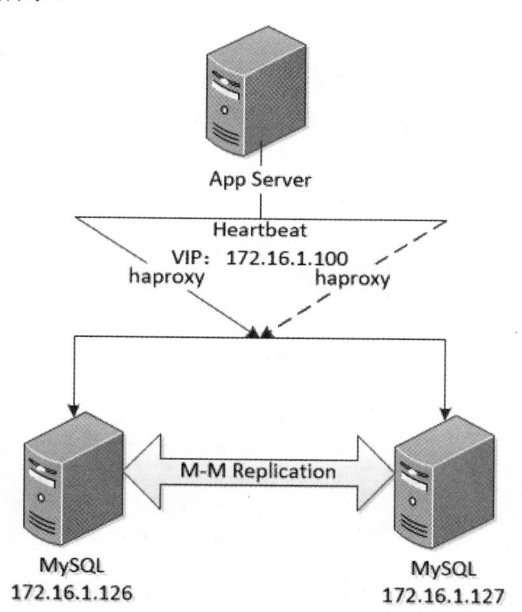

图 12-4　MySQL 双主复制的读写负载均衡及高可用

与上一个示例对比，本架构主要有以下不同点：

- 只使用一个端口 6603 响应全部 MySQL 读写请求。
- MySQL 配置双主复制。在两个等价的 MySQL 服务器之间实现读写负载均衡，即两个 MySQL 实例均可读写，所有客户端请求在两个 MySQL 实例间轮询。
- 配置文件 haresources 中的脚本改变。因为不需要 MySQL 主从切换，也就不再需要执行 remove_slave 脚本。

与上一章中的第二个示例一样，实际应用时，如本例的配置也需要考虑故障转移后，单台MySQL服务器的过载问题。

12.4.2 安装和配置

示例环境各主机的IP地址及其角色如下：

- 172.16.1.126: Heartbeat + HAProxy + MySQL Replication Master
- 172.16.1.127: Heartbeat + HAProxy + MySQL Replication Master
- 172.16.1.100: VIP

注意本示例中的所有Heartbeat、HAProxy相关配置文件以及自定义脚本文件，在两个主机上完全相同。

步骤01 安装和配置MySQL双主复制。

按角色安装和配置MySQL双主复制，具体过程参见9.2.2小节。

步骤02 安装和配置HAProxy。

HAProxy的安装和配置过程参见上一示例的12.3.2小节。配置文件/usr/local/HAProxy/conf/HAProxy.cfg的内容如下：

```
global
    log         127.0.0.1 local2               # 日志定义级别
    chroot      /usr/local/haproxy             # 当前工作目录
    pidfile     /var/run/haproxy.pid           # 进程 id 文件
    maxconn     4000                           # 最大连接数
    user        haproxy                        # 运行该程序的用户
    group       haproxy                        # 运行该程序的组
    daemon                                     # 以后台形式运行
    stats socket /usr/local/haproxy/stats        # socket 文件所在目录

defaults
    mode        tcp                  # HAProxy 运行模式（http | tcp | health）
    log         global               # 采用全局定义的日志
    option      dontlognull          # 不记录健康检查的日志信息
    option      redispatch   # 服务器挂掉后，将其上的请求重新分发到其他健康的服务器
    retries     3            # 三次连接失败则服务器不用
    timeout     http-request    10s      # http 请求超时
    timeout     queue           1m       # 排队超时
    timeout     connect         10s      # 连接超时
    timeout     client          1m       # 客户端超时
    timeout     server          1m       # 服务器超时
    timeout     http-keep-alive 10s      # keep-alive 连接超时
    timeout     check           10s      # 心跳检测
    maxconn     600  # 最大连接数
```

```
listen stats                                          # 配置 HAProxy 状态页（用来查看的页面）
    mode http
    bind :8888
    stats enable
    stats hide-version                                # 隐藏 HAProxy 版本号
    stats uri      /haproxyadmin?stats                # 状态页 uri
    stats realm    Haproxy\ Statistics                # 输入账户和密码时的提示文字
    stats auth     admin:admin                        # 用户名:密码

frontend  main
    bind 0.0.0.0:6603
    # 监听前端端口，表示任何 IP 访问 6603 端口都会将数据轮番转发到 mysql 服务器群组中
    default_backend  mysql                            # 后端服务器组名

backend mysql
    balance        roundrobin                         # 使用轮询方式调度
    server mysql1 172.16.1.126:3306 check port 3306 maxconn 300
    server mysql2 172.16.1.127:3306 check port 3306 maxconn 300
```

这里只定义了一对 frontend 和 backend。前端 main 通过 6603 端口接收所有读写请求，并转发至后端 MySQL 服务器。backend 名为 mysql，其中定义了两个 MySQL 数据库实例，使用轮询策略实现读写负载均衡。

步骤 03 编译安装 Heartbeat。

Heartbeat 编译安装过程参见上一示例在 12.3.2 小节中的说明。

步骤 04 配置 Heartbeat。

Heartbeat 配置过程参见上一示例在 12.3.2 小节中的说明。这里需要注意的是 haresources 的设置。上一示例中，我们建立了一个名为 mysql 的脚本，它调用 remove_slave.sh。当 Heartbeat 主机获得资源时，将自动把 MySQL 主从复制中的从库切换成主库。这次我们用/usr/local/heartbeat/etc/ha.d/haresources 来启动 HAProxy，文件内容只有如下一行：

```
hdp3 172.16.1.100 haproxy
```

而/usr/local/heartbeat/etc/ha.d/resource.d/haproxy 文件中也只有如下一行：

```
/etc/init.d/haproxy restart
```

资源只有 VIP，我们用这种配置保证 HAProxy 的高可用性。当初始启动 Heartbeat 后，VIP 绑定在 172.16.1.126 上。

步骤 05 创建 MySQL 服务检测脚本。

这里保留自定义 mysql_check.sh 脚本用于检查本机 MySQL 服务器的状态。文件内容与部署参见 12.3.2 小节中的示例说明。

步骤 06 启动 Heartbeat 和 HAProxy。

（1）启动两个主机的 Heartbeat 服务

```
chkconfig heartbeat on
systemctl start heartbeat
```

因为两个主机完全对等，所以 Heartbeat 在两主机的启动顺序无关紧要。

（2）在 172.16.1.126 上启动 HAProxy

```
systemctl start haproxy
```

启动完成后，VIP 绑定在 172.16.1.126 上。

12.4.3 功能测试

步骤 01 验证 HAProxy 的负载均衡轮询策略。

用客户端连接 VIP，并多次执行 MySQL 查询，我们可以看到查询请求依次被发送到两个主机上执行。

```
C:\>mysql -utest -p123456 -P6603 -h172.16.1.100 -N -s -e "show variables like 'server_id'"
mysql: [Warning] Using a password on the command line interface can be insecure.
server_id       126

C:\>mysql -utest -p123456 -P6603 -h172.16.1.100 -N -s -e "show variables like 'server_id'"
mysql: [Warning] Using a password on the command line interface can be insecure.
server_id       127

C:\>mysql -utest -p123456 -P6603 -h172.16.1.100 -N -s -e "show variables like 'server_id'"
mysql: [Warning] Using a password on the command line interface can be insecure.
server_id       126

C:\>mysql -utest -p123456 -P6603 -h172.16.1.100 -N -s -e "show variables like 'server_id'"
mysql: [Warning] Using a password on the command line interface can be insecure.
server_id       127
```

步骤 02 验证 MySQL 的高可用性。

在 172.16.1.126 上杀掉 MySQL 进程：

```
pkill -9 mysqld
```

用客户端连接 VIP，并多次执行 MySQL 查询，可以看到查询请求都被发送到 172.16.1.127。一台 MySQL 宕机不影响应用的正常使用，保证了 MySQL 服务的高可用性。

```
C:\>mysql -utest -p123456 -P6603 -h172.16.1.100 -N -s -e "show variables like 'server_id'"
mysql: [Warning] Using a password on the command line interface can be insecure.
```

```
    server_id       127

    C:\>mysql -utest -p123456 -P6603 -h172.16.1.100 -N -s -e "show variables like
'server_id'"
    mysql: [Warning] Using a password on the command line interface can be insecure.
    server_id       127
```

步骤03 验证宕机重新上线后自动添加到 HAProxy 中。

在 172.16.1.126 再次启动 MySQL，并重启 HAProxy 服务：

```
    service mysql start
    systemctl restart haproxy
```

用客户端连接 VIP，并多次执行 MySQL 查询，我们可以看到查询请求依次被发送到两个主机上执行。

```
    C:\>mysql -utest -p123456 -P6603 -h172.16.1.100 -N -s -e "show variables
like 'server_id'"
    mysql: [Warning] Using a password on the command line interface can be insecure.
    server_id       127

    C:\>mysql -utest -p123456 -P6603 -h172.16.1.100 -N -s -e "show variables like
'server_id'"
    mysql: [Warning] Using a password on the command line interface can be insecure.
    server_id       126

    C:\>mysql -utest -p123456 -P6603 -h172.16.1.100 -N -s -e "show variables like
'server_id'"
    mysql: [Warning] Using a password on the command line interface can be insecure.
    server_id       127

    C:\>mysql -utest -p123456 -P6603 -h172.16.1.100 -N -s -e "show variables like
'server_id'"
    mysql: [Warning] Using a password on the command line interface can be insecure.
    server_id       126
```

步骤04 验证 HAProxy 的高可用性。

初始 VIP 在 IP 地址为 172.16.1.126 的主机上，这个 126 主机上的 HAProxy 启动，主机 172.16.1.127 的 HAProxy 停止，客户端轮询访问正常。现在停止 172.16.1.126 上的 MySQL 服务：

```
    pkill -9 mysqld
```

mysql_check 会检测到本机的 MySQL 服务失效，触发停止 Heartbeat 服务。VIP 漂移到 172.16.1.127，同时 172.16.1.127 上的 HAProxy 被启动，客户端访问不受影响：

```
    [root@hdp4~]#ip a
    1: lo: <LOOPBACK,UP,LOWER_UP> mtu 65536 qdisc noqueue state UNKNOWN
        link/loopback 00:00:00:00:00:00 brd 00:00:00:00:00:00
        inet 127.0.0.1/8 scope host lo
           valid_lft forever preferred_lft forever
```

```
    inet6 ::1/128 scope host
       valid_lft forever preferred_lft forever
2: ens160: <BROADCAST,MULTICAST,UP,LOWER_UP> mtu 1500 qdisc mq state UP qlen
1000
    link/ether 00:50:56:a5:49:7f brd ff:ff:ff:ff:ff:ff
    inet 172.16.1.127/24 brd 172.16.1.255 scope global ens160
       valid_lft forever preferred_lft forever
    inet 172.16.1.100/24 brd 172.16.1.255 scope global secondary ens160:0
       valid_lft forever preferred_lft forever
    inet6 fe80::250:56ff:fea5:497f/64 scope link
       valid_lft forever preferred_lft forever
[root@hdp4~]#
[root@hdp4~]#/etc/init.d/haproxy status
haproxy.service - SYSV: HA-Proxy is a TCP/HTTP reverse proxy which is
particularly suited for high availability environments.
   Loaded: loaded (/etc/rc.d/init.d/haproxy)
   Active: active (running) since Thu 2018-07-26 11:25:47 CST; 26s ago
     Docs: man:systemd-sysv-generator(8)
  Process: 532199 ExecStop=/etc/rc.d/init.d/haproxy stop (code=exited,
status=0/SUCCESS)
  Process: 532205 ExecStart=/etc/rc.d/init.d/haproxy start (code=exited,
status=0/SUCCESS)
 Main PID: 532210 (haproxy)
   CGroup: /system.slice/haproxy.service
           └─532210 /usr/sbin/haproxy -D -f /etc/haproxy/haproxy.cfg -p
/var/run/haproxy.pid

Jul 26 11:25:47 hdp4 systemd[1]: Starting SYSV: HA-Proxy is a TCP/HTTP reverse
proxy which is particularly suite...ts....
Jul 26 11:25:47 hdp4 haproxy[532205]: Starting haproxy: [  OK  ]
Jul 26 11:25:47 hdp4 systemd[1]: Started SYSV: HA-Proxy is a TCP/HTTP reverse
proxy which is particularly suited...ents..
Hint: Some lines were ellipsized, use -l to show in full.
[root@hdp4~]#
```

172.16.1.126 再次启动 Heartbeat，VIP 并没有不发生漂移，仍然停留在 172.16.1.127 上：

```
[root@hdp3~]#systemctl start heartbeat
[root@hdp3~]#
[root@hdp3~]#ip a
1: lo: <LOOPBACK,UP,LOWER_UP> mtu 65536 qdisc noqueue state UNKNOWN
    link/loopback 00:00:00:00:00:00 brd 00:00:00:00:00:00
    inet 127.0.0.1/8 scope host lo
       valid_lft forever preferred_lft forever
    inet6 ::1/128 scope host
       valid_lft forever preferred_lft forever
2: ens32: <BROADCAST,MULTICAST,UP,LOWER_UP> mtu 1500 qdisc pfifo_fast state
UP qlen 1000
    link/ether 00:50:56:a5:0f:77 brd ff:ff:ff:ff:ff:ff
    inet 172.16.1.126/24 brd 172.16.1.255 scope global ens32
       valid_lft forever preferred_lft forever
```

```
    inet6 fe80::250:56ff:fea5:f77/64 scope link
       valid_lft forever preferred_lft forever
[root@hdp3~]#
```

从以上测试可以看到，Heartbeat 实现了 HAProxy 的高可用，避免了 HAProxy 的单点故障，出现问题时可以自动切换到正常的节点。HAProxy 服务器起到了负载均衡的作用，将用户请求分发到多个后端。同时，一台 MySQL 故障并不会影响整个集群，因为 HAProxy 会检测后端的状态，并据此自动添加或删除集群中的 MySQL 服务。

还有一点需要注意的是，只采用本例的配置无法处理“脑裂”问题。例如，当心跳线闪断，备机获得 VIP，而此时主机除了与备机失去联系，本身并无任何问题，也绑定同一个 VIP。当心跳恢复正常时，就会出现 VIP 的冲突等问题。“脑裂”问题是一个比较复杂的话题，通常需要引入其他插件，并且“多管齐下”才能彻底解决，参见 15.4.2 小节“测试脑裂”部分的说明。

12.5 小 结

Heartbeat + HAProxy 是另一个常用的高可用负载均衡解决方案，它们提供的功能与 Keepalived + LVS 类似，但更适用于高并发的 Web 服务器。本章通过两个示例详细说明了 Heartbeat 和 HAProxy 的安装和配置，以及如何把它们作为第三方中间件，与 MySQL 复制技术结合在一起使用，实现 MySQL 数据库服务的高可用、读写分离与负载均衡。

第 13 章

InnoDB Cluster

MySQL 作为当今最流行的关系数据库系统，在其高可用特性上也在不断进行着尝试和改进。最初的 MySQL 版本只提供一种简单的主从异步复制，满足最基本的数据同步。为了提高复制性能，从单线程到组提交，再到多线程复制，基本解决了复制延迟问题。为了解决从库与主库的一致性读问题，新增了半同步复制，而为了提供自动故障转移功能，又增加了组复制。要做到真正的高可用，故障转移必须对应用透明，于是在组复制的基础上，又发展出了 InnoDB Cluster。本章说明 InnoDB Cluster 的相关概念、安装部署及管理维护。需要指出的一点是，在 InnoDB Cluster 出现前，实现 MySQL 数据库的高可用性，除了原生的复制功能，通常还需要借助第三方中间件，如 Keepalived、MHA 等等，正如书中第三部分所讲。

13.1 InnoDB Cluster 简介

13.1.1 整体架构

InnoDB Cluster 主要由 MySQL Shell、MySQL Router 和 MySQL 服务器集群组成，三者协同工作，共同为 MySQL 提供完整的高可用性解决方案。图 13-1 所示为 InnoDB Cluster 的整体架构。

InnoDB Cluster 以组复制为基础，集群中的每个 MySQL 服务器实例都是组复制的成员，提供了在 InnoDB Cluster 内复制数据的机制，并且具有内置的故障转移功能。MySQL Shell 在 InnoDB Cluster 中充当控制台角色，使用它包含的 AdminAPI，可以使安装、配置、管理和维护多个 MySQL 组复制实例的工作更加轻松。通过 AdminAPI 的几条交互指令就可以自动完成组复制配置。MySQL Router 可以根据集群部署信息自动生成配置，将客户端应用程序透明地连接到 MySQL 服务器实例。如果服务器实例意外故障，集群将自动重新配置。在默认的单主模式下，InnoDB Cluster 具有单个读写主服务器实例。多个辅助服务器实例是主服务器实例的副本。在主服务器出现故障时，辅助服务器会自动升级为主服务器。MySQL Router 可以检测到这种情况并将客户端应用程序自动转发到新的主服务器。

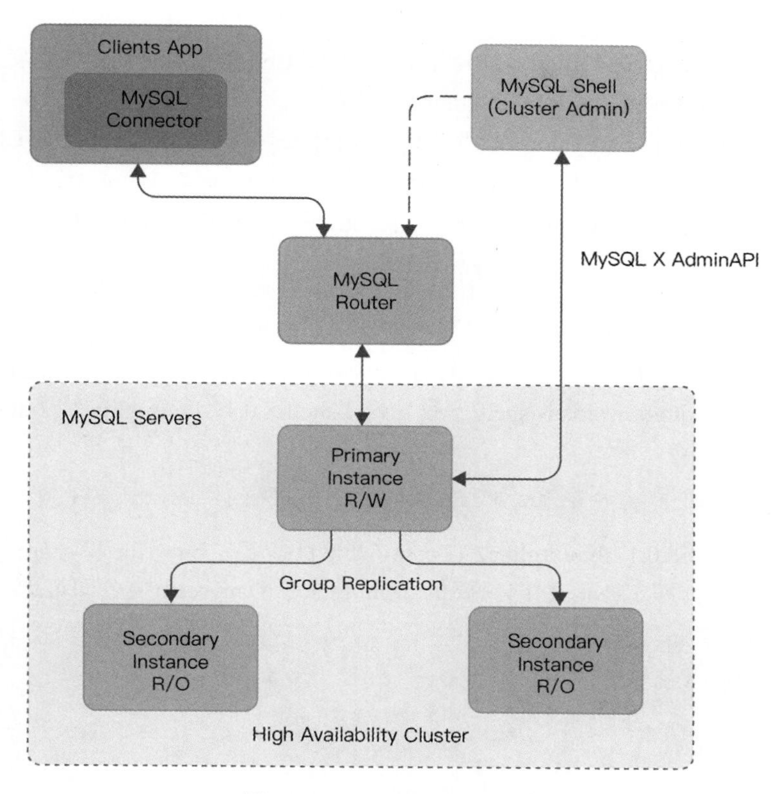

图 13-1 InnoDB Cluster 架构

13.1.2 MySQL Shell

MySQL Shell 是 MySQL Server 的高级客户端和代码编辑器。除了和 mysql 命令行客户端程序一样，执行常规的 SQL 语句外，MySQL Shell 还提供了 JavaScript 和 Python 脚本功能，并包含多个 API，其中的 AdminAPI 用于操作 InnoDB Cluster。

1. MySQL Shell 主要功能

MySQL Shell 中提供了以下主要功能：

- 支持 JavaScript、Python 和 SQL 三种语言，默认为 JavaScript，用\js、\py 和\sql 进行切换。根据当前活动的语言，任何输入的代码都将作为其中一种语言进行处理。
- 除了提供交互式代码执行模式之外，还可以从不同源获取代码并对其进行处理。例如用 mysqlsh 执行一个名为 code.js 的文件：

```
mysqlsh --file code.js
```

- MySQL Shell 包含用 JavaScript 和 Python 实现的 API，可以使用它们来开发与 MySQL 交互的

代码。例如，当 MySQL Shell 使用 X 协议连接到 MySQL 服务器时，X DevAPI 可以将 MySQL 用作文档数据库，提供与 MongoDB 类似的功能；使用 AdminAPI 能够管理 InnoDB Cluster 等。

- 可以使用 JavaScript 或 Python 创建对象，对 MySQL Shell 基本功能进行扩展。
- 能以表格、Tab 分割或 JSON 格式输出返回结果。
- 包含许多 JavaScript 和 Python 模式的内置全局对象：cluster 代表 InnoDB Cluster；dba 使用 AdminAPI 提供对 InnoDB Cluster 管理功能的访问；session 表示全局会话；db 代表默认数据库；shell 提供对 MySQL Shell 函数的访问；util 提供 MySQL Shell 实用程序。

2. 安装 MySQL Shell

MySQL Shell 可谓是“开箱即用”，只要根据操作系统平台选择下载安装包，然后将安装包解压即可直接使用，不需要任何其他额外配置过程。例如，要在 Linux 64 位平台上运行 MySQL Shell，可从 https://dev.mysql.com/downloads/shell/下载 mysql-shell-8.0.17-linux-glibc2.12-x86-64bit.tar.gz 文件，然后执行下面的命令：

```
tar -zxvf mysql-shell-8.0.17-linux-glibc2.12-x86-64bit.tar.gz
```

生成的 mysql-shell-8.0.17-linux-glibc2.12-x86-64bit 目录下有 bin、lib 和 share 三个子目录，用于分别存储可执行文件、库文件和文档。只要执行 bin 目录下的 mysqlsh 文件即可运行 MySQL Shell：

```
[mysql@hdp1~]$~/mysql-shell-8.0.17-linux-glibc2.12-x86-64bit/bin/mysqlsh
MySQL Shell 8.0.17

Copyright (c) 2016, 2019, Oracle and/or its affiliates. All rights reserved.
Oracle is a registered trademark of Oracle Corporation and/or its affiliates.
Other names may be trademarks of their respective owners.

Type '\help' or '\?' for help; '\quit' to exit.
 MySQL  JS >
```

MySQL Shell 包含的 AdminAPI，可通过 dba 全局变量及其相关方法进行访问。dba 对象的方法能够部署、配置和管理 InnoDB Cluster。执行\? dba 指令或 dba.help()函数可以查看 dba 相关的所有方法：

```
 MySQL  JS > \? dba
NAME
      dba - Global variable for InnoDB cluster management.

DESCRIPTION
      The global variable dba is used to access the AdminAPI functionality and
      perform DBA operations. It is used for managing MySQL InnoDB clusters.

PROPERTIES
      verbose
            Enables verbose mode on the dba operations.

FUNCTIONS
      checkInstanceConfiguration(instance[, options])
```

```
        Validates an instance for MySQL InnoDB Cluster usage.

    ...

    For more help on a specific function use: dba.help('<functionName>')

    e.g. dba.help('deploySandboxInstance')
 MySQL  JS >
```

使用常规格式object.help('methodname')可以获得特定方法的联机帮助，例如：

```
 MySQL  JS > dba.help('getCluster')
NAME
     getCluster - Retrieves a cluster from the Metadata Store.

SYNTAX
     dba.getCluster([name][, options])

WHERE
     name: Parameter to specify the name of the cluster to be returned.
     options: Dictionary with additional options.

RETURNS
      The cluster object identified by the given name or the default cluster.

DESCRIPTION
...

EXCEPTIONS
...

 MySQL  JS >
```

13.1.3 MySQL Router

在 InnoDB Cluster 架构中，MySQL Router 起到对应用程序透明路由的作用。当用--bootstrap 选项引导路由时，会根据 InnoDB Cluster 元数据自动生成或更新配置文件。这项功能专为结合 InnoDB Cluster 而设计，简化了对 MySQL Router 的配置维护工作。关于 MySQL Router 的功能、安装、配置和启动等说明，参见第 7 章相关内容。

13.1.4 MySQL 服务器集群

MySQL 服务器集群指组复制中的多个 MySQL 实例，是 InnoDB Cluster 得以实现的核心。关于组复制的概念原理、安装和配置等说明，参见第 6 章。

13.2 创建 InnoDB Cluster

下面使用四台虚拟机搭建一个 InnoDB Cluster，部署如图 13-2 所示。

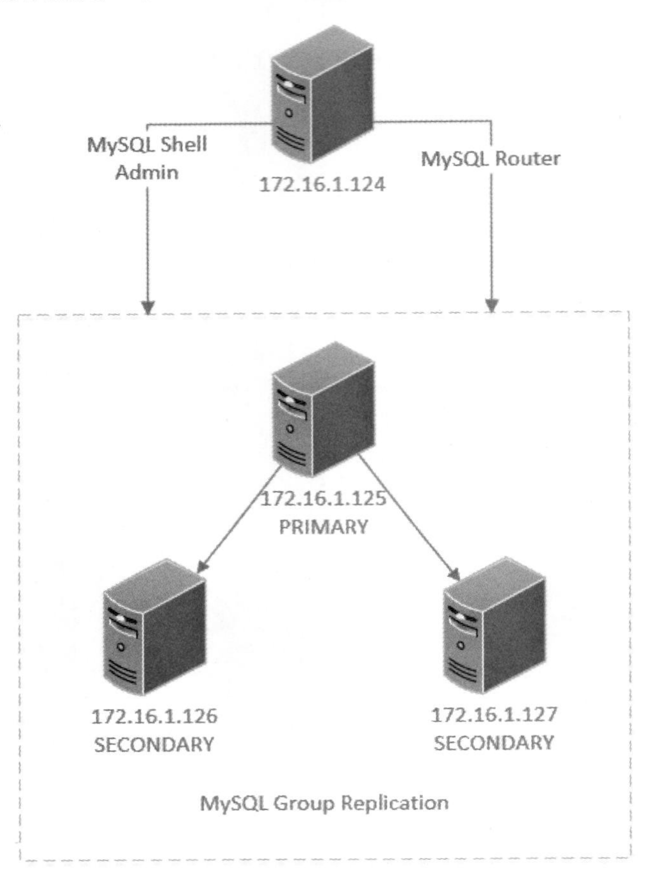

图 13-2　InnoDB Cluster 部署

主机信息如表 13-1 所示。

表 13-1　InnoDB Cluster 主机信息

主机名	IP:端口	角色或组件
hdp1	172.16.1.124	MySQL Shell、MySQL Router
hdp2	172.16.1.125:3306	组复制成员 PRIMARY
hdp3	172.16.1.126:3306	组复制成员 SECONDARY
hdp4	172.16.1.127:3306	组复制成员 SECONDARY

硬件配置为每台主机 CPU 4 核、内存 8GB、硬盘 100GB。软件版本如下：

- 操作系统：CentOS Linux release 7.2.1511 (Core) 64 位。
- MySQL Server：8.0.16（mysql-8.0.16-linux-glibc2.12-x86_64.tar.xz）。
- MySQL Shell：8.0.17（mysql-shell-8.0.17-linux-glibc2.12-x86-64bit.tar.gz）。
- MySQL Router：8.0.17（mysql-router-8.0.17-linux-glibc2.12-x86_64.tar.xz）。

MySQL Server、MySQL Shell 和 MySQL Router 都已经安装好。MySQL Server 和 MySQL Router 的安装分别参见 1.2.2 小节和 7.3.1 小节。

下面使用 MySQL Shell 创建 InnoDB Cluster，并配置 MySQL Router 为应用客户端提供对 InnoDB Cluster 实例的透明连接。在创建 InnoDB Cluster 之前，要确保使用的 MySQL 服务器实例满足以下要求：

- InnoDB Cluster 使用组复制，因此服务器实例必须满足组复制的要求。
- 组复制仅支持 InnoDB 存储引擎，其他如 MyISAM 表不能由组复制写入。
- 必须在 InnoDB Cluster 中的所有实例上启用性能模式（Performance Schema）。
- 所有实例必须具有唯一的 server_id。

MySQL Shell 用于配置 InnoDB Cluster 脚本，需要访问 Python 2.7 版。在 Linux 上必须将 Python 配置成为 Shell 环境的一部分。如果 Python 解释器能够启动，则无需进一步操作，否则可能需要在 /usr/bin/python 和 Python 二进制文件之间创建一个软链接。下面的操作系统命令用于检查是否正确配置了 Python：

```
$ /usr/bin/env python
```

13.2.1 检查实例配置

AdminAPI 提供 dba.checkInstanceConfiguration()方法以验证实例是否满足组复制要求。该方法不会更改实例上的任何配置，也不会检查实例上的任何数据。下面是检查 172.16.1.125:3306 的命令和部分输出的例子：

```
[mysql@hdp1~]$mysqlsh
MySQL Shell 8.0.17
...

 MySQL  JS > dba.checkInstanceConfiguration('wxy@172.16.1.125:3306')
Please provide the password for 'wxy@172.16.1.125:3306': ******
Save password for 'wxy@172.16.1.125:3306'? [Y]es/[N]o/Ne[v]er (default No):
Validating MySQL instance at 172.16.1.125:3306 for use in an InnoDB cluster...

This instance reports its own address as hdp2:3306
Clients and other cluster members will communicate with it through this address
by default. If this is not correct, the report_host MySQL system variable should
be changed.

Checking whether existing tables comply with Group Replication requirements...
No incompatible tables detected
```

```
Checking instance configuration...

NOTE: Some configuration options need to be fixed:
+-----------------+-------+------------+------------------------------------
-----------------+
| Variable        | Cur.. | Required.. | Note                                                  |
+-----------------+-------+------------+------------------------------------
-----------------+
| binlog_checksum | CRC32 | NONE       | Update the server variable            |
| enforce_gtid_.. | OFF   | ON       | Update read-only variable and restart
the server |
| gtid_mode       | OFF   | ON         | Update read-only variable and restart the
server |
| server_id       | 1     | <unique ID> | Update read-only variable and restart
the server |
+-----------------+-------+------------+-------------------------------+

Some variables need to be changed, but cannot be done dynamically on the server.
NOTE: Please use the dba.configureInstance() command to repair these issues.

{
    "config_errors": [
        {
            "action": "server_update",
            "current": "CRC32",
            "option": "binlog_checksum",
            "required": "NONE"
        },
        {
            "action": "server_update+restart",
            "current": "OFF",
            "option": "enforce_gtid_consistency",
            "required": "ON"
        },
        {
            "action": "server_update+restart",
            "current": "OFF",
            "option": "gtid_mode",
            "required": "ON"
        },
        {
            "action": "server_update+restart",
            "current": "1",
            "option": "server_id",
            "required": "<unique ID>"
        }
    ],
    "status": "error"
}
 MySQL  JS >
```

在表格中列出了四个检查项：binlog_checksum、enforce_gtid_consistency、gtid_mode 和 server_id 的当前值和所需值，以及修改它们要采取的措施。同时指出，使用 dba.configureInstance()来修复这些问题。

对集群中的每个服务器实例重复此过程：

```
dba.checkInstanceConfiguration('wxy@172.16.1.126:3306')
dba.checkInstanceConfiguration('wxy@172.16.1.127:3306')
```

运行 dba.checkInstanceConfiguration()之后生成的报告提供了有关所需的任何配置更改的信息，要正确配置这些属性后，才能继续创建集群。

13.2.2 配置实例

AdminAPI 提供了 dba.configureInstance()函数，用于检查实例是否针对 InnoDB Cluster 进行了适当配置，并在发现任何与 InnoDB Cluster 不兼容的设置时配置实例。如果实例不需要更改配置，dba.configureInstance()命令输出确认实例已准备好使用 InnoDB Cluster。根据 MySQL Shell 连接到实例的方式以及在实例上运行的 MySQL 版本，能够通过将这些更改自动保存到远程实例的配置文件（默认为 MySQL 实例数据目录下的 mysqld-auto.cnf 文件）来持久化更改，或者手动更改实例的配置文件。无论进行配置更改的方式如何，必须重新启动实例以确保服务器检测到配置更改。下面是配置 172.16.1.125:3306 的命令和部分输出的例子：

```
MySQL  JS > dba.configureInstance('wxy@172.16.1.125:3306')
Please provide the password for 'wxy@172.16.1.125:3306': ******
Save password for 'wxy@172.16.1.125:3306'? [Y]es/[N]o/Ne[v]er (default No):
Configuring MySQL instance at 172.16.1.125:3306 for use in an InnoDB cluster...

This instance reports its own address as hdp2:3306
Clients and other cluster members will communicate with it through this address
by default. If this is not correct, the report_host MySQL system variable should
be changed.

NOTE: Some configuration options need to be fixed:
+-----------------+-------+-------------+---------------------------------------
-----------------+
| Variable        | Cur.. | Required..  | Note                                  |
+-----------------+-------+-------------+---------------------------------------
-----------------+
| binlog_checksum | CRC32 | NONE        | Update the server variable            |
| enforce_gtid_.. | OFF   | ON          | Update read-only variable and restart the
server |
| gtid_mode       | OFF   | ON          | Update read-only variable and restart the
server |
| server_id       | 1     | <unique ID> | Update read-only variable and restart
the server |
+-----------------+-------+-------------+---------------------------------------
-----------------+
```

```
Some variables need to be changed, but cannot be done dynamically on the server.
Do you want to perform the required configuration changes? [y/n]: y
Do you want to restart the instance after configuring it? [y/n]: y
Configuring instance...
The instance '172.16.1.125:3306' was configured for InnoDB cluster usage.
Restarting MySQL...
NOTE: MySQL server at 172.16.1.125:3306 was restarted.
 MySQL  JS >
```

一旦针对实例发出 dba.configureInstance()，该命令首先检查实例的当前设置是否适合 InnoDB Cluster 使用，并将在一个报告中显示 InnoDB Cluster 所需的设置。这步与 dba.checkInstanceConfiguration 的作用类似。

当针对 MySQL Shell 当前运行的本地 MySQL 实例发出 dba.configureInstance()时，它会尝试自动配置实例。对远程实例发出 dba.configureInstance()时，如果实例支持自动保持配置更改，则可以选择执行此操作，否则必须在本地配置实例。满足以下要求的实例会自动支持持久配置更改：

- 该实例正在运行 MySQL 8.0.11 或更高版本。
- persisted_globals_load 系统变量设置为 ON（默认值）。

对于动态服务器变量，在 dba.configureInstance()后不需要重新启动实例，但对只读服务器变量，需要重启实例。此信息显示在发出 dba.configureInstance()之后生成的报告中。如果实例支持 RESTART 语句，则 MySQL Shell 可以关闭然后启动实例，这可确保 mysqld 检测到对实例配置文件所做的更改。

执行 dba.configureInstance()后再次检查实例配置的部分输出，如下所示：

```
 MySQL  JS > dba.checkInstanceConfiguration('wxy@172.16.1.125:3306')
...

Checking whether existing tables comply with Group Replication requirements...
No incompatible tables detected

Checking instance configuration...
Instance configuration is compatible with InnoDB cluster

The instance '172.16.1.125:3306' is valid for InnoDB cluster usage.

{
    "status": "ok"
}
 MySQL  JS >
```

对集群中的每个服务器实例重复配置过程：

```
dba.configureInstance('wxy@172.16.1.126:3306')
dba.configureInstance('wxy@172.16.1.127:3306')
```

13.2.3 创建集群

准备好实例后，使用 dba.createCluster()函数创建集群，参数为自定义的集群名称。在创建集群之前，MySQL Shell 必须连接到一个实例，并将该实例用作集群的种子实例。种子实例将被复制到集群中的其他实例，从而使它们成为种子实例的副本。当发出 dba.createCluster(name)时，MySQL Shell 会创建连接到服务器实例的会话。下面是以 172.16.1.125:3306 作为种子实例创建名为 testCluster 集群的命令及输出：

```
 MySQL  JS > \connect wxy@172.16.1.125:3306
Creating a session to 'wxy@172.16.1.125:3306'
Please provide the password for 'wxy@172.16.1.125:3306': ******
Save password for 'wxy@172.16.1.125:3306'? [Y]es/[N]o/Ne[v]er (default No):
Fetching schema names for autocompletion... Press ^C to stop.
Your MySQL connection id is 616
Server version: 8.0.16 MySQL Community Server - GPL
No default schema selected; type \use <schema> to set one.
 MySQL  172.16.1.125:3306 ssl  JS > dba.createCluster('testCluster')
A new InnoDB cluster will be created on instance '172.16.1.125:3306'.

Validating instance at 172.16.1.125:3306...

This instance reports its own address as hdp2:3306

Instance configuration is suitable.
Creating InnoDB cluster 'testCluster' on '172.16.1.125:3306'...

Adding Seed Instance...
Cluster successfully created. Use Cluster.addInstance() to add MySQL instances.
At least 3 instances are needed for the cluster to be able to withstand up to
one server failure.

<Cluster:testCluster>
 MySQL  172.16.1.125:3306 ssl  JS >
```

dba.createCluster 函数返回一个 Cluster 对象。千万注意，一旦服务器实例属于一个集群，就只能使用 MySQL Shell 和 AdminAPI 来管理它们。在实例添加到集群后，不支持手动更改组复制的配置。同样，也不支持对 InnoDB Cluster 至关重要的服务器变量进行修改，例如在使用 AdminAPI 配置实例后，不能修改服务器变量 server_uuid。

若要能够容错一个实例的失败，那么集群至少要包含三个 MySQL 实例，添加更多实例会增加对实例失败的容错能力。下面示例添加另外两个实例的情况。

13.2.4 向集群添加实例

使用 Cluster.addInstance(instance)函数向集群添加更多实例，其中 instance 是已配置实例的连接信息。下面是将 172.16.1.126:3306 实例添加到集群的命令及部分输出：

```
 MySQL  172.16.1.125:3306 ssl  JS > var cluster = dba.getCluster()
 MySQL  172.16.1.125:3306 ssl  JS >
cluster.addInstance('wxy@172.16.1.126:3306')
Please provide the password for 'wxy@172.16.1.126:3306': ******
Save password for 'wxy@172.16.1.126:3306'? [Y]es/[N]o/Ne[v]er (default No):
...

Please select a recovery method [I]ncremental recovery/[A]bort (default
Incremental recovery):
...

Adding instance to the cluster...

Monitoring recovery process of the new cluster member. Press ^C to stop
monitoring and let it continue in background.
Incremental distributed state recovery is now in progress.

* Waiting for distributed recovery to finish...
NOTE: '172.16.1.126:3306' is being recovered from 'hdp2:3306'
* Distributed recovery has finished

The instance '172.16.1.126:3306' was successfully added to the cluster.

 MySQL  172.16.1.125:3306 ssl  JS >
```

首先调用 dba.getCluster()函数获取集群对象，并赋予一个变量。如果未指定集群名称，则返回默认集群。默认情况下调用 dba.getCluster()时，MySQL Shell 则会尝试连接到集群的主实例。设置 connectToPrimary 选项以配置此行为。如果 connectToPrimary 为 true 且活动的全局 MySQL Shell 会话没连接到主实例，则会查询主节点的集群，并且集群对象将连接到该集群。如果 connectToPrimary 为 false，则集群对象使用活动会话所属的实例，就是与 MySQL Shell 当前全局会话相同的实例。

注意辅助实例的 super_read_only = ON，因此无法对其进行更改。要在获取集群时强制连接到辅助节点，需要建立与集群辅助成员的连接，并通过发出以下命令来使用 connectToPrimary 选项：

```
mysql-js> shell.connect(secondary_member)
mysql-js> var cluster1 = dba.getCluster(testCluster,
{connectToPrimary:false})
```

如果使用的是 MySQL 8.0.17 或更高版本，则可以选择实例如何恢复与集群同步所需的事务，可选方式有克隆和增量两种，默认为增量。只有当加入实例恢复了先前由集群处理的所有事务时，它才能作为在线实例加入并开始处理事务。同样在 8.0.17 及更高版本中，可以使用 waitRecovery 选项控制 Cluster.addInstance()的行为方式，让恢复操作在后台进行或监视 MySQL Shell 中的实例恢复进度。

如果要在一个已经包含大量数据（如几十 GB 或数百 GB）的集群中添加实例，推荐的方法是使用 xtrabackup 复制一个主节点副本，然后添加该副本实例时使用增量恢复。这样做有两个明显的优点，一是可以在对主节点影响最小的情况下联机执行；二是减少实例恢复所需的时间。如何使用

XtraBackup 进行联机复制可以参考 6.2.1 小节中的第 5 步操作。

以同样的方法添加第三个实例：

```
cluster.addInstance('wxy@172.16.1.127:3306')
```

13.2.5　查看集群状态

集群对象提供了 status()方法，可用于检查集群状态。在检查 InnoDB Cluster 的状态之前，需要通过连接到集群中的任何实例来获取对 InnoDB Cluster 对象的引用。但如果要更改集群的配置，则必须连接到可读写实例。status()从所连接服务器实例的集群视图检索集群的状态，并输出状态报告。所连接实例的状态直接影响状态报告中提供的信息，因此应确保连接实例的状态为 ONLINE。下面是连接 172.16.1.125:3306 获取集群状态及输出：

```
 MySQL  172.16.1.125:3306 ssl  JS > cluster.status()
{
    "clusterName": "testCluster",
    "defaultReplicaSet": {
        "name": "default",
        "primary": "hdp2:3306",
        "ssl": "REQUIRED",
        "status": "OK",
        "statusText": "Cluster is ONLINE and can tolerate up to ONE failure.",
        "topology": {
            "hdp2:3306": {
                "address": "hdp2:3306",
                "mode": "R/W",
                "readReplicas": {},
                "role": "HA",
                "status": "ONLINE",
                "version": "8.0.16"
            },
            "hdp3:3306": {
                "address": "hdp3:3306",
                "mode": "R/O",
                "readReplicas": {},
                "role": "HA",
                "status": "ONLINE",
                "version": "8.0.16"
            },
            "hdp4:3306": {
                "address": "hdp4:3306",
                "mode": "R/O",
                "readReplicas": {},
                "role": "HA",
                "status": "ONLINE",
                "version": "8.0.16"
            }
        },
```

```
        "topologyMode": "Single-Primary"
    },
    "groupInformationSourceMember": "hdp2:3306"
}
 MySQL  172.16.1.125:3306 ssl  JS >
```

Cluster.status()的输出提供以下信息：

- clusterName：在 dba.createCluster 期间分配给此集群的名称。
- defaultReplicaSet：属于 InnoDB Cluster 并包含数据集的服务器实例。
- primary：当前主实例的地址，仅在集群以单主模式运行时显示。如果未显示此字段，则集群将以多主模式运行。
- ssl：集群是否使用安全连接。显示值为 REQUIRED 或 DISABLED，默认是 REQUIRED，具体取决于在 createCluster()或 addInstance()期间如何配置 memberSslMode 选项。此参数返回的值对应于实例上的 group_replication_ssl_mode 服务器变量的值。
- status：集群状态，反映了此集群可提供的高可用性。实例状态为以下之一。
  - ONLINE：实例在线并参与集群。
  - OFFLINE：实例已失去与其他实例的连接。
  - RECOVERING：实例尝试在成为 ONLINE 成员之前检索所需的事务来与集群同步。
  - UNREACHABLE：实例已丢失与集群的通信。
  - ERROR：实例在恢复阶段或应用事务时遇到错误。实例进入 ERROR 状态后，super_read_only 选项设置为 ON，要退出 ERROR 状态，必须手动设置 super_read_only = OFF。
  - MISSING：实例是已配置集群的一部分，但当前不可用。MISSING 状态特定于 InnoDB Cluster，它不是 Group Replication 生成的状态。MySQL Shell 使用此状态来指示在元数据中注册但在实时集群视图中找不到的实例。
- topology：已添加到集群的实例。
- role：此实例在集群中提供的功能，目前只有 HA。
- mode：服务器是读写（“R/W”）还是只读（“R/O”）。从版本 8.0.17 开始，该选项值根据实例中 super_read_only 变量的当前状态以及集群是否具有仲裁来派生。在之前版本中，mode 的值来自实例是作为主实例还是辅助实例。通常，如果是主实例，则模式为“R/W”，如果为辅助实例，则模式为“R/O”。无论 super_read_only 变量的值如何，集群中没有可见仲裁的任何实例都标记为“R/O”。
- groupInformationSourceMember：用于获取有关集群信息的内部连接，显示为类似 URI 的连接字符串，通常为最初用于创建集群的连接。

要显示有关集群的更多信息，可使用扩展选项。如 Cluster.status({'extended'：value})提供附加信息，value 值控制输出：

- 0：默认值，禁用附加信息。
- 1：包括有关组复制报告的组复制协议版本，组名称，集群成员 UUID，集群成员角色和状态以及受保护系统变量列表的信息。

- 2：包括有关连接和应用程序处理的事务信息。

13.2.6 基于已有组复制创建集群

如果已经事先配置好组复制，并且希望使用它来创建 InnoDB Cluster，可将 adoptFromGR 选项传递给 dba.createCluster()函数。创建的 InnoDB Cluster 会匹配复制组是以单主数据库还是多主数据库运行。要采用现有的复制组，使用 MySQL Shell 连接到组成员。以下示例采用单主模式，172.16.1.125:3306 为主实例，172.16.1.126:3306 和 172.16.1.127:3306 为两个辅助实例。这里连接到 172.16.1.125:3306 创建集群。

```
    mysqlsh --uri wxy@172.16.1.125:3306
     MySQL  172.16.1.125:3306 ssl  JS > var cluster =
dba.createCluster('testCluster', {adoptFromGR: true});
    A new InnoDB cluster will be created based on the existing replication group
on instance '172.16.1.125:3306'.

    Creating InnoDB cluster 'testCluster' on '172.16.1.125:3306'...

    Adding Seed Instance...
    Adding Instance 'hdp3:3306'...
    Adding Instance 'hdp4:3306'...
    Adding Instance 'hdp2:3306'...
    Resetting distributed recovery credentials across the cluster...
    Cluster successfully created based on existing replication group.
     MySQL  172.16.1.125:3306 ssl  JS >
```

新集群与组复制的模式匹配。如果组复制以单主模式运行，则会创建单主集群，而如果组复制以多主模式运行，则会创建多主集群。

13.2.7 配置 MySQL Router

MySQL Router 可以使用--bootstrap 选项基于 InnoDB Cluster 的元数据进行自我配置，这会自动配置 MySQL Router 以把连接路由到集群的服务器实例。客户端应用程序连接到 MySQL Router 提供的端口，无需了解 InnoDB Cluster 的拓扑结构。如果发生意外故障，InnoDB Cluster 会自动进行调整，MySQL Router 则会检测到集群配置的变更。这消除了客户端应用程序处理故障转移的需求。MySQL 不建议手动配置 MySQL Router 以重定向到 InnoDB Cluster 的端口，而是建议始终使用--bootstrap 选项，因为这可确保 MySQL Router 从 InnoDB Cluster 的元数据中自动获取其配置。用作生产部署时，建议将 MySQL Router 部署于客户端应用程序所在的每台主机上。

MySQL Router 使用包含的元数据缓存插件来检索 InnoDB Cluster 的元数据，该元数据由构成 InnoDB Cluster 的服务器实例地址列表及其在集群中的角色组成。MySQL Router 从它所连接的实例中检索 InnoDB Cluster 元数据，例如 172.16.1.125:3306：

```
[mysql@hdp1~]$mysqlrouter --bootstrap wxy@172.16.1.125:3306
...
```

```
    # MySQL Router configured for the InnoDB cluster 'testCluster'

    After this MySQL Router has been started with the generated configuration

      $ /etc/init.d/mysqlrouter restart
    or
      $ systemctl start mysqlrouter
    or
      $ mysqlrouter -c
/home/mysql/mysql-router-8.0.17-linux-glibc2.12-x86_64/mysqlrouter.conf

    the cluster 'testCluster' can be reached by connecting to:

    ## MySQL Classic protocol

    - Read/Write Connections: localhost:6446
    - Read/Only Connections: localhost:6447

    ## MySQL X protocol

    - Read/Write Connections: localhost:64460
    - Read/Only Connections: localhost:64470


    Existing configuration backed up to
'/home/mysql/mysql-router-8.0.17-linux-glibc2.12-x86_64/mysqlrouter.conf.bak'

    Existing dynamic state backed up to
'/home/mysql/mysql-router-8.0.17-linux-glibc2.12-x86_64/var/lib/mysqlrouter/st
ate.json.bak'
    [mysql@hdp1~]$
```

基于检索到的InnoDB Cluster元数据，MySQL Router自动配置mysqlrouter.conf文件，包括带有bootstrap_server_addresses的metadata_cache部分，其中包含集群中所有服务器实例的地址。下面是引导时自动生成的/home/mysql/mysql-router-8.0.17-linux-glibc2.12-x86_64/mysqlrouter.conf文件的内容：

```
    # File automatically generated during MySQL Router bootstrap
    [DEFAULT]
    name=system
    keyring_path=/home/mysql/mysql-router-8.0.17-linux-glibc2.12-x86_64/var/li
b/mysqlrouter/keyring
    master_key_path=/home/mysql/mysql-router-8.0.17-linux-glibc2.12-x86_64/mys
qlrouter.key
    connect_timeout=15
    read_timeout=30
    dynamic_state=/home/mysql/mysql-router-8.0.17-linux-glibc2.12-x86_64/var/l
ib/mysqlrouter/state.json
```

```
[logger]
level = INFO

[metadata_cache:testCluster]
router_id=1
user=mysql_router1_c13bmjjayuwr
metadata_cluster=testCluster
ttl=0.5
use_gr_notifications=0

[routing:testCluster_default_rw]
bind_address=0.0.0.0
bind_port=6446
destinations=metadata-cache://testCluster/default?role=PRIMARY
routing_strategy=first-available
protocol=classic

[routing:testCluster_default_ro]
bind_address=0.0.0.0
bind_port=6447
destinations=metadata-cache://testCluster/default?role=SECONDARY
routing_strategy=round-robin-with-fallback
protocol=classic

[routing:testCluster_default_x_rw]
bind_address=0.0.0.0
bind_port=64460
destinations=metadata-cache://testCluster/default?role=PRIMARY
routing_strategy=first-available
protocol=x

[routing:testCluster_default_x_ro]
bind_address=0.0.0.0
bind_port=64470
destinations=metadata-cache://testCluster/default?role=SECONDARY
routing_strategy=round-robin-with-fallback
protocol=x
```

在引导 MySQL Router 之后添加另一个服务器实例导致更改集群拓扑结构时，需要根据更新的元数据修改 bootstrap_server_addresses。可以使用--bootstrap 选项重新引导 MySQL Router 并将其重启。生成的 MySQL Router 配置会创建用于连接到集群的 TCP 端口，包括使用经典 MySQL 协议和 X 协议与集群通信的端口，默认值如下：

- 6446：用于经典 MySQL 协议读写，MySQL Router 将传入连接重定向到主服务器实例。
- 6447：对于经典 MySQL 协议只读，MySQL Router 将传入连接重定向到其中一个辅助服务器实例。
- 64460：用于 X 协议读写，MySQL Router 将传入连接重定向到主服务器实例。
- 64470：用于 X 协议只读，MySQL Router 将传入连接重定向到其中一个辅助服务器实例。

传入连接的重定向方式取决于所使用的集群类型。使用单主集群时，默认情况下，MySQL Router 会发布 X 协议和经典协议端口，客户端连接到这些端口（如 6446 或 64460）进行读写会话，并重定向到集群的单个主节点。使用多主集群时，读写会话将以循环方式重定向到其中一个主实例。例如，到端口 6446 的第一个连接将被重定向到主实例 1，第二个连接将被重定向到主实例 2 等。对于传入的只读连接，MySQL Router 以循环方式将连接重定向到其中一个辅助实例。路由策略由 routing_strategy 选项控制，参见 7.3.2 小节。

使用--bootstrap 选项进行引导后，执行下面的命令后台启动 MySQL Router：

```
[mysql@hdp1~]$mysqlrouter &
```

13.2.8 测试客户端连接

现在可以将 MySQL 客户端，如 mysql 或 MySQL Shell 连接到其中一个 MySQL Router 端口，并查看客户端如何透明地连接到 InnoDB Cluster 中的一个 MySQL 实例。测试脚本文件 router_connect_test.sh 的内容如下：

```
mysql -uwxy -123456 -P6446 --protocol=TCP -N -r -B -e"select @@hostname"
mysql -uwxy -123456 -P6446 --protocol=TCP -N -r -B -e"select @@hostname"
mysql -uwxy -123456 -P6447 --protocol=TCP -N -r -B -e"select @@hostname"
mysql -uwxy -123456 -P6447 --protocol=TCP -N -r -B -e"select @@hostname"
mysql -uwxy -123456 -P6447 --protocol=TCP -N -r -B -e"select @@hostname"
mysqlsh --sql -uwxy -123456 -P64460 -e"select @@hostname"
mysqlsh --sql -uwxy -123456 -P64460 -e"select @@hostname"
mysqlsh --sql -uwxy -123456 -P64470 -e"select @@hostname"
mysqlsh --sql -uwxy -123456 -P64470 -e"select @@hostname"
mysqlsh --sql -uwxy -123456 -P64470 -e"select @@hostname"
```

执行测试脚本并将结果输出到文件：

```
[mysql@hdp1~]$./router_connect_test.sh > result.txt
```

查看测试结果：

```
[mysql@hdp1~]$cat result.txt
hdp2
hdp2
hdp3
hdp4
hdp3
@@hostname
hdp2
@@hostname
hdp2
@@hostname
hdp4
@@hostname
hdp3
@@hostname
hdp4
```

```
[mysql@hdp1~]$
```

13.2.9 测试高可用性

要测试高可用性是否有效，可以通过终止实例来模拟 MySQL 意外故障。集群检测到实例离开并重新配置自身，重新配置的确切方式取决于使用的是单主集群还是多主集群，以及实例在集群中的角色。在单主模式下：

- 如果是当前主节点离开集群，则其中一个辅助实例被选为新主节点，实例由最低 server_uuid 区分优先级。MySQL Router 将读写连接重定向到新选择的主节点。
- 如果是当前辅助节点离开集群，MySQL Router 将停止把只读连接重定向到该实例。

步骤 01 停止 172.16.1.125:3306 实例并测试客户端连接：

```
# 在 hdp2 上执行
mysqladmin -uroot -p123456 shutdown

# 在 hdp1 上执行
router_connect_test.sh > result.txt
cat result.txt
```

测试结果如下：

```
hdp3
hdp3
hdp4
hdp4
hdp4
@@hostname
hdp3
@@hostname
hdp3
@@hostname
hdp4
@@hostname
hdp4
@@hostname
hdp4
```

停止 hdp2 后，选举的新主节点为 hdp3，集群变为了“一主一从”，客户端的访问正常。

步骤 02 重新启动 172.16.1.125:3306 实例并测试客户端连接：

```
# 在 hdp2 上执行
mysqld_safe &

# 在 hdp1 上执行
router_connect_test.sh > result.txt
cat result.txt
```

测试结果如下：

```
hdp3
hdp3
hdp2
hdp4
hdp2
@@hostname
hdp3
@@hostname
hdp3
@@hostname
hdp2
@@hostname
hdp4
@@hostname
hdp2
```

启动 hdp2 后，它自动变为辅助节点，集群又成了“一主二从”，客户端的访问正常。

步骤 03 停止 172.16.1.127:3306 实例并测试客户端连接：

```
# 在 hdp4 上执行
mysqladmin -uroot -p123456 shutdown

# 在 hdp1 上执行
router_connect_test.sh > result.txt
cat result.txt
```

测试结果如下：

```
hdp3
hdp3
hdp2
hdp2
hdp2
@@hostname
hdp3
@@hostname
hdp3
@@hostname
hdp2
@@hostname
hdp2
@@hostname
hdp2
```

停止 hdp4 后，集群变为了“一主一从”，客户端的访问正常。

步骤 04 重新启动 172.16.1.127:3306 实例并测试客户端连接：

```
# 在 hdp4 上执行
```

```
mysqld_safe &

# 在 hdp1 上执行
router_connect_test.sh > result.txt
cat result.txt
```

测试结果如下：

```
hdp3
hdp3
hdp2
hdp4
hdp2
@@hostname
hdp3
@@hostname
hdp3
@@hostname
hdp2
@@hostname
hdp4
@@hostname
hdp2
```

启动 hdp4 后，集群又成为了“一主二从”，客户端的访问正常。

13.3　使用 MySQL Shell 管理 InnoDB Cluster

本节介绍如何使用 MySQL Shell 处理常见的 InnoDB Cluster 管理任务，大部分配置本质上是使用 MySQL Shell 提供的方法和选项配置组复制相关的系统变量。

13.3.1　配置实例自动重新加入

运行 MySQL 8.0.16 及更高版本的实例支持组复制自动重新加入功能，可以将实例配置为在被驱逐后自动重新加入集群。AdminAPI 提供了 autoRejoinTries 选项，用于配置被驱逐后重新加入集群的尝试次数。在默认情况下，实例不会自动重新加入集群。可以使用以下命令在集群级别或单个实例上配置 autoRejoinTries 选项：

- dba.createCluster()
- Cluster.addInstance()
- Cluster.setOption()
- Cluster.setInstanceOption()

autoRejoinTries 选项对应 group_replication_autorejoin_tries 系统变量，例如将重连尝试次数由

默认的 0 改为 10：

```
    [mysql@hdp1~]$mysqlsh --uri wxy@172.16.1.126:3306 -p123456 -e "var cluster = dba.getCluster(); cluster.setOption('autoRejoinTries',10)"
    WARNING: Using a password on the command line interface can be insecure.
    WARNING: Each cluster member will only proceed according to its exitStateAction if auto-rejoin fails (i.e. all retry attempts are exhausted).

    Setting the value of 'autoRejoinTries' to '10' in all ReplicaSet members ...

    Successfully set the value of 'autoRejoinTries' to '10' in the 'default' ReplicaSet.
    [mysql@hdp1~]$mysql -uwxy -p123456 -P6446 --protocol=TCP -N -r -B -e"select @@group_replication_autorejoin_tries"
    Warning: Using a password on the command line interface can be insecure.
    10
    [mysql@hdp1~]$
```

autoRejoinTries 选项接受介于 0 和 2016 之间的正整数值，默认值为 0，即不尝试自动重新加入。自动重连适用于网络不稳定的场景。

运行 MySQL 8.0.12 及更高版本的实例具有 group_replication_exit_state_action 变量，可以使用 AdminAPI 的 exitStateAction 选项配置该变量，用以控制在意外离开集群时实例执行的操作。默认情况下，exitStateAction 选项为 READ_ONLY，这意味着意外离开集群的实例变为只读。如果 exitStateAction 是 ABORT_SERVER，那么在意外离开集群的情况下，MySQL 实例会关闭，并且必须先重启它才能重新加入集群。

```
    [mysql@hdp1~]$mysqlsh --uri wxy@172.16.1.126:3306 -p123456 -e "var cluster = dba.getCluster(); cluster.setOption('exitStateAction','ABORT_SERVER')"
    WARNING: Using a password on the command line interface can be insecure.
    Setting the value of 'exitStateAction' to 'ABORT_SERVER' in all ReplicaSet members ...

    Successfully set the value of 'exitStateAction' to 'ABORT_SERVER' in the 'default' ReplicaSet.
    [mysql@hdp1~]$mysql -uwxy -p123456 -P6446 --protocol=TCP -N -r -B -e"select @@group_replication_exit_state_action"
    Warning: Using a password on the command line interface can be insecure.
    ABORT_SERVER
    [mysql@hdp1~]$
```

需要注意的是，在使用自动重新加入功能时，exitStateAction 选项配置的操作仅在所有尝试重新加入集群失败的情况下发生。

13.3.2 从 InnoDB Cluster 中删除实例

可以使用 Cluster.removeInstance(instance)方法随时从集群中删除实例：

```
 MySQL  172.16.1.126:3306 ssl  JS > var cluster = dba.getCluster();
```

```
 MySQL  172.16.1.126:3306 ssl  JS >
cluster.removeInstance('wxy@172.16.1.126:3306');
    The instance will be removed from the InnoDB cluster. Depending on the instance
    being the Seed or not, the Metadata session might become invalid. If so, please
    start a new session to the Metadata Storage R/W instance.

    Instance '172.16.1.126:3306' is attempting to leave the cluster...

    The instance '172.16.1.126:3306' was successfully removed from the cluster.
```

cluster.removeInstance()操作可确保从 ONLINE 的所有集群成员和实例本身的元数据中删除实例。

当要删除的实例具有仍需要应用的事务时，AdminAPI 会等待 dba.gtidWaitTimeout 选项为要应用的事务（GTID）设置的秒数，默认值为 60 秒：

```
 MySQL  172.16.1.126:3306 ssl  JS > \option dba.gtidWaitTimeout
60
```

更改默认值：

```
 MySQL  172.16.1.126:3306 ssl  JS > shell.options['dba.gtidWaitTimeout']=120
120
 MySQL  172.16.1.126:3306 ssl  JS > \option dba.gtidWaitTimeout
120
```

如果在等待应用事务并且 force 选项为 false（默认）时达到 dba.gtidWaitTimeout 定义的超时值，则会发出错误并中止删除操作。如果在等待应用事务并且 force 选项设置为 true 时达到 dba.gtidWaitTimeout 定义的超时值，则操作将继续而不会出现错误，并从集群中删除该实例。从集群中删除实例时忽略错误可能导致实例与集群不同步，从而阻止其稍后重新加入集群。只有当不再使用实例时再开启强制选项，所有其他情况下，应该始终尝试恢复实例，并仅在状态为 ONLINE 时将其删除。

13.3.3 重启集群

可以使用 dba.rebootClusterFromCompleteOutage()重启整个集群。此操作采用 MySQL Shell 当前连接的实例并使用其元数据来恢复集群。如果集群的实例已完全停止，则必须启动实例后才能启动集群。在生产部署的情况下，要在 MySQL Shell 之外启动实例。实例启动后，需要连接到中断之前应用了最多事务的实例。如果不确定是哪个实例，可连接到任何实例并按照 dba.rebootClusterFromCompleteOutage()中的交互消息进行操作，该消息将检测连接的实例是否应用了最多的事务。

```
 MySQL  JS > dba.rebootClusterFromCompleteOutage();
    Dba.rebootClusterFromCompleteOutage: An open session is required to perform
this operation. (RuntimeError)
 MySQL  JS > \connect wxy@172.16.1.125:3306
    Creating a session to 'wxy@172.16.1.125:3306'
    Please provide the password for 'wxy@172.16.1.125:3306': ******
    Save password for 'wxy@172.16.1.125:3306'? [Y]es/[N]o/Ne[v]er (default No):
```

```
Fetching schema names for autocompletion... Press ^C to stop.
Your MySQL connection id is 165
Server version: 8.0.16 MySQL Community Server - GPL
No default schema selected; type \use <schema> to set one.
 MySQL  172.16.1.125:3306 ssl  JS > dba.rebootClusterFromCompleteOutage();
Reconfiguring the default cluster from complete outage...

The instance 'hdp4:3306' was part of the cluster configuration.
Would you like to rejoin it to the cluster? [y/N]: y

The instance 'hdp3:3306' was part of the cluster configuration.
Would you like to rejoin it to the cluster? [y/N]: y

...

The cluster was successfully rebooted.

<Cluster:testCluster>
```

dba.rebootClusterFromCompleteOutage()遵循以下步骤以确保正确重新配置集群：

- 检查在 MySQL Shell 当前连接的实例上找到的 InnoDB Cluster 元数据，以查看它是否包含最多事务。如果不是，则操作中止。在这种情况下，将 MySQL Shell 连接到错误消息中建议的实例，并从该实例发出 dba.rebootClusterFromCompleteOutage()。
- 如果实例包含最多事务，则会根据实例的元数据来恢复集群。
- 交互模式下会运行一个向导，检查当前可以访问哪个集群实例，并询问是否要将任何已发现的实例重新加入引导的集群。
- 交互模式下向导还会检测当前无法访问的实例，并询问是否要从重新引导的集群中删除此类实例。

手动检测哪个实例具有最多事务，需要检查每个实例上的 gtid_executed 变量：

```
show variables like 'gtid_executed';
```

如果重启过程失败，并且集群元数据已严重损坏，则可能需要删除元数据并从头开始再次创建集群。可以使用 dba.dropMetadataSchema()删除集群元数据。dba.dropMetadataSchema 方法删除的元数据是不可恢复的，它应仅用作无法还原集群时的最后手段。

13.3.4 解散 InnoDB Cluster

解散 InnoDB Cluster 需要连接到状态为 ONLINE 的读写实例，例如单主集群中的主实例，并使用 Cluster.dissolve()命令。这将删除与集群关联的所有元数据和配置，并禁用实例上的组复制，但不会删除在实例之间复制的任何数据。要再次创建集群，使用 dba.createCluster()。

```
 MySQL  JS > \connect wxy@172.16.1.125:3306
Creating a session to 'wxy@172.16.1.125:3306'
Please provide the password for 'wxy@172.16.1.125:3306': ******
```

```
Save password for 'wxy@172.16.1.125:3306'? [Y]es/[N]o/Ne[v]er (default No):
Fetching schema names for autocompletion... Press ^C to stop.
Your MySQL connection id is 2830
Server version: 8.0.16 MySQL Community Server - GPL
No default schema selected; type \use <schema> to set one.
 MySQL  172.16.1.125:3306 ssl  JS > var cluster = dba.getCluster()
 MySQL  172.16.1.125:3306 ssl  JS > cluster.dissolve()
The cluster still has the following registered ReplicaSets:
{
    "clusterName": "testCluster",
    "defaultReplicaSet": {
        "name": "default",
        "topology": [
            {
                "address": "hdp2:3306",
                "label": "hdp2:3306",
                "role": "HA",
                "version": "8.0.16"
            },
            {
                "address": "hdp4:3306",
                "label": "hdp4:3306",
                "role": "HA",
                "version": "8.0.16"
            },
            {
                "address": "hdp3:3306",
                "label": "hdp3:3306",
                "role": "HA",
                "version": "8.0.16"
            }
        ],
        "topologyMode": "Single-Primary"
    }
}
WARNING: You are about to dissolve the whole cluster and lose the high
availability features provided by it. This operation cannot be reverted. All members
will be removed from their ReplicaSet and replication will be stopped, internal
recovery user accounts and the cluster metadata will be dropped. User data will
be maintained intact in all instances.

Are you sure you want to dissolve the cluster? [y/N]: y

Instance 'hdp4:3306' is attempting to leave the cluster...
Instance 'hdp3:3306' is attempting to leave the cluster...
Instance 'hdp2:3306' is attempting to leave the cluster...

The cluster was successfully dissolved.
Replication was disabled but user data was left intact.
```

dba.gtidWaitTimeout 选项用于配置 Cluster.dissolve()操作在从集群中删除目标实例之前等待集群事务的时间，但仅限于目标实例为 ONLINE 的情况。如果在等待要删除的任何实例上应用集群事务超时，则会发出错误（除非使用 force:true）。

13.3.5 配置新主选举权重

可以通过在 dba.createCluster()或 Cluster.addInstance()方法中指定 memberWeight 来影响新主节点的选举结果。memberWeight 选项的值域为 0~100 之间的整数，默认值为 50。该值是故障转移时自动选举主节点的百分比权重，对应 group_replication_member_weight 系统变量。具有较高 memberWeight 值的实例更有可能在单主集群中被选为主节点。如果多个实例具有相同的 memberWeight 值，则根据服务器 UUID 的字典正序，选择第一个实例作为主节点。例如将 wxy@172.16.1.127:3306 配置为故障转移到的首选实例：

```
dba.createCluster('testCluster', {memberWeight:35})
var mycluster = dba.getCluster()
mycluster.addInstance('wxy@172.16.1.126:3306', {memberWeight:25})
mycluster.addInstance('wxy@172.16.1.127:3306', {memberWeight:50})
```

13.3.6 配置故障转移一致性

dba.createCluster()的 consistency 选项用于指定单主模式下故障转移一致性。该选项配置种子实例的 group_replication_consistency 系统变量：consistency = 0 对应 EVENTUAL；consistency = 1 对应 BEFORE_ON_PRIMARY_FAILOVER。该选项仅对单主模式起作用，多主模式下允许使用但选项无效。仅当目标 MySQL 服务器版本为 8.0.14 或更高版本时才支持这个一致性选项，并且添加集群的实例具有相同的 group_replication_consistency 值。

group_replication_consistency 系统变量指定组复制的一致性级别，不同配置对组处理的只读（RO）和读写（RW）事务产生不同的影响。按增加事务一致性保证的顺序，该变量有 EVENTUAL、BEFORE_ON_PRIMARY_FAILOVER、BEFORE、AFTER 和 BEFORE_AND_AFTER 五个可选值，默认为 EVENTUAL。

组复制集群自动检测故障并调整组成员视图，即成员资格配置。如果组以单主模式部署，当成员资格更改时，将执行检查以确定组中是否存在主库。如果没有，则在从库成员列表中选择一个作为新主库，这就是所谓的从库提升。使用默认的 EVENTUAL 一致性级别时，读、写事务都直接执行，不等待之前复制积压的事务，也不等待其他组成员应用这些事务。这也是 MySQL 8.0.14 之前的组复制行为。

一旦发生从库提升，用户期望新主库数据与旧主库数据处于完全相同的状态，在新主库上不能读写旧数据。换句话说，当能够读取和写入新主库时，其上没有积压的复制事务。从 MySQL 8.0.14 版本开始，从库提升后，用户可以指定新主库的行为。新增的 group_replication_consistency 系统参数用于控制新主库是采用之前版本的最终一致性，还是阻止读取或写入，直到完全应用了积压事务。如果在具有 group_replication_consistency='BEFORE_ON_PRIMARY_FAILOVER'设置的新主库上执行事务时，该新主库正在处理积压，则事务将被阻止，直到完全应用待处理的积压事务。这可确保在主库发生故障转移时，无论是自动触发还是手工触发，客户端始终会在新主库上看到最新值，因此防止了以下异常：

- 对于只读和读写事务，没有过时读取。这可以防止新主库将过时读取外部化到应用程序。
- 读写事务没有虚假回滚，因为与复制读写事务产生写-写冲突的事务仍处于待处理状态。
- 读写事务没有读取偏差。

关于组复制中数据一致性的概念、配置、流程、算法和示例的详细讲述，可以参考博文“https://wxy0327.blog.csdn.net/article/details/96482662#二、保证数据一致性”。

13.3.7 更改组复制拓扑

InnoDB Cluster默认以单主模式运行，其中集群中只具有一个接受读写（R/W）的主服务器，所有其余实例仅接受只读（R/O）。将集群配置为多主模式时，集群中的所有实例都是主节点，这意味着它们同时接受读写查询（R/W）。如果集群的所有实例都运行MySQL 8.0.15或更高版本，则可以在集群联机时更改集群的拓扑结构。Cluster.setPrimaryInstance()函数指定一个新的主节点：

```
 MySQL  172.16.1.125:3306 ssl  JS > cluster.setPrimaryInstance
('172.16.1.126:3306')
 Setting instance '172.16.1.126:3306' as the primary instance of cluster
'testCluster'...

 Instance 'hdp2:3306' was switched from PRIMARY to SECONDARY.
 Instance 'hdp4:3306' remains SECONDARY.
 Instance 'hdp3:3306' was switched from SECONDARY to PRIMARY.

 WARNING: The cluster internal session is not the primary member anymore. For
cluster management operations please obtain a fresh cluster handle using
<Dba>.getCluster().

 The instance '172.16.1.126:3306' was successfully elected as primary.
 MySQL  172.16.1.125:3306 ssl  JS >
```

Cluster.switchToMultiPrimaryMode()切换到多主模式：

```
 MySQL  172.16.1.125:3306 ssl  JS > cluster.switchToMultiPrimaryMode()
 Switching cluster 'testCluster' to Multi-Primary mode...

 Instance 'hdp2:3306' was switched from SECONDARY to PRIMARY.
 Instance 'hdp4:3306' was switched from SECONDARY to PRIMARY.
 Instance 'hdp3:3306' remains PRIMARY.

 The cluster successfully switched to Multi-Primary mode.
 MySQL  172.16.1.125:3306 ssl  JS >
```

Cluster.switchToSinglePrimaryMode()切换到单主模式：

```
 MySQL  172.16.1.125:3306 ssl  JS > cluster.switchToSinglePrimaryMode
('172.16.1.125:3306')
```

```
    Switching cluster 'testCluster' to Single-Primary mode...

    Instance 'hdp2:3306' remains PRIMARY.
    Instance 'hdp4:3306' was switched from PRIMARY to SECONDARY.
    Instance 'hdp3:3306' was switched from PRIMARY to SECONDARY.

    WARNING: Existing connections that expected a R/W connection must be
disconnected, i.e. instances that became SECONDARY.

    The cluster successfully switched to Single-Primary mode.
     MySQL  172.16.1.125:3306 ssl  JS >
```

13.3.8 设置 InnoDB Cluster 选项

可以联机查看或修改集群设置。Cluster.options 检查集群的当前设置：

```
 MySQL  172.16.1.125:3306 ssl  JS > cluster.options({all:true})
{
    "clusterName": "testCluster",
    "defaultReplicaSet": {
        "globalOptions": [
            {
                "option": "groupName",
                "value": "c8e80b07-d532-11e9-a1cc-005056a57a4e",
                "variable": "group_replication_group_name"
            },

            ...

        }
    }
}
 MySQL  172.16.1.125:3306 ssl  JS >
```

Cluster.setOption(option, value)用于全局更改所有集群实例的设置或全局设置：

```
  MySQL  172.16.1.125:3306 ssl  JS > cluster.setOption('clusterName',
'procCluster')
  Setting the value of 'clusterName' to 'procCluster' in the Cluster ...

  Successfully set the value of 'clusterName' to 'procCluster' in the Cluster:
'testCluster'.
```

Cluster.setInstanceOption(instance, option, value)用于更改各集群实例的设置：

```
  MySQL  172.16.1.125:3306 ssl  JS > cluster.setInstanceOption
  ('172.16.1.125:3306', 'exitStateAction', 'READ_ONLY')
  Setting the value of 'exitStateAction' to 'READ_ONLY' in the instance:
'172.16.1.125:3306' ...

  Successfully set the value of 'exitStateAction' to 'READ_ONLY' in the 'default'
```

```
ReplicaSet member: '172.16.1.125:3306'.
```

13.4 小 结

InnoDB Cluster 是在 MySQL 组复制基础上，整合了 MySQL Shell 和 MySQL Router 两个产品，形成的一个三位一体的集群解决方案。组复制是核心，MySQL Shell 提供管理控制台，MySQL Router 为应用提供透明路由功能。InnoDB Cluster 提供基本的自动故障转移和对应用程序的透明性，并有多种实例一致性级别供用户选择配置。InnoDB Cluster 显而易见的优点是不需要第三方产品即可实现 MySQL 高可用，缺点是性能比传统的异步或半同步复制存在一定差距，量化方面的具体比较，请参见 6.3 节。本章详细说明了 InnoDB Cluster 的基本概念、安装、配置和管理维护。下一章将介绍与 InnoDB Cluster 类似的另一个 MySQL 集群分支——Galera Cluster。

第14章

Galera Cluster

14.1 基本原理

Galera Cluster 是由 Codership 公司开发的 MySQL 多主集群，包含在 MariaDB 中，同时支持 Percona XtraDB 和 MySQL，是一个易于使用的高可用解决方案，在数据完整性、可扩展性及高性能方面都有可接受的表现。图 14-1 所示为一个三节点 Galera 集群，三个 MySQL 实例是对等的，互为主从，即所谓的多主（Multi-Master）架构。当客户端读写数据时，可连接任一 MySQL 实例。对于读操作，从每个节点读取到的数据都是相同的。对于写操作，当数据写入某一节点后，集群会将其同步到其他节点。这种架构不共享任何数据，是一种高冗余架构。

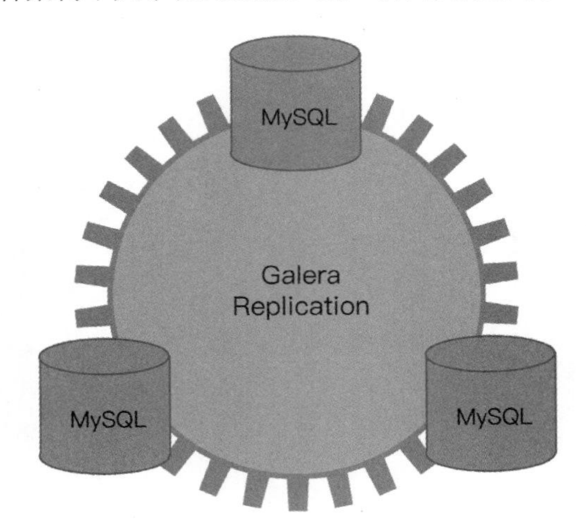

图 14-1　三节点 Galera 集群

Galera 集群具有以下特点：

- 多主架构：真正的多主多活集群，可随时对任何节点进行读写。
- 同步复制：集群不同节点之间数据同步，某节点崩溃时没有数据丢失。
- 数据一致：所有节点保持相同状态，节点之间无数据分歧。
- 并行复制：重放支持多线程并行执行以获得更好的性能。
- 故障转移：故障节点本身对集群的影响非常小，某节点出现问题时无需切换操作，因此不需要使用 VIP，也不会中断服务。客户端通常使用 Galera Load Balancer（负载均衡，功能类似于 MySQL Router）连接到 Galera 集群实现透明故障转移。
- 自动克隆：新增节点会自动拉取在线节点的数据，最终集群所有节点数据一致，而不需要手动备份恢复。
- 应用透明：提供透明的客户端访问，不需要对应用程序进行更改。

Galera 集群复制要求数据库系统支持事务处理，因此仅支持 MySQL 的 InnoDB 存储引擎，并且只能使用可重复读（REPEATABLE-READ）隔离级别。

14.1.1 同步复制

不同于 MySQL 原生的主从异步复制，Galera 采用的是多主同步复制，如图 14-2 所示。

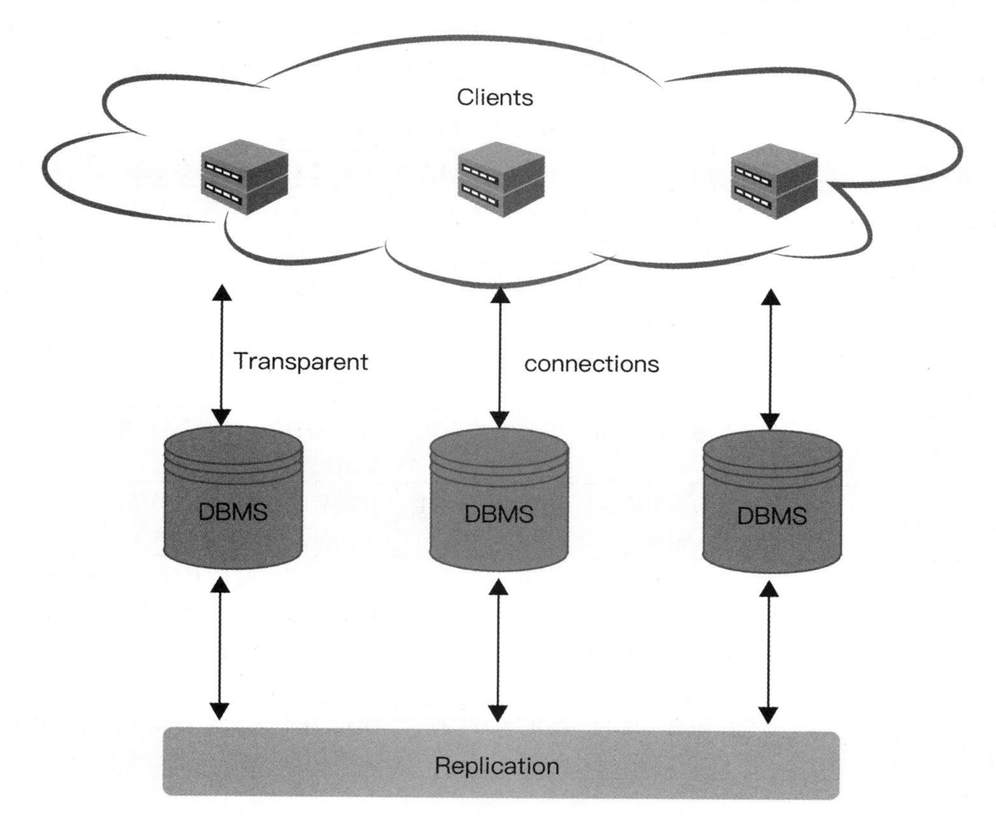

图 14-2 多主同步复制

在异步复制中，主库将数据更新传播给从库后立即提交事务，而不论从库是否成功读取或是否重放数据变化。这种情况下，在主库事务提交后的短时间内，主从库数据并不一致。同步复制时，主库的单个更新事务需要在所有从库上同步更新。换句话说，当主库提交事务时，集群中所有节点的数据保持一致。

相对于异步复制，同步复制的优点主要体现在以下三方面：

- 数据一致：同步复制保证了整个集群的数据一致性，无论何时在任何节点执行相同的 select 查询，结果都一样。
- 高可用性：由于所有节点数据一致，单个节点崩溃不需要执行复杂耗时的故障转移，因此也不会造成丢失数据或停止服务。
- 性能改进：同步复制允许在集群中的所有节点上并行执行事务，从而提高读写性能。

当然，同步复制的缺点也显而易见，这主要源于其实现方式。同步复制协议通常使用两阶段提交或分布式锁协调不同节点的操作。假设集群有 n 个节点，每秒处理 o 个操作，每个操作中包含 t 个事务，则每秒将在网络中产生 n×o×t 条消息。这意味着随着节点数量的增加，事务冲突和死锁的概率将呈指数级增加。这也是 MySQL 默认使用异步复制的主要原因。

为解决传统同步复制的问题，现已提出多种数据库同步复制的替代方法。除理论外，一些原型实现也显示出了很大的希望，如组通信、写集、数据库状态机、事务重排等。Galera 集群就是基于这些方法构建的。可以看到 Galera 复制的原理与实现和 MySQL 组复制有很多相似之处。为了更好地理解 Galera，在深入细节之前，先将它和 MySQL 组复制作一类比，如表 14-1 所示。

表 14-1 Galera 与 MGR 比较

对比项	Galera	MySQL Group Replication
组通信系统	专有组通信系统 GComm，所有节点都必须有 ACK 消息	基于 Paxos，只要求大多数节点有 ACK 消息
二进制日志	不需要二进制日志，将二进制行的事件写入 Gcache	需要二进制日志
节点配置	自动全量同步（SST）与增量同步（IST）	没有自动全量同步，使用异步复制通道
全局事务 ID	使用状态 UUID 和递增序列号 Seqno	依赖 GTID，集群上的写操作产生 GTID 事件
分区控制	分区节点拒绝读写，自动恢复并重新加入集群	分区节点可读，接受写请求但将永久挂起，需要手工重新加入集群
流控	当一个节点慢到一个限制值，阻止所有节点上的写操作	每个节点都有所有成员的统计信息，独立决定该节点写的阈值。如果有节点慢到阈值，其他节点放慢写速度
DDL 支持	总序隔离（TOI），DDL 执行期间，所有写入都将被阻止	DDL 并不会阻塞写，由于没有冲突检测，仅建议在单主模式下使用

14.1.2　Galera 复制架构

Galera 集群的内部架构包含四个组件，如图 14-3 所示。

- 数据库管理系统(DBMS)：在单个节点上运行的数据库服务器。Galera 集群可以使用 MySQL、MariaDB 或 Percona XtraDB。
- wsrep api：Galera 与数据库服务器的接口，为上层提供状态信息和回调函数。wsrep api 由 wsrep hooks 和 dlopen 函数两部分组成。wsrep hooks 钩子程序用于与数据库服务器引擎集成。dlopen 函数使 Galera 插件中的复制程序对 wsrep hooks 可用。
- Galera 复制插件：实现写集复制功能的核心模块。
- 组通信插件：Galera 集群的组通信系统（Group Communication System，GCS），如 GComm。

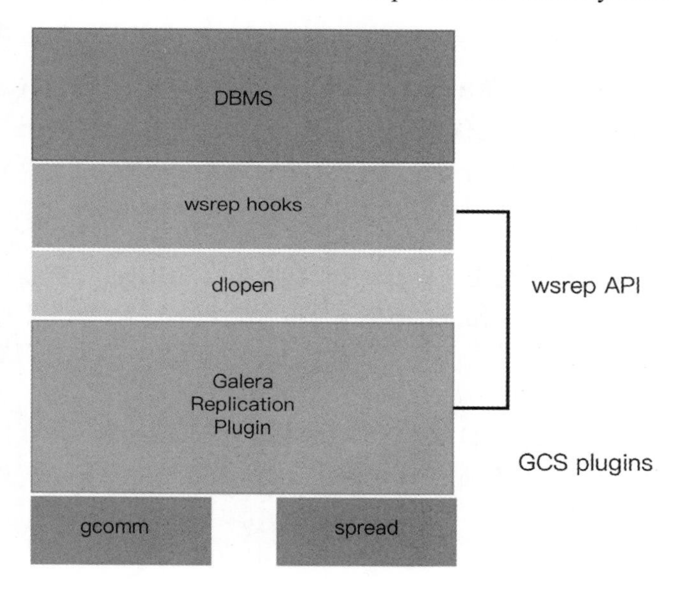

图 14-3　Galera 内部组件

1. wsrep api

wsrep api 是数据库的通用复制插件接口，定义了一组应用程序回调和复制插件调用函数。wsrep api 把数据库中的数据改变视为一种状态变化，当客户端修改数据库内容时，其状态将更改。wsrep api 将数据库状态更改表示为一系列事务。集群中的所有节点始终具有相同状态，它们通过以相同的顺序复制和应用状态更改来实现相互同步。从更改技术角度看，Galera 集群使用以下过程处理状态更改：

（1）一个节点的数据库中发生状态更改。

（2）wsrep 钩子将更改转换为写集。

（3）dlopen 函数连接 wsrep 钩子与 Galera 复制插件。

（4）Galera 复制插件处理写集验证，并将更改复制到集群中的其他节点。

2. 全局事务 ID（Global Transaction ID，GTID）

在 MySQL 社区中，GTID 的概念并不新鲜，MySQL 中的 GTID 由主库生成，是用于标记唯一事务并通过 ID 定位 binlog 位置的一种手段，从而有效解决了级联复制等场景中的各种问题。对于 Galera Cluster 而言，复制不基于 binlog，而是通过 Galera 复制插件来保障。Galera 的 GTID 同样也标记事务唯一性，wsrep api 使用 GTID 来识别状态的更改。Galera 的 GTID 格式如下：

```
45eec521-2f34-11e0-0800-2a36050b826b:94530586304
```

GTID 由两部分组成：

- 状态 UUID：表示当前状态的唯一 ID，可以简单认为是集群的一个唯一标识符。
- 顺序号：一个 64 位有符号整数，表示事务在 Galera Cluster 所有节点中的序号。

3. Galera 复制插件

Galera 复制插件实现 wsrep api，作为 wsrep provider 运行，它由以下两层构成：

- 验证层：该层准备写集，并检测本机以及从其他节点同步过来的事务是否可以提交。
- 复制层：该层的工作包含组通信和并行复制两方面。组通信负责与其他节点同步写集，并为事务分配全局唯一的 GTID。并行复制实现 Galera 事务“乐观”并行控制。

4. 组通信插件

组通信框架为各种 gcomm 系统提供了一个插件体系结构。Galera 集群建立在专有的组通信系统层之上，实现虚拟同步。所谓虚拟同步，简单说是指一个事务在一个节点上执行成功后，保证它在其他节点也一定会被成功执行，但并不能保证实时同步。为了解决实时性问题，Galera 集群实现了自己的运行时可配置的时态流控。

组通信框架还使用 GTID 提供来自多个源的消息总序（Total Order）。在传输层上，Galera 集群是一个对称的无向图，所有节点都通过 TCP 相互连接。在默认情况下，TCP 用于消息复制和集群成员资格服务，但也可以使用 UDP 多播在 LAN 中进行复制。

14.1.3 Galera 工作原理

Galera 复制是一种基于验证的复制，以这两篇论文为理论基础：“Don't be lazy, be consistent”和“Database State Machine Approach”。基于验证的复制使用组通信和事务排序技术实现同步复制。它通过广播并发事务之间建立的全局总序来协调事务提交。简单来说，就是事务必须以相同的顺序应用于集群中的所有实例。事务在本节点乐观执行，然后在提交时运行一个验证过程以保证全局数据一致性。所谓乐观执行是指，事务在一个节点提交时，被认为与其他节点上的事务没有冲突。首先在本地执行，然后再发送到所有节点进行冲突检测，无冲突时在所有节点提交，否则在所有节点回滚。Galera 复制原理如图 14-4 所示。

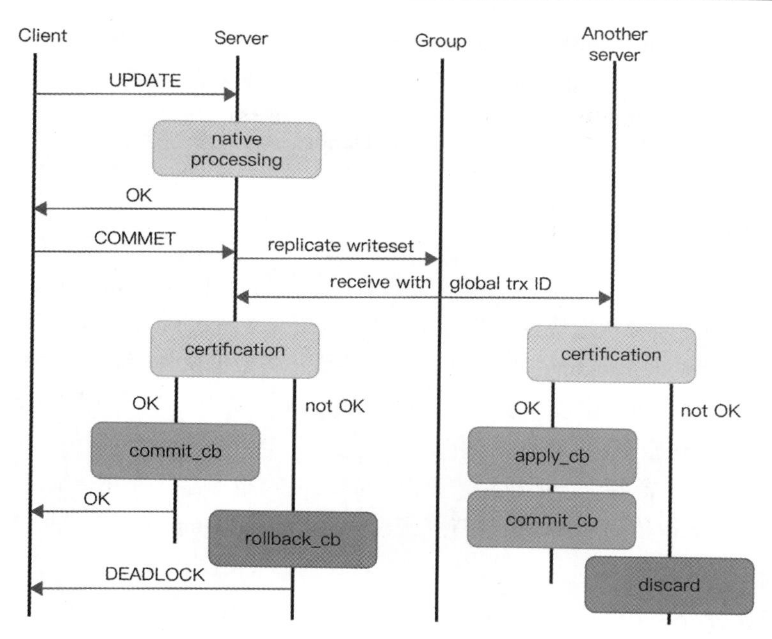

图 14-4 基于验证的复制

当客户端发出 commit 命令时，在实际提交之前，对数据库所做的更改都将被收集到一个写集中，写集中包含事务信息和所更改行的主键。然后，数据库将此写集发送到所有其他节点。节点用写集中的主键与当前节点中未完成事务的所有写集（不仅包括当前节点其他事务产生的写集，还包括其他节点传送过来的写集）的主键相比较，确定节点是否可以提交事务。同时满足以下三个条件则说明存在冲突，验证失败：

- 两个事务来源于不同节点。
- 两个事务包含相同的主键。
- 旧事务对新事务不可见，即旧事务未提交完成。新旧事务的划定依赖于全局事务总序，即 GTID。

验证失败后，节点将删除写集，集群将回滚原始事务。对于所有的节点都是如此，每个节点单独进行验证。因为所有节点都以相同的顺序接收事务，它们对事务的结果都会做出相同的决定，要么全成功，要么都失败。成功后自然就提交了，所有的节点又会重新达到数据一致的状态。节点之间不交换“是否冲突”的信息，各个节点独立异步处理事务。由此可见，Galera 本身的数据也不是严格同步的，很明显在每个节点上的验证是异步的，这也就是前面提到的“虚拟同步”。

最后，启动事务的节点可以通知客户端应用程序是否提交了事务。

14.1.4 状态转移

当一个新节点加入集群时，数据将从集群复制到这个节点，这是一个全自动的过程，Galera 将此称为状态转移。前面介绍 Galera 架构时曾提到，wsrep api 将集群中的数据改变视为状态改变，因此这里将数据同步称作状态转移也就顺理成章了。Galera 集群中有两种状态转移方法：

- 状态快照传输（State Snapshot Transfer，SST），也就是通常所说的全量数据同步。
- 增量状态转移（Incremental State Transfer，IST），指增量数据同步。

当有新节点加入时，集群会选择出一个捐赠者（Donor）节点为新节点提供数据，这点与 MySQL 组复制类似。

1. 状态快照传输（SST）

新节点加入集群时会启动状态快照传输，将其数据与集群同步。Galera 支持 rsync、rsync-wan、xtrabackup 和 mysqldump 四种状态快照传输方法，由系统变量 wsrep_sst_method 指定，默认为 rsync。

rsync、rsync-wan 和 xtrabackup 三种方法是物理备份，将数据文件直接从捐赠者服务器复制到新节点服务器，并在传输后初始化接收服务器，其中 xtrabackup 方式可实现捐赠者无阻塞数据同步。这些方法比 mysqldump 快很多。

mysqldump 方法是逻辑备份，要求用户手动初始化接收服务器，并在传输之前准备好接收连接。这是一种阻塞方法，在传输期间，捐赠节点变为只读。mysqldump 是状态快照传输最慢的方法，不建议在生产环境中使用。

2. 增量状态转移（IST）

增量状态转移只向新节点发送它所缺失的事务。使用 IST 需要满足两个先决条件：

- 新加入节点的状态 UUID 与集群中的节点一致。
- 新加入节点所缺失的写集在捐赠者的写集缓存中存在。这点很容易理解，类比 MySQL 的 binlog，如果所需的 binlog 文件缺失，是无法做增量备份恢复的。

满足这些条件时，捐赠节点单独传输缺失的事务，并按顺序重放它们，直到新节点赶上集群。例如，假设有一个新加入节点落后于集群。新节点和捐赠节点所携带的节点状态如下：

```
# 新加节点
5a76ef62-30ec-11e1-0800-dba504cf2aab:197222
# 捐赠节点
5a76ef62-30ec-11e1-0800-dba504cf2aab:201913
```

集群上的捐赠节点从新加入节点接收状态转移请求。它检查自身写集缓存中的序列号 197223。如果该序号在写集缓存中不可用，则会启动 SST，否则捐赠节点将把从 197223 到 201913 的提交事务发送到新加入节点。增量状态传输的优点是可以显著加快节点合并到集群的速度。另外，这个过程对捐赠者来说是非阻塞的。

增量状态传输最重要的参数是捐赠节点上的 gcache.size，它控制分配多少系统内存用于缓存写集。可用空间越大，则可以存储的写集越多，可以存储的写集越多，通过增量状态传输可以弥合的事务间隙就越大。另一方面，如果写集缓存远大于数据库大小，则增量状态传输开始时的效率低于状态快照传输。

3. 写集缓存（gcache）

Galera 集群将写集存储在一个称为 gcache 的特殊缓存中。gcache 使用三种类型的存储：

- 永久内存存储（Permanent In-Memory Store）：写集使用操作系统的默认内存分配器进行分配，

永久存储于物理内存中。gcache.keep_pages_size 参数指定保留的内存页总大小，默认值为 0。由于硬件的限制，因此默认情况下是禁用的。

- 永久环缓冲区文件（Permanent Ring-Buffer File）：写集在缓存初始化期间预分配到磁盘，生成一个内存映射文件，用作写集存储。文件目录和文件名分别由 gcache.dir 和 gcache.name 参数指定。文件大小由 gcache.size 参数指定，默认值为 128MB。
- 按需页存储（On-Demand Page Store）：根据需要在运行时将写集分配给内存映射页文件。大小由 gcache.page_size 参数指定，默认值为 128MB，可随写集自动变大。页面存储的大小受可用磁盘空间的限制。在默认情况下，Galera 会在不使用时删除页面文件，用户可以设置 gcache.size 参数指定要保留的页面文件总大小。当所有其他存储被禁用时，磁盘上至少保留一个页面的文件。

Galera 集群使用一种分配算法尝试按上述顺序存储写集。也就是说，它首先尝试使用永久内存存储，如果没有足够的空间用于写集，它将尝试存储到永久环缓冲区文件。除非写集大于可用磁盘空间，否则页面存储始终成功。

注意，如果 gcache.recover 参数设置为 yes，则在实例启动时将尝试恢复 gcache，以便该节点可以继续向其他节点提供 IST 服务。如果设置为 no（默认），gcache 将在启动时失效，节点将只能为 SST 提供服务。

14.1.5 流控

Galera 集群内部使用一种称为流控的反馈机制来管理复制过程。流控允许节点根据需要暂停和恢复复制，这可以有效防止任一节点在应用事务时落后其他节点太多。

1. 流控原理

从 Galera 集群同步复制（虚拟同步）原理可知，事务的应用和提交在各个节点上异步发生。节点从集群接收到的但尚未应用和提交的事务将保留在接收队列中。由于不同节点之间执行事务的速度不一样，因此慢节点的接收队列会越积越长。当接收队列达到一定大小时，节点触发流控，作用就是协调各个节点，保证所有节点执行事务的速度大于队列增长速度。流控的实现原理很简单：整个 Galera 集群中，同时只有一个节点可以广播消息，每个节点都会获得广播消息的机会，获得机会后也可以不广播。当慢节点的接收队列超过一定长度后，它会广播一个 FC_PAUSE 消息，所有节点收到消息后都会暂缓广播消息，直到该慢节点的接收队列减小到一定长度后再恢复复制。

流控相关参数如下：

- gcs.fc_limit：接收队列中积压事务的数量超过该值时，流控被触发，默认值为 16。对于主从模式（Master-Slave，只在一个节点写）的 Galera 集群，可以配置一个较大的值，防止主从复制延迟。对启动多写的 Galera 集群，较小的值比较合适，因为较大的接收队列长度意味着更多冲突。
- gcs.fc_factor：当接收队列长度开始小于 gcs.fc_factor × gcs.fc_limit 时恢复复制，默认值为 1。
- gcs.fc_master_slave：Galera 集群是否为主从模式，默认为 no。

2. 理解节点状态

一个节点在 Galera 集群中可能经历的节点状态有 Open、Primary、Joiner、Joined、Synced 和 Donor。可以通过 wsrep_local_state 和 wsrep_local_state_comment 系统变量查看节点的当前状态。节点状态更改如图 14-5 所示。

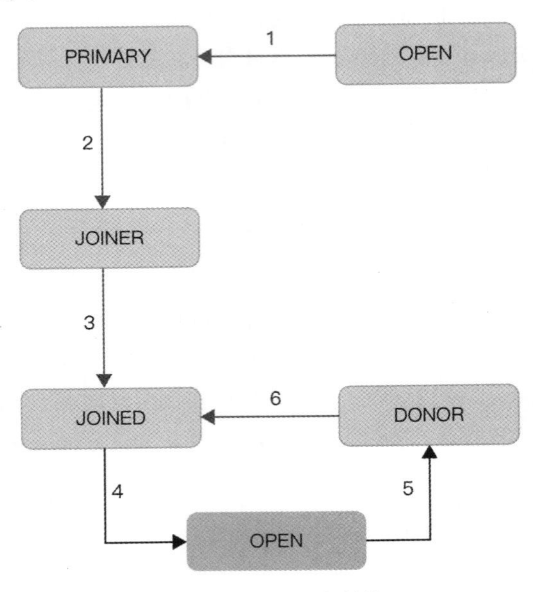

图 14-5　节点状态转换

（1）节点启动并建立到主组件（ Primary Component，PC）的连接。由于网络问题，集群可能被拆分为多个部分，为避免数据差异或“脑裂”问题，此时只能有一部分可以修改数据，这部分称为主组件。

（2）当节点成功执行状态传输请求时，它将开始缓存写集。

（3）节点接收状态快照传输（SST）。它将拥有所有集群数据，并开始应用缓存的写集。

（4）节点完成对集群的追赶。它将 mysql 状态变量 wsrep_ready 设置为值 1，现在允许该节点处理事务。

（5）节点接收状态传输请求，成为捐赠者，缓存它无法应用的所有写集。

（6）节点完成对新加入节点的状态传输。

3. 节点状态与流控

Galera 集群根据节点状态实现多种形式的流控，以保证数据一致性。有四种主要的流控类型：

- 无流控（No Flow Control）：当节点处于 Open 或 Primary 状态时，此流控类型生效。此时节点还不被视为集群的一部分，不允许这些节点复制、应用或缓存任何写集。
- 写集缓存（Write-set Caching）：当节点处于 Joiner 和 Donor 状态时，此流控类型生效。节点在此状态下不能应用任何写集，必须缓存它们以备后用。
- 赶上（Catching Up）：此流控类型在节点处于 Joined 状态时生效，处于该状态的节点可以应用写集。这里的流控确保节点最终能够追赶上集群。由于应用写集通常比处理事务快几倍，因此处于这种状态的节点几乎不会影响集群性能。

- 集群同步（Cluster Sync）：此流控类型在节点处于 Synced 状态时生效。当节点进入该状态时，流控将尝试将接收队列保持最小。

14.1.6 单节点故障与恢复

当一个节点因为硬件、软件和网络等诸多原因与集群失去联系时，都被概括为节点故障。从集群的角度看，若主组件看不到出问题的节点，它将会认为该节点失败。从故障节点本身的角度来看，假设它没有崩溃，那么唯一的迹象是它失去了与主组件的连接。可以通过轮询 wsrep_local_state 状态变量监控 Galera 集群节点的状态。

集群检查从节点最后一次接收到数据包的时间确定该节点是否连接到集群，检查的频率由 evs.inactive_check_period 参数指定，默认值为每隔 0.5 秒检查一次。在检查期间，如果集群发现自上次从节点接收网络数据包以来的时间大于 evs.keepalive_period 参数的值（默认值为 1 秒），则它将开始发出心跳信号。如果集群在 evs.suspect_timeout 参数（默认值为 5 秒）期间没有继续从节点接收到网络数据包，则该节点被声明为 suspect（可疑），表示怀疑该节点已下线。一旦主组件的所有成员都将该节点视为可疑节点，它就被声明为 inactive（失效），即节点失败。如果在大于 evs.inactive_timeout（默认值为 15 秒）的时间内未从节点接收到消息，则无论意见是否一致，都会声明该节点失败。在所有成员同意其成员资格之前，失败节点将保持非操作状态。如果成员无法就节点的活跃性达成一致，说明网络对于集群操作来说太不稳定。这些选项值之间的关系为：

```
    evs.inactive_check_period<=evs.keepalive_period<=evs.suspect_timeout<=
evs.inactive_timeout
```

需要注意，如果网络过于繁忙，以至于无法按时发送消息或心跳信号无响应，也可能被宣布为节点失败，这可以防止集群其余部分的操作被锁。若不希望这样处理，可以增加超时参数。如果用 CAP 原则来衡量，Galera 集群强调的是数据一致性（Consistency），这就导致了集群需要在可用性（Availability）和分区容错性（Partition Tolerance）之间进行权衡。也就是说，当使用的网络不稳定时，低 evs.suspect_timeout 和 evs.inactive_timeout 值可能会导致错误的节点故障检测结果（牺牲 P），而这些参数的较高值可能会导致在实际节点故障的情况下更长的发现时间（牺牲 A）。

集群中的一个节点出现故障不会影响其他节点继续正常工作，单节点故障不会丢失任何数据。失败节点的恢复是自动的，当失败节点重新联机时，它会自动与其他节点同步数据，之后才允许它重新回到集群中。如果在重新同步过程中状态快照传输（SST）失败，会导致接收节点不可用，因为接收节点在检测到状态传输故障时将中止。这种情况下若使用的是 mysqldump 方式的 SST，则需要手动还原。

14.1.7 仲裁

除了单节点故障外，集群还可能由于网络故障而拆分为多个部分。每部分内的节点相互连接，但各部分之间的节点失去连接，这被称为网络分裂（Network Partitioning）。此情况下只有一部分可以继续修改数据库状态，以避免数据差异，这一部分即为主组件。正常情况下主组件就是整个集群。当发生网络分裂时，Galera 集群调用一个仲裁算法选择一部分作为主组件，保证集群中只有一个主组件。

1. 加权法定票数（Weighted Quorum）

集群中的当前节点数量定义了当前集群的大小，集群大小决定达到仲裁所需的票数。Galera 集群在节点不响应并且被怀疑不再是集群的一部分时进行仲裁投票。可以使用 evs.suspect_timeout 参数微调此无响应的超时时间，默认为 5 秒。

发生网络分裂时，断开连接的两侧都有活动节点。主组件要求获得仲裁的多数票，因此具有较多存活节点的部分将成为主组件，而另一部分将进入非主状态并开始尝试与主组件连接，如图 14-6 所示。

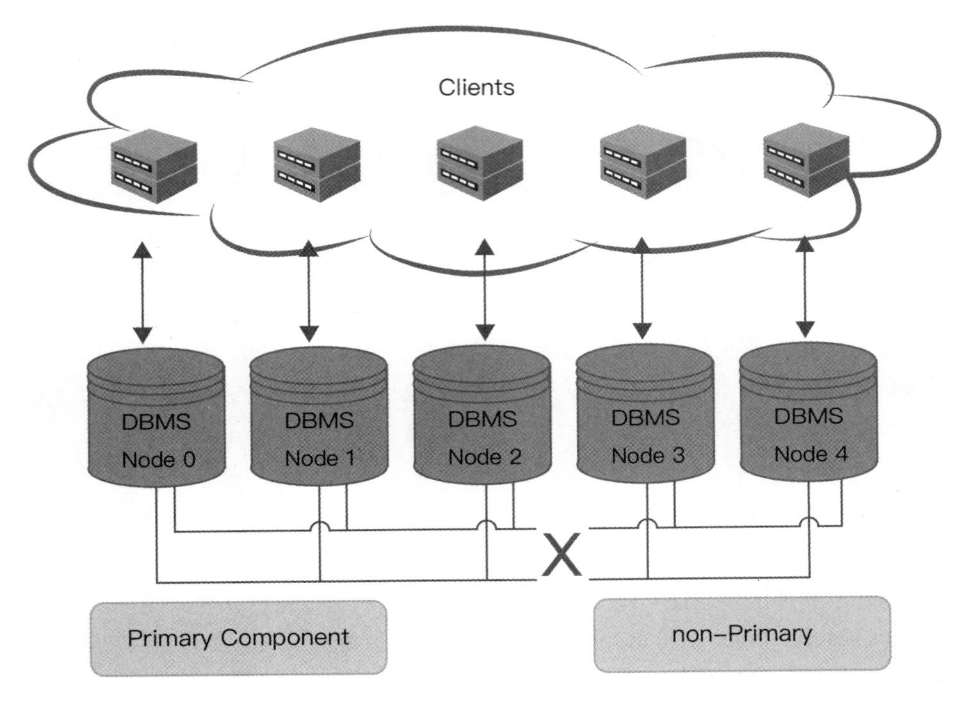

图 14-6 仲裁新主组件

仲裁要求多数，这意味着不能在双节点集群中进行自动故障转移，因为一个节点的故障会导致另一节点自动进入非主状态。而具有偶数个节点的集群则有“脑裂”风险。如果网络分裂导致节点的数量正好分成两半，则两个分区都不能成为主组件，并且都将进入非主状态，如图 14-7 所示。要启用 Galera 集群自动故障切换，至少需要使用三个节点。

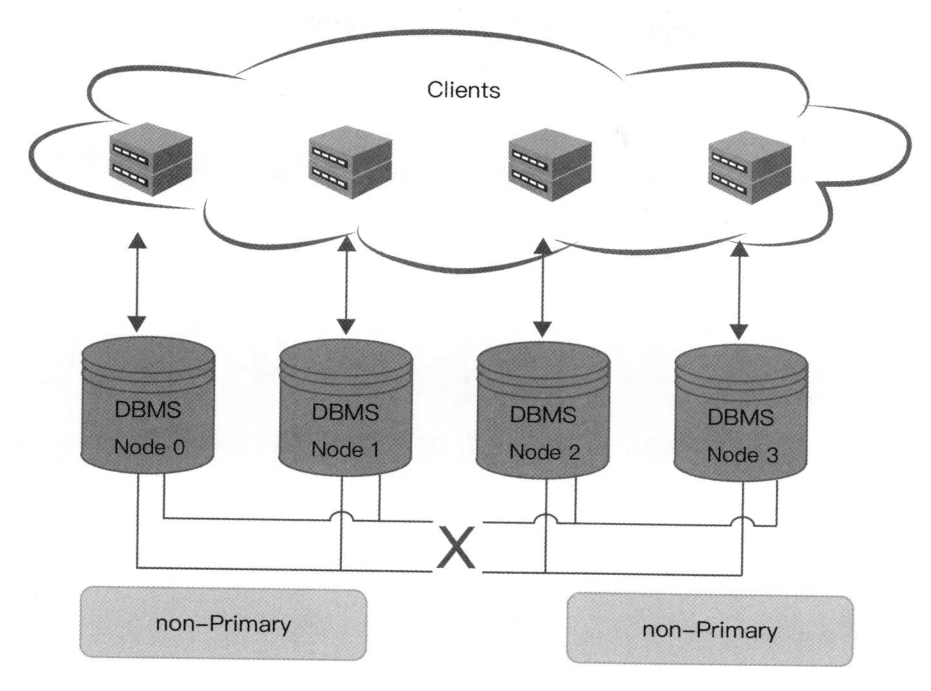

图 14-7　“脑裂”问题示意图

2. 脑裂（Split-Brain）

导致数据库节点彼此独立运行的集群故障称为“脑裂”。这种情况可能导致数据不一致，并且无法修复，例如当两个数据库节点独立更新同一表上的同一行时。与任何基于仲裁的系统一样，当仲裁算法无法选择主组件时，Galera 集群会受到“脑裂”问题的影响。

Galera 设计为避免进入分裂脑状态，如果失败导致将集群分割为两个大小相等的部分，则两部分都不会成为主组件。在节点数为偶数的集群中，为把“脑裂”风险降到最低，可以人为分区将一部分始终划分为集群主组件，例如：

```
4 node cluster -> 3 (Primary) + 1 (Non-primary)
6 node cluster -> 4 (Primary) + 2 (Non-primary)
6 node cluster -> 5 (Primary) + 1 (Non-primary)
```

以上分区示例中，任何中断或失败都很难导致节点完全分成两半。

3. 法定票数计算

Galera 集群支持加权仲裁，其中每个节点可以被分配 0~255 范围内的权重参与计算。法定票数计算公式如下。其中：p_i 表示最后可见的主组件的成员；l_i 为已知正常离开集群的成员；m_i 是当前组件成员；w_i 代表成员权重。

$$\frac{\sum_{p_i \times w_i} - \sum_{l_i \times w_i}}{2} < \sum_{m_i \times w_i}$$

这个公式的含义是：当前节点权重总和大于最后一个主组件节点权重和减去正常离开集群节点权重和的一半时，才会被选为新的主组件。消息传递时带有权重信息，默认的节点权重为 1，此时公式被转换为单纯的节点计数比较。通过设置 pc.weight 参数，可以在运行时更改节点权重，例如：

```
set global wsrep_provider_options="pc.weight=3";
```

4. 加权仲裁示例

在了解了加权仲裁的工作原理后，下面是一些部署模式的示例。

（1）三节点加权仲裁

三个节点配置仲裁权重如下：

```
node1: pc.weight = 2
node2: pc.weight = 1
node3: pc.weight = 0
```

此时如果 node2 和 node3 失效，node1 会成为主组件，而如果 node1 失效，node2 和 node3 都将成为非主组件。

（2）“一主一从”加权仲裁

主节点和从节点配置仲裁权重如下：

```
node1: pc.weight = 1
node2: pc.weight = 0
```

如果主节点失效，node2 将成为非主组件，如果 node2 失效，node1 将继续作为主组件。

（3）“一主多”从加权仲裁

为具有多个从节点的主从方案配置仲裁权重：

```
node1: pc.weight = 1
node2: pc.weight = 0
node3: pc.weight = 0
...
noden: pc.weight = 0
```

如果 node1 失效，所有剩余的节点都将作为非主组件，如果任何其他节点失效，则保留主组件。在网络分裂的情况下，node1 始终作为主组件。

（4）主站点和从站点方案的加权仲裁

为主站点和从站点配置仲裁权重：

```
Primary Site:
  node1: pc.weight = 2
  node2: pc.weight = 2

Secondary Site:
  node3: pc.weight = 1
  node4: pc.weight = 1
```

这种模式下，一些节点位于主站点，而其他节点位于从站点。如果从站点关闭或站点之间的网络连接丢失，则主站点上的节点仍然是主组件。此外，node1 或 node2 其中一个节点崩溃不会让另一个节点成为非主组件。

14.2 安装和配置

本节以搭建三节点 Galera Cluster for MySQL 5.7 为例，说明 Galera 集群的安装步骤与基本配置，示例环境如下：

IP 与主机名：

172.16.1.125 hdp2
172.16.1.126 hdp3
172.16.1.127 hdp4

软件版本：

CentOS Linux release 7.2.1511 (Core)
galera-3.28
mysql-wsrep-5.7.27
Percona-XtraBackup-2.4.15

每台主机的硬件配置：

双核双 CPU，Intel(R) Xeon(R) CPU E5-2420 0 @ 1.90GHz
8GB 物理内存，8GB Swap
100GB 物理硬盘

14.2.1 初始安装

我们从最简单的场景开始，假设在没有任何应用数据和访问的情况下，从头开始安装 Galera 集群。

1. 安装相关软件包

以 root 用户在三台主机执行以下步骤：

步骤 01 安装依赖包：

```
yum install perl-Time-HiRes
yum -y install perl-DBD-MySQL.x86_64
yum -y install libaio*
```

步骤 02 创建 yum 源文件：

```
cat > /etc/yum.repos.d/galera.repo <<-END
[galera]
name = Galera
baseurl = https://releases.galeracluster.com/galera-3.28/centos/7/x86_64
gpgkey = https://releases.galeracluster.com/galera-3.28/
```

```
GPG-KEY-galeracluster.com
    gpgcheck = 1

    [mysql-wsrep]
    name = MySQL-wsrep
    baseurl = https://releases.galeracluster.com/mysql-wsrep-5.7.27-25.19/
centos/7/x86_64
    gpgkey=https://releases.galeracluster.com/mysql-wsrep-5.7.27-25.19/GPG-KEY
-galeracluster.com
    gpgcheck = 1
    END
```

步骤 03 安装 galera-3 与 mysql-wsrep-5.7：

```
yum install -y galera-3 mysql-wsrep-5.7
```

步骤 04 确认相关的 rpm 包：

```
[root@hdp2~]#rpm -qa | grep -E 'galera|wsrep'
mysql-wsrep-client-5.7-5.7.27-25.19.el7.x86_64
galera-3-25.3.28-1.el7.x86_64
mysql-wsrep-common-5.7-5.7.27-25.19.el7.x86_64
mysql-wsrep-libs-5.7-5.7.27-25.19.el7.x86_64
mysql-wsrep-server-5.7-5.7.27-25.19.el7.x86_64
mysql-wsrep-5.7-5.7.27-25.19.el7.x86_64
mysql-wsrep-libs-compat-5.7-5.7.27-25.19.el7.x86_64
[root@hdp2~]#
```

步骤 05 安装 XtraBackup。

如果 SST 使用 XtraBackup 需要执行此步骤。注意 XtraBackup 与 MySQL 服务器的兼容性，如果版本不匹配，则会在 XtraBackup 的日志中报出类似下面的错误：

```
    innobackupex: Error: Unsupported server version: '5.7.27' Please report a bug
 at https://bugs.launchpad.net/percona-xtrabackup
```

对于本示例采用的 mysql-wsrep-5.7.27，可从 https://www.percona.com/downloads/Percona-XtraBackup-2.4/LATEST/下载 xtrabackup 2.4.15 版本。

```
# 安装 XtraBackup
rpm -ivh percona-xtrabackup-24-2.4.15-1.el7.x86_64.rpm
```

至此软件包安装已完成。启动集群只要很少几项必要的配置。

2. 修改配置文件

编辑/etc/my.cnf 文件，增加以下内容：

```
[mysqld]
log-error=/var/log/mysqld.log
wsrep_provider=/usr/lib64/galera-3/libgalera_smm.so
wsrep_cluster_name="mysql_galera_cluster"
wsrep_cluster_address="gcomm://172.16.1.125,172.16.1.126,172.16.1.127"
```

```
    wsrep_sst_method=xtrabackup
    wsrep_sst_auth=wxy:P@sswo2d
    wsrep_node_name=node1                    # 另外两个节点分别为node2、node3
    wsrep_node_address="172.16.1.125"  # 另外两个节点分别为172.16.1.126、
172.16.1.127
```

my.cnf 中系统变量说明：

- log-error：MySQL 错误日志文件，集群初始化后从该文件中查找初始密码。
- wsrep_provider：galera 库文件。
- wsrep_cluster_name：集群名称。
- wsrep_cluster_address：集群节点 IP 地址。
- wsrep_sst_method：SST 方法。
- wsrep_sst_auth：SST 认证信息，XtraBackup 使用此用户名和密码连接数据库实例。
- wsrep_node_name：当前节点名称。
- wsrep_node_address：当前节点地址。

3. 初始化集群

以 root 用户在任一主机执行以下步骤：

步骤 01 启动第一个节点：

```
/usr/bin/mysqld_bootstrap
```

该命令会启动本机的 mysqld 服务，MySQL 默认安装目录为/var/lib/mysql。注意，/usr/bin/mysqld_bootstrap 命令只在集群第一个节点启动时使用，因为该脚本中带有一个参数-wsrep-new-cluster，代表新建集群。

```
# 查看mysqld服务状态
systemctl status mysqld
```

步骤 02 查找并修改初始密码：

```
# 查找初始密码
grep -i 'temporary password' /var/log/mysqld.log

# 修改mysql的root用户密码，需要根据提示输入上一步输出的初始密码
mysqladmin -uroot -p password 'P@sswo2d'
```

步骤 03 创建一个非 root 管理账号：

```
create user wxy identified by 'P@sswo2d';
grant all on *.* to wxy with grant option;
```

4. 启动集群其他节点的 mysqld 服务

在其他两个主机上以 root 用户执行如下命令：

```
systemctl start mysqld
```

5. 验证安装

（1）查看节点数量

```
mysql> show status like 'wsrep_cluster_size';
+--------------------+-------+
| Variable_name      | Value |
+--------------------+-------+
| wsrep_cluster_size | 3     |
+--------------------+-------+
1 row in set (0.00 sec)
```

（2）查看复制情况

三个节点分别建表并插入数据：

```
-- node1
create database test;
use test;
create table t1(a int);
insert into t1 values(1);

-- node2
use test;
create table t2(a int);
insert into t2 values(2);

-- node2
use test;
create table t3(a int);
insert into t3 values(3);
```

在三个节点查询数据，结果一致：

```
mysql> select t1.a,t2.a,t3.a from test.t1,test.t2,test.t3;
+------+------+------+
| a    | a    | a    |
+------+------+------+
|    1 |    2 |    3 |
+------+------+------+
1 row in set (0.00 sec)
```

6. 问题排查

如果在初始化集群或启动 mysqld 服务时，错误日志中出现类似以下的错误：

```
2019-10-05T10:25:29.729981Z 0 [ERROR] WSREP: wsrep_load(): dlopen():
/usr/lib64/galera-3/libgalera_smm.so: symbol SSL_COMP_free_compression_
methods, version libssl.so.10 not defined in file libssl.so.10 with link time
reference
```

这种情况说明 galera 插件没有加载成功。此时 mysqld 仍然可以成功启动，但是以单实例提供读写服务，不会进行节点间数据复制。升级 OpenSSL 软件包即可解决此问题。例如：

```
cd /etc/yum.repos.d/
```

```
wget http://mirrors.163.com/.help/CentOS7-Base-163.repo
yum -y upgrade openssl
```

14.2.2 使用 SST 增加节点

假设 node1 和 node2 是正在使用的 Galera 集群节点，现在要增加第三个节点 node3。目标是在对现有节点无阻塞的前提下，为新增节点进行 SST 数据同步。

（1）将 hdp4 初始化为新节点：

```
# 在 hdp4 上执行
systemctl stop mysqld
rm -rf /var/lib/mysql
rm -rf /var/log/mysqld.log
```

（2）使用 tpcc-mysql 对 hdp2 执行压测，模拟应用负载：

```
# 安装 tpcc-mysql
tar -zxvf tpcc-mysql.tar.gz
cd tpcc-mysql/src
make
cd ..

# 创建压测库表
mysql -uwxy -pP@sswo2d -h172.16.1.125 -e "create database tpcc_test;"
mysql -uwxy -pP@sswo2d -h172.16.1.125 -Dtpcc_test < create_table.sql
mysql -uwxy -pP@sswo2d -h172.16.1.125 -Dtpcc_test < add_fkey_idx.sql

# 准备数据
tpcc_load 172.16.1.125 tpcc_test wxy P@sswo2d 10

# 备份测试库用于重复测试
mysqldump --databases tpcc_test -uwxy -pP@sswo2d -h172.16.1.125 >
tpcc_test.sql

# 执行压测
tpcc_start -h172.16.1.125 -d tpcc_test -u wxy -p "P@sswo2d" -w 10 -c 32 -r 60
-l 300
```

关于 tpcc-mysql 的详细安装及使用说明，请参见 4.2.1 小节。

（3）在压测执行期间，启动 hdp4 上的 MySQL 实例：

```
systemctl start mysqld
```

在 hdp4 启动过程中，hdp2 和 hdp3 都是非阻塞的，但是会报以下错误：

```
Deadlock found when trying to get lock; try restarting transaction
Lock wait timeout exceeded; try restarting transaction
```

在生产系统的 Galera 集群中联机添加节点时需要关注这个问题。当 hdp4 的 MySQL 实例启动成功，后续的压测过程不会再报错。hdp2 的/var/log/mysqld.log 文件中有以下关于 SST 的信息：

```
    2019-10-16T02:04:07.877299Z 2 [Note] WSREP: Assign initial position for
certification: 106026, protocol version: 4
    ...
    2019-10-16T02:04:08.377644Z 0 [Note] WSREP: Running: 'wsrep_sst_xtrabackup
--role 'donor' --address '172.16.1.127:4444/xtrabackup_sst' --socket
'/var/lib/mysql/mysql.sock' --datadir '/var/lib/mysql/' --defaults-file
'/etc/my.cnf' --defaults-group-suffix ''   '' --gtid
'cada8d04-ef2b-11e9-a196-1ea90518b418:106177''
    2019-10-16T02:04:08.380201Z 2 [Note] WSREP: sst_donor_thread signaled with 0
    ...
    2019-10-16T02:05:20.968604Z 0 [Note] WSREP: 0.0 (node3): State transfer from
1.0 (node1) complete.
    2019-10-16T02:09:34.556227Z 0 [Note] WSREP: Member 0.0 (node3) synced with
group.
```

当在 hdp4 上执行 systemctl start mysqld 命令后，集群中的所有节点首先会将已提交的事务刷新到磁盘，之后选择一个捐赠者向新增节点全量同步数据，本例中系统选择了 node1 作为捐赠者。然后 node1 调用 xtrabackup 向 node3 拷贝物理文件，期间产生的写集将被缓存。node1 成为捐赠者时，状态由 SYNCED 转为 DONOR/DESYNCED；当 xtrabackup 备份完成时，其状态变为 JOINED；最后应用完缓存的写集时，状态又从 JOINED 变为 SYNCED。同样从 hdp4 的/var/log/mysqld.log 文件中可以发现，node3 的状态变化过程为：OPEN→PRIMARY→JOINER→JOINED→SYNCED。

14.2.3 使用 IST 增加节点

SST 方法将全量数据从捐赠者节点拷贝到新加入节点，这类似于在 MySQL 主库做一个全备份，然后在从库还原，只不过在 Galera 集群中，该过程依赖于新加入节点的状态而自动触发。对于高并发大库场景下新增节点，SST 方式可能会很痛苦。首先，如果使用 mysqldump 或 rsync 做 SST，捐赠者节点是被阻塞的。其次，对于几百 GB 或更多的数据集，即使网络够快，同步过程也需要几个小时才能完成。所以生产环境新增节点时最好避免使用 SST，而是改用 IST。

IST 只向新节点发送它比捐赠者 Gcache 中少的写集。Gcache 是 Galera 节点保存写集副本的文件。IST 比 SST 快得多，无阻塞，对捐赠者无明显影响。只要可能，这应该是新增节点的首选方案。

有时 SST 是不可避免的，当 Galera 无法确定新增节点状态时会发生这种情况。状态存储在 grastate.dat 文件中，如果发生以下情况将触发 SST：

- MySQL 数据目录下不存在 grastate.dat 文件——节点可能是一个“干净的”新节点。
- grastate.dat 文件中没有 seqno 或 group id——节点可能在 DDL 期间崩溃。
- 由于缺少权限或文件系统损坏，grastate.dat 无法读取。

1. 设置 gcache.size

在上一节介绍 IST 时曾经提到，使用 IST 需要满足两个先决条件：新节点 UUID 与集群相同；Gcache 足够存储增量写集。第一点很好满足，用 XtraBackup 对集群实例做物理备份时会自动为新节点保持集群的 UUID。要满足第二个条件，需要进行一些计算，估计所需 Gcache 的大小。例如，对于 tpcc-mysql 压测，可以按下面的方法估算。

(1)执行压测

```
tpcc_start -h172.16.1.125 -d tpcc_test -u wxy -p "P@sswo2d" -w 10 -c 32 -r 60 -l 300
```

(2)在压测期间执行查询

```
set global show_compatibility_56=on;
set @start := (select sum(variable_value/1024/1024) from information_schema.global_status where variable_name like 'wsrep%bytes');
do sleep(60);
set @end := (select sum(variable_value/1024/1024) from information_schema.global_status where variable_name like 'wsrep%bytes');
select round((@end - @start),2) as `Mb/min`, round((@end - @start),2) * 60 as `Mb/hour`;
```

该查询计算每分钟写的字节数，结果如下：

```
+--------+----------+
| Mb/min | Mb/hour  |
+--------+----------+
| 116.66 | 6999.60  |
+--------+----------+
```

压测总执行时间6分钟（预热1分钟，执行5分钟），gcache.size只要设置为大于117 * 6MB即可，因此这里把gcache.size均设置为1GB，足够演示IST数据同步。在三个节点的配置文件/etc/my.cnf中添加如下参数，然后重启实例使之生效。

```
wsrep_provider_options="gcache.size=1073741824"
```

2. IST测试

同样假设node1和node2是正在使用的Galera集群节点，现在要增加第三个节点node3。为避免SST，将从node1使用XtraBackup创建完整备份，在node3上恢复备份并创建Galera状态文件，以便Galera可以确定节点的状态并跳过SST。为了尽可能接近IST之前的最新数据，还将创建一个增量备份。

(1)将hdp4初始化为新节点

```
# 在hdp4上用root用户执行
systemctl stop mysqld
rm -rf /var/lib/mysql/*
rm -rf /var/log/mysqld.log
rm -rf /tmp/incremental/*
```

(2)向集群重新导入压测库

```
mysql -uwxy -pP@sswo2d -h172.16.1.125 -Dtpcc_test < tpcc_test.sql
```

(3)执行压测模拟生产负载

```
tpcc_start -h172.16.1.125 -d tpcc_test -u wxy -p "P@sswo2d" -w 10 -c 32 -r 60 -l 300
```

接下来的第（4）、（5）、（6）和（7）步均在第（3）步运行期间执行。

（4）手工执行对集群的 XtraBackup 备份

```
# 在 hdp2 上以 mysql 用户执行下面的命令进行全量备份（已经事先配置好了 hdp2 到 hdp4 的免密登录）
innobackupex --defaults-file=/etc/my.cnf --user=wxy --password=P@sswo2d --socket=/var/lib/mysql/mysql.sock --galera-info --no-lock --stream=xbstream ./ | ssh mysql@172.16.1.127 "xbstream -x -C /var/lib/mysql"

# 再执行一个增量备份，这里仅用于演示
scp mysql@172.16.1.127:/var/lib/mysql/xtrabackup_checkpoints /home/mysql/
innobackupex --defaults-file=/etc/my.cnf --user=wxy --password=P@sswo2d --socket=/var/lib/mysql/mysql.sock --incremental --incremental-basedir=/home/mysql --galera-info --no-lock --stream=xbstream ./ | ssh mysql@172.16.1.127 "xbstream -x -C /tmp/incremental"
```

（5）恢复 hdp4 的数据文件

第（4）步执行完成后，在 hdp4 上以 mysql 用户执行以下命令进行恢复：

```
# 恢复全量
innobackupex --apply-log --redo-only /var/lib/mysql/
# 恢复增量
innobackupex --apply-log --redo-only /var/lib/mysql/ --incremental-dir=/tmp/incremental
```

（6）生成 grastate.dat 文件

根据 xtrabackup_galera_info 文件中的内容生成 grastate.dat 文件，用于 IST 增量同步。在 hdp4 上以 mysql 用户执行以下命令：

```
# 查看 xtrabackup_galera_info
cat /var/lib/mysql/xtrabackup_galera_info

# 生成 grastate.dat 文件，uuid 和 seqno 的值来自 xtrabackup_galera_info
tee /var/lib/mysql/grastate.dat <<EOF
# GALERA saved state
version: 2.1
uuid:    650c3acb-eff8-11e9-9905-c73959fd46ca
seqno:   743544
safe_to_bootstrap: 0
EOF
```

（7）启动新节点实例

```
# 以 root 用户在 hdp4 上执行
systemctl start mysqld
```

hdp2 的/var/log/mysqld.log 文件中有以下关于 IST 的信息：

```
2019-10-17T06:37:00.517675Z 1 [Note] WSREP: Assign initial position for certification: 758430, protocol version: 4
2019-10-17T06:37:00.517777Z 0 [Note] WSREP: Service thread queue flushed.
2019-10-17T06:37:00.961803Z 0 [Note] WSREP: Member 2.0 (node3) requested state
```

```
transfer from '*any*'. Selected 1.0 (node2)(SYNCED) as donor.
    2019-10-17T06:37:01.223935Z 0 [Note] WSREP: 1.0 (node2): State transfer to 2.0
(node3) complete.
    2019-10-17T06:37:01.224331Z 0 [Note] WSREP: Member 1.0 (node2) synced with group.
    2019-10-17T06:37:02.301018Z 0 [Note] WSREP: (4ce6e1a5, 'tcp://0.0.0.0:4567')
turning message relay requesting off
    2019-10-17T06:38:14.740957Z 0 [Note] WSREP: 2.0 (node3): State transfer from
1.0 (node2) complete.
    2019-10-17T06:39:31.183193Z 0 [Note] WSREP: Member 2.0 (node3) synced with
group.
```

可以看到，系统选择node2作为捐赠者，它的/var/log/mysqld.log文件中有以下关于IST的信息：

```
    2019-10-17T06:37:00.991588Z 0 [Note] WSREP: Shifting SYNCED ->
DONOR/DESYNCED (TO: 758624)
    2019-10-17T06:37:01.045666Z 2 [Note] WSREP: IST request:
650c3acb-eff8-11e9-9905-c73959fd46ca:743544-758430|tcp://172.16.1.127:4568
    2019-10-17T06:37:01.045701Z 2 [Note] WSREP: wsrep_notify_cmd is not defined,
skipping notification.
    2019-10-17T06:37:01.045885Z 0 [Note] WSREP: Running: 'wsrep_sst_xtrabackup
--role 'donor' --address '172.16.1.127:4444/xtrabackup_sst' --socket
'/var/lib/mysql/mysql.sock' --datadir '/var/lib/mysql/' --defaults-file
'/etc/my.cnf' --defaults-group-suffix ''   '' --gtid
'650c3acb-eff8-11e9-9905-c73959fd46ca:743544' --bypass'
    2019-10-17T06:37:01.046440Z 2 [Note] WSREP: sst_donor_thread signaled with 0
    2019-10-17T06:37:01.048205Z 0 [Note] WSREP: async IST sender starting to serve
tcp://172.16.1.127:4568 sending 743545-758430
```

显示由hdp3向hdp4发送743545-758430之间的写集。可以在hdp4的/var/log/mysqld.log文件中找到node3的IST增量同步和状态变化过程：

```
    2019-10-17T06:37:01.446278Z 2 [Note] WSREP: Receiving IST: 14886 writesets,
 seqnos 743544-758430
    2019-10-17T06:37:01.446519Z 0 [Note] WSREP: Receiving IST...  0.0%
(    0/14886 events) complete.
    ...
    2019-10-17T06:38:14.765552Z 0 [Note] WSREP: Receiving IST...100.0%
(14886/14886 events) complete.
    2019-10-17T06:38:14.765871Z 2 [Note] WSREP: IST received:
650c3acb-eff8-11e9-9905-c73959fd46ca:758430
    2019-10-17T06:38:14.766840Z 0 [Note] WSREP: 2.0 (node3): State transfer from
1.0 (node2) complete.
    2019-10-17T06:38:14.766873Z 0 [Note] WSREP: Shifting JOINER -> JOINED (TO: 777023)
    2019-10-17T06:39:31.208799Z 0 [Note] WSREP: Member 2.0 (node3) synced with group.
    2019-10-17T06:39:31.208825Z 0 [Note] WSREP: Shifting JOINED -> SYNCED (TO:
777023)
    2019-10-17T06:39:31.241137Z 2 [Note] WSREP: Synchronized with group, ready for
connections
    2019-10-17T06:39:31.241155Z 2 [Note] WSREP: wsrep_notify_cmd is not defined,
skipping notification.
```

14.3 管理监控

14.3.1 在线 DDL

MySQL 在线执行 DDL 语句（create table、alter table、create index、grant ...）一直是个令人头疼的操作。一方面大表上的 DDL 语句需要执行很长时间，这是因为 MySQL 的实现，它需要复制一遍表的数据；另一方面在高并发访问的表上执行 DDL 期间会阻塞其上所有 DML（insert、update、delete）语句的执行，直到 DDL 语句执行完。不仅如此，高并发大表上的在线 DDL 还极易产生经典的“Waiting for table metadata lock”等待。

Galera 集群也同样如此，来看下面的例子：

```
-- 创建测试表并装载大量数据
create table t1 as select * from information_schema.tables;
insert into t1 select * from t1;
...
insert into t1 select * from t1;

-- 创建主键
alter table t1 add column id int auto_increment primary key first;

-- 在 session 1 中执行加字段的 DDL
alter table t1 add column c1 int;

-- 在 session 1 的语句执行期间，在 session 2 中执行插入记录的 DML
insert into t1
(table_catalog,table_schema,table_name,table_type,table_comment)
values('a','a','a','a','a');
```

session 2 的 DML 语句会立即返回以下错误，直到 session 1 的 DDL 语句完成，insert 语句才能执行成功。

```
ERROR 1213 (40001): Deadlock found when trying to get lock; try restarting
transaction
```

DDL 语句更改数据库本身，并且是非事务性的（自动提交）。Galera 集群通过两种不同的方法处理 DDL：

- 总序隔离（Total Order Isolation，TOI）：以相同顺序在所有集群节点上执行 DDL，防止在操作期间提交其他事务。
- 滚动升级（Rolling Schema Upgrade，RSU）：在本地执行 DDL，仅影响运行这些更改的节点，更改不会复制到集群的其余部分。

可以配置 wsrep_osu_method 参数指定在线 DDL 方法，默认设置为 TOI。DDL 涉及表锁和 MDL 锁（Meta Data Lock），只要在执行过程中，遇到了 MDL 锁的冲突，所有情况下都是 DDL 优先，将所有使用到这个对象的事务都杀死，被杀的事务都会报出死锁异常，正如前面例子中看到的报错。

1. TOI

如果并不关心集群处理 DDL 语句时其他事务将被阻止，则可使用 TOI 方法。DDL 作为语句复制到集群中的所有节点，节点等待前面的所有事务同时提交，然后单独执行 DDL 更改。在 DDL 处理期间，不能提交其他事务。这种方法的主要优点是它保证了数据的一致性。在使用 TOI 时应考虑以下特性：

- 集群在执行 DDL 之前将其复制为语句。
- 从事务验证的角度看，TOI 模式永远不会与前面的事务冲突，因为它们只在集群提交所有前面的事务之后执行。DDL 更改永远不会使验证失败，并且它们的执行是有保证的。
- DDL 运行时，正在进行的其他事务以及涉及相同数据库资源的事务将在提交时报出死锁错误，并将回滚。
- TOI 可防止单个节点的 DDL 执行出错。

2. RSU

如果要在 DDL 期间保持高可用性，并且避免新旧结构定义之间的冲突，则应该使用 RSU 方法。可以使用 set 语句设置 DDL 执行方法：

```
set global wsrep_osu_method='RSU';
```

RSU 仅在本地节点上处理 DDL。当节点处理表结构更改时，它将与集群解除同步。处理完表结构更改后，它将应用延迟的复制事件并将自身与集群同步。若要在整个集群范围内更改表结构，必须依次在每个节点上手动执行 DDL。在 RSU 期间，集群将继续运行，其中一些节点使用旧表结构，而另一些节点使用新表结构。RSU 的主要优点是一次只阻塞一个节点，主要缺点是可能不安全，如果新旧结构定义在复制事件级别不兼容，则可能会失败。例如：

```
-- 在节点 1 执行
set wsrep_osu_method='RSU';
alter table t1 add column c1 int;
insert into t1(c1) select 1;

-- 在节点 2 执行
alter table t1 add column c1 int;
```

节点 1 向 t1.c1 字段插入值时，节点 2 上并没有 t1.c1 字段，因此数据复制失败。当在节点 2 上手动执行 DDL 添加 t1.c1 字段后，两节点数据不一致。

3. pt-online-schema-change

RSU 只避免了执行 DDL 的节点对其他节点的阻塞，但对于同一节点上 DDL 与 DML 相互影响的问题却无能为力。在当前阶段，解决非阻塞在线 DDL 的终极解决方案是使用 pt-online-schema-change。

pt-online-schema-change 是 percona-toolkit 中的一个工具，功能是无锁定在线修改表结构，要求被修改表具有主键或唯一索引。percona-toolkit 工具包的安装和使用非常简单。例如从 https://www.percona.com/downloads/percona-toolkit/LATEST/下载 percona-toolkit，然后执行下面的命

令进行安装：

```
# 安装依赖包
yum install perl-TermReadKey.x86_64
yum install perl-DBI
yum install perl-DBD-MySQL
yum install perl-Time-HiRes
yum install perl-IO-Socket-SSL

# 安装percona-toolkit
rpm -ivh percona-toolkit-3.1.0-2.el7.x86_64.rpm
```

执行类似下面的命令修改表结构：

```
pt-online-schema-change --alter="add column c1 int;" --execute
D=test,t=t1,u=root,p=P@sswo2d
```

alter 参数指定修改表结构的语句，execute 表示立即执行，D、t、u 和 p 分别指定库名、表名、用户名和密码，执行期间不阻塞其他并行的 DML 语句。pt-online-schema-change 还有许多选项，具体用法可以使用 pt-online-schema-change --help 查看联机帮助。

pt-online-schema-change 工作原理其实很简单：

（1）如果存在外键，根据 alter-foreign-keys-method 参数的值，检测外键相关的表，进行相应设置的处理。如果被修改表存在外键定义但没有使用 --alter-foreign-keys-method 指定特定的值，该工具不予执行。

（2）创建一个新的表，表结构为修改后的数据表，用于从源数据表向新表中导入数据。

（3）创建触发器，用于记录从拷贝数据开始之后，对源数据表继续进行的数据修改。数据拷贝结束后，执行这些操作，保证数据不会丢失。如果表中已经定义了触发器这个工具，就不能工作了。

（4）复制数据，从源数据表复制数据到新表中。

（5）根据修改后的数据，更新外键关联的子表。

（6）把源数据表重新命名为旧表，把新表重新命名为源表名，并将旧表删除。

（7）删除触发器。

14.3.2 恢复主组件

集群节点将主组件状态存储到本地磁盘。节点记录主组件的状态以及连接到它的节点的 UUID。在中断情况下，一旦最后保存状态中的所有节点都实现连接，集群将恢复主组件。如果节点之间的写集位置不同，则恢复过程还需要完成状态快照传输（SST）。

1. 查看主组件状态

节点将主组件状态存储到磁盘时，会将其保存在 MySQL 数据目录下的 gvwstate.dat 文件中，内容类似如下：

```
my_uuid: 4a6cfe9d-f9de-11e9-9ad4-23840b115384
#vwbeg
```

```
view_id: 3 4a6cfe9d-f9de-11e9-9ad4-23840b115384 3
bootstrap: 0
member: 4a6cfe9d-f9de-11e9-9ad4-23840b115384 0
member: 78bdb344-f9de-11e9-bcfa-eb03d339c6d7 0
member: 7d14464b-f9de-11e9-83b3-5b022ee44499 0
#vwend
```

gvwstate.dat 文件分为节点信息和视图信息两部分。节点信息在 my_uuid 字段中提供节点的UUID。视图信息提供有关节点所属主组件视图的信息，该视图包含在 vwbeg 和 vwend 标记之间。view_id 从三个部分构成视图标识符：view_type 始终为 3，表示主视图，view_uuid 和 view_seq 一起构成标识符唯一值。bootstrap 显示节点是否已引导，不影响主组件恢复过程。member 显示此主组件中所有节点的 UUID。

当集群形成或更改主组件时，节点创建并更新此文件，这将确保节点保留其所在的最新主组件的状态。如果节点失去连接，则它具有要引用的文件。如果节点正常关闭，则会删除该文件。

2. 修改主组件状态

如果集群处于需要强制特定节点彼此连接的异常情况，可以通过手动更改保存的主组件状态来执行此操作。注意，正常情况下应完全避免编辑或修改 gvwstate.dat 文件，因为这样做可能会导致意想不到的结果。

当一个节点第一次启动或在正常关机后启动时，它会随机生成一个 UUID 并将其分配给自己，该 UUID 用作集群其余部分的标识符。如果节点在数据目录中找到 gvwstate.dat 文件，它将读取 my_uuid 字段以找到它应该使用的值。通过手动将任意 UUID 值分配给每个节点上的相应字段，可以强制它们在开始时相互连接，形成一个新的主组件。下面看一个例子。

首先停止示例三节点 Galera 集群中的所有 MySQL 实例：

```
systemctl stop mysqld
```

然后在任意节点启动 MySQL 实例都会报错：

```
[root@hdp2/var/lib/mysql]#systemctl start mysqld
Job for mysqld.service failed because the control process exited with error code. See "systemctl status mysqld.service" and "journalctl -xe" for details.
[root@hdp2/var/lib/mysql]#
```

日志中显示如下错误：

```
2019-10-29T00:19:11.470690Z 0 [ERROR] WSREP: failed to open gcomm backend connection: 110: failed to reach primary view: 110 (Connection timed out)
         at gcomm/src/pc.cpp:connect():158
2019-10-29T00:19:11.470710Z 0 [ERROR] WSREP: gcs/src/gcs_core.cpp:gcs_core_open():209: Failed to open backend connection: -110 (Connection timed out)
2019-10-29T00:19:11.470912Z 0 [ERROR] WSREP: gcs/src/gcs.cpp:gcs_open():1458: Failed to open channel 'mysql_galera_cluster' at 'gcomm://172.16.1.125,172.16.1.126,172.16.1.127': -110 (Connection timed out)
2019-10-29T00:19:11.470931Z 0 [ERROR] WSREP: gcs connect failed: Connection timed out
2019-10-29T00:19:11.470941Z 0 [ERROR] WSREP:
```

```
wsrep::connect(gcomm://172.16.1.125,172.16.1.126,172.16.1.127) failed: 7
    2019-10-29T00:19:11.470968Z 0 [ERROR] Aborting
```

很明显，实例启动时取不到主组件视图。由于三个节点中至少要有两个活跃节点才能构成主组件，而现在一个实例都不存在，并且节点找不到集群主组件信息，因而致使节点无法启动。

我们希望三个节点一起启动以形成集群的新主组件。此时需要为每个节点提供一个任意唯一的 UUID 值，例如在其他可用的 MySQL 上执行 select uuid()来获得。下面手工生成三个节点的 gvwstate.dat 文件。节点 1 上的 gvwstate.dat 文件内容如下：

```
my_uuid: 9085dadf-f953-11e9-92e9-005056a50f77
#vwbeg
view_id: 3 9085dadf-f953-11e9-92e9-005056a50f77 3
bootstrap: 0
member: 9085dadf-f953-11e9-92e9-005056a50f77 0
member: 8e2de005-f953-11e9-88b4-005056a5497f 0
member: 7dc3eb7e-f953-11e9-ad17-005056a57a4e 0
#vwend
```

对节点 2 重复该过程：

```
my_uuid: 8e2de005-f953-11e9-88b4-005056a5497f
#vwbeg
view_id: 3 9085dadf-f953-11e9-92e9-005056a50f77 3
bootstrap: 0
member: 9085dadf-f953-11e9-92e9-005056a50f77 0
member: 8e2de005-f953-11e9-88b4-005056a5497f 0
member: 7dc3eb7e-f953-11e9-ad17-005056a57a4e 0
#vwend
```

节点 3 也重复该过程：

```
my_uuid: 7dc3eb7e-f953-11e9-ad17-005056a57a4e
#vwbeg
view_id: 3 9085dadf-f953-11e9-92e9-005056a50f77 3
bootstrap: 0
member: 9085dadf-f953-11e9-92e9-005056a50f77 0
member: 8e2de005-f953-11e9-88b4-005056a5497f 0
member: 7dc3eb7e-f953-11e9-ad17-005056a57a4e 0
#vwend
```

下面启动第一个节点：

```
systemctl start mysqld
```

当节点启动时，Galera 集群将读取每个节点的 gvwstate.dat 文件，从中提取其 UUID 并使用 member 字段的 UUID 来确定它应该连接哪些节点以形成新的主组件。但此时该命令执行的现象是“挂起”。现在集群中还没有主组件，该节点正在等待 SST 完成，正如日志中所显示的：

```
    2019-10-29T00:41:31.058586Z 0 [Note] WSREP: Received NON-PRIMARY.
    2019-10-29T00:41:31.058596Z 0 [Note] WSREP: Waiting for SST to complete.
    2019-10-29T00:41:31.058995Z 2 [Note] WSREP: New cluster view: global state:
4a6db23a-f9de-11e9-ba76-93e71f7c9a45:3, view# -1: non-Primary, number of nodes:
```

```
1, my index: 0, protocol version -1
    2019-10-29T00:41:31.059022Z 2 [Note] WSREP: wsrep_notify_cmd is not defined,
skipping notification.
    2019-10-29T00:41:31.558645Z 0 [Warning] WSREP: last inactive check more than
PT1.5S ago (PT3.50285S), skipping check
```

下面启动第二个节点：

```
systemctl start mysqld
```

依然“挂起”，日志中显示的同节点 1 一样。最后启动第三个节点：

```
systemctl start mysqld
```

此时第三个节点和前面正在启动中的两个节点同时启动成功。只有当主组件包含的所有节点都启动后才能确定 SST 的方向，继而完成整个集群的启动。

14.3.3 重置仲裁

在网络连接出现问题，或超过一半的集群出现故障，或出现“脑裂”等情况时，可能会发现节点不再将自己视为主组件的一部分。可以从 wsrep_cluster_status 变量检查是否发生这种情况：

```
mysql> show global status like 'wsrep_cluster_status';
+----------------------+---------+
| Variable_name        | Value   |
+----------------------+---------+
| wsrep_cluster_status | Primary |
+----------------------+---------+
1 row in set (0.00 sec)
```

返回值 Primary 表示节点是主组件的一部分。当变量返回任何其他值时，表示节点是不可操作组件的一部分。这种情况的节点会向所有应用查询返回未知命令的错误。如果没有任何节点返回 Primary，则意味着需要重置仲裁，这种情况是非常少见的。如果有部分返回 Primary 的节点，则表示是网络连接出现了问题，但不需要重置仲裁。一旦节点连通，它们就会自动与主组件重新同步。

1. 查找最高级别的节点

重置仲裁前需要标识集群中最高级别的节点，即必须找到提交了最后一个事务的节点。无论重置仲裁时使用何种方法，此节点都将作为新主组件的起点。可以使用 wsrep_last_committed 状态变量识别集群中最高级别的节点：

```
mysql> show status like 'wsrep_last_committed';
+----------------------+-------+
| Variable_name        | Value |
+----------------------+-------+
| wsrep_last_committed | 392   |
+----------------------+-------+
1 row in set (0.00 sec)
```

变量返回值是该节点提交的最后一个事务的序号，序号最大的节点是集群中最高级别的节点，将被用作引导新主组件。

2. 重置仲裁

重置仲裁所做的是在可用的最高级别节点上引导主组件，然后该节点作为新的主组件运行，使集群的其余部分与其状态保持一致。有自动和手动两种方法完成仲裁重置，首选方法是自动方法。自动引导在每个节点上保留写集缓存 Gcache，这意味着当新主组件启动时，可以使用增量状态转移（IST）而不是速度慢得多的状态快照转移（SST）进行自我配置。

自动引导是在最高级别节点上动态启用 wsrep_provider_options 参数为 pc.bootstrap 来完成的：

```
set global wsrep_provider_options='pc.bootstrap=yes';
```

该节点现在作为新主组件中的起始节点运行。如果可能，具有网络连接的组件中的节点将尝试启动增量状态传输，否则使用状态快照传输，以使其自己的数据库保持最新。

手动引导时，首先需要关闭集群，然后从最高级别的节点开始重新启动集群。手动引导集群需要完成以下步骤：

步骤01 关闭所有集群节点：

```
systemctl stop mysqld
```

步骤02 使用 wsrep-new-cluster 选项启动最高级别的节点：

```
/usr/bin/mysqld_bootstrap
```

如果命令执行失败，并且日志中显示以下错误信息：

```
    2019-10-29T02:17:41.041493Z 0 [ERROR] WSREP: It may not be safe to bootstrap
 the cluster from this node. It was not the last one to leave the cluster and may
not contain all the updates. To force cluster bootstrap with this node, edit the
grastate.dat file manually and set safe_to_bootstrap to 1 .
    2019-10-29T02:17:41.041501Z 0 [ERROR] WSREP:
wsrep::connect(gcomm://172.16.1.125,172.16.1.126,172.16.1.127) failed: 7
    2019-10-29T02:17:41.041511Z 0 [ERROR] Aborting
```

则说明该节点不是最高级别节点。Galera 认为这种情况下 bootstrap 是不安全的，因为可能丢失事务。如果要强制执行引导，可以编辑 grastate.dat 文件，将 safe_to_bootstrap 设置为 1，然后再执行 mysqld_bootstrap 命令。

首个节点使用前一个集群中的可用数据初始化一个新集群。当其他节点启动时，它们会连接到此节点并请求状态快照传输，以使自己的数据库保持最新。

步骤03 逐次启动集群中的其他节点：

```
systemctl start mysqld
```

14.3.4 管理流控

集群通过全局排序同步复制更改，但从原始节点异步应用这些更改。为了防止任何一个节点落后集群太多，Galera 实现了一种称为流控的反馈机制。节点将接收到的写集按全局顺序排队，并在本地数据库中应用和提交它们。如果接收队列太大，节点将启动流控，此时节点将暂停复制而处理接收队列。一旦接收队列减小到一个阈值，节点就会恢复复制。

1. 查看流控状态

Galera 集群提供全局状态变量用于监视流控，这些变量分为计数流控暂停事件的状态变量和记录暂停复制时长的状态变量。

```
mysql> show status like 'wsrep_flow_control_%';
+------------------------------+----------+
| Variable_name                | Value    |
+------------------------------+----------+
| wsrep_flow_control_paused_ns | 0        |
| wsrep_flow_control_paused    | 0.000000 |
| wsrep_flow_control_sent      | 0        |
| wsrep_flow_control_recv      | 0        |
+------------------------------+----------+
4 rows in set (0.01 sec)
```

流控使用 fc_pause 事件通知集群它正在暂停复制。Galera 集群提供了两个状态变量来监视此事件。

- wsrep_flow_control_sent：显示自上次状态查询以来本地节点发送的流控暂停事件数。
- wsrep_flow_control_recv：显示自上次状态查询以来集群上的流控暂停事件数，包括来自其他节点的事件数和本地节点发送的事件数。

除了跟踪流控暂停事件数之外，Galera 集群还可以跟踪自上次 FLUSH STATUS 以来由于流控而暂停复制的时长。

- wsrep_flow_control_paused：以秒为单位的暂停复制时长。
- wsrep_flow_control_paused_ns：以纳秒为单位的暂停复制时长。

2. 配置流控

Galera 集群提供了两组参数管理节点如何处理复制速率和流控，一组控制写集缓存，另一组控制流控的触发或取消条件。以下三个参数决定节点如何响应复制速率的更改。

- gcs.recv_q_hard_limit：设置最大接收队列的大小，单位是字节。参数值取决于内存、交换区大小和性能考虑。在 32 位系统上，默认值为 ssize_max 减 2GB。64 位系统没有实际限制，默认值为 LLONG_MAX。如果某个节点超过此限制，并且 gcs.max_throttle 未设置为 0.0，则该节点将因内存不足错误而中止。如果 gcs.max_throttle 设置为 0.0，则集群中的复制将停止。
- gcs.max_throttle：限制状态传输期间的复制速率，以避免耗尽内存，默认值为 0.25。如果将参数设置为 1.0，则节点不会限制复制速率。如果将参数设置为 0.0，则可以完全停止复制。
- gcs.recv_q_soft_limit：默认值为 0.25。当复制速率超过软限制时，节点计算在此期间的平均复制速率（以字节为单位）。之后，节点会随着缓存大小线性降低复制速率，以便在 recv_q_hard_limit 下达到 gcs.max_throttle 乘以平均复制速率的值。

以下参数控制节点触发流控的条件以及用于确定何时应断开流控并恢复复制。

- gcs.fc_limit：此参数确定流控触发点。当接收队列中的事务数超过此限制时，节点将暂停复制，默认值为 16。对于多主配置，必须将此限制保持在较低值，以减少验证冲突。如果是主从设置，可以使用更高的值来减少流控干预，降低从库复制延迟。
- gcs.fc_factor：此参数用于确定节点何时可以取消流控，默认值为 1。当节点上的接收队列低于 gcs.fc_limit * gcs.fc_factor 的值时将恢复复制。

虽然使用尽可能小的接收队列对于多主操作来说非常重要，但接收列长度在主从设置中并不是那么重要。根据应用程序和硬件的不同，节点可能在几秒钟内应用 1k 个写集。接收队列长度对故障转移没有影响。

集群节点彼此异步处理事务，节点不能以任何方式预期复制数据的数量，因此流控总是被动的。也就是说，流控只有在节点超过某些限制后才会生效，它并不能防止超过这些限制。

14.3.5 自动逐出

当 Galera 集群发现某个节点出现异常，如很长的响应时间时，可以启动一个进程将该节点从集群中永久删除，此过程称为自动逐出。

1. 配置自动逐出

集群中的每个节点监视集群中所有其他节点的组通信响应时间。当集群在一个节点响应延时，它会向延迟列表中生成一个关于该节点的条目。如果延迟节点在固定时间内再次响应，则该节点的条目将从延迟列表中移除。但如果节点接收到足够多的延迟条目，并且在大多数集群的延迟列表中都可以找到该条目，则会将延迟节点从集群中永久逐出，被逐出的节点重启后才能重新加入集群。

通过 wsrep_provider_options 设置以下选项，可以配置自动逐出的参数：

- evs.delayed_margin：节点响应时间大于该参数定义的时长，则将条目添加到延迟列表，默认为 1 秒。必须设置为大于节点之间往返延时（Round-Trip Time，RTT）的值。
- evs.delayed_keep_period：从被添加到延迟列表，到此参数定义的时间范围内，如果该节点再次响应，则将其从延迟列表条目中删除，默认为 30 秒。
- evs.evict：如果设置为某个节点的 UUID，则该节点将从集群中被逐出。
- evs.auto_evict：定义节点在触发自动逐出协议之前允许的延迟节点条目数，默认值为 0，表示禁用节点上的自动逐出协议，但集群继续监视节点的响应时间。
- evs.version：此参数确定节点使用的 EVS 协议的版本。为了确保向后兼容，默认值为 0。启用自动逐出需要将该参数设置为更高版本，例如在配置文件中添加：

```
wsrep_provider_options="evs.version=1"
```

2. 检查逐出状态

可以通过 Galera 状态变量检查节点逐出状态。

- wsrep_evs_state：提供 evs 协议的内部状态。
- wsrep_evs_delayed：提供延迟列表中以逗号分隔的节点列表。列表中使用的格式是 uuid:address:count。计数是指给定延迟节点的条目数。

- wsrep_evs_evict_list：列出被逐出节点的UUID。

14.3.6 启用 Galera 仲裁员

Galera 仲裁员是参与投票但不参与实际复制的集群成员。虽然 Galera 仲裁员不参与复制，也不存储数据，但它接收的数据与所有其他节点相同，因此必须保证它的网络连接。当集群具有偶数个节点时，仲裁员作为奇数节点发挥作用，以避免出现“脑裂”的情况。具有仲裁员的集群架构如图 14-8 所示。

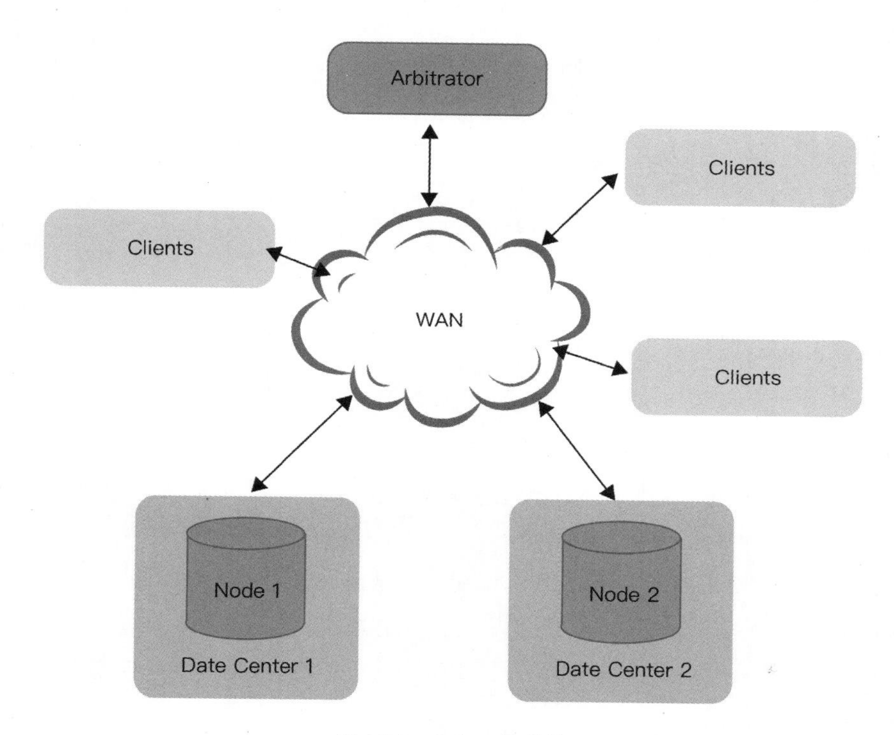

图 14-8 Galera 仲裁员

Galera 仲裁员是 Galera 集群的一个独立守护进程，名为 garbd。这意味着必须从集群单独启动它，并且不能通过 my.cnf 配置文件配置 Galera 仲裁员。可以从 Shell 启动仲裁员，或者作为服务来运行。如何配置 Galera 仲裁员取决于如何启动它。Galera 仲裁员启动时，脚本将在进程中以用户 nobody 身份来执行 sudo 语句。默认的 sudo 配置将阻止没有 tty 访问权限的用户操作。要更正此问题，编辑/etc/sudoers 文件并注释掉此行：

```
Defaults requiretty
```

1. 从 Shell 启动 Galera 仲裁程序

编辑配置文件 arbitrator.config，内容如下：

```
# arbitrator.config
group = mysql_galera_cluster
address = gcomm://172.16.1.125,172.16.1.126,172.16.1.127
```

然后启动仲裁员：

```
garbd --cfg /var/lib/mysql/arbitrator.config &
```

2. 启动 Galera 仲裁员服务

编辑/etc/sysconfig/garb 文件，内容如下：

```
# 已存在的两个节点地址
GALERA_NODES="172.16.1.125:4567 172.16.1.126:4567"
# group 名称保持与两节点的 wsrep_cluster_name 系统变量一致
GALERA_GROUP="mysql_galera_cluster"
# 日志文件
LOG_FILE="/var/log/garb.log"
```

修改日志文件属性后，启动 Galera 仲裁员服务：

```
touch /var/log/garb.log
chown nobody:nobody /var/log/garb.log
chmod 644 /var/log/garb.log
systemctl start garb
```

3. “脑裂”测试

在节点 3 停止 garb 服务。

```
systemctl stop garb
```

在节点 2 drop 丢弃前往第一个节点和仲裁员的数据包。

```
iptables -A OUTPUT -d 172.16.1.125 -j DROP
iptables -A OUTPUT -d 172.16.1.127 -j DROP
```

检查节点 1 和节点 2 的状态，都不是 Synced，于是出现了“脑裂”问题。

```
mysql> show status like 'wsrep_local_state_comment';
+---------------------------+-------------+
| Variable_name             | Value       |
+---------------------------+-------------+
| wsrep_local_state_comment | Initialized |
+---------------------------+-------------+
1 row in set (0.00 sec)
```

清除节点 2 的数据包过滤规则。

```
iptables -F
```

启动节点 3 的 garb 服务。

```
systemctl start garb
```

前两个节点查看集群节点数，结果是 3，说明包括了仲裁节点。

```
mysql> show status like 'wsrep_cluster_size';
+--------------------+-------+
| Variable_name      | Value |
+--------------------+-------+
```

```
| wsrep_cluster_size | 3     |
+--------------------+-------+
1 row in set (0.00 sec)
```

在节点 2 丢弃前往第一个节点和仲裁员的数据包。

```
iptables -A OUTPUT -d 172.16.1.125 -j DROP
iptables -A OUTPUT -d 172.16.1.127 -j DROP
```

这时检查节点 1 的同步状态，仍然是 Synced（即同步的），没有发生“脑裂”。

```
mysql> show status like 'wsrep_local_state_comment';
+---------------------------+--------+
| Variable_name             | Value  |
+---------------------------+--------+
| wsrep_local_state_comment | Synced |
+---------------------------+--------+
1 row in set (0.00 sec)
```

再在节点 1 查看集群节点数，结果是 2，说明节点 1 成为集群主组件。

```
mysql> show status like 'wsrep_cluster_size';
+--------------------+-------+
| Variable_name      | Value |
+--------------------+-------+
| wsrep_cluster_size | 2     |
+--------------------+-------+
1 row in set (0.00 sec)
```

检查节点 2 的同步状态和集群节点数，说明它已和集群主组件断开连接。

```
mysql> show status like 'wsrep_local_state_comment';
+---------------------------+-------------+
| Variable_name             | Value       |
+---------------------------+-------------+
| wsrep_local_state_comment | Initialized |
+---------------------------+-------------+
1 row in set (0.01 sec)

mysql> show status like 'wsrep_cluster_size';
+--------------------+-------+
| Variable_name      | Value |
+--------------------+-------+
| wsrep_cluster_size | 1     |
+--------------------+-------+
1 row in set (0.01 sec)
```

14.3.7 Galera 集群监控

Galera 集群监控主要包括查询状态变量、使用脚本监控和检查数据库服务器日志。

1. 查询状态变量

可以使用标准查询来检查整个集群中写集复制的状态：

```
mysql> show global status like 'wsrep_%';
+-----------------------------+----------------------------------------
----------------------+
| Variable_name              | Value                               |
+-----------------------------+----------------------------------------
+
| wsrep_local_state_uuid       | 4a6db23a-f9de-11e9-ba76-93e71f7c9a45  |
| wsrep_protocol_version       | 9                                   |
| ...                        | ...                             |
| wsrep_ready                | ON                             |
+-----------------------------+----------------------------------------
+
60 rows in set (0.00 sec)
```

2. 使用通知脚本

固然可以通过查询状态变量获得集群状态、节点状态和复制的运行状况，但登录每个节点执行此类查询是何等烦琐。作为更好的替代方法，Galera 集群提供了一种名为通知脚本（Notification Script）的方法，可以通过定制脚本来自动化集群的监控过程。/usr/share/mysql/wsrep_notify 是一个 Galera 自带的示例脚本。该脚本先定义了一个创建数据库表语句的字符串，然后定义了两个函数来维护表数据，最后给出如何处理通知参数。

（1）通知参数

当节点在自身或集群中状态更改时，它将触发通知脚本的执行，并把一些参数传递给通知脚本。下面是参数列表及其基本含义。

- status：节点传递一个指示其当前状态的字符串，值是以下 6 个值之一：
  - Undefined：表示不属于主组件的节点。
  - Joiner：表示作为主组件一部分并正在接收状态快照传输（SST）的节点。
  - Donor：表示作为主组件一部分并正在发送状态快照传输（SST）的节点。
  - Joined：表示作为主组件一部分的节点，该节点处于完成加入集群的状态并正在追赶集群。
  - Synced：表示与集群同步的节点。
  - Error：表示发生了错误。此状态字符串可能会提供一个错误代码，其中包含有关发生情况的详细信息。

通知脚本必须捕获 status 参数的值并执行相应的操作。

- uuid：传一个表示节点 UUID 的字符串。
- primary：传一个 yes 或 no 的字符串，标识它是否认为自己是主组件的一部分。
- members：传送当前集群成员列表，格式为 <node UUID> / <node name> / <incoming address>。Node UUID 指节点唯一标识符，Node Name 指 wsrep_node_name 参数定义的节点名，Incoming Address 指客户端连接的 IP 地址，由参数 wsrep_node_incoming_address 设置，默认值为 auto。

- index：传一个字符串，该字符串指示其在成员资格列表中的索引值。

（2）启用通知脚本

可以通过配置文件中的 wsrep_notify_cmd 参数启用通知脚本。在每个节点上创建可执行的Shell 文件/home/mysql/wsrep_notify.sh，然后设置 wsrep_notify_cmd 参数值为/home/mysql/wsrep_notify.sh。

```
    cp /usr/share/mysql/wsrep_notify /home/mysql/wsrep_notify.sh
    chown mysql:mysql /home/mysql/wsrep_notify.sh
    chmod 755 /home/mysql/wsrep_notify.sh
    sed -i 's/rootpass/P@sswo2d/' /home/mysql/wsrep_notify.sh
    mysql -uroot -pP@sswo2d -e "set global
wsrep_notify_cmd='/home/mysql/wsrep_notify.sh';"
```

节点将为集群成员资格和节点状态的每次更改调用脚本。下面简单说明一下脚本执行情况。在节点 2 丢弃前往第一个和第三个节点的数据包。

```
    iptables -A OUTPUT -d 172.16.1.125 -j DROP
    iptables -A OUTPUT -d 172.16.1.127 -j DROP
```

去掉数据包过滤。

```
    iptables -F
```

在节点 2 上查询 wsrep 库表。

```
    [root@hdp3/var/lib/mysql]#mysql -uroot -pP@sswo2d -e "select * from
wsrep.membership;select * from wsrep.status;"
    mysql: [Warning] Using a password on the command line interface can be insecure.
    +-----+--------------------------------------+-------+-------------------
+
    | idx | uuid                                 | name  | addr              |
    +-----+--------------------------------------+-------+-------------------
+
    |   0 | 605dc61c-fa2e-11e9-8b01-9f8f00127724 | node3 | 172.16.1.127:3306 |
    |   1 | 6c468740-fa1a-11e9-8214-62a1abb74da6 | node2 | 172.16.1.126:3306 |
    |   2 | ac8e4a3f-fadb-11e9-ad70-0a894466f015 | node1 | 172.16.1.125:3306 |
    +-----+--------------------------------------+-------+-------------------
+
    +------+-------+--------+--------------------------------------+------+
    | size | idx   | status | uuid                                 | prim |
    +------+-------+--------+--------------------------------------+------+
    |    3 |     1 | Synced | 4a6db23a-f9de-11e9-ba76-93e71f7c9a45 |    1 |
    +------+------+--------+--------------------------------------+------+
    [root@hdp3/var/lib/mysql]#
```

利用 iptables 数据包过滤，使节点 2 的状态发生变化，此时触发执行了通知脚本。这里只使用了 Galera 自带的示例脚本，可以将它作为编写自定义通知脚本的起点，如加入响应集群更改的警报等。

3. 检查数据库服务器日志

log_error 系统变量指定 MySQL 服务器错误日志文件名，默认将写入数据目录中的 <hostname>.err 文件。Galera 集群提供参数和 wsrep 选项，启用错误日志记录复制冲突事件。

- wsrep_log_conflicts：此参数启用错误日志的冲突日志记录，例如两个节点试图同时写入同一表的同一行，默认为 OFF。
- cert.log_conflicts：此 wsrep_provider_options 选项允许在复制期间记录有关验证失败的信息，默认为 NO。
- wsrep_debug：此参数启用数据库服务器日志的调试信息，默认为 OFF。

可以通过 my.cnf 配置文件启用这些功能：

```
wsrep_log_conflicts=ON
wsrep_provider_options="cert.log_conflicts=ON"
wsrep_debug=ON
```

注意，wsrep_debug 参数会将数据库服务器的连接密码记录到错误日志中。这是一个安全漏洞，因此不要在生产环境中启用它。

除了 MySQL 服务器日志外，如果 wsrep_sst_method 指定为 xtrabackup，Galera 还会将 SST 信息记录到数据目录下的 innobackup.backup.log 文件中。当一个节点无法应用事务时，MySQL 服务器就在数据目录中创建该事件的特殊二进制日志文件，文件命名为 GRA_*.log。该文件的读取可以参考网页：https://community.pivotal.io/s/article/How-to-decode-Galera-GRA-logs-for-MySQL-for-PCF-v1-10。

14.4 性能测试

本节使用 tpcc-mysql 压测工具对示例三节点 Galera 集群进行一系列性能测试，目标为：

- 验证 Galera 的同步复制，检查是否存在复制延迟。
- 对比 Galera 与 MySQL 组复制的每秒事务数（TPS）。
- 验证多线程复制对 Galera 性能的影响。
- 验证流控对 Galera 性能的影响。

14.4.1 测试规划

这里采用与 MySQL 组复制性能测试相同的环境和方法，不过组复制与 Galera 使用的 MySQL 版本不同。组复制测试用的是 MySQL 8.0.16 版本，而 Galera 测试用的是 Galera Cluster for MySQL 5.7。之所以版本不同，是因为在测试时我们使用的都是当时的最新版本。具体测试思路与环境参见 6.3.2 小节。

需要注意 tpcc-mysql 与 Galera 集群的兼容性，并不是所有版本的 tpcc-mysql 安装包都支持

Galera。在我们的示例环境中，最新的tpcc-mysql-master在Galera Cluster for MySQL 5.7上执行数据加载（tpcc_load）时就报错。

Galera是多主复制，集群中的多个节点对等。为了便于与MySQL组复制推荐的单主模式进行比较，我们只在Galera集群的节点1上执行压测。Galera使用自己定义的GTID，当前版本也没有提供master_pos_wait或wait_for_executed_gtid_set类似功能的函数，因此需要修改测试脚本以获得压测在节点2和节点3上的执行时间。修改后的tpcc_test.sh文件内容如下：

```
    # 初始化tpcc数据
    mysql -uwxy -pP@sswo2d -h172.16.1.125 -Dtpcc_test < tpcc_test.sql

    # 获取节点1的last_committed用于比较
    read last_committed < <(mysql -uwxy -pP@sswo2d -h172.16.1.125 -e "show status
like 'wsrep_last_committed';" --skip-column-names | awk '{print $2}' | sed
"s/\\\n//g")

    # 等待其他两个节点执行完初始化复制
    read last_committed_1 < <(mysql -uwxy -pP@sswo2d -h172.16.1.126 -e "show status
like 'wsrep_last_committed';" --skip-column-names | awk '{print $2}' | sed
"s/\\\n//g")
    read last_committed_2 < <(mysql -uwxy -pP@sswo2d -h172.16.1.127 -e "show status
like 'wsrep_last_committed';" --skip-column-names | awk '{print $2}' | sed
"s/\\\n//g")

    while [ $last_committed_1 -lt $last_committed -o $last_committed_2 -lt
$last_committed ]
    do
        read last_committed_1 < <(mysql -uwxy -pP@sswo2d -h172.16.1.126 -e "show
status like 'wsrep_last_committed';" --skip-column-names | awk '{print $2}' | sed
"s/\\\n//g")
        read last_committed_2 < <(mysql -uwxy -pP@sswo2d -h172.16.1.127 -e "show
status like 'wsrep_last_committed';" --skip-column-names | awk '{print $2}' | sed
"s/\\\n//g")
    done

    # 开始时间
    start_time=`date '+%s'`

    # 开始事务序号
    read start_last_committed < <(mysql -uwxy -pP@sswo2d -h172.16.1.125 -e "show
status like 'wsrep_last_committed';" --skip-column-names | awk '{print $2}' | sed
"s/\\\n//g")

    # 节点1执行压测，10个仓库，32个并发线程，预热1分钟，压测5分钟
    tpcc_start -h172.16.1.125 -d tpcc_test -u wxy -p "P@sswo2d" -w 10 -c 32 -r 60
-l 300 > tpcc_test.log 2>&1

    # 获取节点1的last_committed用于比较
    read last_committed < <(mysql -uwxy -pP@sswo2d -h172.16.1.125 -e "show status
like 'wsrep_last_committed';" --skip-column-names | awk '{print $2}' | sed
```

```
"s/\\\n//g")

    # 等待其他两个节点执行完复制
    read last_committed_1 < <(mysql -uwxy -pP@sswo2d -h172.16.1.126 -e "show status
like 'wsrep_last_committed';" --skip-column-names | awk '{print $2}' | sed
"s/\\\n//g")
    read last_committed_2 < <(mysql -uwxy -pP@sswo2d -h172.16.1.127 -e "show status
like 'wsrep_last_committed';" --skip-column-names | awk '{print $2}' | sed
"s/\\\n//g")

    while [ $last_committed_1 -lt $last_committed -o $last_committed_2 -lt
$last_committed ]
    do
        if [ $last_committed_1 -lt $last_committed ]
        then
            read last_committed_1 < <(mysql -uwxy -pP@sswo2d -h172.16.1.126 -e "show
status like 'wsrep_last_committed';" --skip-column-names | awk '{print $2}' | sed
"s/\\\n//g")
        else
            end_time1=`date '+%s'`
        fi

        if [ $last_committed_2 -lt $last_committed ]
        then
            read last_committed_2 < <(mysql -uwxy -pP@sswo2d -h172.16.1.127 -e "show
status like 'wsrep_last_committed';" --skip-column-names | awk '{print $2}' | sed
"s/\\\n//g")
        else
            end_time2=`date '+%s'`
        fi
    done

    if [ ! $end_time1 ]; then
        end_time1=`date '+%s'`
    fi

    if [ ! $end_time2 ]; then
        end_time2=`date '+%s'`
    fi

    # 复制执行时长
    elapsed1=$(($end_time1 - $start_time))
    elapsed2=$(($end_time2 - $start_time))

    # 执行的事务数
    trx=$(($last_committed - $start_last_committed))

    # 计算三个节点的 TPS
    Master_TPS=`expr $trx / 360`
    Slave1_TPS=`expr $trx / $elapsed1`
```

```
Slave2_TPS=`expr $trx / $elapsed2`

# 打印输出
echo "TRX: $trx"
echo "Node1 TPS: $Master_TPS"
echo "Elapsed1: $elapsed1" "Node2 TPS: $Slave1_TPS"
echo "Elapsed2: $elapsed2" "Node3 TPS: $Slave2_TPS"
```

当三个节点的 last_committed 相等时，它们执行了相同的事务数。如果存在复制延迟，则节点 2 或节点 3 会比节点 1 稍后执行到 last_committed 点。

14.4.2　测试过程

每次测试只需要执行 tpcc_test.sh 即可。

1. 默认配置

获得默认配置的测试结果，作为后面不同配置的对比基准。测试结果如下：

```
TRX: 76472
Node1 TPS: 212
Elapsed1: 360 Node2 TPS: 212
Elapsed2: 360 Node3 TPS: 212
```

可以看到，虽然 Galera 只是虚拟同步复制，每个节点上的事务验证是异步的，但实际测试中没有复制延迟，压测节点 1 与复制节点 2、节点 3 的执行时间和 TPS 相同。这一点与组复制大相径庭。单主模式的组复制中，相同压测主库比从库的 TPS 高一倍。另一方面，默认配置组复制中主库的 TPS 比 Galera 高一倍，也就是说 Galera 的性能与单主模式组复制中的从库大致相当。因为两者 MySQL 版本不同，这里的测试结果只作参考。

2. 多线程

上一节提到 wsrep_cert_deps_distance 状态变量用于指定并行提交的事务序号之差，可用作 wsrep_slave_threads 的参考值。

```
mysql> show status like 'wsrep_cert_deps_distance';
+--------------------------+-----------+
| Variable_name            | Value     |
+--------------------------+-----------+
| wsrep_cert_deps_distance | 56.673657 |
+--------------------------+-----------+
1 row in set (0.00 sec)
```

wsrep_cert_deps_distance 的值为 56，因此在三个节点执行下面的 SQL 命令，指定复制线程数为 60。

```
set global wsrep_slave_threads=60;
```

然后再次执行测试，TPS 提高了 40%，结果如下：

```
TRX: 106848
```

```
Node1 TPS: 296
Elapsed1: 360 Node2 TPS: 296
Elapsed2: 360 Node3 TPS: 296
```

3. 流控

查看流控相关的状态变量：

```
mysql> show status like 'wsrep_flow_control_%';
+------------------------------+--------------+
| Variable_name                | Value        |
+------------------------------+--------------+
| wsrep_flow_control_paused_ns | 947249076448 |
| wsrep_flow_control_paused    | 0.037075     |
| wsrep_flow_control_sent      | 0            |
| wsrep_flow_control_recv      | 932          |
+------------------------------+--------------+
4 rows in set (0.00 sec)
```

接收队列中的事务数超过 gcs.fc_limit 时触发流控，节点将暂停复制，gcs.fc_limit 的默认值 16 显然太小。在三个节点上执行下面的 SQL 命令，指定流控限制为 1000。

```
set global wsrep_provider_options = 'gcs.fc_limit = 1000';
```

再次执行测试，流控状态变量的值如下，确认没有触发流控。

```
mysql> show status like 'wsrep_flow_control_%';
+------------------------------+--------------+
| Variable_name                | Value        |
+------------------------------+--------------+
| wsrep_flow_control_paused_ns | 947249076448 |
| wsrep_flow_control_paused    | 0.000000     |
| wsrep_flow_control_sent      | 0            |
| wsrep_flow_control_recv      | 0            |
+------------------------------+--------------+
4 rows in set (0.00 sec)
```

测试结果如下：

```
TRX: 103760
Node1 TPS: 288
Elapsed1: 361 Node2 TPS: 287
Elapsed2: 361 Node3 TPS: 287
```

可见，是否触发流控对于 TPS 并没有明显影响。

由以上测试过程和结果我们得出：

- Galera 是同步复制，节点间无延迟。
- Galera 比单主模式组复制中的主库，TPS 相差一半。
- wsrep_slave_threads 参数对 TPS 影响较大。
- 是否触发流控对 TPS 没有明显影响。

14.5 负载均衡

Galera 负载均衡器（Galera Load Balancer，GLB）为客户端的请求提供了一个简单的 TCP 连接平衡功能，它与 Galera Cluster 的关系类似于 MySQL Router 之于组复制。它从另一个轻量级负载均衡器 Pen 中汲取灵感，开发中考虑了可伸缩性和性能，但仅限于平衡 TCP 连接。GLB 提供了以下几个功能：

- 支持在运行时配置后端服务器。
- 支持服务器排除（Draining）。
- 支持用于改善路由性能的 epoll API。
- 支持多线程操作。
- 可选的看门狗模块，用于监视目的地址和调整路由表。

14.5.1 安装

与 Galera Cluster 不同，GLB 没有提供二进制安装包，需要以 root 用户执行下面的命令从源代码文件来构建 GLB 的可执行文件。

```
# 安装依赖包
yum install gcc* libtool
# 下载 GLB 源代码文件
git clone https://github.com/codership/glb
# 在 git 创建的 glb 目录中，运行 bootstrap 脚本
cd glb/
./bootstrap.sh
# 配置
./configure
# 编译
make
# 安装
make install
```

成功执行了上述所有操作后，就可以使用 glbd 命令从命令行启动 GLB。除了系统守护程序外，还安装了 libglb 共享库，用于对使用 C 标准库中 connect()调用的任何 Linux 应用程序进行连接平衡。

上面的过程只安装了 GLB 软件，以便从命令行手动启动它，不过，有时将应用程序作为系统服务来运行更利于维护。若要将 GLB 安装为服务，则只需把 files 目录下的两个文件复制到相应位置即可。

```
# GLB 脚本文件
cp files/glbd.sh /etc/init.d/glb
# GLB 配置文件
cp files/glbd.cfg /etc/sysconfig/glbd
```

14.5.2 配置

编辑/etc/sysconfig/glbd 配置文件，内容如下：

```
LISTEN_ADDR="8010"
CONTROL_ADDR="127.0.0.1:8011"
CONTROL_FIFO="/var/run/glbd.fifo"
THREADS="4"
MAX_CONN=256
DEFAULT_TARGETS="172.16.1.125:3306 172.16.1.126:3306 172.16.1.127:3306"
OTHER_OPTIONS="--round-robin"
```

配置项的说明：

- LISTEN_ADDR：监听客户端连接的地址。这里只给出了端口，意为绑定本机所有网卡的地址。
- CONTROL_ADDR：控制地址，用于显示 GLB 信息。
- CONTROL_FIFO：控制 FIFO 文件，它总是打开的。
- THREADS：连接池线程数，通常为每个 CPU 内核分配一些线程。
- MAX_CONN：最大连接数，可能需要修改系统打开文件的限制（ulimit 的 open file）。
- DEFAULT_TARGETS：空格分隔的目标服务器地址，格式为 IP[:PORT[:WEIGHT]]，权重 WEIGHT 和端口 PORT 可选。WEIGHT 默认值为 1，PORT 默认为 LISTEN_ADDR 指定的端口。
- OTHER_OPTIONS：其他 glbd 命令行选项，这里指定了目标选择策略为轮询。

如果在 glbd 命令行中没有指定目标选择策略，则必须将参数添加到 glbd 配置文件里的 OTHER_OPTIONS 选项中。GLB 支持以下 5 种目标选择策略：

- Least Connected：最少连接，这是默认策略。将新连接定向到当前连接最少的服务器。它将根据服务器权重进行调整。
- Round Robin：轮询，使用 - round 选项来启用。将新连接循环定向到 DEFAULT_TARGETS 列表中的下一个目标地址。
- Single：新连接指向具有最大权重的单台服务器，使用 - single 选项来启用。路由将持续定向到该服务器，直到它失败，或者直到具有更高权重的服务器可用为止。
- Random：随机连接到可用服务器，使用 - random 选项来启用。
- Source Tracking：把来自同一地址的连接定向到同一台服务器，使用 - source 选项来启用。

14.5.3 启动

```
[root@manager~/glb]#service glb start
[Sat Feb 29 14:16:15 CST 2020] glbd: starting...
glb v1.0.1 (epoll)
Incoming address: 0.0.0.0:8010, control FIFO: /var/run/glbd.fifo
```

```
    Control  address:  127.0.0.1:8011
    Number of threads: 4, max conn: 256, nodelay: ON, keepalive: ON, defer accept: OFF, linger: OFF, daemon: YES, lat.count: 0, policy: 'round-robin', top: NO, verbose: NO
    Destinations: 3
       0:    172.16.1.125:3306 , w: 1.000
       1:    172.16.1.126:3306 , w: 1.000
       2:    172.16.1.127:3306 , w: 1.000
       INFO: glb_daemon.c:44: Changing effective user to 'daemon'
    [Sat Feb 29 14:16:15 CST 2020] glbd: started, pid=542577
    [root@manager~/glb]#
```

查看状态：

```
[root@manager~/glb]#service glb getinfo
Router:
------------------------------------------------------------------
        Address       :    weight   usage    map  conns
   172.16.1.125:3306  :     1.000   0.000    N/A     0
   172.16.1.126:3306  :     1.000   0.000    N/A     0
   172.16.1.127:3306  :     1.000   0.000    N/A     0
------------------------------------------------------------------
Destinations: 3, total connections: 0 of 256 max
[root@manager~/glb]#
```

结果中显示出可用服务器列表、它们的权重和使用率，以及与它们建立的连接数。

以下是 glb 服务脚本支持的操作：

- start 启动 GLB。
- stop 停止 GLB。
- restart 重启 GLB
- getinfo 检索当前路由信息。
- getstats 提供与集群相关的性能统计信息。
- add<IP Address>从路由表中添加 IP 地址。
- remove<IP Address>从路由表中删除指定的 IP 地址。
- drain<IP Address>将指定服务器设置为 drain。执行此操作时，GLB 不会向给定服务器发送新连接，但也不会终止现有连接，而是会等待到指定服务器的连接正常结束。

运行时添加 IP 地址，必须遵循的格式为：IP 地址:端口:权重。可以使用主机名代替 IP 地址。

14.5.4 测试

```
    [root@manager~/glb]#mysql -uwxy -pP@sswo2d -h127.0.0.1 -P8010 -N -s -e "select @@wsrep_node_name;"
    Warning: Using a password on the command line interface can be insecure.
    node1
    [root@manager~/glb]#mysql -uwxy -pP@sswo2d -h127.0.0.1 -P8010 -N -s -e "select @@wsrep_node_name;"
```

```
    Warning: Using a password on the command line interface can be insecure.
    node2
    [root@manager~/glb]#mysql -uwxy -pP@sswo2d -h127.0.0.1 -P8010 -N -s -e "select
@@wsrep_node_name;"
    Warning: Using a password on the command line interface can be insecure.
    node3
    [root@manager~/glb]#mysql -uwxy -pP@sswo2d -h127.0.0.1 -P8010 -N -s -e "select
@@wsrep_node_name;"
    Warning: Using a password on the command line interface can be insecure.
    node1
    [root@manager~/glb]#
```

每个通过 8010 端口的客户端新连接都循环指向下一个有效服务器，可见已成功使用 GLB 完成 Galera 集群的负载均衡。

14.6 小　结

Galera Cluster 是一个 MySQL 高可用集群解决方案，包含在 MySQL 的一个分支 MariaDB 中，同时也支持 Percona XtraDB 和 Oracle 的 MySQL。Galera Cluster 的主要特点是真正的“多主多活”。集群中任何一个节点均可读写，并且在任何节点上的查询结果都保持一致，有效地避免了 MySQL 传统复制的主从延迟，以及主库出现问题时的故障转移过程。为了实现这样的功能，同时达到可接受的性能，Galera Cluster 使用了组通信、写集、数据库状态机和事务重排等多种技术。它与 MGR 在实现上有相似之处，但从两者较新版本的性能对比测试来看，Galera Cluster 比 MGR 还是略逊一筹。无论如何，Galera Cluster 依然是 MySQL 高可用集群的候选方案之一。

第 15 章

DRBD

大多数 MySQL 高可用解决方案是基于 MySQL 自带的各种复制技术。本质上是将一个实例上的数据更新或事务，在其他实例上进行重放，从而完成数据同步。当生产实例出现问题无法提供服务时，应用连接切换到其他实例，实现数据库服务持续可用。从这个角度看，使用 DRBD 实现 MySQL 高可用的方式则完全不同，它与 MySQL 本身的复制毫无关系，而是利用存储层面的块复制，可以说是另辟蹊径。

15.1 DRBD 简介

DRBD 的全称是 Distributed Replicated Block Device，即分布式复制块设备，是一个用软件实现的、无共享的、服务器之间以镜像块设备对内容进行存储复制的解决方案。DRBD 镜像块设备，是按数据位镜像成的数据块。简单来说，DRBD 是实现活动节点存储数据更新后自动复制到备用节点相应存储位置的一款软件，它采用的是一种数据块级别的物理复制。

15.1.1 工作原理

图 15-1 所示为 DRBD 的工作栈模型。可以看到 DRBD 需要运行在各个节点上，而且是运行在节点主机的内核中，所以 DRBD 是内核模块，从 Linux 2.6.33 版本起开始整合到内核中。

图 15-1 中假设左节点为活动节点（Primary，也就是主节点），右节点为备用节点（Secondary，也就是从节点或辅助节点）。左节点接收到数据发往内核的数据通路，DRBD 在数据通路中注册钩子程序来检查数据，当发现接收到的数据是发送到自己管理的存储位置，就复制另一份，一份存储到本机的 DRBD 存储设备，另一份发给 TCP/IP 协议栈，通过网卡网络传输到右节点主机的 TCP/IP 协议栈。而对端节点运行的 DRBD 模块同样在数据通路上检查数据，当发现传输过来的数据时，

就保存到 DRBD 存储设备对应的位置。

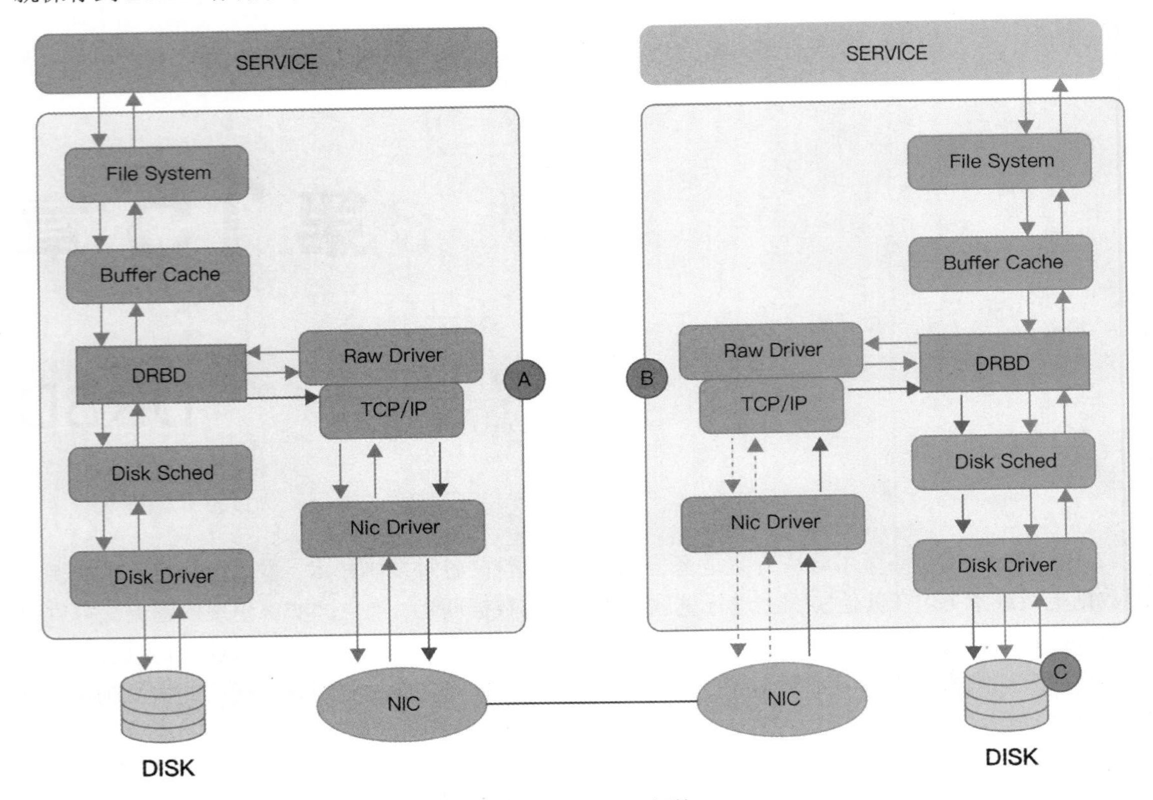

图 15-1 DRBD 架构

如果左节点宕机，右节点可以在高可用集群中成为活动节点，把接收到的数据先存储到本地，当左节点恢复上线时，再把宕机后右节点变动的数据镜像到左节点。镜像过程完成后还需要返回成功或失败的回应消息，这个回应消息可以在传输过程中的不同位置返回。如上图 A、B 和 C 所标识的位置，可以分为三种复制模式：

- A：Async，异步。本地写成功后立即返回，数据放在发送缓冲区中，数据可能会丢失，但传输性能好。
- B：Semi Sync，半同步。对方接收到数据后，但在数据还没有落盘前就返回。
- C：Sync，同步。本地和对方写成功落盘确认后再返回，数据可靠性高，生产系统中一般都采用这种方式。

15.1.2 基本概念

1. DRBD 支持的底层设备

DRBD 需要在底层设备上构建出一个块设备。对于用户来说，一个 DRBD 设备，就像是一块物理的磁盘，可以在 DRBD 设备内创建文件系统。DRBD 所支持的底层设备类别包括：磁盘或者是磁盘的一个分区、Soft Raid 设备（软 RAID 设备）、LVM 的逻辑卷、EVMS（Enterprise Volume

Management System，企业卷管理系统）卷，或其他任何块设备。

2. DRBD 资源

DRBD 资源定义 DRBD 管理的存储空间及相关信息，主要配置以下四个选项：

- 资源名称：可以是除了空白字符外的任意 ACSII 码字符。
- DRBD 设备：在双方节点上 DRBD 的设备文件，一般为/dev/drbdN，主设备号是 147。
- 磁盘：在双方节点上，各自提供的存储设备。
- 网络配置：双方数据同步时所使用的网络属性。

3. DRBD 配置工具

DRBD 主要有以下三个配置工具：

- drbdadm：高级管理工具，管理/etc/drbd.conf，向 drbdsetup 和 drbdmeta 发送指令。
- drbdsetup：配置加载进操作系统内核的 DRBD 模块，平时很少直接使用。
- drbdmeta：管理元数据结构，平时很少直接使用。

4. DRBD 与 RAID1 区别

RAID1 也是实现不同存储设备间的数据镜像，不同的是 RAID1 的各个存储设备是通过连接一个 RAID 控制器而接入到一台主机中，而 DRBD 是通过网络实现不同节点主机存储设备数据的复制。因此也有把 DRBD 叫作网络 RAID1 的说法。

5. DRBD 与共享存储

如果各节点访问的一个数据文件都在同一个存储空间内，就是说共享的数据就一份，则这个存储空间就是共享存储。而 DRBD 定义上就指出了“无共享”，不同节点存储设备空间是镜像关系。DRBD 可以直接应用在主备模型的集群中，也可通过高可用软件如 corosycn 应用在双主模型集群中，不过这就需要 DML/OCFS2/GFS2 等分布式集群文件系统为双主读写的时候分配锁。还有就是 DRBD 为 NFS 共享存储提供高可用镜像备份。

15.2　DRBD 安装和配置

15.2.1　测试环境

IP 地址与主机名：

172.16.1.125　node1

172.16.1.126　node2

软件版本：

CentOS Linux release 7.2.1511 (Core)

DRBD 9.0.20

MySQL 5.6.14

DRBD：
磁盘分区：/dev/sdb
DRBD 设备：/dev/drbd0
DRBD 资源：mysql
挂载点（Mount Point）：/mnt

15.2.2 安装前的准备工作

在安装和配置 DRBD 前，需要做一些准备工作，以下步骤需要在两个节点中都执行。

1. 设置静态 IP

编辑网卡的配置文件，将 BOOTPROTO 设置为 static，如 node1 的网卡配置文件/etc/sysconfig/network-scripts/ifcfg-ens32 内容如下：

```
# Generated by dracut initrd
NAME="ens32"
DEVICE="ens32"
ONBOOT=yes
NETBOOT=yes
UUID="adb62466-2361-405e-ada9-b48fe7c09546"
IPV6INIT=yes
BOOTPROTO=static
TYPE=Ethernet
IPADDR=172.16.1.125
NETMASK=255.255.255.0
GATEWAY=172.16.1.254
DNS1=172.16.1.10
```

然后重启网络服务使配置生效：

```
service network restart
```

2. 配置域名解析

编辑/etc/hosts 文件配置域名解析，如 node1 的内容如下：

```
127.0.0.1   localhost localhost.localdomain localhost4
localhost4.localdomain4
::1         localhost localhost.localdomain localhost6 localhost6.localdomain6

172.16.1.125 node1
172.16.1.126 node2
```

3. 禁用 SELINUX 与防火墙

编辑/etc/selinux/config 文件，设置 SELINUX=disabled。

执行下面的命令禁用防火墙：

```
systemctl stop firewalld
systemctl disable firewalld
```

4. 设置双向 ssh 免密

```
# 在 node1 上执行：
ssh-keygen -t rsa
ssh-copy-id 172.16.1.126

# 在 node2 上执行：
ssh-keygen -t rsa
ssh-copy-id 172.16.1.125
```

5. 设置 NTP 时钟同步

```
yum install ntp
systemctl enable ntpd.service
service ntpd start
```

15.2.3 下载安装 DRBD

在两个节点执行以下命令：

```
# 导入 GPG-KEY，用于验证签名
rpm --import https://www.elrepo.org/RPM-GPG-KEY-elrepo.org
# 安装 yum 源
yum install https://www.elrepo.org/elrepo-release-7.0-4.el7.elrepo
.noarch.rpm
# 安装 DRBD 和内核相关软件包
yum install -y drbd90-utils kmod-drbd90 kernel*
# 因为升级了内核，需要重启系统
reboot
```

然后向内核加载 DRBD 模块并确认：

```
[root@node2~]#modprobe drbd
[root@node2~]#lsmod | grep drbd
drbd_transport_tcp    22144  1
drbd               568697  3 drbd_transport_tcp
libcrc32c            12644  4 xfs,drbd,ip_vs,nf_conntrack
[root@node2~]#
```

15.2.4 配置 DRBD

DRBD 的主配置文件为/etc/drbd.conf。为了管理的便捷性，一般会将这些配置文件分成多个部分，并保存到/etc/drbd.d 目录中，主配置文件中仅使用“include”指令将这些配置文件的各个片断整合起来。通常，/etc/drbd.d 目录中的 DRBD 配置文件为 global_common.conf 以及所有以.res 结尾的文件。其中 global_common.conf 中主要定义 global 段和 common 段，而每一个.res 的文件用于定义一个资源。

在配置文件中，global 段仅能出现一次，并且如果所有的配置信息都保存至同一个配置文件中而不分开为多个文件的话，global 段则必须位于配置文件的最开始处，common 段则用于定义被每一个资源默认继承的参数，而且能够在资源定义中使用的参数都可以在 common 段中定义。在实际应用中，common 段并非必须的，不过我们建议将多个资源共享的参数定义为 common 段中的参数，以降低配置文件的复杂度。

resource 段则用于定义 DRBD 资源，每个资源通常定义在一个单独的位于/etc/drbd.d 目录下的以.res 结尾的文件中。资源在定义时必须为其命名，名字可以由非空白的 ASCII 字符组成。每一个资源段的定义里至少要包含两个 host 子段，以便定义此资源关联到的节点，其他参数均可以从 common 段或 DRBD 的默认值中进行继承而无须定义。

在 node1 上完成下面的 1~4 步骤的操作。

步骤 01 查看主配置文件/etc/drbd.conf 内容。

```
[root@node1~]#cat /etc/drbd.conf
# You can find an example in  /usr/share/doc/drbd.../drbd.conf.example

include "drbd.d/global_common.conf";
include "drbd.d/*.res";
[root@node1~]#
```

/etc/drbd.conf 主配置文件里面配置了包含关系，不需要做任何修改，直接使用即可。

步骤 02 配置/etc/drbd.d/global_common.conf 文件。

文件内容如下：

```
global {
    usage-count no;             # 不统计 DRBD 的使用情况
}

common {
        protocol C;             # 使用 DRBD 的同步协议，数据可靠性高
        disk {
        on-io-error detach; # 配置 I/O 错误处理策略为分离
    }
}
```

on-io-error 策略可能为以下选项之一：

- detach：这是默认和推荐的选项。如果在节点上发生了底层的硬盘 I/O 错误，它会将设备运行在 Diskless 无盘模式下。
- pass_on：DRBD 将 I/O 错误报告到上层。在主节点上，它会将其报告给挂载的文件系统。
- local-in-error：调用本地磁盘 I/O 处理程序定义的命令。这需要有相应的 local-io-error 调用的资源处理程序，给管理员权利调用 local-io-error 处理 I/O 错误。

/etc/drbd.d/global_common.conf 是全局配置文件，这里只设置了使用统计量、同步协议、I/O 错误处理策略三项。该文件中还有一些其他选项，每个选项的含义在安装后的样例文件中都有相应说明。

步骤 03 配置资源文件/etc/drbd.d/mysql.res。

文件内容如下，资源文件名前缀要和资源名相同：

```
resource mysql {                              # 资源名称
  disk /dev/sdb;                              # 磁盘分区
  device /dev/drbd0;                          # DRBD 设备
  meta-disk internal;                         # 元数据存储方式
  on node1 {
    device /dev/drbd0;
    disk /dev/sdb;
    address 172.16.1.125:7789;               # 节点 1 的地址
  }
  on node2 {
    device /dev/drbd0;
    disk /dev/sdb;
    address 172.16.1.126:7789;               # 节点 2 的地址
  }
}
```

Metadata（元数据）有 internal 和 external 两种存储方式，在每个 resource 配置段中指定。配置成使用 internal 意味着 DRBD 把它的 metadata 和实际生产数据存储于相同的底层物理设备中。该存储方式是在设备的最后位置留出一个区域来保存 metadata。

步骤 04 将配置文件远程复制到 node2。

```
scp -rp /etc/drbd.d/* node2:/etc/drbd.d/
```

步骤 05 初始化设备元数据。

分别在两节点上创建 DRBD 设备元数据：

```
drbdadm create-md mysql
```

步骤 06 启动 DRBD 资源。

分别在两节点上启动 DRBD 资源：

```
drbdadm up mysql
```

步骤 07 查看节点角色。

当前两台均为备机：

```
[root@node1~]#drbdadm role mysql
Secondary
```

步骤 08 将 node1 设置 primary（主设备）。

在 node1 执行如下命令：

```
[root@node1~]#drbdadm --force primary mysql
[root@node1~]#drbdadm role mysql
Primary
```

primary 表示资源目前为主设备（或主节点），并且可能正在被读写，如果不是双主则只会出现在其中一个节点上。secondary 表示资源目前为从设备（从节点），正常接收对等节点的更新。

步骤 09 在主节点对磁盘进行格式化。

```
[root@node1~]#mkfs.xfs /dev/drbd0
meta-data=/dev/drbd0             isize=512    agcount=4, agsize=327668 blks
     =                       sectsz=512   attr=2, projid32bit=1
     =                       crc=1        finobt=0, sparse=0
data     =                       bsize=4096   blocks=1310671, imaxpct=25
     =                       sunit=0      swidth=0 blks
naming   =version 2              bsize=4096   ascii-ci=0 ftype=1
log      =internal log           bsize=4096   blocks=2560, version=2
     =                       sectsz=512   sunit=0 blks, lazy-count=1
realtime =none                   extsz=4096   blocks=0, rtextents=0
```

因为文件系统的挂载只能在主节点进行，所以也只有在设置了主节点后才能对 DRBD 设备进行格式化，这样格式化后的全部信息都会镜像到另一个节点。

步骤 10 在主节点挂载设备。

```
mount /dev/drbd0 /mnt
```

步骤 11 查看状态。

查看两个节点的磁盘状态和资源状态：

```
[root@node1/mnt]#drbdadm dstate mysql
UpToDate/UpToDate
[root@node1/mnt]#drbdadm cstate mysql
Connected
```

节点的磁盘有可能为下列状态之一：

- Diskless：本地没有块设备分配给 DRBD 使用。这表示没有可用的设备，或者使用 drbdadm 命令手工分离，或是底层的 I/O 错误导致自动分离。
- Attaching：读取无数据时候的瞬间状态。
- Failed：本地块设备报告 I/O 错误的状态，其下一个状态为 Diskless 无盘。
- Negotiating：在已经连接的 DRBD 设置进行 Attach 读取无数据前的瞬间状态。
- Inconsistent：数据是不一致的，出现在两个节点上初始的完全同步之前。这种状态出现后立即创建一个新的资源。此外，同步期间在目标节点上出现这种状态。
- Outdated：数据资源是一致的，但是已经过时。
- DUnknown：当对等节点网络连接不可用时出现这种状态。
- Consistent：一个没有连接的节点数据一致。当建立连接时，它决定数据是 UpToDate 或是 Outdated。
- UpToDate：一致的最新数据状态，这个状态为正常状态。

一个资源可能有以下连接状态中的一种：

- StandAlone：网络配置不可用，资源还没有被连接或是被管理断开（使用 drbdadm disconnect 命令），或是由于出现认证失败或“脑裂”的情况。
- Disconnecting：断开只是临时状态，下一个状态是 StandAlone。
- Unconnected：是尝试连接前的临时状态，下一个状态可能为 WFconnection 或 WFReportParams。
- Timeout：与对等节点连接超时，是临时状态，下一个状态为 Unconnected。
- BrokerPipe：与对等节点连接丢失，是临时状态，下一个状态为 Unconnected。
- NetworkFailure：与对等节点进行连接后的临时状态，下一个状态为 Unconnected。
- ProtocolError：与对等节点进行连接后的临时状态，下一个状态为 Unconnected。
- TearDown：临时状态，对等节点关闭，下一个状态为 Unconnected。
- WFConnection：等待和对等节点建立网络连接。
- WFReportParams：已经建立 TCP 连接，本节点等待从对等节点传来的第一个网络包。
- Connected：DRBD 已经建立连接，数据镜像现在可用，节点处于正常状态。
- StartingSyncS：完全同步，由管理员发起的同步刚开始，未来可能的状态为 SyncSource 或 PausedSyncS。
- StartingSyncT：完全同步，由管理员发起的同步刚开始，下一状态为 WFSyncUUID。
- WFBitMapS：部分同步刚开始，下一步可能的状态为 SyncSource 或 PausedSyncS。
- WFBitMapT：部分同步刚开始，下一步可能的状态为 WFSyncUUID。
- WFSyncUUID：同步即将开始，下一步可能的状态为 SyncTarget 或 PausedSyncT。
- SyncSource：以本节点为源的同步正在进行。
- SyncTarget：以本节点为目标的同步正在进行。
- PausedSyncS：以本地节点为持续同步的源，但是目前同步已经暂停，可能是因为另外一个同步正在进行或是使用命令 drbdadm pause-sync 暂停了同步。
- PausedSyncT：以本地节点为持续同步的目标，但是目前同步已经暂停，可能是因为另外一个同步正在进行或是使用命令 drbdadm pause-sync 暂停了同步。
- VerifyS：以本地节点为验证源的线上设备验证正在执行。
- VerifyT：以本地节点为验证目标的线上设备验证正在执行。

15.3 测试 MySQL 数据同步

两个节点都已事先安装了 MySQL 数据库服务器。

步骤 01 将 MySQL 数据目录设置为 DRBD 挂载点。

在两个节点的 my.cnf 配置文件中修改设置：

```
datadir = /mnt/
```

步骤 02 在主节点上启动 MySQL 服务。

```
# 停止MySQL服务
service mysql stop
# 将数据目录拷贝到挂载点
cp -r /data/* /mnt/
# 将数据目录的属主改为mysql
chown -R mysql:mysql /mnt/
# 启动MySQL服务
service mysql start
```

步骤 03 在主节点上建立数据库表。

```
create database db1;
use db1;
create table t1 (a int);
insert into t1 select 1;
commit;
```

步骤 04 手工执行主从切换，验证数据同步。

（1）将 node1 降为从节点

```
service mysql stop
umount /mnt
drbdadm secondary mysql
```

（2）将 node2 升为主节点，并启动 MySQL 服务

```
drbdadm primary mysql
mount /dev/drbd0 /mnt
chown -R mysql:mysql /mnt
service mysql start
```

（3）验证数据同步

```
mysql> select * from db1.t1;
+-------+
| a     |
+-------+
|     1 |
+-------+
1 row in set (0.00 sec)
```

可以看到数据已经同步，在 node2 上已经有了新建的库表数据。

15.4 Heartbeat+DRBD+MySQL 高可用方案

前面的 DRBD 环境充其量只是一个容灾配置，还算不上高可用方案。设想当主节点的 MySQL 数据库出现问题时，需要人为发现故障，并手工执行主从切换，这个过程不可避免地会导致长时间停止服务。要保证主服务器不间断地提供服务，需要实现自动故障转移，与 DRBD 一起使用比较

多的是 Heartbeat。顾名思义，Heartbeat 像是在主从两个 DRBD 之间接了一条心跳线，当从节点接收不到主节点的心跳时接管服务。Heartbeat 双机热备软件用来保证数据库服务的稳定性和连续性，数据一致性由 DRBD 来保证。在默认情况下，只有一台 MySQL 实例在工作，当主 MySQL 服务器出现问题后，系统将自动切换到备机上继续提供服务。这个方案的优点显而易见：安全性、稳定性、可用性高，出现故障可自动切换；但缺点也彰明较著：只有一台服务器提供服务，成本相对较高，不方便扩展，可能会发生“脑裂”。

15.4.1　配置

实验环境的架构如图 15-2 所示。

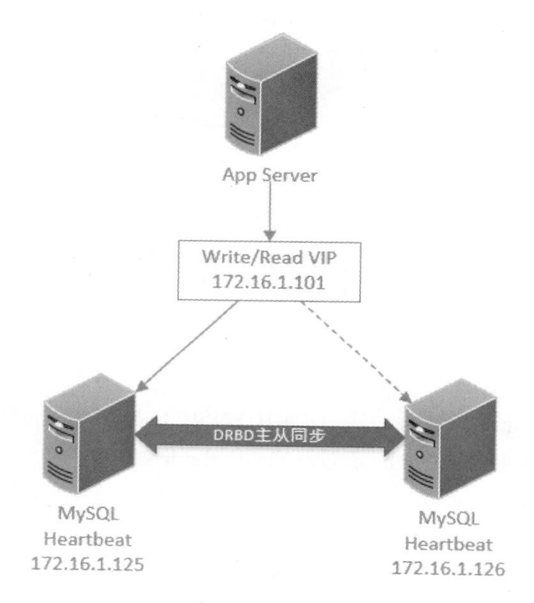

图 15-2　DRBD+Heartbeat+MySQL 高可用架构

前面已经完成了 DRBD+MySQL 的安装部署，下面配置 Heartbeat。Heartbeat 的安装参见 12.3.2 小节。

步骤 01　编辑配置文件 ha.cf。

node1 上的/usr/local/heartbeat/etc/ha.d/ha.cf 文件内容如下：

```
debugfile /var/log/ha-debug
logfile /var/log/ha-log
logfacility local0
keepalive 1
deadtime 30
warntime 10
initdead 120
udpport 694
bcast ens32
```

```
ucast ens32 172.16.1.126
auto_failback off
node node1
node node2
ping 172.16.1.254
respawn hacluster /usr/local/heartbeat/libexec/heartbeat/ipfail
apiauth ipfail gid=haclient uid=hacluster
```

节点 2 上的/usr/local/heartbeat/etc/ha.d/ha.cf 文件只有以下一个配置项与节点 1 不同：

```
ucast ens32 172.16.1.125
```

步骤 02 编辑认证文件 authkeys。

节点 1 与节点 2 上的/usr/local/heartbeat/etc/ha.d/authkeys 文件内容相同：

```
auth 1
1 crc
```

步骤 03 编辑资源配置文件 haresources。

所有节点的/usr/local/heartbeat/etc/ha.d/haresources 文件内容相同：

```
node1 IPaddr::172.16.1.101/24/ens32 drbddisk::mysql
Filesystem::/dev/drbd0::/mnt::xfs mysql
```

- node1：主节点的主机名。
- IPaddr::172.16.1.101/24/ens32：用 IPaddr 脚本配置对外服务的浮动 VIP。
- drbddisk::mysql：用 drbddisk 脚本实现 DRBD 主从节点资源组的挂载和卸载。该脚本文件为 heartbeat 自带，默认安装在/etc/ha.d/resource.d/drbddisk。
- Filesystem::/dev/drbd0::/mnt::xfs：用 Filesystem 脚本实现磁盘挂载和卸载。
- mysql：MySQL 启动脚本。

步骤 04 建立插件软链接。

在两个节点执行以下命令：

```
cd /usr/local/heartbeat/etc/ha.d
chmod 600 authkeys
ln -svf /usr/local/heartbeat/lib64/heartbeat/plugins/RAExec/*
/usr/local/heartbeat/lib/heartbeat/plugins/RAExec/
ln -svf /usr/local/heartbeat/lib64/heartbeat/plugins/*
/usr/local/heartbeat/lib/heartbeat/plugins/
```

步骤 05 创建脚本文件。

在两个节点的 Heartbeat 资源目录下创建脚本文件，说明如下。

- 创建 drbddisk 文件：

```
cp /etc/ha.d/resource.d/drbddisk /usr/local/heartbeat/etc/ha.d/resource.d/
```

- 创建如下内容的/usr/local/heartbeat/etc/ha.d/resource.d/mysql 文件：

```
chown -R mysql:mysql /mnt/
service mysql start
```

- 创建如下内容的 MySQL 服务检测脚本文件/home/mysql/mysql_check.sh:

```
#!/bin/bash
. /home/mysql/.bashrc

count=1

while true
do

mysql -uroot -p123456 -S /data/mysql.sock -e "show status;" > /dev/null 2>&1
i=$?
ps aux | grep mysqld | grep -v grep > /dev/null 2>&1
j=$?
if [ $i = 0 ] && [ $j = 0 ]
then
   sleep 3
else
   if [ $i = 1 ] && [ $j = 0 ]
   then
       sleep 3
   else
        if [ $count -gt 5 ]
        then
              break
        fi
   let count++
   continue
   fi
fi

done

/etc/init.d/heartbeat stop
```

步骤 06 启动 Heartbeat。

在两个节点上启动 HeartBeat 服务，先启动 node1，再启动 node2。

```
systemctl start heartbeat
systemctl enable heartbeat
systemctl status heartbeat
```

如下所示，此时 VIP 绑定在 node1 上：

```
[root@node1 ~]# ip a
1: lo: <LOOPBACK,UP,LOWER_UP> mtu 65536 qdisc noqueue state UNKNOWN qlen 1000
```

```
    link/loopback 00:00:00:00:00:00 brd 00:00:00:00:00:00
    inet 127.0.0.1/8 scope host lo
      valid_lft forever preferred_lft forever
    inet6 ::1/128 scope host
      valid_lft forever preferred_lft forever
2: ens32: <BROADCAST,MULTICAST,UP,LOWER_UP> mtu 1500 qdisc pfifo_fast state
UP qlen 1000
    link/ether 00:50:56:a5:7a:4e brd ff:ff:ff:ff:ff:ff
    inet 172.16.1.125/24 brd 172.16.1.255 scope global ens32
      valid_lft forever preferred_lft forever
    inet 172.16.1.101/24 brd 172.16.1.255 scope global secondary ens32:0
      valid_lft forever preferred_lft forever
    inet6 fe80::250:56ff:fea5:7a4e/64 scope link
      valid_lft fo  preferred_lft forever
[root@node1 ~]#
```

客户端通过 VIP 可以正常访问 MySQL：

```
C:\Users\wxy>mysql -h172.16.1.101 -uroot -p123456 -e "select * from db1.t1"
mysql: [Warning] Using a password on the command line interface can be insecure.
+------+
| a    |
+------+
|    1 |
+------+
```

当启动 Heartbeat 时同时启动 ipfail 来检测心跳。hacluster 为启动 Heartbeat 时的用户，也就是说 ipfail 和 Heartbeat 都是用 hacluster 这个用户来运行的。Heartbeat 和 ipfail 进程如下所示：

```
[root@node1 ~]# pstree
systemd─┬─abrt-watch-log
        ├─abrtd
        ├─2*[agetty]
        ├─atd
        ├─chronyd
        ├─cmagent───25*[{cmagent}]
        ├─crond
        ├─dbus-daemon
        ├─heartbeat─┬─7*[heartbeat]
        │           ├─ipfail
        ├─irqbalance
        ├─lsmd
        ├─lvmetad
...
        ├─sshd─┬─sshd───bash───tailf
        │      └─sshd───bash───pstree
        ├─systemd-journal
        ├─systemd-logind
        ├─systemd-udevd
        └─tuned───4*[{tuned}]
[root@manager~]#
```

15.4.2 测试

步骤 01 停掉 node1 上的 mysqld，查看服务是否切换到 node2。

```
# 首先启动 mysql 检测脚本，因为 Heartbeat 不检查服务的可用性，需要通过自定义脚本实现
nohup /home/mysql/mysql_check.sh &
# 停止 mysqld
service mysql stop
```

node1 的 Heartbeat 进程停止，VIP 和 DRBD 都切换到 node2，客户端可以正常访问 MySQL，自动切换成功。

步骤 02 停掉 node2 的 Heartbeat 服务，查看服务是否切换回 node1。

```
# 先启动 node1 的 Heartbeat
systemctl start heartbeat
# 停止 node2 的 Heartbeat，也可以使用 iptables -I INPUT -p icmp -j DROP 禁用 ping
systemctl stop heartbeat
```

node2 的 Heartbeat 进程停止，VIP 和 DRBD 都切换到 node1，客户端可以正常访问 MySQL，自动切换成功。

步骤 03 node1 重启，同时查看服务是否切换到 node2。

```
# 先启动 node2 的 Heartbeat
systemctl start heartbeat
# node1 重启
reboot
```

VIP 和 DRBD 都切换到 node2，客户端可以正常访问 MySQL，自动切换成功。

步骤 04 测试“脑裂”。

```
# 停止 node1 的 Heartbeat 服务
systemctl stop heartbeat

# 停止 node2 的 Heartbeat 服务
systemctl stop heartbeat

# 在 node1 上添加防火墙策略，拒绝 node2 的广播包
iptables -A INPUT -i ens32 -p udp -s 172.16.1.126 --dport 694 -j DROP

# 在 node2 上添加防火墙策略，拒绝 node1 的广播包
iptables -A INPUT -i ens32 -p udp -s 172.16.1.125 --dport 694 -j DROP

# 启动 node1 的 Heartbeat 服务
systemctl start heartbeat

# 启动 node2 的 Heartbeat 服务
systemctl start heartbeat
```

在超过 initdead 定义的时间后，两个节点都绑定了同一个 VIP，都挂载了 DRDB 盘，都启动了 MySQL 服务，都可以进行 MySQL 的读写。它们之间各自为政，不知道对方的存在，出现了“脑裂”问题。

由于两节点之间在指定的时间内，无法互相检测到对方心跳而各自启动故障转移功能，取得了资源以及服务的所有权，此时的两个节点都还活着并作正常运行，这样会导致资源冲突，最严重的就是两台主机同时占用一个 VIP 地址，当用户写入数据的时候可能会分别写入到两端，这样可能会导致两节点数据不一致或造成数据丢失，这种情况被称为“脑裂”。

一般来说，“脑裂”的发生，主要是由以下的几个原因导致的：

- 节点对之间心跳线路故障，导致无法正常的通信。
- 节点对上开启的防火墙阻挡了心跳消息的传输。
- 节点对上的心跳网卡地址等信息配置错误，导致发送心跳失败。
- 其他服务配置不当等原因，如心跳的方式不同，心跳广播冲突，软件出现了 BUG 等。

发生“脑裂”的时候，对业务的影响极其严重，有时甚至是致命的。例如：两个节点之间发生“脑裂”，导致互相竞争同一个 IP 资源，就如同局域网内常见的 IP 地址冲突一样，两个机器就会有一个不正常或者都不正常，影响用户正常访问服务器。如果是应用在数据库或存储服务这种重要的高可用系统上，那就会导致用户发布的数据间断地写在两台服务器上的恶果，最终使得数据难以恢复。

实际的生产环境中，可以从以下几个方面来防止“脑裂”问题的发生：

- 使用串行电缆和以太网电缆连接，同时启用两条心跳线路。这样一条线路坏了，另一条线路还是好的，依然能传送消息。这是最简单的一个方案，也是推荐的防“脑裂”方法。
- 检测到“脑裂”的时候强行关闭一个心跳节点（需要特殊的节点支持，如 Stonith、fence 等），相当于程序上备节点发现心跳线故障，发送关机命令到主节点。
- 做好对“脑裂”的监控报警，如邮件以及手机短信等，在问题发生的时候能够人为介入到仲裁，降低损失。当然，在实施高可用方案的时候，要根据业务的实际需求确定是否能够容忍这样的损失。对于一般的网站业务，这个损失是可控的。
- 启用磁盘锁。正在服务一方锁住共享磁盘，“脑裂”发生的时候，让对方完全抢不走共享的磁盘资源。但使用锁磁盘也会有问题，如果占用共享盘的一方不主动解锁，另一方就永远得不到共享磁盘。现实中如果服务节点突然死机或者崩溃，另一方就永远不可能执行解锁命令，后备节点也就接管不了共享的资源和应用服务。于是有人在高可用（HA）中设计了“智能”锁，正在服务的一方只在发现心跳线全部断开时才启用磁盘锁，平时就不上锁。
- 报警报在服务器接管之前，给人员处理留出足够的时间。例如，1 分钟内报警，但服务器在 5 分钟后接管。这样数据不易丢失，但是会导致一段时间内用户无法写入数据。
- 报警后，不直接执行自动服务器接管，而是由人工处理的。
- 增加仲裁机制，确定谁该获得资源。例如设置参考的 IP，当心跳完全断开的时候，两个节点各自 ping 一下参考 IP，不通则表明断点就出现在本段。这样就主动放弃竞争，让能够 ping 通参考 IP 的一端去接管服务，或者通过第三方软件仲裁谁该获得资源。

15.5 小 结

DRBD是一个用软件实现的、无共享的、以镜像块设备来存储内容的存储复制解决方案。DRBD模块被集成到Linux内核中，功能上类似于网络RAID1。在将数据写入本地文件系统时，数据还将通过网卡发送到网络中另一台主机上，以相同的形式记录在一个文件系统中，主节点与备节点（也称为从节点）的数据可以保证实时同步。当本地系统出现故障时，远程主机上还保留有一份相同的数据，可以继续使用。在高可用系统中使用DRBD，可以代替一个共享盘阵。因为数据同时存在于本地主机和远程主机上，切换时远程主机只要使用它上面的那份备份数据就可以继续进行服务。

对于MySQL来说，只需要将数据目录配置在DRBD的块设备上，就可利用DRBD提供的功能实现数据库服务高可用。与前面章节介绍的所有MySQL高可用技术不同，DRBD是存储级别的数据块镜像，因此完全不依赖MySQL的任何复制技术。为了能让MySQL实施自动故障转移，通常要将DRBD与其他工具（如Heartbeat等）配合使用。该方案的主要优点是安全性高，对应用透明，主要缺点是容易出现“脑裂”的情况，需要采取其他手段来防范。

第 16 章

优化案例

保持高可用环境下的 MySQL 数据库性能是一项具有挑战性的工作。数据库的并发访问量和每秒查询数（Queries Per Second，QPS）等指标决定了它的维护难度，主要体现在维护的同时要将对现有数据库服务的影响最小化。在一个很高负载的生产数据库上，执行任何操作都必须小心翼翼，因为一条 SQL 语句出现性能问题，就可能出现卡库，甚至让整个数据库停止运行。本章分析 4 个具体的优化案例，说明一些提高性能的技巧和方法。

16.1　快速生成数字辅助表

数字辅助表只有一个整数列，包含从 1 到 N 个整数序列值，N 通常很大。对 MySQL 来讲，数字辅助表是一个强大的工具，编写 SQL 语句时经常用数据表与数字辅助表做笛卡尔积连接来创建额外的行。建议创建一个持久的数据辅助表，并根据需要填充一定数据量的值。实际上如何填充数字辅助表无关紧要，因为只需要运行这个过程一次，不过，还是可以对此过程进行优化。假设需要为如下数字辅助表生成数据：

```
create table nums (a bigint unsigned not null primary key) engine=innodb;
```

1. 方法一：平凡低效

```
drop procedure if exists pcreatenums;
delimiter //
create procedure pcreatenums(cnt bigint)
begin
    declare s int default 1;
    set session autocommit=0;
    while s<=cnt do
        insert into nums values(s);
        set s=s+1;
```

```
    end while;
    commit;
end;
//
```

这个存储过程很简单，就是一个循环，每次插入一条数据，以生成的数据行数作为循环次数。在实验环境中执行这个过程生成100万行数据需要执行的时间接近1分24秒。效率不高的原因在于insert语句被执行了100万次。

```
mysql> call pcreatenums(1000000);
Query OK, 0 rows affected (1 min 24.39 sec)
```

2. 方法二：高效迭代

```
drop procedure if exists pcreatenums;
delimiter //
create procedure pcreatenums(cnt int)
begin
    declare s int default 1;
    set session autocommit=0;
    insert into nums select s;
    while s<=cnt do
        insert into nums select a+s from nums where a+s <=cnt;
        set s=s*2;
    end while;
    commit;
end;
//
```

执行该存储过程生成同样的数据只用了不到17秒：

```
mysql> call pcreatenums(1000000);
Query OK, 0 rows affected (16.53 sec)
```

在这个改进的存储过程中，变量s保存插入nums表的行数。循环开始前先插入1条数据，然后当s小于等于所要生成的数据行数时执行循环。在每次迭代中，该过程把nums表当前所有行的值加上s后再插入到nums表中。这样每次循环插入的行数以2的幂次方递增，insert语句只被执行了21次，其中还包括作为种子数据的第一次插入。因此这个过程的执行速度很快。

3. 方法三：一次生成

```
set session cte_max_recursion_depth=1000000;
insert into nums
with recursive temp (n) as (select 1 union all select n+1 from temp where n
< 1000000)
select n from temp;
```

这种方法利用MySQL 8提供的CTE（Common Table Expressions）功能，用递归一次性生成所有数据，只需要不到13秒，性能进一步提高了四分之一。

```
mysql> insert into nums
    ->   with recursive temp (n) as (select 1 union all select n+1 from temp where
n < 1000000)
```

```
    -> select n from temp;
Query OK, 1000000 rows affected (12.28 sec)
Records: 1000000  Duplicates: 0  Warnings: 0
```

CTE 类似 Oracle 中的 connect by，经常用于实现递归查询，但功能更强大，能够解决非常复杂的查询问题。https://dev.mysql.com/doc/refman/8.0/en/with.html 是 MySQL 官方文档对 CTE 的详细说明。

16.2 将 MySQL 去重操作优化到极致

16.2.1 问题描述

源表 t_source 结构如下：

```
item_id int,
created_time datetime,
modified_time datetime,
item_name varchar(20),
other varchar(20)
```

要求：

（1）源表中有 100 万条数据，其中有 50 万 created_time 和 item_name 重复。

（2）要把去重后的 50 万数据写入到目标表。

（3）重复 created_time 和 item_name 的多条数据，可以保留任意一条，不做规则限制。

实验环境：

CentOS release 6.4；8GB 物理内存（MySQL 的 innodb_buffer_pool_size 配置为 4GB）；100GB 机械硬盘；双核双物理 CPU，共 4 个处理器；MySQL 8.0.16 版本。

建立测试表和数据：

```
-- 建立源表
create table t_source
(
  item_id int,
  created_time datetime,
  modified_time datetime,
  item_name varchar(20),
  other varchar(20)
);

-- 建立目标表
create table t_target like t_source;

-- 生成 100 万条测试数据，其中有 50 万 created_time 和 item_name 重复
delimiter //
```

```
create procedure sp_generate_data()
begin
    set @i := 1;

    while @i<=500000 do
        set @created_time := date_add('2017-01-01',interval @i second);
        set @modified_time := @created_time;
        set @item_name := concat('a',@i);
        insert into t_source
        values (@i,@created_time,@modified_time,@item_name,'other');
        set @i:=@i+1;
    end while;
    commit;

    set @last_insert_id := 500000;
    insert into t_source
    select item_id + @last_insert_id,
           created_time,
           date_add(modified_time,interval @last_insert_id second),
           item_name,
           'other'
      from t_source;
    commit;
end
//
delimiter ;

call sp_generate_data();

-- 源表没有主键或唯一性约束，有可能存在两条完全一样的数据，所以再插入一条记录模拟这种情况
insert into t_source select * from t_source where item_id=1;
```

源表中有 1000001 条记录，去重后的目标表应该有 500000 条记录。

```
mysql> select count(*),count(distinct created_time,item_name) from t_source;
+----------+----------------------------------------+
| count(*) | count(distinct created_time,item_name) |
+----------+----------------------------------------+
| 1000001 |                                  500000 |
+----------+----------------------------------------+
1 row in set (1.92 sec)
```

16.2.2　巧用索引与变量

1. 无索引对比测试

（1）使用相关子查询

```
truncate t_target;
insert into t_target
select distinct t1.* from t_source t1 where item_id in
```

```
    (select min(item_id) from t_source t2 where t1.created_time=t2.created_time
and t1.item_name=t2.item_name);
```

这个语句很长时间都出不来结果，就先看一下执行计划吧：

```
    mysql> explain select distinct t1.* from t_source t1 where item_id in
        -> (select min(item_id) from t_source t2 where
t1.created_time=t2.created_time
        -> and t1.item_name=t2.item_name)\G
    *************************** 1. row ***************************
              id: 1
      select_type: PRIMARY
            table: t1
       partitions: NULL
             type: ALL
    possible_keys: NULL
              key: NULL
          key_len: NULL
              ref: NULL
             rows: 997282
         filtered: 100.00
            Extra: Using where; Using temporary
    *************************** 2. row ***************************
              id: 2
      select_type: DEPENDENT SUBQUERY
            table: t2
       partitions: NULL
             type: ALL
    possible_keys: NULL
              key: NULL
          key_len: NULL
              ref: NULL
             rows: 997282
         filtered: 1.00
            Extra: Using where
    2 rows in set, 3 warnings (0.00 sec)
```

主查询和相关子查询都是全表扫描，一共要扫描 100 万*100 万数据行，难怪不出结果。

（2）使用表连接

```
    truncate t_target;
    insert into t_target
    select distinct t1.* from t_source t1,
    (select min(item_id) item_id,created_time,item_name from t_source group by
created_time,item_name) t2
    where t1.item_id = t2.item_id;
```

这种方法用时 14 秒，查询计划如下：

```
    mysql> explain select distinct t1.* from t_source t1,
        -> (select min(item_id) item_id,created_time,item_name
        ->    from t_source group by created_time,item_name) t2
```

```
    ->   where t1.item_id = t2.item_id\G
*************************** 1. row ***************************
           id: 1
  select_type: PRIMARY
        table: t1
   partitions: NULL
         type: ALL
possible_keys: NULL
          key: NULL
      key_len: NULL
          ref: NULL
         rows: 997282
     filtered: 100.00
        Extra: Using where; Using temporary
*************************** 2. row ***************************
           id: 1
  select_type: PRIMARY
        table: <derived2>
   partitions: NULL
         type: ref
possible_keys: <auto_key0>
          key: <auto_key0>
      key_len: 5
          ref: test.t1.item_id
         rows: 10
     filtered: 100.00
        Extra: Distinct
*************************** 3. row ***************************
           id: 2
  select_type: DERIVED
        table: t_source
   partitions: NULL
         type: ALL
possible_keys: NULL
          key: NULL
      key_len: NULL
          ref: NULL
         rows: 997282
     filtered: 100.00
        Extra: Using temporary
3 rows in set, 1 warning (0.00 sec)
```

- 内层查询扫描t_source表的100万行，建立临时表，找出去重后的最小item_id，生成导出表derived2，此导出表有50万行。
- MySQL会在导出表derived2上自动创建一个item_id字段的临时索引auto_key0。
- 外层查询也要扫描t_source表的100万行数据，在与导出表做链接时，对t_source表每行的item_id，使用auto_key0索引查找导出表中匹配的行，并在此时优化distinct操作，在找到第一个匹配的行后即停止查找同样值的动作。

（3）使用变量

```
set @a:='1000-01-01 00:00:00';
set @b:=' ';
set @f:=0;
truncate t_target;
insert into t_target
select item_id,created_time,modified_time,item_name,other
  from
(select t0.*,if(@a=created_time and @b=item_name,@f:=0,@f:=1) f,
@a:=created_time,@b:=item_name
  from (select * from t_source order by created_time,item_name) t0) t1 where
f=1;
```

这种方法用时 13 秒，查询计划如下：

```
mysql> explain select item_id,created_time,modified_time,item_name,other
    ->   from (select t0.*,if(@a=created_time and @b=item_name,@f:=0,@f:=1) f,
    ->               @a:=created_time,@b:=item_name
    ->   from (select * from t_source order by created_time,item_name) t0) t1
where f=1\G
*************************** 1. row ***************************
           id: 1
  select_type: PRIMARY
        table: <derived2>
   partitions: NULL
         type: ref
possible_keys: <auto_key0>
          key: <auto_key0>
      key_len: 4
          ref: const
         rows: 10
     filtered: 100.00
        Extra: NULL
*************************** 2. row ***************************
           id: 2
  select_type: DERIVED
        table: <derived3>
   partitions: NULL
         type: ALL
possible_keys: NULL
          key: NULL
      key_len: NULL
          ref: NULL
         rows: 997282
     filtered: 100.00
        Extra: NULL
*************************** 3. row ***************************
           id: 3
  select_type: DERIVED
        table: t_source
```

```
   partitions: NULL
         type: ALL
possible_keys: NULL
          key: NULL
      key_len: NULL
          ref: NULL
         rows: 997282
     filtered: 100.00
        Extra: Using filesort
3 rows in set, 5 warnings (0.00 sec)
```

- 最内层的查询扫描 t_source 表的 100 万行，并使用文件排序，生成导出表 derived3。
- 第二层查询要扫描 derived3 的 100 万行，生成导出表 derived2，完成变量的比较和赋值，并自动创建一个导出列 f 上的临时索引 auto_key0。
- 最外层使用 auto_key0 索引扫描 derived2 得到去重的结果行。

与上面方法 2 比较，总的扫描行数不变，都是 200 万行。只存在一点微小的差别，这次自动生成的索引是在常量列 f 上，而表关联自动生成的索引是在 item_id 列上，所以查询时间几乎相同。

至此，我们还没有在源表上创建任何索引。无论使用哪种写法，要查重都需要对 created_time 和 item_name 字段进行排序，因此很自然地想到，如果在这两个字段上建立联合索引，就可利用索引本身有序的特性消除额外排序，从而提高查询性能。

2. 建立 created_time 和 item_name 上的联合索引进行对比测试

```
-- 建立 created_time 和 item_name 字段的联合索引
create index idx_sort on t_source(created_time,item_name,item_id);
analyze table t_source;
```

（1）使用相关子查询

本次用时 19 秒，查询计划如下：

```
mysql> explain select distinct t1.* from t_source t1 where item_id in
    -> (select min(item_id) from t_source t2 where
t1.created_time=t2.created_time and
    -> t1.item_name=t2.item_name)\G
*************************** 1. row ***************************
           id: 1
  select_type: PRIMARY
        table: t1
   partitions: NULL
         type: ALL
possible_keys: NULL
          key: NULL
      key_len: NULL
          ref: NULL
         rows: 997281
     filtered: 100.00
        Extra: Using where; Using temporary
*************************** 2. row ***************************
           id: 2
```

```
  select_type: DEPENDENT SUBQUERY
        table: t2
   partitions: NULL
         type: ref
possible_keys: idx_sort
          key: idx_sort
      key_len: 89
          ref: test.t1.created_time,test.t1.item_name
         rows: 2
     filtered: 100.00
        Extra: Using index
2 rows in set, 3 warnings (0.00 sec)
```

- 外层查询的 t_source 表是驱动表，需要扫描 100 万行。
- 对于驱动表每行的 item_id，通过 idx_sort 索引查询出两行数据。

（2）使用表连接

本次用时 13 秒，查询计划如下：

```
mysql> explain select distinct t1.* from t_source t1,
    -> (select min(item_id) item_id,created_time,item_name
    ->    from t_source group by created_time,item_name) t2
    ->   where t1.item_id = t2.item_id\G
*************************** 1. row ***************************
           id: 1
  select_type: PRIMARY
        table: t1
   partitions: NULL
         type: ALL
possible_keys: NULL
          key: NULL
      key_len: NULL
          ref: NULL
         rows: 997281
     filtered: 100.00
        Extra: Using where; Using temporary
*************************** 2. row ***************************
           id: 1
  select_type: PRIMARY
        table: <derived2>
   partitions: NULL
         type: ref
possible_keys: <auto_key0>
          key: <auto_key0>
      key_len: 5
          ref: test.t1.item_id
         rows: 10
     filtered: 100.00
        Extra: Distinct
*************************** 3. row ***************************
```

```
          id: 2
 select_type: DERIVED
       table: t_source
  partitions: NULL
        type: index
possible_keys: idx_sort
         key: idx_sort
     key_len: 94
         ref: NULL
        rows: 997281
    filtered: 100.00
       Extra: Using index
3 rows in set, 1 warning (0.00 sec)
```

和没有索引相比，子查询虽然从全表扫描变为了全索引扫描，但还是需要扫描 100 万行记录。因此查询性能并没有明显提升。

（3）使用变量

本次用时 13 秒，查询计划与没有索引时的完全相同。可见索引对这种写法没有作用。能不能消除嵌套，只用一层查询出结果呢？

（4）使用变量，并且消除嵌套查询

```
set @a:='1000-01-01 00:00:00';
set @b:=' ';
truncate t_target;
insert into t_target
select * from t_source force index (idx_sort)
 where (@a!=created_time or @b!=item_name)
   and (@a:=created_time) is not null and (@b:=item_name) is not null
 order by created_time,item_name;
```

本次用时 12 秒，查询计划如下：

```
mysql> explain select * from t_source force index (idx_sort)
    ->   where (@a!=created_time or @b!=item_name)
    ->     and (@a:=created_time) is not null and (@b:=item_name) is not null
    ->   order by created_time,item_name\G
*************************** 1. row ***************************
           id: 1
  select_type: SIMPLE
        table: t_source
   partitions: NULL
         type: index
possible_keys: NULL
          key: idx_sort
      key_len: 94
          ref: NULL
         rows: 997281
     filtered: 99.00
        Extra: Using where
```

```
1 row in set, 3 warnings (0.00 sec)
```

该语句具有以下特点：

- 消除了嵌套子查询，只需要对 t_source 表进行一次全索引扫描，查询计划已达最优。
- 无须 distinct 二次查重。
- 变量判断与赋值只出现在 where 子句中。
- 利用索引消除了 filesort。

在 MySQL 8 之前，该语句是单线程去重的最佳解决方案。仔细分析这条语句，发现它巧妙地利用了 SQL 语句的逻辑查询处理步骤和索引特性。一条 SQL 查询的逻辑步骤为：

（1）执行笛卡尔乘积（交叉连接）
（2）应用 ON 筛选器（连接条件）
（3）添加外部行（outer join）
（4）应用 where 筛选器
（5）应用 group by 分组
（6）应用 cube 或 rollup
（7）应用 having 筛选器
（8）处理 select 列表
（9）应用 distinct 子句
（10）应用 order by 子句
（11）应用 limit 子句

每条 select 查询语句的逻辑执行步骤都是这 11 步的子集。拿这条查询语句来说，其执行顺序为：强制通过索引 idx_sort 查找数据行→应用 where 筛选器→处理 select 列表→应用 order by 子句。

为了使变量能够按照 created_time 和 item_name 的排序顺序进行赋值和比较，必须按照索引顺序查找数据行。这里的 force index (idx_sort)提示就起到了这个作用，必须这样写才能使整条查重语句成立。否则，因为先扫描表才处理排序，所以不能保证变量赋值的顺序，也就不能确保查询结果的正确性。order by 子句同样不可忽略，否则即使有 force index 提示，MySQL 也会使用全表扫描而不是全索引扫描，从而使结果出现错误。索引同时保证了 created_time,item_name 的顺序，避免了文件排序。force index (idx_sort)提示和 order by 子句缺一不可，索引 idx_sort 在这里可谓恰到好处、一举两得。

查询语句开始前，先给变量初始化为数据中不可能出现的值，然后进入 where 子句从左向右判断。先比较变量和字段的值，再将本行 created_time 和 item_name 的值赋给变量，按 created_time、item_name 的顺序逐行处理。item_name 是字符串类型，(@b:=item_name)不是有效的布尔表达式，因此要写成(@b:=item_name) is not null。

最后补充一句，我们在本例中忽略了“insert into t_target select * from t_source group by created_time,item_name;”的写法，因为它受“sql_mode='ONLY_FULL_GROUP_BY'”的限制，不符合 SQL 标准。

16.2.3　利用窗口函数

MySQL 8 中新增的窗口函数使得原来麻烦的去重操作变得很简单。

```
truncate t_target;
insert into t_target
select item_id, created_time, modified_time, item_name, other
 from (select *, row_number() over (partition by created_time,item_name) as
rn
from t_source) t where rn=1;
```

这个语句执行只需要 12 秒，而且写法清晰易懂，其查询计划如下：

```
mysql> explain select item_id, created_time, modified_time, item_name, other
    ->  from (select *, row_number() over (partition by created_time,item_name)
as rn
    ->       from t_source) t where rn=1;
*************************** 1. row ***************************
           id: 1
  select_type: PRIMARY
        table: <derived2>
   partitions: NULL
         type: ref
possible_keys: <auto_key0>
          key: <auto_key0>
      key_len: 8
          ref: const
         rows: 10
     filtered: 100.00
        Extra: NULL
*************************** 2. row ***************************
           id: 2
  select_type: DERIVED
        table: t_source
   partitions: NULL
         type: ALL
possible_keys: NULL
          key: NULL
      key_len: NULL
          ref: NULL
         rows: 997281
     filtered: 100.00
        Extra: Using filesort
2 rows in set, 2 warnings (0.00 sec)
```

该查询对 t_source 表进行一次全表扫描，同时用 filesort 对表按分区字段 created_time、item_name 进行了排序。外层查询从每个分区中保留一条数据。因为重复 created_time 和 item_name 的多条数据中可以保留任意一条，所以 over 中不需要使用 order by 子句。

从执行计划看，窗口函数去重语句似乎没有消除嵌套查询的变量去重的效果好，但此方法实际执行是最快的。MySQL 窗口函数的文档链接是：https://dev.mysql.com/doc/refman/8.0/en/window-functions.html。

16.2.4 多线程并行

前面已经将单条查重语句调整到最优，但还是以单线程的方式执行。能否利用多处理器，让去重操作多线程并行执行，从而进一步提高速度呢？比如我们的实验环境是 4 颗处理器，如果使用 4 个线程同时执行查重 SQL，理论上应该接近 4 倍的性能提升。

在生成测试数据时，created_time 采用每条记录加一秒的方式，也就是最大和最小的时间差为 50 万秒，而且数据均匀分布，因此先把数据平均分成 4 份。

（1）查询出 4 份数据的 created_time 边界值：

```
mysql> select date_add('2017-01-01',interval 125000 second) dt1,
    ->        date_add('2017-01-01',interval 2*125000 second) dt2,
    ->        date_add('2017-01-01',interval 3*125000 second) dt3,
    ->        max(created_time) dt4
    ->    from t_source;
+---------------------+---------------------+---------------------+-------
--------------+
| dt1                 | dt2                 | dt3                 | dt4                 |
+---------------------+---------------------+---------------------+-------
--------------+
| 2017-01-02 10:43:20 | 2017-01-03 21:26:40 | 2017-01-05 08:10:00 | 2017-01-06 
18:53:20 |
+---------------------+---------------------+---------------------+-------
--------------+
1 row in set (0.00 sec)
```

（2）查看每份数据的记录数，确认数据平均分布：

```
mysql> select case when created_time >= '2017-01-01'
    ->          and created_time < '2017-01-02 10:43:20'
    ->         then '2017-01-01'
    ->         when created_time >= '2017-01-02 10:43:20'
    ->          and created_time < '2017-01-03 21:26:40'
    ->         then '2017-01-02 10:43:20'
    ->         when created_time >= '2017-01-03 21:26:40'
    ->          and created_time < '2017-01-05 08:10:00'
    ->         then '2017-01-03 21:26:40'
    ->         else '2017-01-05 08:10:00'
    ->      end min_dt,
    ->      case when created_time >= '2017-01-01'
    ->          and created_time < '2017-01-02 10:43:20'
    ->         then '2017-01-02 10:43:20'
    ->         when created_time >= '2017-01-02 10:43:20'
    ->          and created_time < '2017-01-03 21:26:40'
```

```
    ->            then '2017-01-03 21:26:40'
    ->            when created_time >= '2017-01-03 21:26:40'
    ->             and created_time < '2017-01-05 08:10:00'
    ->            then '2017-01-05 08:10:00'
    ->            else '2017-01-06 18:53:20'
    ->        end max_dt,
    ->        count(*)
    ->   from t_source
    ->  group by case when created_time >= '2017-01-01'
    ->              and created_time < '2017-01-02 10:43:20'
    ->             then '2017-01-01'
    ->             when created_time >= '2017-01-02 10:43:20'
    ->              and created_time < '2017-01-03 21:26:40'
    ->             then '2017-01-02 10:43:20'
    ->             when created_time >= '2017-01-03 21:26:40'
    ->              and created_time < '2017-01-05 08:10:00'
    ->             then '2017-01-03 21:26:40'
    ->             else '2017-01-05 08:10:00'
    ->          end,
    ->         case when created_time >= '2017-01-01'
    ->              and created_time < '2017-01-02 10:43:20'
    ->              then '2017-01-02 10:43:20'
    ->              when created_time >= '2017-01-02 10:43:20'
    ->              and created_time < '2017-01-03 21:26:40'
    ->              then '2017-01-03 21:26:40'
    ->              when created_time >= '2017-01-03 21:26:40'
    ->              and created_time < '2017-01-05 08:10:00'
    ->              then '2017-01-05 08:10:00'
    ->              else '2017-01-06 18:53:20'
    ->          end;
+---------------------+---------------------+----------+
| min_dt              | max_dt              | count(*) |
+---------------------+---------------------+----------+
| 2017-01-01          | 2017-01-02 10:43:20 |   249999 |
| 2017-01-02 10:43:20 | 2017-01-03 21:26:40 |   250000 |
| 2017-01-03 21:26:40 | 2017-01-05 08:10:00 |   250000 |
| 2017-01-05 08:10:00 | 2017-01-06 18:53:20 |   250002 |
+---------------------+---------------------+----------+
4 rows in set (4.86 sec)
```

4 份数据的并集应该覆盖整个源数据集，并且各份数据之间无重叠。也就是说 4 份数据的 created_time 要连续且互斥，连续保证处理全部数据，互斥确保了不需要二次查重。实际上这和时间范围分区的概念类似，或许用分区表会更好些，只是这里省略了重建表的步骤。

步骤 01 建立查重的存储过程。

有了以上信息，我们就可以写出 4 条语句处理全部数据。为了使调用接口尽量简单，创建下面的存储过程：

```
delimiter //
```

```
    create procedure sp_unique(i smallint)
    begin
        set @a:='1000-01-01 00:00:00';
        set @b:=' ';
        if (i<4) then
            insert into t_target
            select * from t_source force index (idx_sort)
             where created_time >= date_add('2017-01-01',interval (i-1)*125000
second)
               and created_time < date_add('2017-01-01',interval i*125000 second)
               and (@a!=created_time or @b!=item_name)
               and (@a:=created_time) is not null
               and (@b:=item_name) is not null
             order by created_time,item_name;
        else
        insert into t_target
            select * from t_source force index (idx_sort)
             where created_time >= date_add('2017-01-01',interval (i-1)*125000
second)
               and created_time <= date_add('2017-01-01',interval i*125000
second)
               and (@a!=created_time or @b!=item_name)
               and (@a:=created_time) is not null
               and (@b:=item_name) is not null
             order by created_time,item_name;
        end if;
    end
    //
```

查询语句的执行计划如下：

```
    mysql> explain select * from t_source force index (idx_sort)
        ->       where created_time >= date_add('2017-01-01',interval (1-1)*125000
second)
        ->         and created_time < date_add('2017-01-01',interval 1*125000
second)
        ->         and (@a!=created_time or @b!=item_name)
        ->         and (@a:=created_time) is not null
        ->         and (@b:=item_name) is not null
        ->       order by created_time,item_name;
    *************************** 1. row ***************************
               id: 1
      select_type: SIMPLE
            table: t_source
       partitions: NULL
             type: range
    possible_keys: idx_sort
              key: idx_sort
          key_len: 6
              ref: NULL
             rows: 498640
```

```
     filtered: 100.00
        Extra: Using index condition
1 row in set, 3 warnings (0.00 sec)
```

MySQL 优化器进行索引范围扫描，并且使用索引条件下推（ICP）优化查询。

步骤 02　并行执行。

下面分别使用 Shell 后台进程和 MySQL Schedule Event 两种方法实现并行。

（1）Shell 后台进程

- 建立 duplicate_removal.sh 文件，内容如下：

```
#!/bin/bash
mysql -vvv -u root -p123456 test -e "truncate t_target" &>/dev/null
date '+%H:%M:%S'
for y in {1..4}
do
  sql="call sp_unique($y)"
  mysql -vvv -u root -p123456 test -e "$sql" &>par_sql1_$y.log &
done
wait
date '+%H:%M:%S'
```

- 执行脚本文件，输出如下：

```
[mysql@hdp2~]$./duplicate_removal.sh
14:27:30
14:27:35
```

这种方法用时不到 5 秒，并行执行的 4 个过程调用分别用时为 4.87 秒、4.88 秒、4.91 秒和 4.73 秒：

```
[mysql@hdp2~]$cat par_sql1_1.log | sed '/^$/d'
mysql: [Warning] Using a password on the command line interface can be insecure.
--------------
call sp_unique(1)
--------------
Query OK, 124999 rows affected (4.87 sec)
Bye
[mysql@hdp2~]$cat par_sql1_2.log | sed '/^$/d'
mysql: [Warning] Using a password on the command line interface can be insecure.
--------------
call sp_unique(2)
--------------
Query OK, 125000 rows affected (4.88 sec)
Bye
[mysql@hdp2~]$cat par_sql1_3.log | sed '/^$/d'
mysql: [Warning] Using a password on the command line interface can be insecure.
--------------
call sp_unique(3)
```

```
---------------
Query OK, 125000 rows affected (4.91 sec)
Bye
[mysql@hdp2~]$cat par_sql1_4.log | sed '/^$/d'
mysql: [Warning] Using a password on the command line interface can be insecure.
---------------
call sp_unique(4)
---------------
Query OK, 125001 rows affected (4.73 sec)
Bye
[mysql@hdp2~]$
```

4 个过程的平均执行时间为 4.85 秒，因为是并行执行，总的过程执行时间为最慢的 4.91 秒，比单线程速度提高了 2.5 倍。

（2）MySQL Schedule Event（调度事件）

- 建立事件历史日志表，如下所示:

```
-- 用于查看事件执行时间等信息
create table t_event_history  (
   dbname  varchar(128) not null default '',
   eventname  varchar(128) not null default '',
   starttime  datetime(3) not null default '1000-01-01 00:00:00',
   endtime  datetime(3) default null,
   issuccess  int(11) default null,
   duration  int(11) default null,
   errormessage  varchar(512) default null,
   randno  int(11) default null
);
```

- 为每个并发线程创建一个事件，如下所示:

```
delimiter //
create event ev1 on schedule at current_timestamp + interval 1 hour on completion
preserve disable do
begin
    declare r_code char(5) default '00000';
    declare r_msg text;
    declare v_error integer;
    declare v_starttime datetime default now(3);
    declare v_randno integer default floor(rand()*100001);

    insert into t_event_history (dbname,eventname,starttime,randno)
    # 作业名
    values(database(),'ev1', v_starttime,v_randno);

    begin
        # 异常处理段
        declare continue handler for sqlexception
        begin
```

```
            set v_error = 1;
            get diagnostics condition 1 r_code = returned_sqlstate , r_msg =
message_text;
          end;

          # 此处为实际调用的用户程序过程
          call sp_unique(1);
      end;

    update t_event_history set
          endtime=now(3),
          issuccess=isnull(v_error),
          duration=timestampdiff(microsecond,starttime,now(3)),
          errormessage=concat('error=',r_code,', message=',r_msg),
          randno=null
     where starttime=v_starttime and randno=v_randno;

    end
    //

    create event ev2 on schedule at current_timestamp + interval 1 hour on completion
preserve disable do
    begin
        declare r_code char(5) default '00000';
        declare r_msg text;
        declare v_error integer;
        declare v_starttime datetime default now(3);
        declare v_randno integer default floor(rand()*100001);

        insert into t_event_history (dbname,eventname,starttime,randno)
        # 作业名
        values(database(),'ev2', v_starttime,v_randno);

        begin
            # 异常处理段
            declare continue handler for sqlexception
            begin
              set v_error = 1;
              get diagnostics condition 1 r_code = returned_sqlstate , r_msg =
message_text;
            end;

            # 此处为实际调用的用户程序过程
            call sp_unique(2);
        end;

    update t_event_history set
          endtime=now(3),
          issuccess=isnull(v_error),
          duration=timestampdiff(microsecond,starttime,now(3)),
```

```
        errormessage=concat('error=',r_code,', message=',r_msg),
        randno=null
        where starttime=v_starttime and randno=v_randno;
    end
    //

    create event ev3 on schedule at current_timestamp + interval 1 hour on completion
preserve disable do
    begin
        declare r_code char(5) default '00000';
        declare r_msg text;
        declare v_error integer;
        declare v_starttime datetime default now(3);
        declare v_randno integer default floor(rand()*100001);

        insert into t_event_history (dbname,eventname,starttime,randno)
        # 作业名
        values(database(),'ev3', v_starttime,v_randno);

        begin
            # 异常处理段
            declare continue handler for sqlexception
            begin
              set v_error = 1;
              get diagnostics condition 1 r_code = returned_sqlstate, r_msg =
message_text;
            end;

            # 此处为实际调用的用户程序过程
            call sp_unique(3);
        end;

        update t_event_history set
               endtime=now(3),
               issuccess=isnull(v_error),
               duration=timestampdiff(microsecond,starttime,now(3)),
               errormessage=concat('error=',r_code,', message=',r_msg),
               randno=null
          where starttime=v_starttime and randno=v_randno;

    end
    //

    create event ev4 on schedule at current_timestamp + interval 1 hour on completion
preserve disable do
    begin
        declare r_code char(5) default '00000';
        declare r_msg text;
        declare v_error integer;
        declare v_starttime datetime default now(3);
```

```
      declare v_randno integer default floor(rand()*100001);

      insert into t_event_history (dbname,eventname,starttime,randno)
      # 作业名
      values(database(),'ev4', v_starttime,v_randno);

      begin
        # 异常处理段
        declare continue handler for sqlexception
        begin
          set v_error = 1;
          get diagnostics condition 1 r_code = returned_sqlstate , r_msg =
message_text;
        end;

        # 此处为实际调用的用户程序过程
        call sp_unique(4);
      end;

    update t_event_history set
        endtime=now(3),
        issuccess=isnull(v_error),
        duration=timestampdiff(microsecond,starttime,now(3)),
        errormessage=concat('error=',r_code,', message=',r_msg),
        randno=null
        where starttime=v_starttime and randno=v_randno;

  end
  //
```

为了记录每个事件执行的时间，在事件定义中增加了操作日志表的逻辑，因为每个事件中只多执行了一条 insert 和一条 update，4 个事件总共多执行了 8 条很简单的语句，对测试的影响可以忽略不计。执行时间精确到毫秒。

- 触发事件执行:

```
mysql -vvv -u root -p123456 test -e "truncate t_target;
alter event ev1 on schedule at current_timestamp enable;
alter event ev2 on schedule at current_timestamp enable;
alter event ev3 on schedule at current_timestamp enable;
alter event ev4 on schedule at current_timestamp enable;"
```

该命令行依序触发了 4 个事件，但不会等前一个执行完才执行下一个，而是立即向下执行。这可从命令的输出清楚地看到：

```
[mysql@hdp2~]$mysql -vvv -u root -p123456 test -e "truncate t_target;
> alter event ev1 on schedule at current_timestamp enable;
> alter event ev2 on schedule at current_timestamp enable;
> alter event ev3 on schedule at current_timestamp enable;
> alter event ev4 on schedule at current_timestamp enable;"
mysql: [Warning] Using a password on the command line interface can be insecure.
```

```
--------------
truncate t_target
--------------

Query OK, 0 rows affected (0.06 sec)

--------------
alter event ev1 on schedule at current_timestamp enable
--------------

Query OK, 0 rows affected (0.02 sec)

--------------
alter event ev2 on schedule at current_timestamp enable
--------------

Query OK, 0 rows affected (0.00 sec)

--------------
alter event ev3 on schedule at current_timestamp enable
--------------

Query OK, 0 rows affected (0.02 sec)

--------------
alter event ev4 on schedule at current_timestamp enable
--------------

Query OK, 0 rows affected (0.00 sec)

Bye
[mysql@hdp2~]$
```

- 查看事件执行日志:

```
mysql> select * from test.t_event_history;
...
+------+-----+-------------------------+-------------------------+---+-----
----+-----+------+
| test | ev1 | 2019-07-31 14:38:04.000 | 2019-07-31 14:38:09.389 | 1 | 5389000
| NULL | NULL |
| test | ev2 | 2019-07-31 14:38:04.000 | 2019-07-31 14:38:09.344 | 1 | 5344000
| NULL | NULL |
| test | ev3 | 2019-07-31 14:38:05.000 | 2019-07-31 14:38:09.230 | 1 | 4230000
| NULL | NULL |
| test | ev4 | 2019-07-31 14:38:05.000 | 2019-07-31 14:38:09.344 | 1 | 4344000
| NULL | NULL |
+------+-----+-------------------------+-------------------------+---+-----
----+-----+------+
4 rows in set (0.00 sec)
```

4 个过程的平均执行时间为 4.83 秒，又因为是并行执行，因此总的执行时间为最慢的 5.3 秒，优化效果和 Shell 后台进程方式几乎相同。

16.3　重叠时间段问题之算法优化

16.3.1　问题描述与分析

这是一个实际的业务需求，某一直播表中记录了如下格式的用户进出直播间的日志数据：

```
+--------+--------+---------------------+---------------------+
| roomid | userid | s                   | e                   |
+--------+--------+---------------------+---------------------+
|      1 |      1 | 2018-01-01 01:01:01 | 2018-01-01 01:10:01 |
|      1 |      1 | 2018-01-01 01:01:02 | 2018-01-01 01:11:01 |
|      1 |      1 | 2018-01-01 01:01:05 | 2018-01-01 01:10:01 |
|      1 |      1 | 2018-01-01 01:11:02 | 2018-01-01 01:11:05 |
|      1 |      2 | 2018-01-01 01:01:02 | 2018-01-01 01:01:05 |
|      1 |      3 | 2018-01-01 01:01:05 | 2018-01-01 01:02:05 |
|      2 |      1 | 2018-01-01 01:01:03 | 2018-01-03 01:11:01 |
|      2 |      4 | 2018-01-01 01:03:02 | 2018-01-01 01:12:05 |
|      2 |      5 | 2018-01-01 01:11:02 | 2018-01-01 01:12:05 |
|      2 |      6 | 2018-01-01 01:15:02 | 2018-01-01 01:16:05 |
|      2 |      7 | 2018-01-01 01:01:03 | 2018-01-01 01:11:05 |
|      2 |      8 | 2018-01-01 23:01:03 | 2018-01-02 01:11:01 |
|      3 |      1 | 2018-01-05 01:01:01 | 2018-01-10 01:01:01 |
|      3 |      2 | 2018-01-05 01:01:01 | 2018-01-06 01:01:01 |
|      3 |      3 | 2018-01-06 01:01:01 | 2018-01-06 02:01:01 |
...
```

上面的 4 个字段分别表示直播间 ID、用户 ID、进入时间和退出时间。要统计出每天每个活跃房间的峰值人数和总时长。活跃房间的定义是：以每秒为时间粒度，如果在某一时刻同时有两个及两个以上的用户在房间内，该房间当天即为活跃房间。峰值人数是指一天内同时在一个活跃房间的最大人数。总活跃时长是指一天内活跃时长的总和。

这是一个典型的重叠时间段的统计问题。具体来说，该需求可以细分为这样几个需要解决的问题：

（1）一个房间内同一用户的重叠时间段的合并。

（2）拆分起止时间段跨天的时段。

（3）取得活跃的时段。

（4）按天计算每个房间活跃时段内的不同用户数及其活跃时段的长度。

（5）选取活跃时段内的最大人数，并汇总活跃时长。

1. 一个房间内同一用户的重叠时段问题

理论上同一用户进出房间的时间段是不存在重叠的。表数据是由移动端程序上报的，做过移动应用的开发者应该都理解，类似数据统计类的需求不能直接依赖移动端上报的数据，因为有各种原因造成上报数据不准确。此案例中，任意给定的一个房间，用户在其内的时间存在重叠部分，而重叠又分同一用户的重叠与不同用户之间重叠两种情况。对于前一种情况，在判断房间是否活跃时，不应该对用户重复计数，因此这部分的重叠时段需要进行合并。例如，2018-01-01，用户 1 在房间 1 有 4 条日志记录：

```
+--------+--------+---------------------+---------------------+
| roomid | userid | s                   | e                   |
+--------+--------+---------------------+---------------------+
|      1 |      1 | 2018-01-01 01:01:01 | 2018-01-01 01:10:01 |
|      1 |      1 | 2018-01-01 01:01:05 | 2018-01-01 01:10:01 |
|      1 |      1 | 2018-01-01 01:01:02 | 2018-01-01 01:11:01 |
|      1 |      1 | 2018-01-01 01:11:02 | 2018-01-01 01:11:05 |
+--------+--------+---------------------+---------------------+
```

为了判断房间 1 在'2018-01-01 01:01:01'和'2018-01-01 01:11:05'之间是否存在活跃时间段，需要将 4 条记录合并为如下 2 条记录：

```
+--------+--------+---------------------+---------------------+
| roomid | userid | s                   | e                   |
+--------+--------+---------------------+---------------------+
|      1 |      1 | 2018-01-01 01:01:01 | 2018-01-01 01:11:01 |
|      1 |      1 | 2018-01-01 01:11:02 | 2018-01-01 01:11:05 |
+--------+--------+---------------------+---------------------+
```

2. 起止时段跨天的问题

由于是按天进行统计，对于进出时间点跨天的情况，因此要进行拆分。例如，用户 1 在房间 2 的进出时间跨越了 3 天：

```
+--------+--------+---------------------+---------------------+
| roomid | userid | s                   | e                   |
+--------+--------+---------------------+---------------------+
|      2 |      1 | 2018-01-01 01:01:03 | 2018-01-03 01:11:01 |
+--------+--------+---------------------+---------------------+
```

为了统计'2018-01-01'、'2018-01-02' 和 '2018-01-03' 3 天的数据，需要将这条记录拆分为如下 3 条记录：

```
+--------+--------+---------------------+---------------------+
| roomid | userid |   s                 |   e                 |
|--------+--------+---------------------+---------------------|
|      2 |      1 | 2018-01-01 01:01:03 | 2018-01-01 23:59:59 |
|      2 |      1 | 2018-01-02 00:00:00 | 2018-01-02 23:59:59 |
|      2 |      1 | 2018-01-03 00:00:00 | 2018-01-03 01:11:01 |
+--------+--------+---------------------+---------------------+
```

拆分的起止时间相差 1 秒，不能相同。在后面解析计算活跃时间段内的不同用户数及其活跃时长的算法时，会看到这点非常重要。

3. 统计活跃时段

经过了前两步的数据预处理便可以统计活跃时段。这一步是一个令人头疼的问题，关键在于如何高效地获取活跃时段。我们尝试了多种解决方案，后面将介绍其中两种，它们的性能有着天壤之别。

下面建立测试表并生成数据，用于演示各种 SQL 的执行结果：

```
create table test1 (roomid int, userid int, s datetime, e datetime);
insert into test1 values
(1, 1, '2018-01-01 01:01:01', '2018-01-01 01:10:01'),
(1, 2, '2018-01-01 01:01:02', '2018-01-01 01:01:05'),
(1, 3, '2018-01-01 01:01:05', '2018-01-01 01:02:05'),
(2, 4, '2018-01-01 01:03:02', '2018-01-01 01:12:05'),
(2, 5, '2018-01-01 01:11:02', '2018-01-01 01:12:05'),
(2, 6, '2018-01-01 01:15:02', '2018-01-01 01:16:05'),
(2, 7, '2018-01-01 01:01:03', '2018-01-01 01:11:05'),
(1, 1, '2018-01-01 01:01:05', '2018-01-01 01:10:01'),
(1, 1, '2018-01-01 01:01:02', '2018-01-01 01:11:01'),
(1, 1, '2018-01-01 01:11:02', '2018-01-01 01:11:05'),
(2, 1, '2018-01-01 01:01:03', '2018-01-03 01:11:01'),
(2, 8, '2018-01-01 23:01:03', '2018-01-02 01:11:01'),
(3, 1, '2018-01-05 01:01:01', '2018-01-10 01:01:01'),
(3, 2, '2018-01-05 01:01:01', '2018-01-06 01:01:01'),
(3, 3, '2018-01-06 01:01:01', '2018-01-06 02:01:01');

commit;
```

为了验证不同方案在实际数据集上的执行性能，采集了 3 天的 2505495 条业务数据，存储在 u_room_log 表中。u_room_log 与 test1 表结构相同，并且都没有任何索引。

16.3.2 优化重叠查询

如前所述，我们需要解决的第一个问题是合并一个房间内同一用户的重叠时间段。下面讨论自关联和游标两种实现方案。

1. 自关联

在重叠问题的 SQL 解决方案中，最容易想到的是自关联。先求出每个分组的开始时间，并用 DISTINCT 去重，然后用同样的方法得到每组结束的时间，最后把前两步的结果集合并，并通过 MIN 函数取得结束的时间。完整的查询如下面的 SQL 语句所示：

```
select distinct roomid, userid,
       if(date(s)!=date(e) and id>1,date(s+interval id-1 day),s) s,
       if(date(s+interval id-1 day)=date(e),e,
          date_format(s+interval id-1 day,'%Y-%m-%d 23:59:59')) e
  from (select distinct s.roomid, s.userid, s.s,
                -- 合并后每个区间的结束时间
                (select min(e)
                   from (select distinct roomid, userid, e
```

```
                    from test1 a
                   where not exists (select * from test1 b
                                     where a.roomid = b.roomid
                                       and a.userid = b.userid
                                       and a.e >= b.s
                                       and a.e < b.e)) s2
          where s2.e > s.s
            and s.roomid = s2.roomid
            and s.userid = s2.userid) e
      from
           -- 每个房间每个用户的开始时间
           (select distinct roomid, userid, s
               from test1 a
             where not exists (select * from test1 b
                               where a.roomid = b.roomid
                                 and a.userid = b.userid
                                 and a.s > b.s
                                 and a.s <= b.e)) s,
           -- 每个房间每个用户的结束时间
           (select distinct roomid, userid, e
              from test1 a
             where not exists (select * from test1 b
                               where a.roomid = b.roomid
                                 and a.userid = b.userid
                                 and a.e >= b.s
                                 and a.e < b.e)) e
     where s.roomid = e.roomid
       and s.userid = e.userid) t1,
    (select id from nums where id<=100) nums
 where nums.id<=datediff(e,s)+1;
```

最外层的查询用于处理跨天时段。关联数字辅助表将单行数据分解为多行，id<=100 表示单个时段跨越的天数最多是 100。对于按天统计的直播业务，这个跨度足够了。为了提高查询性能，该值应该为满足需求的最小值。下面是该查询的执行结果：

```
+--------+--------+---------------------+---------------------+
| roomid | userid | s                   | e                   |
+--------+--------+---------------------+---------------------+
|      1 |      1 | 2018-01-01 01:01:01 | 2018-01-01 01:11:01 |
|      1 |      2 | 2018-01-01 01:01:02 | 2018-01-01 01:01:05 |
|      1 |      3 | 2018-01-01 01:01:05 | 2018-01-01 01:02:05 |
|      2 |      4 | 2018-01-01 01:03:02 | 2018-01-01 01:12:05 |
|      2 |      5 | 2018-01-01 01:11:02 | 2018-01-01 01:12:05 |
|      2 |      6 | 2018-01-01 01:15:02 | 2018-01-01 01:16:05 |
|      2 |      7 | 2018-01-01 01:01:03 | 2018-01-01 01:11:05 |
|      1 |      1 | 2018-01-01 01:11:02 | 2018-01-01 01:11:05 |
|      2 |      1 | 2018-01-01 01:01:03 | 2018-01-01 23:59:59 |
|      2 |      8 | 2018-01-01 23:01:03 | 2018-01-01 23:59:59 |
|      3 |      1 | 2018-01-05 01:01:01 | 2018-01-05 23:59:59 |
|      3 |      2 | 2018-01-05 01:01:01 | 2018-01-05 23:59:59 |
```

```
|    3 |    3 | 2018-01-06 01:01:01 | 2018-01-06 02:01:01 |
|    2 |    1 | 2018-01-02 00:00:00 | 2018-01-02 23:59:59 |
|    2 |    8 | 2018-01-02 00:00:00 | 2018-01-02 01:11:01 |
|    3 |    1 | 2018-01-06 00:00:00 | 2018-01-06 23:59:59 |
|    3 |    2 | 2018-01-06 00:00:00 | 2018-01-06 01:01:01 |
|    2 |    1 | 2018-01-03 00:00:00 | 2018-01-03 01:11:01 |
|    3 |    1 | 2018-01-07 00:00:00 | 2018-01-07 23:59:59 |
|    3 |    1 | 2018-01-08 00:00:00 | 2018-01-08 23:59:59 |
|    3 |    1 | 2018-01-09 00:00:00 | 2018-01-09 23:59:59 |
|    3 |    1 | 2018-01-10 00:00:00 | 2018-01-10 01:01:01 |
+--------+--------+---------------------+---------------------+
22 rows in set (0.01 sec)
```

原表的 15 行数据，经过重叠合并与跨天拆分后变为 22 条数据。自关联的写法比较易懂，在小数据集上的性能尚可，但如果表很大，这种写法就会突显性能的问题。将查询中的 test1 表改为 u_room_log 表，慢到没有等到出结果，慢的原因从查询计划中就可得到直观的反映：

```
+----+---------+------------+-------+-------------+-------------+------+...
.+----------+...
| 1 | PRIMARY | nums       | range | PRIMARY     | PRIMARY     | 8    |...|     100
|...
| 1 | PRIMARY | <derived2> | ALL   | NULL        | NULL        | NULL |...| 24980980
|...
| 2 | DERIVED | <derived6> | ALL   | NULL        | NULL        | NULL |...|  2498089
|...
| 2 | DERIVED | <derived8> | ref   | <auto_key0> | <auto_key0> | 14   |...|
10 |...
| 8 | DERIVED | a          | ALL   | NULL        | NULL        | NULL |...|  2498089 |...
| 9 | DEPE... | b          | ALL   | NULL        | NULL        | NULL |...|  2498089 |...
| 6 | DERIVED | a          | ALL   | NULL        | NULL        | NULL |...|  2498089 |...
| 7 | DEPE... | b          | ALL   | NULL        | NULL        | NULL |...|  2498089 |...
| 3 | DEPE... | <derived4> | ref   | <auto_key0> | <auto_key0> | 14   |...|
249809 |...
| 4 | DERIVED | a          | ALL   | NULL        | NULL        | NULL |...|  2498089 |...
| 5 | DEPE... | b          | ALL   | NULL        | NULL        | NULL |...|  2498089
|...
+----+---------+------------+-------+-------------+-------------+------+...
.+----------+...
```

要对一个 250 万行的表多次进行相关子查询，总计要扫描的行数是多个 250 万的乘积，从执行时间看基本没有意义，因此这个写法被否定了。我们希望找到只扫描一遍表的实现方法，这是最优的解决方案，因为无论如何也要扫描一遍表。

2. 游标+内存临时表

在数据库优化中有一条基本原则，就是尽量使用集合操作而避免使用游标。下面来看一个最简单的例子，nums 是单列 100 万行的数字辅助表，select 查询时间为 0.41 秒。

```
mysql> select @id:=id from nums;
...
1000000 rows in set, 1 warning (0.41 sec)
```

而游标遍历的时间为 3.05 秒，比单条 select 语句慢了 7.4 倍。

```
mysql> delimiter //
mysql> create procedure p_cursor()
    -> begin
    ->     declare done int default 0;
    ->     declare v_id bigint;
    ->
    ->     declare cur_nums cursor for select id from nums;
    ->     declare continue handler for not found set done = 1;
    ->
    ->     open cur_nums;
    ->     repeat
    ->         fetch cur_nums into v_id;
    ->     until done end repeat;
    ->     close cur_nums;
    -> end//
Query OK, 0 rows affected (0.01 sec)

mysql>
mysql> call p_cursor()//
Query OK, 0 rows affected (3.05 sec)
```

下面案例中的情况却有所不同。有可能通过业务数据表上的游标，在逐行遍历表时编写复杂的应用逻辑，避免大表之间的关联，极大减少扫描行数，性能会比表关联好很多。下面是用游标合并重叠时间段的存储过程。

```
drop procedure if exists sp_overlap;
delimiter //

create procedure sp_overlap()
begin
    declare done int default 0;
    declare v_roomid bigint;
    declare v_userid bigint;
    declare v_start datetime;
    declare v_end datetime;

    declare v_prev_roomid int;
    declare v_prev_userid bigint;
    declare v_max_end datetime;

    declare cur_t1 cursor for select roomid,userid,s,e from test1 order by
roomid,userid,s,e;

    declare continue handler for not found set done = 1;

    drop table if exists t;
    drop table if exists t1;
    drop table if exists tmp_s;
```

```
create temporary table t(
   roomid bigint,
   userid bigint,
   s datetime,
   e datetime,
   broken int
) engine=memory;

create temporary table t1 (
   roomid int,
   userid bigint,
   s datetime,
   e datetime
) engine=memory;

create temporary table tmp_s(
   roomid bigint,
   userid bigint,
   s datetime,
   e datetime,
   i int
) engine=memory;

open cur_t1;
repeat
   fetch cur_t1 into v_roomid,v_userid,v_start,v_end;
   if done !=1 then
     if(v_roomid=v_prev_roomid and v_userid=v_prev_userid) then
        if(v_start<=v_max_end) then
           insert into t values(v_roomid,v_userid,v_start,v_end,0);
        else
           insert into t values(v_roomid,v_userid,v_start,v_end,1);
        end if;
        if(v_end>=v_max_end) then
           set v_max_end:=v_end;
        end if;
        set v_prev_roomid:=v_roomid;
        set v_userid:=v_userid;
     else
        set v_max_end:=v_end;
        set v_prev_roomid:=v_roomid;
        set v_prev_userid:=v_userid;
        insert into t values(v_roomid,v_userid,v_start,v_end,1);

     end if;
   end if;
until done end repeat;
close cur_t1;

insert into tmp_s
```

```
        select roomid,userid,min(s) s,max(e) e,datediff(max(e),min(s))+1 i
          from (select roomid,userid,s,e,
                       case when @flag=flag then @rn:=@rn+broken
                            when @flag:=flag then @rn:=broken end ran
                  from (select roomid,userid,s,e,broken,concat(roomid,',',userid)
flag
                          from t,(select @flag:='',@rn:=0) vars) a
                  order by roomid,userid,s,e) b
         group by roomid,userid,ran;

        select max(i) into @c from tmp_s;

        insert into t1(roomid,userid,s,e)
        select roomid, userid,
            if(date(s)!=date(e) and id>1,date(s+interval id-1 day),s) s,
            if(date(s+interval id-1 day)=date(e) ,e,
                date_format(s+interval id-1 day,'%y-%m-%d 23:59:59')) e
          from tmp_s t1, (select id from nums where id<=@c) nums
         where (nums.id<=t1.i);
    end
    //
```

定义游标的查询需要按房间 ID、用户 ID、起始时间和终止时间排序。v_roomid、v_userid、v_start 和 v_end 这 4 个变量存储游标当前行 4 个字段的数据。由于要按房间和用户分组，v_prev_roomid 和 v_prev_userid 分别存储前一行的房间 ID 和用户 ID，用于与当前行进行比较，判断哪些行属于同一组。

v_max_end 变量存储同一分组中当前最大的结束时间。在当前行的开始时间小于等于 v_max_end 时，说明当前行与同组中前面的时间段存在重叠，用 0 标识该行，否则表示当前行与同组中前面的时间段不存在重叠，用 1 标识该行。将游标遍历结果存储在临时表 t 中，t 只比原表多了 broken 字段，用于存储所在行是否需要合并的标识：

```
+--------+--------+---------------------+---------------------+--------+
| roomid | userid | s                   | e                   | broken |
+--------+--------+---------------------+---------------------+--------+
|      1 |      1 | 2018-01-01 01:01:01 | 2018-01-01 01:10:01 |      1 |
|      1 |      1 | 2018-01-01 01:01:02 | 2018-01-01 01:11:01 |      0 |
|      1 |      1 | 2018-01-01 01:01:05 | 2018-01-01 01:10:01 |      0 |
|      1 |      1 | 2018-01-01 01:11:02 | 2018-01-01 01:11:05 |      1 |
|      1 |      2 | 2018-01-01 01:01:02 | 2018-01-01 01:01:05 |      1 |
|      1 |      3 | 2018-01-01 01:01:05 | 2018-01-01 01:02:05 |      1 |
|      2 |      1 | 2018-01-01 01:01:03 | 2018-01-03 01:11:01 |      1 |
|      2 |      4 | 2018-01-01 01:03:02 | 2018-01-01 01:12:05 |      1 |
|      2 |      5 | 2018-01-01 01:11:02 | 2018-01-01 01:12:05 |      1 |
|      2 |      6 | 2018-01-01 01:15:02 | 2018-01-01 01:16:05 |      1 |
|      2 |      7 | 2018-01-01 01:01:03 | 2018-01-01 01:11:05 |      1 |
|      2 |      8 | 2018-01-01 23:01:03 | 2018-01-02 01:11:01 |      1 |
|      3 |      1 | 2018-01-05 01:01:01 | 2018-01-10 01:01:01 |      1 |
|      3 |      2 | 2018-01-05 01:01:01 | 2018-01-06 01:01:01 |      1 |
|      3 |      3 | 2018-01-06 01:01:01 | 2018-01-06 02:01:01 |      1 |
```

```
+--------+--------+--------------------+--------------------+--------+
15 rows in set (0.00 sec)
```

临时表 tmp_s 存储合并行后的结果。除了原有的 4 列外，该表还增加了表示开始时间和结束时间之间跨越天数的一列。在生成该表数据的查询语句中：

```
case when @flag=flag then @rn:=@rn+broken when @flag:=flag then @rn:=broken
end
```

这句的含义是按房间和用户分组（@flag 相同的表示为同一组），并且累加同一组中的 broken，因为需要合并行的 broken=0，所以所有需要合并行的累加 broken 都是 1。外层查询就按 min(s)、max(e)和 datediff(max(e),min(s))+1 这 3 列 group by 分别得到合并后的开始时间、结束时间和跨越天数。

然后用下面的查询取得最大跨越天数：

```
select max(i) from tmp_s;
```

最后将 tmp_s 与数字辅助表连接，进行跨天时间段的拆分，并将拆分后的结果存入临时表 t1。

本过程使用游标仅扫描一遍原始数据表，将中间处理结果存储到内存临时表中，对于处理重叠问题具有一定的通用性。之所以用到了三个临时表，是为了增加代码的可读性。每步产生的中间结果都存储于内存临时表，逻辑比较清晰。在性能优化时也要进行可读性、灵活性、易维护性等多方面权衡，避免“优化强迫症”。本例是可以不用写三个临时表的，去掉一个临时表可能会提高一些性能，但若将此复杂的处理步骤合并为单一查询，必然使 SQL 语句变得极为晦涩难懂，更不易维护，最终结果是得不偿失。

此存储过程在 u_room_log 表上执行，生成 2557836 行数据，用时 2 分 26 秒，这是一个可以接受的性能度量。

```
mysql> set max_heap_table_size=268435456;
Query OK, 0 rows affected (0.00 sec)
mysql> set tmp_table_size=268435456;
Query OK, 0 rows affected (0.00 sec)
mysql> call sp_overlap();
Query OK, 2557836 rows affected (2 min 26.36 sec)
```

16.3.3 改进取得活跃时段的算法

经过了前两步的数据处理，得到了结果集 t1，其中同一房间同一用户不存在重叠时间段，包括开始和结束的两个时间点也不重合，并且每行的开始时间和结束时间都不跨天。下面要依据活跃时段的定义，以 t1 作为输入，找到不同用户的重叠时间段。这里使用了“最小范围”和“正负计数器”两种不同算法来实现，但在大数据量的生产环境中，只有后者在性能上是可行的。

1. 最小范围算法（表连接）

该算法步骤如下：

步骤 01 将进出同一房间的所有时间点（不分用户）统一排序。例如，roomid=1 的进出房间记录如下：

```
+--------+--------+---------------------+---------------------+
| roomid | userid | s                   | e                   |
+--------+--------+---------------------+---------------------+
|      1 |      1 | 2018-01-01 01:01:01 | 2018-01-01 01:11:01 |
|      1 |      2 | 2018-01-01 01:01:02 | 2018-01-01 01:01:05 |
|      1 |      3 | 2018-01-01 01:01:05 | 2018-01-01 01:02:05 |
|      1 |      1 | 2018-01-01 01:11:02 | 2018-01-01 01:11:05 |
+--------+--------+---------------------+---------------------+
```

这一步处理完成后的输出为：

```
+--------+---------------------+
| roomid | timepoint           |
+--------+---------------------+
|      1 | 2018-01-01 01:01:01 |
|      1 | 2018-01-01 01:01:02 |
|      1 | 2018-01-01 01:01:05 |
|      1 | 2018-01-01 01:01:05 |
|      1 | 2018-01-01 01:02:05 |
|      1 | 2018-01-01 01:11:01 |
|      1 | 2018-01-01 01:11:02 |
|      1 | 2018-01-01 01:11:05 |
+--------+---------------------+
```

步骤 02 对于上一步输出中同一 roomid 的数据，将当前行的时间点作为结束时间，前一行的时间点作为开始时间，并且过滤掉开始时间为空或开始时间等于结束时间的数据。输出为每个房间最小时间范围的间隔。例如，roomid=1 的最小时间范围间隔为：

```
+--------+---------------------+---------------------+
| roomid | starttime           | endtime             |
+--------+---------------------+---------------------+
|      1 | 2018-01-01 01:01:01 | 2018-01-01 01:01:02 |
|      1 | 2018-01-01 01:01:02 | 2018-01-01 01:01:05 |
|      1 | 2018-01-01 01:01:05 | 2018-01-01 01:02:05 |
|      1 | 2018-01-01 01:02:05 | 2018-01-01 01:11:01 |
|      1 | 2018-01-01 01:11:01 | 2018-01-01 01:11:02 |
|      1 | 2018-01-01 01:11:02 | 2018-01-01 01:11:05 |
+--------+---------------------+---------------------+
```

这步是算法的核心，实际上就是把同一房间的所有进出时间点串行化到一个连续的时间轴上，输出的每个时间段首尾相接但不重叠。

步骤 03 将上一步的输出与 t1 表进行内连接。如果用户的在线时间和最小范围重叠，就将重叠的最小范围、userid 和 roomid 输出。结果包含了某个房间某个用户一个或者多个的最小范围。例如，roomid=1 的房间，每个用户对应的最小时间范围的间隔为：

```
+--------+--------+---------------------+---------------------+
| roomid | userid | s                   | e                   |
+--------+--------+---------------------+---------------------+
|      1 |      1 | 2018-01-01 01:01:01 | 2018-01-01 01:01:02 |
|      1 |      1 | 2018-01-01 01:01:02 | 2018-01-01 01:01:05 |
```

```
|      1 |      2 | 2018-01-01 01:01:02 | 2018-01-01 01:01:05 |
|      1 |      3 | 2018-01-01 01:01:05 | 2018-01-01 01:02:05 |
|      1 |      1 | 2018-01-01 01:01:05 | 2018-01-01 01:02:05 |
|      1 |      1 | 2018-01-01 01:02:05 | 2018-01-01 01:11:01 |
|      1 |      1 | 2018-01-01 01:11:02 | 2018-01-01 01:11:05 |
+--------+--------+---------------------+---------------------+
```

步骤 04 按上一步输出中的 roomid 和最小时间范围分组，过滤出每组中 userid 个数大于 1 的数据，结果为每个房间对应的活跃时间段。例如，roomid=1 的房间输出为：

```
+--------+---------------------+---------------------+---+
| roomid | s                   | e                   | c |
+--------+---------------------+---------------------+---+
|      1 | 2018-01-01 01:01:02 | 2018-01-01 01:01:05 | 2 |
|      1 | 2018-01-01 01:01:05 | 2018-01-01 01:02:05 | 2 |
+--------+---------------------+---------------------+---+
```

步骤 05 统计每个房间每天活跃时段内的最大人数，并汇总活跃时长（舍入到分钟）。例如，roomid=1 的房间输出为：

```
+--------+------------+------+------+
| roomid | dt         | ts   | c    |
+--------+------------+------+------+
|      1 | 2018-01-01 |    1 |    2 |
+--------+------------+------+------+
```

下面是实现最小范围算法的存储过程：

```
drop procedure if exists sp_active_duration;
delimiter //

create procedure sp_active_duration()
begin
    declare done int default 0;
    declare v_roomid bigint;
   declare v_start datetime;
   declare v_end datetime;

   drop table if exists tmp_time_point;
   create temporary table tmp_time_point(
      roomid bigint,
      timepoint datetime
   ) engine=memory;

   insert into tmp_time_point select roomid,s from t1;
   insert into tmp_time_point select roomid,e from t1;

   select roomid,date(s) dt,round(sum(timestampdiff(second,s,e))/60)
ts,max(c) c
     from (select roomid,s,e ,count(distinct userid) c
         from (select distinct v6.roomid,v6.userid,starttime s,endtime e
```

```
                from (select distinct roomid,cast(starttime as datetime)
starttime,
                                                 cast(endtime as datetime) endtime
                       from (select if(@roomid=roomid,@d,'') as starttime,
                                              @d:=timepoint,@roomid:=roomid,
                                              p.roomid,p.timepoint endtime
                             from tmp_time_point p,
                                              (select @d:='',@roomid:=-1) vars
                            order by roomid,timepoint) v4
                      where starttime!='' and date(starttime)=date(endtime)
                                and starttime <> endtime) v5
                inner join t1 v6 on(v5.starttime between v6.s and v6.e
                                             and v5.endtime between v6.s and v6.e
                                             and v5.roomid=v6.roomid)) v6
          group by roomid,s,e having count(distinct userid)>1) v7
     group by roomid,date(s);

   end
   //

   delimiter ;
```

tmp_time_point 表即为步骤 01 的输出结果。MySQL 限制在一条查询中只能引用临时表一次，否则会报 ERROR 1137 (HY000): Can't reopen table: 't1' 错误，所以生成 tmp_time_point 表数据时执行了两次 insert 语句。中间结果集 v5、v6 和 v7 分别为步骤 02、步骤 03 和步骤 04 的输出结果。

最小范围算法获取活跃时段的逻辑没问题，但在步骤 03 中需要表关联，当数据量很大时，这一步需要花费非常多的时间，因为要扫描大量的数据行。存储过程中最后的 select 语句在 u_room_log 表上的执行计划如下：

```
   +-----+---------+------------+------+------------+------------+------+..
.+------------+...
   |  1 | PRIMARY | <derived2> | ALL  | NULL        | NULL        | NULL |...|
1308213650 |...
   |  2 | DERIVED | <derived3> | ALL  | NULL        | NULL        | NULL |...|
1308213650 |...
   |  3 | DERIVED | v6         | ALL  | roomid     | NULL       | NULL |...|   2557836
|...
   |  3 | DERIVED | <derived4> | ref  | <auto_key0> | <auto_key0>
|9    |...|    41436 |...
   |  4 | DERIVED | <derived5> | ALL  | NULL       | NULL       | NULL |...|    5115672
|...
   |  5 | DERIVED | <derived6> | sy.. | NULL       | NULL       | NULL |...|          1
|...
   |  5 | DERIVED | p          | ALL  | NULL       | NULL       | NULL |...|    5115672
|...
   |  6 | DERIVED | NULL       | NULL | NULL       | NULL       | NULL |...|       NULL
|...
   +-----+---------+------------+------+------------+------------+------+..
.+------------+...
```

可以看到，步骤 03 需要关联两个几百万行的大表，因此在 u_room_log 表上执行 sp_active_duration()过程没有等到出结果。

2. 正负计数器算法（一次扫描）

与重叠时间段优化思想类似，我们希望只扫描一遍表数据，去掉表关联以提高性能。实际上，经过 sp_overlap 过程处理后，可以用一种高效的方式得到活跃时段。该算法的核心思想是：将所有的进出时间点统一排序，同时记录每个时间点的进出用户数。这样我们可以将在线时间分成多个互斥的时间段，并把当前时间点前面所有累计进出的用户数作为前一个时间点到当前时间点的重叠度，也就是不同用户数。用户进入房间标记为+1，离开房间标记为-1，因此不妨称之为正负计数器算法，具体步骤如下。

步骤 01 将同一房间的所有进入时间点和退出时间点合并成一列，进入时间标记为 1，退出时间标记为-1。实际上，1 表示在对应的时间点有一个用户进入，-1 表示在对应的时间点有一个用户退出。这步处理后 roomid=1 的记录变为：

```
+--------+---------------------+------+
| roomid | timepoint           | type |
+--------+---------------------+------+
|      1 | 2018-01-01 01:01:01 |    1 |
|      1 | 2018-01-01 01:01:02 |    1 |
|      1 | 2018-01-01 01:01:05 |   -1 |
|      1 | 2018-01-01 01:01:05 |    1 |
|      1 | 2018-01-01 01:02:05 |   -1 |
|      1 | 2018-01-01 01:11:01 |   -1 |
|      1 | 2018-01-01 01:11:02 |    1 |
|      1 | 2018-01-01 01:11:05 |   -1 |
+--------+---------------------+------+
```

步骤 02 按房间和时间点分组，对标志位汇总聚合，目的是去除重复的时间点。重复时间点表示在同一秒有多个用户进入、或者退出、或者进入退出同一个房间。汇总的目的就是确定在该时间点，最终进出的用户数。这一步是必须的，原因有两个：①我们必须保证对于一个房间每个时间点是唯一的；②必须确定某一时间点的进出方向和进出数量。这两点是保证算法成立的充要条件。出于同样的理由，在拆分跨天记录时，为保持时间点的唯一性，起止时间相差一秒。这一步处理后 roomid=1 的记录变为：

```
+--------+---------------------+------+
| roomid | timepoint           | type |
+--------+---------------------+------+
|      1 | 2018-01-01 01:01:01 |    1 |
|      1 | 2018-01-01 01:01:02 |    1 |
|      1 | 2018-01-01 01:01:05 |    0 |
|      1 | 2018-01-01 01:02:05 |   -1 |
|      1 | 2018-01-01 01:11:01 |   -1 |
|      1 | 2018-01-01 01:11:02 |    1 |
|      1 | 2018-01-01 01:11:05 |   -1 |
+--------+---------------------+------+
```

步骤 03 按房间分组，时间点排序，取得当前时间点的前一个时间点对应的进出用户数。如果没有前一个时间点，说明是该房间的第一次进入，前一个时间点对应的进出用户数设为 0。这一步处理后的记录 roomid=1 变为：

```
+--------+---------------------+------+----------+
| roomid | timepoint           | type | prevType |
+--------+---------------------+------+----------+
|      1 | 2018-01-01 01:01:01 |    1 |        0 |
|      1 | 2018-01-01 01:01:02 |    1 |        1 |
|      1 | 2018-01-01 01:01:05 |    0 |        1 |
|      1 | 2018-01-01 01:02:05 |   -1 |        0 |
|      1 | 2018-01-01 01:11:01 |   -1 |       -1 |
|      1 | 2018-01-01 01:11:02 |    1 |       -1 |
|      1 | 2018-01-01 01:11:05 |   -1 |        1 |
+--------+---------------------+------+----------+
```

步骤 04 取当前时间点的前一个时间点作为起始时间，当前时间点作为终止时间，将房间的在线时间区间划分成互斥时段。用当前时间点前面的所有累计进出用户数作为该时段的重叠度。这一步处理后 roomid=1 的记录如下，rn 即为 starttime 和 endtime 这段时间内的不同用户数：

```
+--------+---------------------+---------------------+------+
| roomid | starttime           | endtime             | rn   |
+--------+---------------------+---------------------+------+
|      1 | NULL                | 2018-01-01 01:01:01 |    0 |
|      1 | 2018-01-01 01:01:01 | 2018-01-01 01:01:02 |    1 |
|      1 | 2018-01-01 01:01:02 | 2018-01-01 01:01:05 |    2 |
|      1 | 2018-01-01 01:01:05 | 2018-01-01 01:02:05 |    2 |
|      1 | 2018-01-01 01:02:05 | 2018-01-01 01:11:01 |    1 |
|      1 | 2018-01-01 01:11:01 | 2018-01-01 01:11:02 |    0 |
|      1 | 2018-01-01 01:11:02 | 2018-01-01 01:11:05 |    1 |
+--------+---------------------+---------------------+------+
```

步骤 05 按天统计每个房间活跃时长（重叠度大于 1 的时段汇总），并求出活跃时段的峰值人数（最大重叠度）。最终 roomid=1 的结果如下，其中 dur 为活跃时长（单位舍入为分钟），c 是峰值人数：

```
+--------+------------+------+------+
| roomid | dt         | dur  | c    |
+--------+------------+------+------+
|      1 | 2018-01-01 |    1 |    2 |
+--------+------------+------+------+
```

采用正负计数器算法后的 sp_active_duration 如下：

```
drop procedure if exists sp_active_duration;
delimiter //

create procedure sp_active_duration()
begin
   declare done int default 0;
```

```
    declare v_roomid bigint;
    declare v_start datetime;
    declare v_end datetime;

    declare cur_test cursor for select roomid,s,e from t1;

    declare continue handler for not found set done = 1;

    drop table if exists tmp_time_point;
    create temporary table tmp_time_point(
     roomid bigint,
      timepoint datetime,
      type smallint
    ) engine=memory;

    -- 开始点+1, 结束点-1
    insert into tmp_time_point(roomid,timepoint,type) select roomid,s,1 from
t1;
    insert into tmp_time_point(roomid,timepoint,type) select roomid,e,-1 from
t1;

    select roomid,date(s) dt,round(sum(timestampdiff(
           second,date_format(s,'%Y-%m-%d %H:%i:%s'),
           date_format(e,'%Y-%m-%d %H:%i:%s')))/60) ts,max(rn) c
     from (select if(@roomid=roomid,@d,'') as s,
             @d:=str_to_date(timepoint,'%Y-%m-%d %H:%i:%s.%f'),
             @roomid:=roomid,
             p.roomid,
             str_to_date(timepoint,'%Y-%m-%d %H:%i:%s.%f') e,
             rn
         from (select round(case when @roomid=roomid then @rn:=@rn+prevType
                                 when @roomid:=roomid then @rn:=prevType end)
                          rn,b.prevType,roomid,timepoint,type
               from (select a.roomid,timepoint,type,
                               if(@roomid=roomid,@type,0) prevType,
                               @roomid:=roomid, @type:=type
                     from (select *
                           from (select roomid,timepoint,sum(type) type
                                       from tmp_time_point
                                      group by roomid,timepoint)
tmp_time_point,
                               (select @roomid:=-1,@rn:=0,@type:=0) vars
                          order by roomid ,timepoint) a) b
              order by roomid ,timepoint) p,
             (select @d:='',@roomid:=-1) vars
        order by roomid,timepoint) v4
    where rn>=2
    group by roomid,date(s);

  end
```

```
//

delimiter ;
```

tmp_time_point 表存储步骤 01 的结果。b 和 v4 分别是步骤 03 和步骤 04 的输出结果。过程中最后的查询只扫描一遍 tmp_time_point 表，处理速度大为提高。u_room_log 表上 sp_active_duration 过程的执行时间为 1 分 13 秒。

为满足原始需求，只需要在一个会话中连续调用两个存储过程即可。250 万的业务日志数据，总执行时间约为 3 分 40 秒。

```
set max_heap_table_size=268435456;
set tmp_table_size=268435456;
call sp_overlap();
call sp_active_duration();
```

16.3.4 MySQL 8 的单条查询解决方案

MySQL 8 提供了丰富的窗口函数，使复杂分析查询成为可能。更进一步，旧版 MySQL 的行级变量用法已经不再推荐使用：

```
mysql> select @a:=id from nums limit 1;
+--------+
| @a:=id |
+--------+
|      1 |
+--------+
1 row in set, 1 warning (0.00 sec)

mysql> show warnings;
+-------+----+------------------------------------------------------------
------------------+
| Level
|Code|Message                                                            |
+-------+----+------------------------------------------------------------
------------------+
|Warning|1287|Setting user variables within expressions is deprecated and will
be removed ...|
+-------+----+------------------------------------------------------------
------------------+
1 row in set (0.00 sec)
```

没有提供窗口函数之前，为了处理复杂逻辑，使用行级变量也是不得已而为之。本身就不是标准 SQL，可读性很差，如果需要换 RDBMS，比重做一遍还麻烦。而 MySQL 8 在 SQL 功能上已经接近 Oracle，重叠时间段问题用一句查询即可解决：

```
with c1 as  -- 合并同一房间同一用户的重叠时间段，用于统计峰值人数
(
  select distinct roomid,userid,min(s) s,max(e) e
   from (select roomid,userid,s,e,
```

```
            sum(broken) over (partition by roomid, userid order by s,e) flag
          from (select *,
                  (case when s <= max(e)
                          over (partition by roomid, userid order by s,e
                          rows between unbounded preceding and 1 preceding)
then 0
                    else 1
                  end) as broken
              from test1
            ) t
        ) t
     group by roomid,userid,flag
   ),
   c2 as  -- 拆分跨天的时间段
   (
     select *
      from (select roomid,userid,s,e
          from c1
         where date(s) = date(e)  -- 不跨天
         union all
        select roomid,userid,
            case when id = 1 then s
                 else date_add(date(s),interval id-1 day)
              end s,
            case when id = m2 then e
                 else date_add(date(s),interval id*3600*24 -1 second)
              end e
         from (select roomid,userid,s,e,id,
                 max(id) over (partition by roomid,userid,s) m2
             from c1,(select id from nums where id<=100) n
            where date(s) <> date(e) -- 跨天
             and id <= date(e)-date(s)+1) t1) t1
   ),
   c3 as -- 在计算最小范围的同时，计算区间用户数
   (
    select roomid,ts endtime,
         sum(prevtype) over (partition by roomid order by ts) rn,
         lag(ts) over (partition by roomid order by ts) starttime
     from (select a.*,
             ifnull(lag(type) over (partition by roomid order by ts),0)
prevtype
         from (select roomid,ts,sum(type) type
              from (select roomid,e ts, -1 type from c2
                   union all
                  select roomid,s ts, 1 type from c2
                  ) t1 group by roomid,ts
            ) a
        ) c
   )
   select roomid,dt,round(sum(dur)/60) ts,max(rn) c
```

```
    from (select roomid,date(starttime)
dt,timestampdiff(second,starttime,endtime) dur,rn
            from c3 where rn>=2
         ) t
    group by roomid,dt
    order by roomid,dt;
```

该查询处理逻辑和存储过程完全相同，只是大部分复杂工作都交给窗口函数完成了，写法更简练，但执行时间没有存储过程快。在相同环境下，with 查询在 u_room_log 上的执行时间为 4 分 10 秒左右，比自定义的存储过程执行还要慢半分钟。

16.4 快速安全删除 MySQL 大表

在一个高负载的生产数据库上删除大表需要一些技巧，倘若直接使用 drop table 来删除大表，将产生大量磁盘 I/O，严重的会卡库，这是高可用服务所不能接受的。要优化删除表，需要了解其内部执行过程。

16.4.1 表删除过程

表删除原理上分为内存和磁盘两部分的操作：

- 清除表相关的缓冲池（buffer pool）页面。
- 删除表相关的磁盘文件。

1. 清除缓冲池

在删除表时，MySQL 会将表在缓冲池中对应的页面清除，这个过程中会对缓冲池加上一个全局的互斥锁。在 MySQL 5.5.23 以后的版本，实现了一个 lazy drop table 方式，以减轻清理内存结构的操作对系统吞吐量产生的波动，具体过程为：

（1）持有缓冲池互斥锁。

（2）持有缓冲池中的刷新列表互斥锁。

（3）开始扫描刷新列表。

①如果脏页面属于删除表，那么就直接从刷新列表中删除掉。

②如果删除的页面数超过了“#define BUF_LRU_DROP_SEARCH_SIZE 1024”定义的这个数：

- 释放刷新列表互斥锁。
- 释放缓冲池互斥锁。
- 强制通过 pthread_yield 进行一次操作系统上下文切换，释放剩余的 CPU 时间片。

③重新持有缓冲池互斥锁。

④重新持有刷新列表互斥锁。

（4）释放刷新列表互斥锁。

（5）释放缓冲池互斥锁。

上面的过程中持有了缓冲池互斥锁和刷新列表互斥锁，从锁的保护范围来看，缓冲池互斥锁直观上瓶颈会比较明显。如果缓冲池很大，或者表中有很多脏页面需要清除，则持有互斥锁的时间会相对较长，导致其他事务在用到相应缓冲池实例时被阻塞，从而影响整个数据库的性能。

这部分优化的难点在于涉及源代码，作为应用能做的是先要保证被删除表上没有活动事务，最好是没有任何访问，然后在业务低峰期执行表删除操作。

2. 删除表相关的磁盘文件

这里只讨论采用独立表空间（innodb_file_per_table=1）的 innodb 表删除。独立表空间在性能和运维上都大大强于共享表空间，也是当前绝大多数情况下的表存储方式。相对于内存扫描，删除磁盘文件对系统的影响要大得多。问题在于如果表文件过大，直接删除会瞬时占用大量 I/O，造成 I/O 阻塞。通常可以使用以下三个步骤删除大表：

（1）创建表文件的硬链接。

（2）用 drop table 删除表。

（3）删除表文件释放磁盘空间。

16.4.2　创建硬链接

一个磁盘上的存储文件，可以由多个文件名引用。这多个文件是完全相同的，都指向同一个磁盘上的 inode index。当我们删除任何一个文件时，都不会影响真实的存储文件，只是会将其引用数据减 1，只有当被引用数目变为 1 时，再次删除文件，文件才会真正被删除。

例如，在没有硬链接时，t1 表对应的磁盘文件如下：

```
-rw-r----- 1 mysql mysql       17973 Jul  3  2018 t1.frm
-rw-rw---- 1 mysql mysql 498115543040 Oct  7 17:14 t1.ibd
```

其中第二个字段的“1”表示文件只有一个 inode 引用，此时删除 t1 表会实际删除磁盘文件。创建硬链接：

```
ln t1.frm t1.frm.h
ln t1.ibd t1.ibd.h
```

之后每个表文件有两个 inode 引用：

```
-rw-r----- 2 mysql mysql       17973 Jul  3  2018 t1.frm
-rw-r----- 2 mysql mysql       17973 Jul  3  2018 t1.frm.h
-rw-rw---- 2 mysql mysql 498115543040 Oct  7 17:14 t1.ibd
-rw-rw---- 2 mysql mysql 498115543040 Oct  7 17:14 t1.ibd.h
```

再删除表时，只是删除了 inode 引用，此操作非常快，不会影响正常服务。可以使用以下脚本对一个数据库内的所有表创建硬链接：

```
# MySQL 数据目录
datadir=`mysql -uroot -p123456 -S /data/mysqldata/mysql.sock -e "show
```

```
variables like 'datadir'" -N -B | awk '{print $2}'`
    # 数据库名
    dbname='dbname'

    cd $datadir/$dbname
    # 创建硬链接
    ls *.{ibd,frm} | awk '{print "ln "$0" "$0".h"}' | bash
```

16.4.3 删除表

```
    drop table t1;
```

MySQL 的 drop table 操作会删除系统表中 t1 的相关记录，同时会删除表的 t1.frm 与 t1.ibd 文件。如上所述，此操作可以瞬间完成，之后剩下两个.h 文件：

```
    -rw-r----- 1 mysql mysql      17973 Jul  3  2018 t1.frm.h
    -rw-rw---- 1 mysql mysql 498115543040 Oct  7 17:14 t1.ibd.h
```

16.4.4 删除文件释放空间

t1 表虽然秒删，但并没有释放磁盘空间。下面我们需要删除 t1.frm.h 与 t1.ibd.h 文件以真正释放空间。此时不能直接用 rm 命令删除这个接近 500GB 的文件，原因同样是会造成 I/O 冲击。相应的对策就是使用 coreutils 的 truncate 工具，逐次缩减文件大小，最终执行一个不会对系统产生影响的小文件删除操作。例如要实际删除 t1 的两个剩余.h 文件，命令如下：

```
    rmtablefile.sh t1
```

rmtablefile.sh 脚本文件内容如下：

```
    #!/bin/bash
    # 表定义文件很小，可直接删除
    rm $1.frm.h

    # 表数据文件大小，单位 MB
    filesize=`ls -l $1.ibd.h | awk '{print int($5/1024/1024)}'`
    if (( $filesize < 100 ))
    then
        # 小于 100MB 直接删除
        rm $1.ibd.h
    else
        # 大于等于 100，每次截断 100MB
        for i in `seq $filesize -100 0`
        do
            sleep 2
            echo $i
            truncate -s ${i}M $1.ibd.h
        done

        # 删除小于 100MB 的文件
```

```
    rm $1.ibd.h
fi
```

使用 truncate 每次将文件大小缩小 100MB，当文件小于 100MB 时再将其删除。需要说明的是，这一步实际删除磁盘文件的操作，目标是不影响在线的服务，在此基础上只要能够安全删除文件即可。

16.5　小　结

对于任何数据库系统，性能优化都是一个重中之重的课题，同时也是一项艰巨的任务。本书并非专门介绍 MySQL 优化，但希望用 MySQL 高可用环境中的实际案例，使读者能够管中窥豹，掌握一些常用的 MySQL 性能优化的方法。

本章讲述的 4 个案例各有特点。第 1 个案例说明了快速生成数据的通用方法，即使用 insert into...select，每次以指数级增长插入数据，包括严格控制插入行数的技巧。第 2 个案例是关于 MySQL 去重，通过恰到好处地组合利用索引有序、用提示强制使用索引、并行执行存储过程等特性，一步步将去重操作优化到最佳。本例虽然需求简单，但最终实现却需要了解很多 MySQL 的基础知识。第 3 个案例是经典的重叠时间段问题，也是一个实际业务需求。该案例理解起来有一定难度，但当清楚了所有实现细节之后，就会体会到算法的巧妙之处。第 4 个案例说明如何利用硬链接技术，安全快速地删除 MySQL 大表，对于高负载生产数据库表的维护，具有通用的借鉴意义。